AMERICAN MATHEMATICAL SOCIETY TRANSLATIONS

Series 2
Volume 30

10 papers on topology

——————————— AUTHORS ———————————

Aleksandrov, P. S.	Gluskin, L. M.	Sobolev, S. L.
Bognar, M.	Ponomarev, V. I.	Taĭmanov, A. D.
Čogošvili, G. S.	Rohlin, V. A.	Vladimirov, D. A.

Published by the

AMERICAN MATHEMATICAL SOCIETY

190 Hope Street, Providence, Rhode Island

under a grant from the

NATIONAL SCIENCE FOUNDATION

1963

Copyright © 1963 by the American Mathematical Society
Printed in the United States of America

All Rights Reserved
No portion of this book may be
reproduced without the written
permission of the publisher

Library of Congress Catalog Card Number A51-5559

PHOTOLITHOPRINTED BY CUSHING - MALLOY, INC.
ANN ARBOR, MICHIGAN, UNITED STATES OF AMERICA

TABLE OF CONTENTS

Page

iii

TOPOLOGICAL DUALITY THEOREMS. I
CLOSED SETS

P. S. ALEKSANDROV

CONTENTS

Preface

In this paper we systematically present the topological duality theorems for sets lying in Euclidean spaces. The first part is devoted to the case of closed sets (the duality theorem of Pontrjagin in ∇- and $\triangle$-forms). In the second part we investigate the general case of non-closed sets. There we present the general duality theorem proved by me in 1947, and the newest duality theorem of Sitnikov, which in a well-known sense completes the solution of the problem of topological duality for sets lying in Euclidean spaces.

All the auxiliary theorems of combinatorial topology are presented in the first chapters of the first part. Thus, the first part of this monograph becomes at the same time a general introduction to combinatorial topology.

CHAPTER I

Elementary concepts of combinatorial topology

§1. Complexes

1. **Simplicial complexes. Triangulations. Polyhedra.** A simplicial complex is a set of simplexes. Simplexes, if not specified otherwise, will always be assumed open; i.e., the points of the boundary of the simplex will not be counted as being in the simplex. The closure of a simplex will be called a closed simplex.

In the first part of this work we shall consider exclusively finite complexes; i.e., we shall assume that the number of simplexes appearing as elements of a given complex is finite. The *condition of completeness* will be quite essential: *each face of each simplex appearing as an element of the given complex will be an element of that complex.*[*] If the elements of a given finite complete complex are open simplexes of Euclidean or spherical space, not having common points, then such a complex is said to be a (finite) *triangulation*. It is easy to see that the corresponding closed simplexes, i.e., the closures of the simplexes appearing

---

[*] If we do not specify otherwise, we shall suppose that the complexes considered by us satisfy this condition; *however, in complete complexes we shall frequently consider subcomplexes not satisfying the completeness condition* (in particular, the so-called open subcomplexes; see below).

For each noncomplete complex K one may define its *combinatorial closure* $[K]$ consisting of all the simplexes $T \in K$ and of all the faces of these simplexes. Obviously $[K]$ is a complete complex.

as elements of the triangulation, satisfy the following condition: any two of these closed simplexes either do not intersect or else their intersection is a common closed face of these simplexes.

The set-theoretic sum of the simplexes appearing as elements of a given complex K is called the *body* of that complex and will be denoted by $\widetilde{K}$. Sets which are bodies of triangulations will be called *polyhedra*; the triangulation K is one of the triangulations of the polyhedron $\widetilde{K}$.

2. **Abstract complexes; nerves.** In every simplicial complex the simplexes correspond in a one-one way to their skeletons.* This circumstance leads to the concept of an *abstract simplicial complex;* we imagine a finite set X of some kind of elements, conventionally called *vertices,* and we suppose that in it there are defined certain nonempty subsets, called *skeletons* or *abstract simplexes.*

The set of these skeletons is called an abstract complex. The condition of completeness transforms in the following way: every nonempty subset of a skeleton is a skeleton. The number of elements in a given skeleton, diminished by unity, is called its *dimension*; the dimension of a complex is the largest of the dimensions of the skeletons entering into it.

The most important example of an abstract complex is the *nerve* of a finite system of sets. Let α be such a system, and put into correspondence with each of its elements e some vertex (the simplest of all, but not always the most convenient, is to take as the vertex the set e itself). A set of these vertices forms a "skeleton" if the sets corresponding to them have a nonempty intersection. The abstract complex obtained in this way is called the *nerve* of the given system of sets. The concept of nerve is fundamental to this entire paper.

3. **Stars.** We shall call the *combinatorial star* of some simplex T of the complex K the set of all simplexes of the complex K having T as a *proper* or *improper* face (the simplex itself is called an *improper face* of itself).**

A subcomplex K_0 of the complex K will be said to be *open* in K if, along with each simplex T in K_0, K_0 contains also the whole star of that simplex. The open subcomplexes of a triangulation form the most important example of noncomplete complexes.

A subcomplex K_0 of the complex K, whose complement $K - K_0$ is open, is

---

*The skeleton of a simplex is the set of all its vertices.

**The skeletons of the simplicial complex K form a set which is partially ordered by inclusion. This ordering, naturally, carries over to the simplexes of our set. For two simplexes of our complex K we write $T' \geq T$ if T is a face (possibly improper) of the simplex T'. $T' > T$ means that T is a proper face of the simplex T'. The combinatorial star of the simplex $T \in K$ consists of all the simplexes $T' \geq T$ of the complex K. Evidently *stars, generally speaking, are noncomplete complexes.*

said to be *closed*. If K is a complete complex, then its closed subcomplexes coincide with its complete subcomplexes.

The star of the simplex T in the complex K will be denoted by $O_K T$ or simply by OT.

The stars $O_K T_1, \cdots, O_K T_r$ of a complete simplicial complex have a nonempty intersection if and only if the simplexes $T_1, \cdots, T_r$ are faces of some simplex of the complex K, i.e., when the sum of the skeletons of the simplexes $T_1, \cdots, T_r$ is the skeleton of some simplex $T \in K$ (this simplex is called the *combinatorial sum* of the simplexes $T_1, \cdots, T_r$). In this case

$$O_K T_1 \cap \cdots \cap O_K T_r = O_K T.$$

The proof consists in the automatic verification of the fact that a simplex included on one side of this equality is included also on the other.

In particular, *the stars $O_K e_0, \cdots, O_K e_r$ of the vertices $e_0, \cdots, e_r$ of the complex K have a nonempty intersection if and only if these vertices form a simplex of the complex K.* In other words, *the complex K is the nerve of the system of stars of its vertices.*

If K is a triangulation, then in the majority of cases we shall consider not combinatorial stars of the complex K, but rather their bodies, i.e., the sum of the simplexes constituting them. These we shall call *open stars* and denote them by $o_K T$; this nomenclature results from the fact that open stars of the complex K are open sets of the polyhedron $\widetilde{K}$. Moreover, *the body $\widetilde{K}_0$ of the subcomplex K_0 is open in the polyhedron $\widetilde{K}$ if and only if K_0 is an open subcomplex of the complex K.*

Since the intersection of the bodies of any set of subcomplexes of the complex K is the body of the intersection of these subcomplexes, the properties proved above for the intersection of combinatorial stars are valid also for open stars. In particular, *the open stars $o_K e_0, \cdots, o_K e_r$ of any vertices $e_0, \cdots, e_r$ of the complex K have a nonempty intersection if and only if these vertices form the skeleton of some simplex of the complex K.*

The open stars of the vertices of a triangulation K are called the *principal stars* of that triangulation, and we have the theorem:

A triangulation K is the nerve of the system of its principal stars.

4. **Simplicial mappings.** Suppose that to each vertex e of a simplicial complex X we make correspond some vertex $e' = f(e)$ of a simplicial complex X', and that the following condition is satisfied: if any group of vertices $e_0, \cdots, e_r$ of the complex X forms in X the skeleton of some simplex, then also the vertices $f(e_0), \cdots, f(e_r)$, among which some may coincide, form the skeleton of some simplex

of the complex X' ("the mapping does not break simplexes").

In view of this condition, to each skeleton, and accordingly to each simplex $T = |e_0 \cdots e_r|$ of the complex X, there corresponds some simplex $T' = fT$ of the complex X', and we obtain a mapping of the complex X into the complex X'. The mapping f thus defined of the complex X into the complex X' will be called a *simplicial* mapping.

If a simplicial mapping of the complex X onto the complex X' is 1-1, then it is called an *isomorphic mapping* (isomorphism), and complexes X and X' which admit of an isomorphism of one onto the other are said to be *isomorphic* to each other. It is easy to prove that *each finite n-dimensional simplicial complex is isomorphic to some triangulation lying in* R^{2n+1}.

In fact, let $a_1, \cdots, a_s$ be all the vertices of the complex X. We choose a collection of points $b_1, \cdots, b_s$ in R^{2n+1} in general position, i.e., such that no $2n + 2$ of them lie on a $2n$-dimensional plane in the space R^{2n+1}. We span a simplex across the points $b_{i_0}, \cdots, b_{i_r}$ in R^{2n+1} if and only if $a_{i_0}, \cdots, a_{i_r}$ form a skeleton of the complex X. The simplexes $|b_{i_0} \cdots b_{i_r}|$ thus obtained are disjoint. Indeed, consider the simplexes

$$|b_{i_0} \cdots b_{i_h}| \text{ and } |b_{j_0} \cdots b_{j_k}|$$

corresponding to the simplexes

$$|a_{i_0} \cdots a_{i_h}| \text{ and } |a_{j_0} \cdots a_{j_k}|$$

of the complex X. Since the dimension of the complex X is equal to n, therefore $h \leq n$ and $k \leq n$, so that the sum of the skeletons of the simplexes $|b_{i_0} \cdots b_{i_h}|$ and $|b_{j_0} \cdots b_{j_k}|$ consists of not more than $(h + 1) + (k + 1) \leq 2n + 2$ points and thus is the skeleton of some simplex of the space R^{2n+1}. This last simplex has $|b_{i_0} \cdots b_{i_h}|$ and $|b_{j_0} \cdots b_{j_k}|$ as faces. These simplexes therefore have no common point. *

Thus the simplexes we have constructed in R^{2n+1} form a triangulation, obviously isomorphic to the complex X.

Let f be a simplicial mapping of the complex X into the complex X'. Then the vertices of each simplex T in X are mapped into the vertices of some simplex T' in X'. But then there is determined an affine mapping of the simplex T into the simplex T'. It is easy to see that this affine mapping, defined on each simplex of the complex X, in its entirety yields a single continuous ("piecewise affine") mapping of the polyhedron $\widetilde{X}$ into the polyhedron $\widetilde{X}'$, which we shall denote by $\widetilde{f}$. **

---

* Translator's note: or else they are identical.
** Frequently simply by f.

Thus, *a simplicial mapping* f *of the triangulation* X *into the triangulation* X' *generates a continuous mapping* $\tilde{f}$ *of the polyhedron* $\tilde{X}$ *into the polyhedron* $\tilde{X}'$. *The mapping* $\tilde{f}$ *is also called a simplicial mapping.*

5. **Successors of triangulations and subdivisions of them.** By definition, *the triangulation* K' *succeeds the triangulation* K *if each simplex* T' *of the triangulation* K' *is contained in some* (evidently, unique) *simplex* T *of the triangulation* K. *The simplex* T *is called the carrier of the simplex* T' *in the triangulation* K. If K' succeeds K, then evidently $\tilde{K}' \subseteq \tilde{K}$; if $\tilde{K}' = \tilde{K}$, i.e., if each simplex $T \in K$ is the union of those simplexes of the triangulation K' lying in it, then K' is said to be a *subdivision of the triangulation* K.[*]

6. **Barycentric subdivisions.** One of the simplest subdivisions of a given triangulation of K, and doubtless the most important subdivision, is the barycentric subdivision. Take in each simplex $T \in K$ a point, which we shall call the *center* of the simplex T. Usually one takes as this point the center of mass of the simplex T. A barycentric subdivision of the complex K is defined as a triangulation K_1 whose vertices are the centers of simplexes of the complex K (so that the vertices of the complex K_1 fall into 1-1 correspondence with the simplexes of the complex K). A set of vertices of the complex K_1 forms by definition a skeleton of that complex if it may be written in an order

$$e_0, e_1, \cdots, e_r$$

such that the simplexes $T_i \in K$ corresponding to them form a decreasing chain

$$T_0 > T_1 > \cdots > T_r \tag{1}$$

in the sense that T_1 is a face of the simplex T_0, T_2 is a face of the simplex T_1, and so forth.

The proof of the fact that the triangulation K_1 succeeds K presents no difficulties. The proof of the fact that K_1 is a subdivision of the complex K is based on the important concept of a pyramid over a complex. A pyramid with the vertex o over the complex X is an abstract (non-complete) complex oX, whose simplexes are the vertex o itself and all simplexes of the form $|oe_1 \cdots e_r|$, where $|e_1 \cdots e_r|$ is a simplex of X. One may construct a barycentric subdivision K_1 of the triangulation K by applying induction: a zero-dimensional simplex of the complex K enters into the complex K_1 unchanged. Suppose that the barycentric subdivisions of all simplexes of dimension $\leq r$ of the complex K have been defined. Then one obtains the barycentric subdivision of an $(r+1)$-dimensional simplex $T^{r+1} \in K$

---

[*] For what follows it is important to observe that the subdivision of a non-complete complex K is defined as the set of those simplexes of a subdivision of the combinatorial closure $[K]$ of the complex K which lie on the simplexes of that complex.

by constructing, over the already subdivided boundary of the simplex T^{r+1}, a pyramid with its vertex at the center of the simplex T^{r+1}. This pyramid turns out immediately to be a subdivision of the simplex T^{r+1}.

Remark. The barycentric subdivision of the triangulation K has, as is clear from examples, a very graphic geometric sense. But it is very convenient to regard K_1 also as an abstract complex. As vertices of this abstract complex, one may take the simplexes of the complex K. Then a decreasing sequence (1) of simplexes $T_i \in K$ is defined to be a skeleton of the complex K_1. Thus the sole basis for the definition of the barycentric subdivision lies in the circumstance that each complex is a partially ordered set with the so-called geometrical order relation: $T' > T$ means that T is a proper face of the simplex T'.

7. **Star complexes adjoint to a given triangulation K.** In each simplex $\tau = |e_0 \cdots e_r|$ of a barycentric subdivision K_1 of a triangulation K there is given a certain natural order of vertices: this is the decreasing order of those simplexes

$$T_0 > T_1 > \cdots > T_r \tag{1}$$

of the complex K whose centers are the vertices $e_0, \cdots, e_r$ of the simplex τ. Therefore, among the vertices of each simplex $\tau \in K_1$ there is a leading vertex e_0, corresponding to the simplex T_0 of largest dimension, and a least vertex, corresponding to the simplex T_r of smallest dimension, in the sequence (1). Taking the set of all simplexes of the complex K_1 having one and the same leading vertex, namely, the center e of the simplex T, we obtain the set of all simplexes of the complex K_1 lying on the simplex T. We find that they form a barycentric subdivision of that (open) simplex. Taking, conversely, the set of all simplexes of the complex K_1 having the given vertex e, i.e., the center of a given simplex T, as their least vertex, we obtain a certain non-complete subcomplex U of the complex K_1 called the *open barycentric star adjoint to the simplex T of the complex K.* The combinatorial closure of a given barycentric star is called a closed barycentric star $[U]$,* adjoint to the given simplex T of the complex K.

The complete complex $[U] - U$, consisting of the simplexes which complete a given open barycentric star to its combinatorial closure, is called *the boundary of the barycentric star U.* The barycentric star U' is contained in the boundary of the barycentric star U if and only if the simplex T adjoint to the star U is a face of the simplex T' adjoint to the star U'. *In this case the barycentric star U' is called a (proper) face of the barycentric star U.*

The boundary of a barycentric star is entirely formed from its faces. *A bary-*

---

* The open barycentric star U forms an open subcomplex of the closed star $[U]$.

centric star adjoint to some vertex of the complex K *is said to be a principal barycentric star*; such a star is not the face of any other barycentric star.[*] The set of barycentric stars adjoint to an element of the triangulation K, with the partial ordering defined by $U > U'$ if the star U' is a face of the star U (i.e., the simplex T is a face of the simplex T'), is said to be the *star complex* K^* adjoint to the triangulation K.

A barycentric subdivision K_1 of a triangulation may be decomposed into pairwise nonintersecting complexes in two natural ways. First, K_1 is the sum of the barycentric subdivisions of (open) simplexes of the complex K; secondly, K_1 is the sum of (open) barycentric stars, adjoint to the simplexes of the complex K. Thus the body of the complex K^* is the same polyhedron $\tilde{K}_1 = \tilde{K} = \tilde{K}^*$. The complex K_1 itself is to the same degree a subdivision of the complex K and of the complex K^*.

For what follows it is important to observe that the intersection of any simplex $T_i \in K$ with any star $U_j \in K^*$ consists of all those simplexes $\tau \in K_1$ which have the center e_i of the simplex T_i as their leading vertex, and the center e_j of the star U_j (i.e., the center of the simplex T_j adjoint to it) as their least vertex. *It therefore immediately follows that the simplex T_i and the star U_j intersect if and only if the simplex T_j is a face of the simplex T_i.* The same condition is necessary and sufficient for the closed simplex $\bar{T}_i$ to intersect with the closed star $\bar{U}_j$.[**]

Indeed, if $\bar{T}_i$ intersects $\bar{U}_j$, then there exist a $T_{i_1} \le T_i$ and a $U_{j_1} \le U_j$ such that T_{i_1} intersects U_{j_1}. But then $T_i \ge T_{i_1} \ge T_{j_1} \ge T_j$, i.e., $T_i \ge T_j$.

In particular: 1) The intersection of a simplex T_i^r with the star U_i adjoint to it consists only of the common center of the simplex T_i^r and of the star U_i, while the closed simplex $\bar{T}_i^r$ does not intersect with closed barycentric stars adjoint to r-dimensional simplexes of the complex K other than itself. 2) Each closed simplex $\bar{T}$ of the complex K is contained in the union of the bodies of the closed barycentric stars adjoint to its vertices and does not intersect any principal barycentric star whose center is not a vertex of the simplex T. Hence it follows that *the complex K is the nerve of the system of its closed barycentric stars.*

8. **Connectivity of complexes.** A complex is called connected if it can not be represented as the sum of two nonempty nonintersecting closed subcomplexes. If

------

[*]As is well known, a simplex T of a complex K is said to be principal if it is not the face of any simplex of the same complex other than itself.

[**]Here we diverge from the conventions accepted above and understand by a closed star $\bar{U}_j$ the body of the star $[U_j]$.

the complex is not connected, then it uniquely decomposes into its *components*, namely, into pairwise nonintersecting maximal connected subcomplexes, each of which is at the same time a closed and open subcomplex of the given complex.

It is easy to verify that *a complete simplicial complex is connected if and only if any two of its vertices* e_1 *and* e_s *may be connected in that complex by a "broken line", i.e., a finite sequence of one-dimensional simplexes*

$$(e_1 e_2), \ (e_2 e_3), \ \cdots, \ (e_{s-1} e_s),$$

each two successive elements of which have a common vertex.

9. **The extent of the concept of a complex required in the sequel. Pseudo-manifolds.** Along with complete simplicial complexes we shall find it necessary in the sequel to consider also their open subcomplexes or, what amounts to the same thing, simplicial complexes open in their combinatorial closures. Along with these we shall have to consider star complexes adjoint to triangulations (in the first instance, triangulations of the n-sphere).

A very important class of complexes, which may be noncomplete, are the so-called *combinatorial n-dimensional pseudomanifolds.*

A simplicial complex K open in its combinatorial closure is called an n-dimensional (combinatorial) pseudomanifold if it satisfies the following three conditions:

1) *The complex K is a dimensionally homogeneous n-dimensional complex;* this means that there are no simplexes of dimension $> n$ in it and that every simplex of dimension $< n$ is the face of some n-dimensional simplex.

2) *The complex K is strongly connected;* this means that any two n-dimensional simplexes T_1^n and T_s^n of the complex K may be connected in K by a finite succession of simplexes

$$T_1^n, \ T_2^n, \ \cdots, \ T_{s-1}^n, \ T_s^n,$$

in which each two successive links T_i and T_{i+1} have a common n-dimensional face.

3) *Each $(n-1)$-dimensional simplex of the complex K is the common face of two and only two n-dimensional simplexes of that complex.*

Complete pseudomanifolds are usually called *closed* pseudomanifolds.

Remark. We defined combinatorial pseudomanifolds as *complexes* (satisfying certain conditions); the adjective "combinatorial", which will frequently be dropped, is indeed intended to emphasize this circumstance.

The bodies of combinatorial manifolds are also called pseudomanifolds (geometrical).

10. **Curved polyhedra.** Every set (topological space) homeomorphic to a

polyhedron is called a curved polyhedron. If one is given a polyhedron $P = \tilde{K}$ and a topological mapping of it onto a curved polyhedron Q, then under this mapping the triangulation K of the polyhedron P goes into a "curved triangulation" of the curved polyhedron Q. For a similar construction, see [1], Chapter 4, §6.

§2. **The simplest algebraic concepts of combinatorial topology.**

Δ- and ∇-groups of a complex (Betti groups)

1. **Orientation.** To give an orientation to a simplex, or to orient it, means to arrange its vertices in a definite order, with the agreement that two orderings which transform into each other under an even substitution define the same orientation. Thus, although there are $(n+1)!$ orderings which may be given to the vertices of an n-dimensional simplex, there are only two orientations. A simplex along with a particular one of its orientations is called an *oriented simplex*. We shall denote one or the other of the two orientations of T^n by the symbol t^n. The other will then be denoted by $-t^n$, and we shall frequently write $T^n = |t^n| = |-t^n|$. An oriented zero-dimensional simplex (in this case the preceding definition loses its meaning) is simply the given point ("vertex") along with one of the two coefficients $+1$ or -1. If the oriented simplex t^n is given by the ordering

$$e_0, e_1, \cdots, e_n$$

of its vertices, we write

$$t^n = (e_0 e_1 \cdots e_n).$$

Then

$$t^n = \epsilon(e_{i_0} e_{i_1} \cdots e_{i_n}),$$

where ϵ is the sign of the substitution $\begin{pmatrix} 0 & 1 & \cdots & n \\ i_0 & i_1 & \cdots & i_n \end{pmatrix}$.

If $t^n = (e_0 \cdots e_n)$ and $t'^n = (c'_0 \cdots e'_n)$ are two oriented n-dimensional simplexes in R^n, then there exists a unique affine mapping of R^n onto itself, carrying the points $e_0, \cdots, e_n$, respectively, into $e'_0, \cdots, e'_n$; if this affine mapping is positive (i.e., has a positive determinant),[*] then the simplexes t^n and t'^n are said to be identically *oriented* in R^n. In the opposite case they are said to be *oppositely oriented*. This definition is correct, since any even permutation of the vertices of a given simplex $T^n \subset R^n$ is realized by a positive affine mapping of the space R^n containing the simplex, and an odd permutation by a negative such mapping. An affine mapping of the space onto itself is positive if it may be continuously deformed into the identity mapping, and it is negative if it may be carried by a continuous deformation into the mirror transformation. Therefore the following

---

[*]As is known, the sign of the determinant of a mapping does not depend on the choice of a system of affine coordinates in R^n. See, for example [1], second supplement.

proposition holds:

Suppose that in an n-dimensional plane $R^{n-1} \subset R^n$ there is given a system of n independent points $e_1, \cdots, e_n$; suppose moreover that two points e_0' and e_0'' are given, not lying in the plane R^{n-1}. Then the affine mapping leaving the points $e_1, \cdots e_n$ in place and carrying e_0' into e_0'' is positive if the points e_0' and e_0'' lie on the same side of the plane R^{n-1}, and is negative if they lie on opposite sides.

It follows from this that the choice of an orientation of any n-dimensional simplex t^n of the space R^n defines the identical orientation for every other simplex of the same R^n (namely, that orientation into which the given one goes under a positive affine mapping). In this sense we shall say that the orientation of any n-dimensional simplex of the space R^n *determines the orientation of the whole space R^n.*

2. **Incidence coefficients.** Let t^n be an oriented simplex, and let t^{n-1} be an $(n-1)$-dimensional face of it oriented in some way. The simplex t has a single vertex e not a vertex of the face $|t^{n-1}|$ (the vertex opposite to this face). We choose any order of the vertices $e_1, \cdots, e_n$ of the face $|t^{n-1}|$, defining by this choice an orientation of t^{n-1}. Then, putting the vertex e in front of all the vertices of the face $|t^{n-1}|$, we obtain a definite order

$$e, e_1, \cdots, e_n$$

of all the vertices of the simplex $|t^n|$, and this order defines either the orientation of t^n or the orientation of $-t^n$.* In the first case we say that the incidence coefficient $(t^n : t^{n-1})$ of the oriented simplexes t^n and t^{n-1} is equal to $+1$. In the second case we say that it is equal to -1. In other words, the incidence coefficient $\epsilon = (t^n : t^{n-1})$ is defined from the equation

$$t^n = \epsilon(et^{n-1}).$$

Evidently, we have

$$(-t^n : t^{n-1}) = (t^n : -t^{n-1}) = -(t^n : t^{n-1}). \tag{1}$$

Further, suppose that $t^n = (e_0 \cdots e_n)$ and $t_i^{n-1} = (e_0 \cdots \hat{e}_i \cdots e_n)$, where $\hat{e}_i$ denotes that the vertex e_i is omitted. Then

$$(t^n : t_i^{n-1}) = (-1)^i,$$

so that in particular

---

*The orientation $(ee_1 \cdots e_n) = \pm t^n$ depends only on the orientation of the simplex $t^{n-1} = (e_1 \cdots e_n)$, and not on the order of the vertices $e_1, \cdots, e_n$ defining this orientation. Indeed, an even permutation of the vertices $e_1, \cdots, e_n$ defines an even permutation of the vertices $e, e_1, \cdots, e_n$ (since e stays in place). Because of the circumstance just mentioned, it is all right to denote the oriented simplex $(ee_1 \cdots e_n)$ by (et^{n-1}).

$$((e_0 e_1 e_2 \cdots e_n) : (e_1 e_2 \cdots e_n)) + ((e_0 e_1 e_2 \cdots e_n) : (e_0 e_2 \cdots e_n)) = 0. \quad (2)$$

Now we take an n-dimensional simplex, two of its $(n-1)$-dimensional faces, and an $(n-2)$-dimensional face common to the two chosen $(n-1)$-dimensional ones. All of these four simplexes we take in quite general orientations, which we denote by t^n, t_0^{n-1}, t_1^{n-1}, t^{n-2}.

Without loss of generality we may assume

$$t^{n-2} = (e_2 \cdots e_n); \quad t_0^{n-1} = \pm (e_1 e_2 \cdots e_n); \quad t_1^{n-1} = \pm (e_0 e_2 \cdots e_n).$$

We shall prove the equality

$$(t^n : t_0^{n-1})\, (t_0^{n-1} : t^{n-2}) + (t^n : t_1^{n-1})\, (t_1^{n-1} : t^{n-2}) = 0. \quad (3)$$

Without loss of generality we may assume that the orientations of t_0^{n-1} and t_1^{n-1} are such that $(t_0^{n-1} : t^{n-2}) = +1$, $(t_1^{n-1} ; t^{n-2}) = +1$. Indeed, if one of the two latter equalities were not satisfied and if, for example, $(t_0^{n-1} : t^{n-2}) = -1$, we could replace the orientation t_0^{n-1} in (3) by $-t_0^{n-1}$, and the left side of the equality (3) would not change.

So it is sufficient to prove the equality (3) for the case when $(t_0^{n-1} : t^{n-2}) = (t^{n-1} : t^{n-2}) = 1$. But in this case, with the notation chosen, we have

$$t_0^{n-1} = (e_1 e_2 \cdots e_n), \quad t_1^{n-1} = (e_0 e_2 \cdots e_n),$$

and the equality (3) which we are proving turns into the equality (2) already proved.

We now assume that for the two oriented simplexes t^n, t^{n-1} we always have $(t^n : t^{n-1}) = 0$ if the simplex $|t^{n-1}|$ is not a face of the simplex $|t^n|$.

For $i = 0, 1, \cdots, n$ we now let t_i^{n-1} denote arbitrary orientations of the $(n-1)$-dimensional faces of the simplex $|t^n|$, so that we may write the same equality (3) in the form

$$\sum_i (t^n : t_i^{n-1})\, (t_i^{n-1} : t^{n-2}) = 0.$$

All the terms in this sum, except the two corresponding to the faces t_0^n and t_1^{n-1} adjoining the face t^{n-2}, are equal to zero.

If K is any complete simplicial complex, $|t^n|$ any n-dimensional simplex, and $|t^{n-2}|$ any $(n-2)$-dimensional simplex in it, then, taking for each $(n-1)$-dimensional simplex of the complex K a definite orientation t_i^{n-1}, we have again equation (4), where the summation is taken now over all $(n-1)$-dimensional simplexes of the complex K.

3. **Chains. The operator Δ. Betti groups.** The frontier Δt^n of an oriented simplex t^n is obtained, by definition, if to each arbitrarily oriented face t_i^{n-1} of the simplex t^n one assigns in the form of a "weight" or "coefficient" (with which that face appears in the boundary of the simplex t^n) its incidence coefficient

$(t^n : t_i^{n-1})$. In other words, *the boundary Δt^n of the oriented simplex t^n is defined in the form of an algebraic expression*

$$\Delta t^n = \sum_i (t^n : t_i^{n-1}) t_i^{n-1} .$$

In particular, if

$$t^n = (e_0 \cdots e_n), \quad t_i^{n-1} = (e_0 \cdots \hat{e}_i \cdots e_n),$$

then

$$\Delta t^n = \sum (-1)^i \, t_i^{n-1} .$$

The boundary of an oriented simplex is a special case of the most important general concept of combinatorial topology: the concept of a *chain*. An r-dimensional chain of a given complex K is obtained if to each oriented r-dimensional simplex t_i^r one attaches a certain integer-valued weight, with the understanding that if one has attached the weight a_i to the simplex t_i^r, one attaches the weight $-a_i$ to the simplex $-t_i^r$. In other words, *an r-dimensional chain is an integer-valued function, defined on the set of all oriented r-dimensional simplexes of a given complex K, subjected to the sole requirement that to two opposite orientations of the same simplex of the complex K one attaches values which are equal in absolute value and opposite in sign.* Let x^r be an r-dimensional chain of the complex K. Suppose that for each r-dimensional simplex of this complex we have defined an orientation denoted by t_i^r; denoting by a_i the value taken by the chain x^r on the simplex t_i^r, and putting this value in the form of a coefficient for the simplex t_i^r, we write the chain x^r in the form

$$x^r = \sum a_i t_i .$$

Here, writing the chain as a linear form means nothing more than that the function x^r takes the value a_i on the simplex t_i^r. The same thing is expressed by saying that the simplex t_i^r enters into the chain x^r with coefficient a_i. This way of writing a chain is motivated as follows. A chain that assigns to a given oriented simplex t_i^r the value 1, and assigns the value 0 to all remaining simplexes, is naturally identified with the oriented simplex t_i^r itself, and written in the form $x^r = t_i^r$. But then the chain that assigns to a given t_i^r the value a_i and to all remaining simplexes the value 0 must now be written in the form $a_i t_i^r$, and the chain x^r that assigns to an arbitrary t_i^r, $|t_i^r| \in K$, the value a_i must be written in the form $\sum_i a_i t_i^r$.

Chains are added as linear forms, and with this addition the r-dimensional chains of a complex K form a group denoted by $L^r K$.

Now we may say: *the boundary of an r-dimensional oriented simplex t^r of the complex K is an $(r-1)$-dimensional chain that takes on each oriented $(r-1)$-dimensional simplex t^{r-1} of the complex K a value equal to the incidence*

14 P. S. ALEKSANDROV

coefficient $(t^r : t^{r-1})$.

For any chain $x^r = \Sigma\, a_i t_i^r$ of the complex K, the boundary Δx^r is defined as an $(r-1)$-dimensional chain by the formula

$$\Delta x^r = \sum_i a_i \Delta t_i^r.$$

This means, on the basis of formula (5), that

$$\Delta x^r = \sum_j b_j t_j^{r-1},$$

where $b_j = \sum_i (t_i^r : t_j^{r-1})\, a_i$. Evidently, for two chains x_1^r and x_2^r,

$$\Delta(x_1^r + x_2^r) = \Delta x_1^r + \Delta x_2^r,$$

so that Δ turns out to be a linear operator that homomorphically maps the group $L^r K$ of all r-dimensional chains of the complex K into the group $L^{r-1} K$ of all $(r-1)$-dimensional chains of the same complex. The kernel of this homomorphism is the subgroup $Z^r K$ of the group $L^r K$ consisting of all chains whose boundaries equal zero. Such chains are called *cycles*. The image of the group $L^r K$ under the homomorphism Δ is a subgroup $H^{r-1} K$ of the group $L^{r-1} K$ consisting of all those $(r-1)$-dimensional chains of the complex K which are boundaries of r-dimensional chains of the same complex. Such chains are said to be *homologous to zero*. We shall prove that chains which are homologous to zero in the complex K are always cycles, i.e., that the group $H^r K$, defined as a subgroup of the group $L^r K$, in fact is a subgroup of the group $Z^r K$.

For this it is evidently sufficient to prove that the boundary for any chain x is a cycle, and this means that

$$\Delta\Delta x = 0 \tag{6}$$

for any chain x whatever. It is sufficient to prove formula (6) for the case when the chain consists of one oriented simplex, $x = t^r$, which is carried out by a simple calculation with an application of formula (4):

$$\Delta\Delta t^r = \Delta \sum_i (t^r : t_i^{r-1}) t_i^{r-1} = \sum_i (t^r : t_i^{r-1})\, \Delta t_i^r = \sum_i \sum_j (t^r : t_i^{r-1})(t_i^{r-1} : t_j^{r-2})\, t_j^{r-2}\,;$$

i.e.,

$$\Delta\Delta t^r = \sum_j d_j t_j^{r-2},$$

where

$$d_j = \sum_i (t^r : t_i^{r-1})(t_i^{r-1} : t_j^{r-2}) = 0,$$

which gives the desired result.

The factor-group $Z^r K - H^r K$ is called the r-dimensional Δ-group of the complex K (or the r-dimensional lower Betti group of the complex K) and is denoted by $\Delta^r K$.

Remark. Each zero-dimensional chain is a cycle,[*] so that the group $\Delta^0 K$ is the factor-group $L^0 K - H^0 K$.

However, it is sometimes convenient to distinguish in the group $L^0 K = Z^0 K$ the subgroup $Z^{00} K$ of those cycles, i.e., linear forms $\Sigma c_i t_i^0$, the sum of whose coefficients Σc_i is equal to zero. It is easy to see that $H^0 K \subseteq Z^{00} K$. The factor-group $Z^{00} K - H^0 K$ will be denoted by $\Delta^{00} K$. Frequently, by the zero-dimensional Δ-group of the complex K we understand indeed the group $\Delta^{00} K$, and not $\Delta^0 K$.

If k is the number of connected components of the complex K, then $\Delta^0 K$ is the direct sum of k terms, and $\Delta^{00} K$ is the direct sum of $k - 1$ terms, each of which is an infinite cyclic group.

4. **The operator** ∇. **The groups** $\nabla^r K$. The operator Δ is frequently called the *lower boundary operator*. Along with it, in combinatorial topology a fundamental role is played by the so-called *upper boundary operator* ∇, discovered in 1934 by A. N. Kolmogorov and J. Alexander. Its definition is quite analogous to the definition of the operator Δ: for any oriented simplex t^r of the complex K the chain Δt^r is an $(r - 1)$-dimensional chain that takes on each oriented $(r - 1)$-dimensional simplex t^{r-1} of the complex K a value equal to the incidence coefficient $(t^r : t^{r-1})$. Analogously, ∇t^r *is defined as an* $(r + 1)$-*dimensional chain that takes on each oriented* $(r + 1)$-*dimensional simplex* t^{r+1} *of the complex* K *a value equal to the incidence coefficient* $(t^{r+1} : t^r)$. For any chain $x^r = \Sigma a_i t_i^r$, the chain ∇x^r is defined as

$$\nabla x^r = \Sigma a_i \nabla t_i^r.$$

This means that

$$\nabla x^r = \Sigma b_h t_h^{r+1}, \tag{7a}$$

where

$$b_h = \sum_i (t_h^{r+1} : t_i^r) a_h. \tag{7b}$$

The chain x is called a ∇-*cycle* if $\nabla x = 0$. The chain x is said to be ∇-*homologous to zero* in K if it is the ∇-boundary of some chain (of one less dimension) of the complex K. Again it is easily verified, by a simple calculation, that $\nabla\nabla x = 0$, i.e., that the ∇-boundary of any chain is a ∇-cycle.[**] So, in the group $Z_\nabla^r K$ of all r-dimensional ∇-cycles of the complex K there is contained a subgroup $H_\nabla^r K$ consisting of all the ∇-cycles which are homologous to zero in K; this subgroup is the image of the group $L^{r-1} K$ under the homomorphism ∇, whose kernel is the

---

[*] The boundary of a zero-dimensional simplex is by definition zero (i.e., the null element which is the single element of the group $L^{-1} K$).

[**] Indeed, $\nabla\nabla t^r = \nabla \Sigma (t_i^{r+1} : t^r) t_i^{r+1} = \Sigma (t_i^{r+1} : t^r) \nabla t_i^{r+1} = \Sigma_i \Sigma_h (t_i^{r+1} : t^r)(t_h^{r+2} : t_i^{r+1}) t_h^{r+2} = \Sigma d_h t_h^{r+2}$, where $d_h = \sum_i (t_h^{r+2} : t_i^{r+1})(t_i^{r+1} : t^r) = 0$.

group $Z_\nabla^{r-1}K$. The factor group

$$Z_\nabla^r K - H_\nabla^r K$$

is denoted by $\nabla^r K$ *and is called the* r-*dimensional* ∇-*group of the complex* K (or r-dimensional upper Betti group of that complex).

5. **Observation on the scalar product of chains. Proof that the operators** Δ **and** ∇ **are adjoint.** Sometimes it turns out to be convenient to consider the scalar product $(x^r \cdot y^r)$ of two chains $x^r = \sum_i a_i t_i^r$ and $y^r = \sum_i b_i t_i^r$ of the complex K, defined naturally in this case by the formula

$$(x^r \cdot y^r) = \sum_i a_i b_i.$$

We observe in particular that if $y^r = t_i^r$, we have $(x \cdot t_i^r) = a_i$, so that *the value of a chain on a given oriented simplex* t_i^r *is the scalar product* $(x^r \cdot t_i^r)$; we shall make constant use of this remark in what follows.

Now it is easy to verify that for any two chains x^r and y^{r-1} we have the equality

$$(x^r \cdot \nabla y^{r-1}) = (\Delta x^r \cdot y^{r-1}), \tag{8}$$

which expresses the fact that the operators Δ *and* ∇ *are adjoint.*

It is sufficient to verify this equality in the special case $x^r = t^r$, $y^{r-1} = t^{r-1}$, where it follows easily from the fact that both the left and the right side of the equality are equal to the incidence coefficient $(t^r : t^{r-1})$.

6. **Cell complexes.** The algebraic construction introduced above makes use of two concepts: orientation and incidence coefficients. But both of these concepts may be introduced not only for simplexes but, for example, also for any convex polyhedra (this has been done in [1], supplement to Chapter 7).

Still more frequently, especially when K is a triangulation of the n-dimensional sphere S^n (or generally, any manifold M^n), one has to consider the adjoint star complex K^*. It turns out in those cases that the barycentric stars also admit orientations, and that one may define for them the incidence coefficients. We shall at the proper time make essential use of this. In order to take in as many such special cases as possible, to permit the construction of a common theory, the general notion of a *cell complex* was proposed, by A. N. Kolmogorov in our country and by A. W. Tucker in America.

Definition. *A cell complex* K *is a set of elements of arbitrary character, called cells,* * *satisfying the following conditions:*

1°. *To each cell there corresponds a non-negative integer, called the dimension*

---

* Cells serve as an analog of *oriented* simplexes.

of that cell (denoted by a superscript).

2°. *For each cell there is a unique cell of the same dimension, called the cell opposite to the given one. Passage from the given cell to the opposite cell is denoted by the sign* $-$; *here* $-(-t^r) = t^r$ *for any cell* t^r.

3°. *To each two cells* t^r *and* t^{r-1} *of successive dimensions there corresponds an integer, denoted by* $(t^r : t^{r-1})$, *and called the incidence coefficient of these cells. One assumes that the incidence coefficients satisfy the following two conditions:*

(a) $$(-t^r : t^{r-1}) = (t^r : t^{r-1}) = -(t^r : t^{r-1});$$

(b) *Let* t^r *and* t^{r-2} *be any two cells of given dimensions* r *and* $r-2$. *Suppose that of any pair of mutually opposite* $(r-1)$-*dimensional cells, one is denoted by* t_i^{r-1} *and the other by* $-t_i^{r-1}$. *Then*

$$\sum_i (t^r : t_i^{r-1}) \, (t_i^{r-1} : t^{r-1}) = 0.$$

7. Chains and Betti groups in cell complexes. In any cell complex one may define an r-dimensional chain as a function that is defined on the set of all r-dimensional cells of the given cell complex and is *odd*, i.e., which takes opposite signs on each two mutually opposite r-dimensional cells. A chain is said to be integer-valued if (as we have assumed up till now) its values are whole numbers. More generally, indeed, there may be given an arbitrary (additively written) abelian group $\mathfrak{A}$, called the *domain* (group) *of values* or *coefficient domain;* then an r-dimensional chain (over the coefficient domain $\mathfrak{A}$) is a function x^r with values which are elements of the group $\mathfrak{A}$, defined on the set of all r-dimensional cells of the given cell complex and taking on opposite cells opposite values. Supposing now that for each pair of mutually opposite cells one of them has once and for all been denoted by t_i^r and the other by $-t_i^r$. Denoting the value of the chain x^r on the cell t_i^r by a_i, we write this chain in the form $x^r = \Sigma \, a_i t_i^r$.* With respect to addition the r-dimensional chains of a cell complex K form a group denoted by $L^r(K, \mathfrak{A})$ (here $\mathfrak{A}$ is the coefficient domain; frequently one does not write it).

The Δ-boundary Δt^r is an $(r-1)$-dimensional chain that takes on each $(r-1)$-dimensional cell t^{r-1} the value $(t^r : t^{r-1})$. The ∇-boundary ∇t^r of the cell t^r is an $(r+1)$-dimensional chain that takes on each $(r+1)$-dimensional cell t^{r+1} the value $(t^{r+1} : t^r)$.

$$\Delta t^r = \Sigma \, (t^r : t_j^{r-1}) \, t_j^{r-1} \, ; \quad \Delta t^r = \Sigma \, (t_h^{r+1} : t^r) \, t_n^{r+1}.$$

---

* This way of writing is again based on the fact that we identify the cell t^r with the integer-valued chain taking on t^r the value 1 (and thus taking on $-t^r$ the value -1), and taking on all the cells other than $\pm t^r$ the value zero.

After this, one may for any chain $x^r = \Sigma\, a_i t_i^r$ over the coefficient domain $\mathfrak{A}$ define the boundary chains

$$\Delta x^r = \Sigma\, a_i \Delta t_i^r; \quad \nabla x^r = \Sigma\, a_i \nabla t_i^r,$$

while from property (b) of the incidence coefficients we again have $\Delta\Delta x^r = 0$, $\nabla\nabla x^r = 0$. Therefore, word for word as for simplicial complexes, we arrive at the groups $Z^r(K,\ \mathfrak{A}) \supseteq H^r(K,\ \mathfrak{A})$ and $Z^r_\nabla(K,\ \mathfrak{A}) \supseteq H^r_\nabla(K,\ \mathfrak{A})$ of cycles and of cycles homologous to zero in K.

The Betti groups (Δ- and ∇-) are defined as the factor-groups

$$\Delta^r(K,\ \mathfrak{A}) = Z^r(K,\ \mathfrak{A}) - H^r(K,\ \mathfrak{A});$$

$$\nabla^r(K,\ \mathfrak{A}) = Z^r_\nabla(K,\ \mathfrak{A}) - H^r_\nabla(K,\ \mathfrak{A}).$$

Remark. If $x^r = \Sigma\, a_i t_i^r$ is a chain over the coefficient domain $\mathfrak{A}$, then its scalar product with the chain $y^r = \Sigma b_i t_i^r$ is defined by the formula

$$(x^r \cdot y^r) = \Sigma\, a_i b_i$$

in the following cases: first, if $\mathfrak{A}$ is a ring and both chains are taken over the coefficient domain $\mathfrak{A}$; secondly, if $\mathfrak{A}$ is any abelian group, and y^r is an integer-valued chain (in particular, the value of the chain x^r on the chain t^r may always be written in the form of a scalar product $(x^r t^r)$, a notation which we shall constantly use); thirdly (we shall not come to these cases until Chapter VI), if the group of coefficients $\mathfrak{A}$ over which the chain x^r is taken, and the group $\mathfrak{B}$, over which the chain y^r is taken, are mutually dual in the sense of the Pontrjagin theory of character groups, and accordingly each term $a_i b_i$ is defined in the sense of the theory of characters.

8. Remark on open and closed subcomplexes. Two cells t^n and t^{n-1} of successive dimensions are said to be *incident* to each other in the given cell complex if their incidence coefficient $(t^n : t^{n-1})$ is different from zero. The set $K_0 \subseteq K$ is said to be an *open (closed) subcomplex of the cell complex K* if along with each cell $t^n \in K_0$ the set K_0 also contains, first, the cell t^n, opposite to the cell t^n, and secondly, every cell t^{n+1} (every cell t^{n-1}), incident to the cell t^n.

It is easy to prove the following

Theorem. *Every open (closed) subcomplex K_0 of the cell complex K is a cell complex (if in K_0 for all elements one takes the dimension, the relation of oppositeness, and the incidence coefficients the same as in K).*

Indeed, we need only convince ourselves of the fact that the identity

$$\sum_i (t^n : t_i^{n-1})\, (t_i^{n-1} : t^n) = 0$$

holds in K_0 for any two cells t^n, t^{n-2} of K_0. But from the definition of an open (closed) subcomplex it follows that in the sum just written those terms which can be different from zero are the same whether we sum over all* $t_i^{r-1} \in K$ or over all $t_i^{r-1} \in K_0$; and since in the first sum we get zero (K is a cell complex!), then we get zero also in the second summation.

The most important application of this theory is the fact that *the set of all oriented simplexes of any open subcomplex of a complete simplicial complex is a cell complex.* We shall say for simplicity that *an open subcomplex of a complete simplicial complex forms a cell complex.* In particular, *every pseudomanifold* (being open in its combinatorial closure) *forms a cell complex.*

9. **Inclusion and excision operators.** These operators will take an important place in many of the constructions we make in the sequel.

Let A be a subcomplex of the complex K. Then each chain x of the complex corresponds in a tirival way to a chain $E_K^A x$ of the complex K, equal to the chain x on the elements of A and equal to zero on $K - A$. We may say that $E_K^A x$ is the same chain x, but considered as a chain of the whole complex K. Therefore we shall frequently drop the sign E_K^A. Sometimes we shall simply write E instead of E_K^A. Evidently the operator is an isomorphic mapping of the group $L^r A$ into the group $L^r K$.

If A is a closed subcomplex of the complex K, then the operator E_K^A commutes with the operator Δ:
$$\Delta E_K^A x = E_K^A \Delta x,$$
where, naturally, the boundary Δ is taken on the left in K and on the right in A. Therefore the operator E maps (isomorphically) the group $Z^r A$ into the group $Z^r K$ and the group $H^r A$ into the group $H^r K$. Accordingly, the operator E generates a homomorphic mapping of the group $\Delta^r A$ into the group $\Delta^r K$. This homomorphism is denoted also by E_K^A and is called the *inclusion homomorphism.*

Remark. If A is an open subcomplex of the complex K, then the operator E_K^A, as the simplest examples show, generally speaking does not commute with the operator Δ. But if A is open, the operator E_K^A commutes with the operator ∇ and therefore generates an inclusion homomorphism of the group $\nabla^r A$ into the group $\nabla^r K$.

We turn to the excision homomorphism J_A^K. For any subcomplex A of the complex K the operator J_A^K puts into correspondence with each chain x of the complex K the chain J_A^K of the complex A, which on any element of the complex

---

* Here, as we stipulated above, in each pair of mutually opposite $(n-1)$-dimensional cells a definite one of them is denoted by t^{n-1}.

A takes the same value as the chain x. If the chain x is written in the form of a linear *form* $x = \Sigma\, a_i t_i^r$, then for the resulting chain $J_A^K x$ we need in this linear form to retain only those of its terms which correspond to elements t_i^r of the complex A. The chain $J_A^K x$ is called the *portion, or piece, of the chain x excised by the subcomplex A, or the portion (piece) lying on the complex A.* Sometimes instead of $J_A^K x$ we shall write simply Ax.

Evidently *the operator J_A^K is a homomorphic mapping* of the group $L^r K$ onto the group $L^r A$, while the operators E_K^A and J_A^K are adjoint: for two chains x_A^r and x_K^r of the complexes A and K, respectively, we have[*]

$$(E_K^A x_A^r, x_K^r) = (x_A^r, J_A^K x_K^r). \tag{9}$$

This formula need only be verified for the case, for example, when the chain x_K^r is a simplex t^r. Then both scalar products in the formula (9) are equal to the value of x_A^r on that simplex if it lies in the complex A, and equal to zero in the opposite case.

We automatically verify the following assertion:

If A is a closed subcomplex of the complex K, then the operator J_A^K commutes with the operator ∇:

$$\nabla J_A^K x^r = J_A^K \nabla x^r,$$

and therefore generates a homomorphism J_A^K of the group $\nabla^r K$ into the group $\nabla^r A$, called the excision homomorphism. If A is an open subcomplex of the complex K, then the operator J_A^K commutes with the operator Δ and therefore generates an excision homomorphism J_A^K of the group $\Delta^r K$ into the group $\Delta^r A$.

10. **The case of integer coefficients.** In this case the group $L^r K$ is a free abelian group with a finite number of generators.[**] Since the subgroups and factor-groups of a group with a finite number of generators are also groups with finite numbers of generators, then both the groups $\Delta^r K$ and $\nabla^r K$ are groups with finite numbers of generators. Therefore $\Delta^r K$ is the direct sum of a finite group $\Theta^r K$, consisting of all the elements of finite order of the group $\Delta^r K$, and a free abelian group, whose rank is equal to the rank of the group $\Delta^r K$. This rank is called *the r-dimensional Betti number* of the complex K and is denoted by $\pi^r K$. The group

---

[*] See the remark in subsection 8.

[**] Moreover, $L^r K$ is a so-called *grating* of rank a^r, where $2a^r$ is the number of r-dimensional cells of the cell complex K (if K is a simplicial complex, then a^r is the number of r-dimensional simplexes of the complex K). By a grating we understand a free abelian group with an elementary basis singled out once and for all, i.e., a system of independent generators. In the grating $L^r K$ as a basis one chooses the system of cells $t_1^r, \cdots, t_a^r$ (denoting in each pair of opposite cells one by t_i^r, the other by $-t_i^r$). On the subject of gratings, abelian groups with a finite number of generators, and so forth, see the supplement to the book [1], where detailed proofs of the algebraic theorems mentioned here are given.

$\Theta^r K$ is called the *r-dimensional torsion group of the complex* K.

For further study of the homology groups of a finite complex we shall need the following algebraic theorem:

Theorem A. Let X be a free abelian group, U any subgroup of the group X, $\hat{U}$ its quotient closure.[*] Then there exists a basis $z_1, \cdots, z_\pi; u_1, \cdots, u_\tau; v_1, \cdots, v_\sigma$ of the group X, satisfying the following conditions:

(a) the elements z_i have infinite order relative to the subgroup U;[**]

(b) $v_1, \cdots, v_\sigma$ are elements of the group U; the order of the element u_i relative to the subgroup U is a natural number $\theta_i > 1$ (i.e., the smallest natural number θ_i for which $\theta_i u_i \in U$), while θ_i divides θ_{i+1};

(c) the elements $\theta_1 u_1, \cdots, \theta_\tau u_\tau; v_1, \cdots, v_\sigma$ form a basis for the group U;

(d) the elements $u_1, \cdots, u_\tau; v_1, \cdots, v_\sigma$ form a basis for the group $\hat{U}$.

Every basis of a free group X satisfying these conditions is called a basis of the group X *canonical with respect to the subgroup* U.

We return to the complex K and consider the grating $L^r = L^r K$ whose elementary basis is the set of cells

$$t_1^r, \cdots, t_{\alpha^r}^r.$$

Along with this grating we shall consider also the gratings L^{r-1} and L^{r+1}.

With the elementary bases thus selected for these gratings, the matrix of the homomorphism Δ of the grating L^r into the grating L^{r-1} is given by the incidence matrix

$$E^{r-1} = \|\epsilon_{ij}^{r-1}\|, \text{ where } \epsilon_{ij}^{r-1} = (t_i^r : t_j^{r-1}).$$

Now suppose that in the grating L^r there have been chosen two adjoint bases:[***]

$$W^r = \{w_1^r, \cdots, w_{\alpha^r}^r\}$$

and

$$\overline{W}^r = \{\overline{w}_1^r, \cdots, \overline{w}_{\alpha^r}^r\}.$$

Analogously, suppose that

$$W^{r-1} = \{w_1^{r-1}, \cdots, w_{\alpha^{r-1}}^{r-1}\};$$
$$\overline{W}^{r-1} = \{\overline{w}_1^{r-1}, \cdots, \overline{w}_{\alpha^{r-1}}^{r-1}\}$$

---

[*] A *quotient closure* of a subgroup U of a group X is a subgroup $\hat{U}$ of the group X, consisting of all elements x of the group X, for which there exists an integer θ such that $\theta x \in U$. Evidently $U \subseteq \hat{U}$. If $\hat{U} = U$, then U is said to be a *quotient-closed* subgroup of the group. The quotient closure of a subgroup U is completely characterized by the fact that the factor-group $X - U$ is a free abelian group.

[**] This means that for every integer $c \neq 0$, the element cz_i is not contained in U.

[***] The bases $\{x_1, \cdots, x_n\}$ and $\{\overline{x}_1, \cdots, \overline{x}_n\}$ of the grating X are said to be adjoint if $(x_i, \overline{x}_j) = \delta_{ij}$ (the scalar product in the grating is taken with respect to its elementary basis). The elements x_i and $\overline{x}_i$ of adjoint bases are said to be adjoint.

For each basis of the grating there exists a unique basis adjoint to it (see [1], Supplement 1, §5).

are two adjoint bases of the grating L^{r-1}. Suppose that $H^{r-1} = \|\eta_{ij}^{r-1}\|$ is the matrix of the homomorphism Δ of the group L^r into the group L^{r-1}, taken in the bases W^r, W^{r-1}:

$$\Delta w_i^r = \sum_j \eta_{ij}^{r-1} w_j^{r-1}.$$

Since the homomorphisms Δ and ∇ are adjoint, then[*]

$$\nabla \overline{w}_j^{r-1} = \sum_i \eta_{ij}^{r-1} \overline{w}_i^r.$$

The number η_{ij}^{r-1} is thus the coefficient with which w_j^{r-1} enters into ∇w_i^r and the coefficient with which $\overline{w}_i^r$ enters into $\nabla \overline{w}_j^{r-1}$.

We now apply Theorem A to the group $Z^r K$ and its subgroup $H^r K$. We obtain:

Theorem B. The group $Z^r = Z^r K$ has a basis $z_1^r, \cdots, z_{\pi^r}^r$; $u_1^r, \cdots, u_{\tau^r}^r$; $v_1^r, \cdots, v_{\sigma^r}^r$ satisfying the following conditions:

(a) the orders of the elements z_i^r relative to the subgroup $H^r = H^r K$ are equal to ∞;[**]

(b) $v_1^r, \cdots, v_{\sigma^r}^r$ are elements of the group H^r; the order of the element u_i^r with respect to H^r is an integer θ_i^r, while θ_i^r divides θ_{i+1}^r;

(c) the cycles $\theta_1^r u_1^r, \cdots, \theta_{\tau^r}^r u_{\tau^r}^r$ and $v_1^r, \cdots, v_{\sigma^r}^r$ form a basis of the group H^r, and the cycles u_i^r and v_i^r form a basis of the group $\hat{H}^r$.

A basis of the group Z^r satisfying these conditions is said to be *canonical*.

Remark. Denoting by Z_h^r, U_i^r, V_j^r infinite cyclic groups whose generators are, respectively, the elements z_h^r, u_i^r, v_j^r, we have the decomposition into direct sums:

$$Z^r = \sum_{h=1}^{\pi^r} Z_h + \sum_{i=1}^{\tau^r} U_i + \sum_{j=1}^{\sigma^r} V_j;$$

$$H^r = \sum_{i=1}^{\tau^r} \theta_i^r U_i + \sum_{j=1}^{\sigma^r} V_j,$$

from which it follows that the group $\Delta^r K = Z^r - H^r$ is isomorphic to the direct sum of infinite cyclic groups, taken in the number π^r, and the sum of τ^r cyclic groups of orders θ_i^r, $i = 1, \cdots, \tau^r$.

Now suppose that for all $r = 0, 1, \cdots, n$, where n is the dimension of the complex K, canonical bases have been chosen for the groups $Z^r K$. In view of the

---

[*] This results from the following theorem, whose proof is given for example in the book [1], page 643. Suppose that $\{x_1, \cdots, x_p\}$ and $\{\overline{x}_1, \cdots, \overline{x}_p\}$ are adjoint bases of the grating X, and that $\{y_1, \cdots, y_q\}$ and $\{\overline{y}_1, \cdots, \overline{y}_q\}$ are adjoint bases of the grating Y; suppose that f and g are adjoint homomorphisms, respectively, of X into Y and of Y into X; suppose, finally, $fx_i = \sum_j c_{ij} y_j$. Then $g\overline{y}_j = \sum_i c_{ij} \overline{x}_i$.

[**] I.e., for any integer $c \neq 0$ whatever, the cycle cz_i^r is not homologous to zero in K.

isomorphism Δ between the groups $L^r - z^r$ and H^{r-1}, there corresponds to the basis

$$\theta_1^{r-1} u_1^{r-1}, \cdots, \theta_{\tau^{r-1}}^{r-1} u_{\tau^{r-1}}^{r-1}; \ v_1^{r-1}, \cdots, v_{\sigma^{r-1}}^{r-1}$$

of the group H^{r-1} the basis

$$\xi_1^r, \cdots, \xi_{\tau^{r-1}}^r; \ \eta_1^r, \cdots, \eta_{\sigma^{r-1}}^r$$

of the group $L^r - Z^r$, while, for any $x_i^r \in \xi_i^r$, $y_j^r \in \eta_j^r$,

$$\Delta x_i^r = \theta_i^{r-1} u_i^{r-1}; \ \Delta y_j^r = v_j^{r-1}.$$

Choose once and for all the chains $x_i^r \in \xi_i^r$, $y_j^r \in \eta_j^r$. It is easy to see that then the chains

$$x_1^r, \cdots, x_{\tau^{r-1}}^r; \ y_1^r, \cdots, y_{\sigma^{r-1}}^r; \ z_1^r, \cdots, z_{\pi^r}^r; \ u_1^r, \cdots, u_{\tau^r}^r; \ v_1^r, \cdots, v_{\sigma^r}^r \quad (10)$$

form a basis W^r of the group L^r. These bases, constructed for $r = 0, 1, \cdots, n$, *also form a system of canonical bases of the complex K.* Evidently the homomorphism Δ (of the group L^r into the group L^{r-1}) has in these bases the matrix (in which each row expresses the boundary of the element at its head)

	$x_1^{r-1} \dots x_{\tau^{r-2}}^{r-1}$	$v_1^{r-1} \dots y_{\sigma^{r-2}}^{r-1}$	$z_1^{r-1} \dots z_{\pi^{r-1}}^{r-1}$	$u_1^{r-1} \dots u_{\tau^{r-1}}^{r-1}$	$v_1^{r-1} \dots v_{\sigma^{r-1}}^{r-1}$
x_1^r ⋮ $x_{\tau^{r-1}}^r$	0	0	0	$\begin{smallmatrix}\theta_1^{r-1} & & \\ & \ddots & \\ & & \theta_{\tau^{r-1}}^{r-1}\end{smallmatrix}$	0
y_1^r ⋮ $y_{\sigma^{r-1}}^r$	0	0	0	0	$\begin{smallmatrix}1 & & \\ & \ddots & \\ & & 1\end{smallmatrix}$
z_1^r ⋮ $z_{\pi^r}^r$	0	0	0	0	0
u_1^r ⋮ $u_{\tau^r}^r$	0	0	0	0	0
v_1^r ⋮ $v_{\sigma^r}^r$	0	0	0	0	0

Now we choose bases $\overline{W}^r$ adjoint to the bases (10); the elements of these bases, adjoint, respectively, to the elements

$$x_1^r, \cdots, x_{\tau^{r-1}}^r; \; y_1^r, \cdots, y_{\sigma^{r-1}}^r; \; z_1^r, \cdots, z_{\pi^r}^r; \; u_1^r, \cdots, u_{\tau^r}^r; \; v_1^r, \cdots, v_{\sigma^r}^r,$$

we denote by

$$\overline{u}_1^r, \cdots, \overline{u}_{\tau^{r-1}}^r; \; \overline{v}_1^r, \cdots, \overline{v}_{\sigma^{r-1}}^r; \; \overline{z}_1^r, \cdots, \overline{z}_{\pi^r}^r; \; \overline{x}_1^r, \cdots, \overline{x}_{\tau^r}^r; \; \overline{y}_1^r, \cdots, \overline{y}_{\sigma^r}^r. \quad (10^*)$$

In view of what has been said above about matrices of adjoint homomorphisms in adjoint bases, the homomorphism ∇ (of the group L^{r-1} into L^r) has in the bases (10^*) the matrix

	$\overline{u}_1^{r-1} \cdots \overline{u}_{\tau^{r-2}}^{r-1}$	$\overline{v}_1^{r-1} \cdots \overline{v}_{\sigma^{r-2}}^{r-1}$	$\overline{z}_1^{r-1} \cdots \overline{z}_{\tau^{r-1}}^{r-1}$	$\overline{x}_1^{r-1} \cdots \overline{x}_{\tau^{r-1}}^{r-1}$	$\overline{y}_1^{r-1} \cdots \overline{y}_{\sigma^{r-1}}^{r-1}$
$\overline{u}_1^r \; \vdots \; \overline{u}_{\tau^{r-1}}^r$	0	0	0	$\theta_1^{r-1} \cdots \theta_{\tau^{r-1}}^{r-1}$ (diagonal)	0
$\overline{v}_1^r \; \vdots \; \overline{v}_{\tau^{r-1}}^r$	0	0	0	0	$1 \cdots 1$ (diagonal)
$\overline{z}_1^r \; \vdots \; \overline{z}_{\pi^r}^r$	0	0	0	0	0
$\overline{x}_1^r \; \vdots \; \overline{x}_{\tau^r}^r$	0	0	0	0	0
$\overline{y}_1^r \; \vdots \; \overline{y}_{\sigma^r}^r$	0	0	0	0	0

in which each column expresses the ∇-boundary of the element at its head.

Now we may without difficulty express the ∇-groups of the complex K by means of the Δ-groups of that complex, which constitutes the fundamental object of this section of our exposition. Indeed we prove the following proposition.

Fundamental theorem. *For any cell complex K, the group $\nabla^r K$ is isomorphic to the direct sum of the $(r-1)$-dimensional torsion group $\Theta^{r-1} K$ of the complex*

K and of a free abelian group of rank $\pi^r = \pi^r K$.

Proof. Since (10^*) is the basis of the group L^r, each chain x^r may be represented uniquely in the form

$$x^r = \Sigma a_i \bar{u}^r_i + \Sigma b_j \bar{v}^r_j + \Sigma c_h \bar{z}^r_h + \Sigma d_k \bar{x}^r_k + \Sigma e_l \bar{y}^r_l,$$

while

$$\nabla x^r = \Sigma d_k \nabla \bar{x}^r_k + \Sigma e_l \nabla \bar{y}^r_l = \sum_1^{\tau^r} d_k \theta^r_k \bar{u}^{r+1}_k + \sum_1^{\sigma^r} e_l \bar{v}^{r+1}_l, \tag{11}$$

Therefore the chain x^r is a ∇-cycle if and only if in the representation (11) we have

$$d_k = 0, \ e_l = 0 \quad \begin{cases} k = 1, \cdots, \tau^r; \\ l = 1, \cdots, \sigma^r. \end{cases}$$

In other words, the ∇-cycles $\bar{u}^r_i, \bar{v}^r_j, \bar{z}^r_h$ form a basis for the group Z^r_∇.

Further, since $\bar{W}^r$ and $\bar{W}^{r-1}$ are respectively bases of the groups L^r and L^{r-1}, each equality of the form ∇x^{r-1} is a linear combination of equalities of the type

$$\nabla \bar{x}^{r-1}_i = \theta^{r-1}_i \bar{u}^r_i, \ \nabla \bar{y}^{r-1}_j = \bar{v}^r_j.$$

This means that the ∇-cycle

$$x^r = \Sigma a_i \bar{u}^r_i + \Sigma b_j \bar{v}^r_j + \Sigma c_h \bar{z}^r_h \tag{12}$$

lies in the group H^r_∇ if and only if in (12) all the c_h are equal to zero and all the a_i may be divided by θ^{r-1}_i. In other words, the ∇-cycles

$$\theta^{r-1}_1 \bar{u}^r_1, \cdots, \theta^{r-1}_{\tau^{r-1}} \bar{u}^r_{\tau^{r-1}}; \ \bar{v}^r_1, \cdots, \bar{v}^r_{\sigma^{r-1}}$$

form a basis for the group H^r_∇. Thus, denoting the cyclic subgroups of the groups L^r, generated by the elements $\bar{x}^r_k, \bar{y}^r_l, \bar{z}^r_h, \bar{u}^r_i, \bar{v}^r_j$, respectively, by $\bar{X}^r_k, \bar{Y}^r_l, \bar{Z}^r_h, \bar{U}^r_i, \bar{V}^r_j$, we have a decomposition into direct sums:

$$L^r = \sum_1^{\tau^r} \bar{X}^r_k + \sum_1^{\sigma^r} \bar{Y}^r_l + \sum_1^{\pi^r} \bar{Z}^r_h + \sum_1^{\tau^{r-1}} \bar{U}^r_i + \sum_1^{\sigma^{r-1}} \bar{V}^r;$$

$$Z^r_\nabla = \sum_1^{\pi^r} \bar{Z}^r_h + \sum_1^{\tau^{r-1}} \bar{U}^r_i + \sum_1^{\sigma^{r-1}} \bar{V}^r_j;$$

$$H^r_\nabla = \sum_1^{\tau^{r-1}} \theta^{r-1}_i \bar{U}^r_i + \sum_1^{\sigma^{r-1}} V^r_j,$$

whence (writing isomorphisms in the form of equalities)

$$\nabla^r K = Z^r_\nabla - H^r_\nabla = \sum_{k=1}^{\pi^r} \bar{Z}^r_h + \sum_{i=1}^{\tau^{r-1}} (\bar{U}^r_i - \theta^{r-1}_i \bar{U}^r_i).$$

Here the group $U^r_i - \theta^{r-1}_i U^r_i$ is a finite cyclic group of order θ^{r-1}_i, so that, in view of the remark on page 22, the group $\Sigma(\bar{U}^r_i - \theta^{r-1}_i \bar{U}^{r-1}_i)$ is isomorphic to the group $\Theta^{r-1} K$ and the fundamental theorem is proved.

CHAPTER II

Direct and inverse group spectra [†]

1. Directed sets and cofinal sections. As is well known, the set X is said to be partially ordered if for certain pairs x_0, x_1 of its elements there has been established an ordering relation, expressed by the words "one element (e.g., x_1) follows another element (e.g., x_0)". One writes this as follows: $x_1 > x_0$. Here we assume that the condition of transitivity is satisfied: if $x_1 > x_0$ and $x_2 > x_1$, then $x_2 > x_0$.

A partially ordered set is said to be *directed* if for any two of its elements x_0 and x_1 there exists an element x_2 following both x_0 and x_1.

A directed set X' is said to be a *section* of the directed set X if the two following conditions are satisfied:

$1°$. $X' \subseteq X$.

$2°$. If $x_0' \in X'$, $x_1' \in X'$ and if $x_1' > x_0'$ in X', it follows that $x_1' > x_0'$ in X.

A directed set X' which is a section of a directed set X is said to be a *cofinal section* of X if each element x of the set X is followed by at least one element $x' \in X'$ (this is the *cofinality condition*).

Special cases. If the directed set X' is a section of the directed set X and if X and X' contain the same set of elements, then X' is always a cofinal section of X.

In this case one says that X' is obtained from X by *weakening the ordering*.

At the same time suppose that X' is a section of the directed set X preserving the order of X (i.e., if x_0, $x_1 \in X'$, then $x_1 > x_0$ in X' if and only if $x_1 > x_0$ in X). If X' satisfies the condition of cofinality, then X' is automatically directed and, accordingly, is a cofinal section of the set X.

2. On groups. All the groups considered in this book are commutative, as will be assumed without further mention. A finite cyclic group of order m will be denoted by I_m; the group of integers by I.

The additive group of real numbers, taken modulo 1 in its natural topology, is denoted by K. Every group topologically isomorphic to the group K is called a *continuous cyclic group*. Topological groups are considered only in the sixth chapter; in these discussions, if the opposite is not specified, subgroups are taken to be closed and homomorphisms to be continuous. By $\mathfrak{A}$ we denote discrete, and by $\mathfrak{B}$ bicompact, groups; if $\mathfrak{A}$ and $\mathfrak{B}$ are considered simultaneously, then we

---

[†] This chapter has an auxiliary character, to be referred to only as necessary. This remark refers especially to the text in small print, whose study we recommend be deferred to Chapter VI.

suppose that they are dual to one another in the sense of the theory of characters.*

3. **Direct spectra of groups.** At the base of the definition of the spectrum of a group lies some directed set A, whose elements will be called *indices* (and denoted by α, β, γ, and so forth).

We shall suppose that to each index $\alpha \in A$ there corresponds a group X_α, while we suppose that if $\alpha \neq \beta$ the groups X_α and X_β have no common elements.** We suppose moreover that for any two indices α, β, where β follows α, there is given an isomorphism π_β^α of the group X_α into the group X_β, while for $\gamma > \beta > \alpha$ we always have $\pi_\gamma^\alpha = \pi_\gamma^\beta \pi_\beta^\alpha$. Under these assumptions we say that there is given a *direct spectrum*

$$\{X_\alpha, \pi_\beta^\alpha\}$$

of the groups X_α with projections π_β^α. A direct spectrum defines a limit group $X = \lim_{\longrightarrow}(X_\alpha, \pi_\beta^\alpha)$ (or simply $X = \lim X_\alpha$) as follows. Two elements $x_\alpha \in X_\alpha$, $x_\beta \in X_\beta$ are said to be *equivalent* in the given spectrum if there exists a γ following both α and β such that $\pi_\gamma^\alpha x_\alpha = \pi_\gamma^\beta x_\beta$.

It is easy to verify that this definition of equivalence satisfies the conditions of reflectivity, symmetry and transitivity, and accordingly leads to a decomposition of the set $\underset{\alpha}{U} X_\alpha$ *** into *classes* or *aggregates* of elements which are equivalent to one another. These aggregates are by definition the elements of the limit group X. The sum of two aggregates ξ' and ξ'' is defined as follows: we choose an α such that there are representatives x'_α and x''_α lying in ξ' and ξ'', respectively, and both lying in the group X_α.****

The aggregate ξ containing the element $x'_\alpha + x''_\alpha$ is by definition the sum of the of the aggregates ξ' and ξ''. The fact that this definition of the sum does not depend on the choice of the elements $x'_\alpha \in \xi'$, $x''_\alpha \in \xi''$ is easily verified. The

---

*Two groups $\mathfrak{A}$ and $\mathfrak{B}$ are said to be dual to one another (in the sense of L. S. Pontrjagin) if each of them is the character group of the other. The duality of the groups $\mathfrak{A}$ and $\mathfrak{B}$ is denoted as follows: $\mathfrak{A} | \mathfrak{B}$. Here we restrict ourselves to the case when one of these groups, namely in our notation the group $\mathfrak{A}$, is discrete and the other, $\mathfrak{B}$, is bicompact. We shall not encounter the theory of characters until the sixth chapter, in which it will be supposed that the fundamental concepts of characters is known to the reader. We shall recall them in subsections 6 and 7 of the present chapter.

**It is easy to satisfy this condition, replacing, if necessary, the group X_α by a group X'_α isomorphic to it with elements (a, x), where x runs through the whole group X_α and the isomorphism between X'_α and X_α is given by the formula $(a, x) \rightleftarrows x$.

***This set is called the *spectral set* of the given spectrum.

****Such an α always exists. Choose, indeed, some $x'_{\alpha'} \in \xi'$, $x''_{\alpha''} \in \xi''$, with α following α' and α''. Then the elements $x'_\alpha = \pi_\alpha^{\alpha'} x'_{\alpha'}$ and $x''_\alpha = \pi_\alpha^{\alpha''} x''_{\alpha''}$ of the group belong respectively to the aggregates ξ' and ξ''.

null element of the group X is the aggregate containing the zeros of all the groups X_α.

Definition. Two direct spectra

$$\{X_\alpha,\ \rho_\beta^\alpha\}, \tag{1}$$

$$\{Y_\alpha,\ \sigma_\beta^\alpha\}, \tag{2}$$

ordered by the same indices α, are said to be isomorphic if for each α there exists an isomorphism f_α of the group X_α onto the group Y_α, commuting with the projections in the sense that

$$f_\beta \rho_\beta^\alpha x_\alpha = \sigma_\beta^\alpha f_\alpha x_\alpha$$

for any $\alpha,\ \beta,\ \beta > \alpha$.

Theorem 1. *Isomorphic direct spectra have isomorphic limit groups.*

Proof. In view of the isomorphisms f_α we have a $1-1$ mapping f of the spectral set $\underset{\alpha}{U} X_\alpha$ of one spectrum onto the spectral set $\underset{\alpha}{U} Y_\alpha$ of the other. Here two elements x_α and $x'_{\alpha'}$, equivalent in the spectrum (1), go over into the elements $y_\alpha = f_\alpha x_\alpha$ and $y'_{\alpha'} = f_{\alpha'} x'_{\alpha'}$, equivalent in the spectrum (2) (since, from $\rho_\beta^\alpha x_\alpha = \rho_\beta^{\alpha'} x'_{\alpha'}$ it follows that $f_\beta \rho_\beta^\alpha x_\alpha = f_\beta \rho_\beta^{\alpha'} x'_{\alpha'}$, i.e., that $\sigma_\beta^\alpha y_\alpha = \sigma_\beta^{\alpha'} y'_{\alpha'}$). In other words, we have a $1-1$ mapping of the group $\lim X_\alpha$ onto the group $\lim Y_\alpha$. This mapping is obviously an isomorphism.

4. **Inverse spectra of (bicompact) groups.**[*] Suppose that to each element α of a given directed set of indices one puts in correspondence a (bicompact) group Y_α, while for $\beta > \alpha$ there is defined a (continuous) homomorphism ϖ_α^β of the group Y_β into the group Y_α satisfying the transitivity condition: for each triple $\gamma > \beta > \alpha$ we have $\varpi_\alpha^\gamma = \varpi_\alpha^\beta \varpi_\beta^\gamma$.

In this case we say that there is given an inverse spectrum

$$\{Y_\alpha,\ \varpi_\alpha^\beta\}$$

of the (bicompact) groups Y_α with the projections ϖ_α^β. The limit group

$$Y = \varprojlim\ (Y_\alpha, \varpi_\alpha^\beta)$$

(or simply $Y = \lim Y_\alpha$) of the inverse spectrum is defined as follows. Its elements are so-called threads, i.e., a system $\eta = \{y_\alpha\}$ (with one element y_α from each group Y_α), satisfying the condition: *if $\beta > \alpha$, then* $\varpi_\alpha^\beta y_\beta = y_\alpha$. The elements y_α of a thread $\eta = \{y_\alpha\}$ are called its coordinates. If $\eta' = \{y'_\alpha\}$ and $\eta'' = \{y''_\alpha\}$ are two threads, then for each α the quantity

$$y_\alpha = y'_\alpha + y''_\alpha$$

is defined. As the sum of η' and η'' we take the thread $\eta = \{y_\alpha\}$. The null element of the group $\lim Y_\alpha$ is the thread all of whose coordinates are zero.

The topology in the group Y is introduced as follows. In order to define a neighbor-

---

[*] Inverse spectra are considered also for groups without a topology, as well as for topological groups, and in the latter case for bicompact groups. Our exposition is adapted to both of these cases; the *bicompact* case is put in parentheses.

hood of any element $\eta^0 = \{y_\alpha^0\}$ of the group Y, we fix a finite number of coordinates $y_{\alpha_1}^0, \cdots, y_{\alpha_s}^0$ of this element and choose arbitrarily neighborhoods $u_{\alpha_1} = Oy_{\alpha_1}^0, \cdots, u_{\alpha_s} = Oy_{\alpha_s}^0$ in $Y_{\alpha_1}, \cdots, Y_{\alpha_s}$, respectively. A neighborhood of η^0 is the neighborhood $O_{u_{\alpha_1}, \cdots, u_{\alpha_s}}$ consisting of all the threads $\eta = \{y_\alpha\}$ satisfying the condition

$$y_{\alpha_1} \in u_{\alpha_1}, \cdots, y_{\alpha_s} \in u_{\alpha_s}$$

(evidently, it would have been sufficient to restrict ourselves to defining neighborhoods of the null element).

Theorem 2. *The topological group $Y = \lim Y_\alpha$ is bicompact.*

Indeed, the group Y, considered as a topological space, is obviously a subset of the topological (Tihonov) product Π of the given bicompacta Y_α,[*] and therefore it is sufficient to prove that Y is closed in Π. Take an arbitrary point $\xi = \{x_\alpha\} \in \Pi - Y$. Then there exist α, β, with $\beta > \alpha$, such that $y_\alpha = \omega_\alpha^\beta x_\beta \neq x_\alpha$. We choose nonintersecting neighborhoods u_α and v_α of the points x_α and y_α in Y_α. Because of the continuity of the mapping ω_α^β there exists a neighborhood v_β of the point x_β such that $\omega_\alpha^\beta v_\beta \subseteq u_\alpha$. Then the neighborhood $O_{u_\alpha v_\beta}$ of the point ξ in Π contains no thread, since $\omega_\alpha^\beta v_\alpha$ and u_α do not intersect. Thus, no point $\xi \in \Pi - Y$ is a point of the adherence of the set Y in the set Π. Thus the fact that Y is closed in Π and thus bicompact is proved.

5. Cofinal sections of spectra. If A' is a cofinal section of a directed set of indices A, then, if we retain from the given direct or inverse spectrum only those groups and those projections which correspond to elements $\alpha \in A'$ (or to pairs α, β of elements of A'), we obtain a spectrum, called a cofinal section of the initial spectrum. If S' is a section of the direct spectrum S, then every two elements of the spectral set S' which are equivalent in S' will also be equivalent in the spectrum S. But if S' forms a cofinal section of the spectrum S, then it is easily seen that the inverse assertion will also be true: any two elements of the spectral set S' equivalent in S will also be equivalent in S'. From the first assertion it follows that each coset of the spectrum S' is contained in one (evidently in only one) coset of the spectrum S. But if S' is a cofinal section of the spectrum S, then, as is easily seen, each coset of the spectrum S contains at least one, and in view of the second assertion, only one coset of the spectrum S'. Thus:

Theorem 3. *If the direct spectrum S' is a cofinal section of the spectrum S, then, putting into correspondence with each coset of the spectrum S' the unique coset of the spectrum S containing it, we obtain an isomorphism of the limit group of the spectrum S' with the limit group of the spectrum S.*

---

[*] The elements of the Tihonov product space are all possible systems $\xi = \{x_\alpha\}$ (with one x_α in each Y_α) having the same topology as above. The Tihonov product of any number of bicompacta is a bicompact (see [2], p. 394). If, as in our case, we take the Tihonov product of bicompact groups, it turns out to be, with coordinatewise addition, a (bicompact) group. The group Y, together with its algebra and topology, indeed turns out to be a subgroup of this group.

We turn to the case of inverse spectra. If Σ' is a section of the inverse spectrum Σ, then each thread of the spectrum Σ contains (uniquely) a thread of the spectrum Σ', which establishes a homomorphism of the limit group of the spectrum Σ into the limit group of the spectrum Σ'. If Σ' is a cofinal section of the spectrum Σ, then, supplementing each thread $\eta' = \{y_{\alpha'}\}$ of the spectrum Σ' by means of the projections $\omega_{\alpha}^{\beta'} y_{\beta'}$ of its elements to all the Y_{α}, for each $\alpha \in A - A'$ we obtain a unique thread of the spectrum Σ, containing the given thread of the spectrum Σ'. Thus:

Theorem 3$'$. *If Σ' is a cofinal section of the inverse spectrum Σ, then, putting into correspondence with each thread of the spectrum Σ the unique thread of the spectrum Σ' contained in it, we obtain an isomorphic mapping of the limit group of the spectrum Σ onto the limit group of the spectrum Σ'.*

6. **On the theory of characters.** We have already mentioned that in Chapter VI of this monograph we shall assume an acquaintance with the theory of characters.

We shall recall the fundamental concepts and facts of the theory of characters which we shall need in the sequel.

A character of a group X (discrete or bicompact) is a homomorphism of that group into a group K. Here, if the group X is bicompact, we consider only continuous characters. The sum of two characters ϕ_1 and ϕ_2 is the character ϕ, defined by the formula $\phi(x) = \phi_1(x) + \phi_2(x)$ for every $x \in X$. Thus it turns out that the set of all characters of a given group X is itself a group.

If X is a discrete group, then in the group X^* of its characters one introduces a topology as follows. In order to obtain an arbitrary neighborhood of the null element of the group X^*, we choose a neighborhood W of the null element of the group K and a finite number of elements $x_1, \cdots, x_s$ of the group X. These define a neighborhood of the null element of the group X^* as follows: the neighborhood consists of all those characters ϕ for which $\phi(x_i) \in W$ for $i = 1, 2, \cdots, s$. In this topology the group X^* turns out to be bicompact.

If X is a bicompact group, then its character group will not be provided with a topology, i.e., it remains discrete.

If x^* is a character of the group X, then the value $x^*(x)$ of the function x^* for a given element $x \in X$ will for convenience be written in the form xx^* and called the *scalar product* of x by x^*. Here we must keep in mind that $xx^* \in K$. If we choose a definite element $x \in X$, then, writing $x(x^*) = xx^*$ for any $x^* \in X^*$, we obtain a character of the group X^*. Here, if X^* is bicompact (i.e., if X is discrete), then the character xx^* turns out to be continuous. Thus each element x of the group X is, in view of the formula $x(x^*) = xx^*$, a character of the group X^*. The fundamental fact of the theory of characters is the fact that *every* character ϕ of the group X^* is obtained in this way (i.e., realized by the formula $\phi(x^*) = xx^*$) for some element x of the group X, and that different elements of the group X are different characters of the group X^*.[*] Moreover, if the group X is bicompact (hence X^* discrete), then the topology given in X coincides with the topology introduced in this group as a group of characters of the discrete group X^*. All of this

---

* This last condition may evidently be reformulated as follows: for any $x \in X$, $x \neq 0$, there is an $x^* \in X^*$ such that $xx^* \neq 0$; by the very definition of the group X^*, for each $x^* \in X^*$, $x^* \neq 0$, there exists an $x \in X$ such that $xx^* \neq 0$.

is expressed in short form by the words:

If X^* is a group of characters of the group X, then, inversely, the group X is a group of characters of the group X^* under the formula

$$x^*(x) = xx^* = x(x^*).$$

This important theorem, which we have formulated only for the case of discrete and bicompact groups, is the fundamental theorem of the theory of characters and is known as the *algebraic duality law of Pontrjagin*. For its proof we refer the reader to the book of Pontrjagin [3].

Regarding this theorem, we give in addition the following remarks, which are important in the sequel.

Let X and Y be two groups, of which the first is discrete and the second bicompact. Suppose that for any two elements $x \in X$, $y \in Y$ the scalar product $xy \in K$ is defined, distributive with respect to x and y, continuous with respect to y and satisfying the following condition: the zero of each of the two groups is the only element in its group whose scalar product with an arbitrary element of the second group is equal to the zero of the group K.

In this case each of our two groups is by formula (1) (in which we need to put y in place of x^*) the character group of the second group, and we may say that *the groups X and Y are dual to each other under the given definition of the scalar product*.

We again stress the importance of the so-called existence theorem in the theory of characters (whose proof constitutes an important part of the proof of the whole algebraic duality law): if $X \mid Y$, then for each element of one of the two groups X, Y, not equal to zero, one may find an element of the other group for which the scalar product of these elements will be different from zero.

A slight strengthening of this existence theorem is given by the following proposition:

Strong existence theorem. If $X \mid Y$ and A is a subgroup of X, not coinciding with the whole group X, then for $x_0 \in X - A$ one can always find an element y of the group Y such that, for all $x \in A$,

$$xy = 0,$$
$$x_0 y \neq 0.$$

For the proof it is sufficient to observe that the so-called annihilator* B of the group A in the group Y is a group dual to the factor-group $X - A$ of the group X modulo the subgroup A. Denoting by $\xi_0 \in X - A$ the coset containing the element x_0, we have by hypothesis $\xi_0 \neq 0$. Therefore there exists an element $y_0 \in B$, such that $x_0 y_0 = \xi_0$, $y_0 \neq 0$; while $xy_0 = 0$ for all $x \in A$ by definition of the group B.

7. Adjoint homomorphisms. Adjoint spectra. Suppose we are given two pairs of dual groups:

$$X_\alpha \mid Y_\alpha;$$
$$X_\beta \mid Y_\beta$$

(X_α and X_β discrete, Y_α and Y_β bicompact).

---

* I.e., the subgroup $B \subseteq Y$ consisting of all elements having null scalar products with arbitrary elements of the group A.

The homomorphisms f_β^α of the group X_α into the group X_β and ϕ_α^β of the group Y_β into the group Y_α are said to be *adjoint* if for any $x_\alpha \in X_\alpha$, $y_\beta \in Y_\beta$ we have

$$x_\alpha \phi_\alpha^\beta y_\beta = f_\beta^\alpha x_\alpha y_\beta.$$

Now we suppose that for the same set of indices we have two spectra:

the direct spectrum

$$S = \{X_\alpha, \pi_\beta^\alpha\}$$

and

the inverse spectrum

$$\Sigma = \{Y_\alpha, \varpi_\alpha^\beta\},$$

while for any α the groups X_α and Y_α are dual to each other, and for any pair $\beta > \alpha$ the homomorphisms π_β^α and ϖ_α^β are adjoint. Under these conditions the spectra S and Σ are called adjoint. Write $X = \varinjlim X_\alpha$, $Y = \varprojlim Y_\alpha$.

For any two elements

$$x \in X, \ y \in Y$$

one defines the scalar product xy in the following way: take any $x_\alpha \in x$ and set $xy = x_\alpha y_\alpha$, where $y_\alpha \in y$.

We shall show first of all that this definition is correct, i.e., that it does not depend on the choice of $x_\alpha \in x$. Indeed, suppose that $x_\beta' \in x$. Take $\gamma > \alpha$, β such that $\pi_\gamma^\alpha x_\alpha = \pi_\gamma^\beta x_\beta' = x_\gamma$. It suffices to prove that $x_\alpha y_\alpha = x_\gamma y_\gamma$ and $x_\beta' y_\beta = x_\gamma y_\gamma$. But

$$x_\gamma y_\gamma = \pi_\gamma^\alpha x_\alpha \cdot y_\gamma = x_\alpha \cdot \varpi_\alpha^\gamma y_\gamma = x_\alpha y_\alpha.$$

The equality $x_\beta' y_\beta = x_\gamma y_\gamma$ is proved in the same way.

Distributivity of the scalar product introduced here is verified without difficulty. Accordingly, each element $y \in Y$ defines a character $f_y(x) = xy$ (where x runs through the whole group X) of the group X, i.e., there is an algebraic homomorphism ϕ, namely, a mapping $y \to f_y$ of the group Y into the group of characters X^* of the group X. We shall show that for any element $y \in Y$ different from zero, the character $\phi(y) = f_y$ is also different from zero. Thus we shall have proved that ϕ is an (algebraic) isomorphism of the group Y with the group Y^*. But if $y \neq 0$, then there exists an element $y_\alpha \in y$, differing from the zero of the group Y_α, and accordingly an element $x_\alpha \in X_\alpha$ such that $x_\alpha y_\alpha \neq 0$. But then, denoting by $x \in X$ the coset containing the element x_α, we have

$$f_y(x) = xy = x_\alpha y_\alpha \neq 0,$$

which proves our assertion. We prove further that the mapping ϕ is a mapping onto the whole group X^*, i.e., we construct for any character f of the group X a $y \in Y$ such that $f(x) = f_y(x) = xy$ for all $x \in X$. In order to do this we consider a homomorphism π^α of the group X_α in the group X, putting into correspondence with each $x_\alpha \in X_\alpha$ that coset $x \in X$ which contains the element x_α. Evidently $\pi^\beta \pi_\beta^\alpha = \pi^\alpha$ for any $\beta > \alpha$. The mapping $f\pi^\alpha$ is a homomorphism of the group X_α into the group K, i.e., a character of the group X_α, which means $f\pi^\alpha = y_\alpha \in Y_\alpha$. We shall show that the set y of all y_α constructed in this way forms a thread; then, taking arbitrarily $x \in X$ and $x_\alpha \in x$, we shall have

$$xy = x_\alpha y_\alpha = f\pi^\alpha(x_\alpha) = f(x),$$

i.e., our object will be attained. Let us then consider any $\beta > \alpha$. We shall prove that $\omega_\alpha^\beta y_\beta = y_\alpha$, which will prove that $y = y_\alpha$ is a thread. Choose $x_\alpha \in X_\alpha$ arbitrarily. We have

$$x_\alpha y_\alpha = f\pi^\alpha(x_\alpha) = f\pi^\beta(\pi_\beta^\alpha x_\alpha) = \pi_\beta^\alpha x_\alpha \cdot y_\beta = x_\alpha \cdot \omega_\alpha^\beta y_\beta.$$

Since the equality $x_\alpha y_\alpha = x_\alpha \omega_\alpha^\beta y_\beta$ is satisfied for all $y_\alpha \in Y_\alpha$, then $y_\alpha = \omega_\alpha^\beta y_\beta$.

Thus ϕ is an algebraic isomorphism of the group Y onto the group X^*. It remains to prove that this isomorphism is topological. But since Y and X^* are bicompact it suffices to prove that ϕ is a continuous mapping of the group Y onto the group X^*. It is sufficient to prove the continuity at the null point. Suppose we are given an arbitrary neighborhood U of the null element of the group X^*. This means that we are given a certain neighborhood W of the null element in the group K and a certain finite number of elements $x^1, \cdots, x^s$ of the group X; then U (by definition of the topology in the character group) consists of all those $f_y \in X^*$ for which $f_y(x^j) = x^j y \in W$, $j = 1, \cdots, s$. Now we fix on an α for which we may take x_α^j to be an element of the jth coset, $j = 1, \cdots, s$. To these x_α^j we assign a neighborhood v_α of the null element in Y_α such that $x_\alpha^j v_\alpha \subseteq W$ for all $j = 1, \cdots s$. Then in the neighborhood O_{v_α} of the null element in Y we obtain

$$x^j O_{v_\alpha} = x_\alpha^j v_\alpha \subseteq W \quad \text{(for all } j = 1, \cdots, s\text{)},$$

which is to say,

$$\phi(O_{v_\alpha}) \subseteq U.$$

Thus we have proved:

Theorem 4. *Under the definition of the scalar product introduced above for the elements of the limit groups of two adjoint spectra, these limit groups are dual to each other.*

CHAPTER III

Betti groups of compacta

In this chapter we define the Δ- and ∇-groups for compacta, which are the subject of the duality rules proved further on. The definition of Δ-groups is given in two forms: first, by nerves of coverings, and second, in metric form through ϵ-cycles and homologies. In passing we consider the special case of polyhedra and prove the theorem on the topological invariance of their Betti groups.[*] In the first two sections we study homomorphisms of Betti groups generated by simplicial mappings.

§1. The homomorphisms ω_α^β and π_β^α generated by a simplicial mapping ω_α^β. Subdivisions of triangulations and their canonical shifts

1. Suppose we are given a simplicial mapping ω_α^β of the complex[**] β into the complex α.

----

[*] By this we mean, as usual, that the Betti groups of the triangulations of two homeomorphic polyhedra are isomorphic to each other.

[**] Complexes are assumed simplicial and complete.

Let T_α^r be any simplex of the complex α. We consider the set

$$T_{\beta 1}^r, \cdots, T_{\beta s}^r$$

of all the r-dimensional simplexes of the complex β which map onto T_α^r.

If t_α^r is any orientation of the simplex T_α^r, then each of the simplexes $T_{\beta i}^r$ may be assigned an orientation $t_{\beta i}^r$, $i = 1, \cdots, s$, such that $\omega_\alpha^\beta t_{\beta i}^r = t_\alpha^r$.* Noting this, we choose any chain y^r of the complex β and define a chain $\omega_\alpha^\beta y^r$ of the complex α, giving it on the simplex t_α^r a value equal to the sum of the values of the chain y^r on the simplexes $t_{\beta 1}^r, \cdots, t_{\beta s}^r$. Since t_α^r is any oriented simplex of the complex α, the chain $\omega_\alpha^\beta y^r$ is completely determined.

Thus we have defined a homomorphism ω_α^β of the group $L^r \beta$ into the group $L^r \alpha$, for any coefficient domain.

Evidently the homomorphism ω_α^β may also be defined as follows. Let $t_\beta^r = (e_0 \cdots e_r)$ be any oriented simplex of the complex β. If it does not degenerate under the simplicial mapping ω_α^β, i.e., if all the vertices $\omega_\alpha^\beta e_0, \cdots, \omega_\alpha^\beta e_r$ are distinct, then the image $\omega_\alpha^\beta t_\beta^r$ of the simplex t_β^r is the oriented simplex

$$\omega_\alpha^\beta t_\beta^r = (\omega_\alpha^\beta e_0 \cdots \omega_\alpha^\beta e_r)$$

of the complex α; but if among the vertices $\omega_\alpha^\beta e_0, \cdots, \omega_\alpha^\beta e_r$ at least two coincide, then we put $\omega_\alpha^\beta t_\beta^r = 0$. After this we put, for any chain $y^r = \sum_i c_i t_{\beta i}^r$ of the complex β,

$$\omega_\alpha^\beta y^r = \sum_i c_i \omega_\alpha^\beta t_{\beta i}^r.$$

2. It is easily verified that the homomorphism ω_α^β commutes with the operator Δ, i.e., that for any chain y^r of the complex β we have

$$\omega_\alpha^\beta \Delta y^r = \Delta \omega_\alpha^\beta y^r; \tag{1}$$

this verification may be realized either by going back to the original definition of the chain $\omega_\alpha^\beta y^r$, or making use of the second definition. It is sufficient to prove that for any simplex $t_\beta^r = (e_0 \cdots e_r)$ we have $\omega_\alpha^\beta \Delta t_\beta^r = \Delta \omega_\alpha^\beta t_\beta^r$. This last identity is evident if the simplex t_β^r does not degenerate under the mapping ω_α^β. If it does degenerate, then $\omega_\alpha^\beta t_\beta^r = 0$ and $\Delta \omega_\alpha^\beta t_\beta^r = 0$. We need only to prove that also $\omega_\alpha^\beta \Delta t_\beta^r = 0$. In order to do this we must consider separately two cases: *when the number of vertices of the simplex $\omega_\alpha^\beta t_\beta^r$ is less that $r - 1$* (in this case not only the image of the simplex t_β^r, but also the image of each or its oriented $(r - 1)$-dimensional faces, is equal to zero), *and when the number of vertices of the simplex $\omega_\alpha^\beta t_\beta^r$ is equal to $r - 1$*; in this case there is only one pair of vertices which merge under the mapping ω_α^β, for example, e_0 and e_1:

---

*I.e., such an orientation $t_{\beta i}^r = (e_0 \cdots e_r)$ that $(\omega_\alpha^\beta e_0 \cdots \omega_\alpha^\beta e_r) = t_\alpha^r$.

$$\omega_\alpha^\beta e_0 = \omega_\alpha^\beta e_1.$$

Then the images of all the $(r-1)$-dimensional faces of the simplex other than the faces $(e_1 e_2 \cdots e_r)$ and $(e_0 e_2 \cdots e_r)$ are equal to zero and

$$\omega_\alpha^\beta \Delta t_\beta^r = (\omega_\alpha^\beta e_1, \ \omega_\alpha^\beta e_2, \ \cdots, \ \omega_\alpha^\beta e_r) - (\omega_\alpha^\beta e_0, \ \omega_\alpha^\beta e_2, \ \cdots, \ \omega_\alpha^\beta e_r) = 0.$$

It follows immediately from the relation (1) that under the mapping ω_α^β every Δ-cycle of the complex β goes into a Δ-cycle of the complex α, and that a cycle homologous to zero in β goes into a cycle homologous to zero in α. In other words, *under the homomorphism ω_α^β of the group $L^r\beta$ into the group $L^r\alpha$, the group $Z^r\beta$ maps into $Z^r\alpha$, and the group $H^r\beta$ into the group $H^r\alpha$, so that the group $\Delta^r\beta$ maps into the group $\Delta^r\alpha$.*

This last homomorphism is called the homomorphism induced by the simplicial mapping ω_α^β.

3. As an application we shall compute the Δ-groups, first, for the combinatorial closure $[T^n]$ of the simplex T^n, and, secondly, for the n-dimensional combinatorail sphere S^n, i.e., for the complex consisting of all the proper faces of the $(n+1)$-dimensional simplex. As the coefficient domain we take for simplicity the group of integers. Since both complexes are connected, their null-dimensional Δ-group is an infinite cyclic group. Further, since S^n is an oriented n-dimensional psuedomanifold, its n-dimensional Δ-group is an infinite cyclic group. We shall prove that

$$\Delta^r[T^n] = 0 \ \text{ for } \ 0 < r \le n;$$

$$\Delta^r S^n = 0 \ \text{ for } \ 0 < r < n.$$

Both assertions follow from the following proposition:

Theorem. *If in a complete simplicial complex Q any $r+2$ vertices form a skeleton, then each r-dimensional cycle in Q is homologous to zero.*

The proof is based on the concept of a pyramid oK over the given complex K (Chapter I, §1.6).

The pyramid ot, of one higher dimension than t, over the oriented simplex t of the complex K, will be called the pyramid over t.

The pyramid over the chain $x^r = \sum_i a_i t_i^r$ is the chain

$$ox^r = \sum_i a_i (ot_i^r).$$

It is easy to see that each cycle z^r in K bounds in $[oK]^*$ a chain $x^{r+1} = oz^r$. For $r = 0$ we need to assume that the cycle z^0 has the sum of its coefficients

---

* The combinatorial closure $[oK] = oK \cup K$ of the pyramid oK is called a closed pyramid (with base K and vertex o).

equal to zero.

We shall now prove the theorem formulated above. We denote by $|z^r| \subset Q$ the *modulus of the cycle* z^r, i.e., the complex consisting of all simplexes entering into z^r with coefficients differing from zero, and of all the faces of these simplexes.

We shall define a simplicial mapping of the closed pyramid $o|z^r| \cup |z^r|$ into the complex Q, putting into correspondence with the vertex o some vertex e_0 of the complex Q, and with each vertex e_i of the complex Q that same vertex e_i. In view of the fact that each $r+2$ vertices in Q form a skeleton, the mapping thus constructed indeed defines a simplicial mapping of the complex $o|z^r| \cup |z^r|$ into Q, under which all the simplexes $|z^r|$ stay in place. Under this mapping the cycle z^r therefore goes into itself, and the chain in the pyramid $o|z^r|$ bounded by this cycle goes over into some chain of the complex Q, bounded by the same cycle, so that $z^r \sim 0$ in Q.

4. The homomorphism π_β^α adjoint to the homorphism $\tilde{\omega}_\alpha^\beta$ (and to the simplicial mapping $\tilde{\omega}_\alpha^\beta$), is a homomorphism of the group L_α^r into the group L_β^r defined as follows: if x_α^r is a chain of the complex α, then the chain $\pi_\beta^\alpha x_\alpha^r$ on each oriented simplex t_β^r of the complex β takes on the value

$$(\pi_\beta^\alpha x_\alpha^r \cdot t_\beta^r) = (x_\alpha^r \cdot \tilde{\omega}_\alpha^\beta t_\beta^r). \tag{2}$$

This homomorphism commutes with the operator ∇, i.e.,

$$\pi_\beta^\alpha \nabla x_\alpha^r = \nabla \pi_\beta^\alpha x_\alpha^r. \tag{3}$$

Indeed, for any simplex t_β^{r+1} of β we have (on the basis of formula (2) and formula (8), Chapter 1, §1.5)

$$(\pi_\beta^\alpha \nabla x_\alpha^r \cdot t_\beta^{r+1}) = (\nabla x_\alpha^r \cdot \tilde{\omega}_\alpha^\beta t_\beta^{r+1}) = (x_\alpha^r \cdot \nabla \tilde{\omega}_\alpha^\beta t_\beta^{r+1}) =$$

$$= (x_\alpha^r \cdot \tilde{\omega}_\alpha^\beta \Delta t_\beta^{r+1}) = (\pi_\beta^\alpha x_\alpha^r \cdot \Delta t_\beta^{r+1}) = (\nabla \pi_\beta^\alpha x_\alpha^r \cdot t_\beta^{r+1}).$$

Therefore, under the homomorphism π_β^α, the group Z_∇^r maps into $Z_\nabla^r \beta$, and the group $H_\triangle^r \alpha$ into $H_\nabla^r \beta$, so that we have also defined a homomorphism π_β^α of the group $\nabla^r \alpha$ into the group $\nabla^r \beta$. This homomorphism is called the homomorphism of ∇-groups *adjoint to the simplicial mapping* $\tilde{\omega}_\alpha^\beta$.

5. **Example: canonical shifts of triangulations and the homomorphisms adjoint to them: subdivisions.** Let α and β be triangulations with β following α. Each simplex, in particular each vertex, of the triangulation β, lies on its carrier, a definite simplex of the triangulation α. If we put into correspondence with each vertex e_β of the triangulation β some vertex of the carrier of the vertex e_β, we obtain a simplicial mapping σ_α^β of the triangulation β into the triangulation α,

called the *canonical shift of the triangulation* β *into the triangulation* α.[*] Since the vertex e_β may be put into correspondence with any vertex of its carrier in α, there exist many canonical shifts, generally speaking, of the triangulation β into the triangulation α. We shall denote by $\bar{\sigma}^\alpha_\beta$ the homomorphism of the group $L^r\alpha$ into $L^r\beta$ adjoint to the canonical shift σ^β_α.

Now let the triangulation β be a subdivision of the triangulation α. Suppose that t^r_α is any oriented simplex of the triangulation α. Taking all the r-dimensional simplexes into which the simplex $|t^r_\alpha|$ was subdivided, with the same orientation $t^r_{\beta j}$ as that of the simplex t^r_α which is their common carrier,[**] we obtain an integer-valued chain $s^\alpha_\beta t^r_\alpha = \sum_j t^r_{\beta j}$, called a *subdivision* (in the triangulation β) *of the oriented simplex* t^r_α. If $x^r = \sum_i c_i t^r_{\alpha i}$ is some chain of the complex α, then the chain

$$s^\alpha_\beta x^r_\alpha = \sum_i c_i s^\alpha_\beta t^r_{\alpha i}$$

of the complex β is called a *subdivision of the chain* x^r_α *in the triangulation* β. Here it is easy to verify that $\Delta s^\alpha_\beta t^r_\alpha = s^\alpha_\beta \Delta t^r_\alpha$,[***] which means that, for any chain $x^r_\alpha \in L^r_\alpha$

$$\Delta s^\alpha_\beta x^r_\alpha = s^\alpha_\beta \Delta x^r_\alpha,$$

from which it follows that the subdivision of a cycle is a cycle, and that the subdivision of a cycle homologous to zero in α is a cycle homologous to zero in β. In other words, *the subdivision operator* s^α_β *realizing an isomorphic mapping of the group* L^r_α *into* L^r_β *generates a homomorphism with the same name of the group* $\Delta^r \alpha$ *into the group* $\Delta^r \beta$.

For the proof of the fact that s^α_β *is an isomorphism of the group* Δ^r_α *onto the group* Δ^r_β, we need to prove two theorems:

(A) *Every cycle* z^r_β *of the complex* β *is homologous in* β *to the subdivision* $s^\alpha_\beta z^r_\alpha$ *of some cycle* z^r_α *of the complex* α.

(B) *If the cycle* $s^\alpha_\beta z^r_\alpha$ *is homologous to zero in* β, *then* $z^r_\alpha \sim 0$ *in* α.

The second of these assertions means that the mapping s^α_β of the group $\Delta^r \alpha$ into $\Delta^r \beta$ is an isomorphism, and the first, that it is an isomorphism onto the whole group Δ^r_β.

Theorem B will be proved now, and Theorem A somewhat later, but in this same chapter (§5).

---

[*] Here the image $\sigma^\beta_\alpha T_\beta$ of any simplex $T_\beta \in \beta$ is a proper or improper face of the carrier $T_\alpha \in \alpha$ of the simplex T_β.

[**] For the identity of the orientations in the space R^r carrying the simplex $|t^r_\alpha|$, and the orientation of t^r_α orienting it, see Chapter I, §2.1.

[***] See, for example, [1], Theorem [2.12] on page 373.

Theorem B is an obvious consequence of the significant and important formula

$$\sigma^{\beta}_{\alpha} s^{\alpha}_{\beta} x^r_{\alpha} = x^r_{\alpha}, \tag{4}$$

which holds for any chain x^r_{α} of the complex α and for any canonical shift σ^{α}_{β}. Formula (4), which we shall call the Sperner-Alexander formula, needs only to be proved for the case when the chain x^r_{α} consists of just one oriented simplex t^r_{α}:

$$\sigma^{\beta}_{\alpha} s^{\alpha}_{\beta} t^r_{\alpha} = t^r_{\alpha}. \tag{4'}$$

Since for zero-dimensional chains x^0_{α} the formula (4) is obvious, we may prove (4') by induction on the dimension r, under the assumption that (4) has been proved for the dimension $r - 1$. The proof is based on the following lemma:

Lemma. *Suppose that we are given a simplicial map ω of some complex β into the complex $[T^n]$ consisting of the n-dimensional simplex T^n and all of its faces; suppose that in β we are given an n-dimensional chain y^n whose boundary has under the mapping ω as image the cycle Δt^n (where t^n is an orientation of the simplex T^n):*

$$\omega \Delta y^n = \Delta t^n.$$

Then

$$\omega y^n = t^n.$$

Indeed, under the mapping ω the image of the chain y^n is some n-dimensional chain of the complex $[T^n]$. But all the n-dimensional chains of this complex have the form ct^n, since in $[T^n]$ there is only one n-dimensional simplex. So, $\omega y^n = ct^n$. But then $\omega \Delta y^n = c\Delta t^n$; however we know that $\omega \Delta y^n = \Delta t^n$, which means that $c = 1$ and the lemma is proved.

We shall now prove formula (4') for the dimension r. The canonical shift σ^{β}_{α} maps the chain $y^r = s^{\alpha}_{\beta} t^r_{\alpha}$ into $[t^r]$. Since formula (4) is assumed to have been proved for $r - 1$ and $\Delta s^{\alpha}_{\beta} t^r_{\alpha} = s^{\alpha}_{\beta} \Delta t^r_{\alpha}$, therefore $\omega^{\beta}_{\alpha} \Delta s^{\alpha}_{\beta} t^r_{\alpha} = \Delta t^r_{\alpha}$; but then by the lemma

$$\omega^{\beta}_{\alpha} s^{\alpha}_{\beta} t^r_{\alpha} = t^r_{\alpha}.$$

We have proved the Sperner-Alexander lemma for integer chains. But in exactly the same way we may prove it also in the case of chains over any coefficient domain, in particular for chains modulo 2. The formula (4') in this last case has the following content:

Suppose we are given a subdivision β of the closed simplex $\overline{T}^n$ (i.e., of the complex $[T^n]$). If one puts into correspondence with each vertex of the subdivision β some vertex of its carrier in $[T^n]$, we obtain a simplicial mapping of the complex β into $[T^n]$ (canonical shift) under which an odd number of simplexes of the complex map onto T^n.

This last is a well-known proposition, called the lemma of Sperner.

From this lemma it follows at once that *under a canonical shift σ_α^β of the subdivision β of the triangulation α, the image of the triangulation β is the whole triangulation α:*

$$\sigma_\alpha^\beta \beta = \alpha.$$

§2. Coverings. Dimension

1. A covering of a given set A (lying in some space R) is a system of sets lying in R whose sum contains the set A. Most frequently one considers coverings whose elements are subsets of A itself. A covering is said to be *open* (respectively, closed) if its elements are open (respectively, closed) sets in R or in A. In the entire first part of this work it will be always supposed that the coverings are *finite* (i.e., consist of a finite number of elements), and this will not be specially mentioned.

The *multiplicity* of a covering (more generally the *multiplicity of a system of sets*) α is the largest number k such that in the system α there exist k sets having a nonempty intersection. Evidently, *the multiplicity of a system of sets is one larger than the dimension number of the nerve of this system of sets.*

If all the elements of a covering α have a diameter less than some $\epsilon > 0$, then the covering is said to be an ϵ-*covering*.

We remark finally that two coverings $\alpha = \{A_i\}$ and $\beta = \{B_i\}$ are said to be *similar* if their elements may be labeled in such a way that any set of elements $A_{i_1}, \cdots, A_{i_r}$ of the one covering have a nonempty intersection if and only if the corresponding elements $B_{i_1}, \cdots, B_{i_r}$ of the other covering also have a nonempty intersection. Evidently, *similar coverings have the same nerve.*

The consideration of open or closed coverings of compacta, and in particular of the nerves of these coverings, furnish the fundamental method for carrying over to compacta the concepts of combinatorial topology.

Lebesgue's lemma for open coverings (Lemma 1). *For every open covering*

$$\alpha = \{O_1, \cdots, O_s\}$$

of a compactum Φ one may find a positive number η with the following property: every set $M \subseteq \Phi$, having a diameter $< \eta$, is entirely contained in at least one element of the covering α; every number η satisfying this condition is called a Lebesgue number of the open covering α.

Proof. We shall call *too big* any set $M \subseteq \Phi$ not contained in any element of the system α.

We denote by η the lower bound of diameters of all the too big sets $M \subseteq \Phi$, and we shall prove that it is positive. This will prove the lemma.

Let

$$M_1, M_2, \cdots, M_\nu, \cdots$$

be any sequence of too big sets whose diameters converge to η. In each M_ν we choose a point p_ν; passing if necessary to a subsequence we may suppose that the points p_ν converge to some point $p \in \Phi$. The point p lies in some element O_i of the system α and has accordingly a positive distance ϵ from the closed set $R - O_i$, so that

$$O(p, 2\epsilon) \subseteq O_i;$$

by their very definition none of the sets M_ν is contained in O_i, and *a fortiori* not in $O(p, 2\epsilon)$. Meanwhile, for all sufficiently large ν we have $p_\nu \in O(p, \epsilon)$, so that for all such ν the diameter of M_ν certainly has to be larger than ϵ. Since η is the limit of the diameters of the M_ν, therefore $\eta \geq \epsilon$, as we were required to prove.

2. **Dimension of compacta.** We may now define the *dimension of the compactum* as the smallest positive number n such that for every $\epsilon > 0$ there is an open ϵ-covering of the compactum Φ having multiplicity $n + 1$. If there is no such number (i.e., for sufficiently small ϵ every open covering of the compactum Φ has a multiplicity larger than any number chosen in advance), then the dimension of the compactum is taken to be infinite.

We shall denote the dimension of the compactum Φ by $\dim \Phi$.

Remark. We may easily prove that in the definition of dimension we may replace open sets by closed ones.*

---

* This follows from the two following propositions:

Lemma 2. (The lemma of Lebesgue for closed coverings.) To every closed covering $\{A_1, \cdots, A_s\}$ of a compactum there exists an $\epsilon > 0$ (*the Lebesgue number of the closed covering*) such that the ϵ-neighborhoods $O_1 = O(A_1, \epsilon), \cdots, O_s = O(A_s, \epsilon)$ of the sets $A_1, \cdots, A_s$ form a covering similar to the covering $\{A_1, \cdots, A_s\}$ (i.e., the intersection $O_{i_1} \cap \cdots \cap O_{i_r}$ is nonempty only when $A_{i_1} \cap \cdots \cap A_{i_r}$ is nonempty).

This lemma is proved easily by contradiction.

Lemma 3. To each open covering

$$\alpha = \{O_1, \cdots, O_s\}$$

of the compactum Φ there corresponds a closed covering

$$\beta = \{A_1, \cdots, A_s\}$$

of that compactum such that $A_1 \subseteq O_1, \cdots, A_s \subseteq O_s$.

Indeed, we choose an open set $O_i' \supseteq \Phi - O_i$, $i = 1, 2, \cdots, s$, such that the system $\alpha' = \{O_1', \cdots, O_s'\}$ is similar to the system $\beta' = \{\Phi - O_1, \cdots, \Phi - O_s\}$.

From the similarity of the systems α' and β' it follows in particular (since β is a covering) that $O_1' \cap \cdots \cap O_s'$ is empty, which means that the sets

$$A_1 = \Phi - O_1', \cdots, A_s = \Phi - O_s'$$

form a covering of the compactum Φ. Since evidently $A_i \subseteq O_i$, this is the desired covering.

We shall prove the following fundamental proposition.

Theorem. *If Φ is an n-dimensional (in the elementary sense of the word) polyhedron, i.e., if it is the body of an n-dimensional triangulation, then $\dim \Phi = n$.*

Proof. Since the principal stars of a sufficiently fine triangulation of an n-dimensional polyhedron Φ form a system of order $n + 1$ (the nerve of this system is the given triangulation itself), we are only required to prove that for sufficiently small $\epsilon > 0$ every open ϵ-covering of the given n-dimensional polyhedron Φ has multiplicity $\geq n + 1$.

Thus, let α be a triangulation of the polyhedron Φ. The triangulation α is the nerve of the system $\{O_1, \cdots, O_\nu\}$ of its principal stars. Let ϵ be a Lebesgue number of the covering $\{O_1, \cdots, O_\nu\}$. We shall prove that every open ϵ-covering $\beta = \{o_1, \cdots, o_s\}$ of the compactum Φ has multiplicity $\geq n + 1$, i.e., that the nerve of the covering has a dimension number $\geq n$. To this end we choose a Lebesgue number δ of the covering β and a subdivision α' of the complex α so fine that its simplexes are less than $\frac{\delta}{2}$ in diameter, which means that its stars are less than δ in diameter.

For each star of the triangulation α' we choose an element of the covering β containing this star; in turn, for each $o_i \in \beta$ we choose a star of the triangulation α containing the set o_i.

Since each triangulation is the nerve of the system of its principal stars, we have established:

(a) a simplicial mapping σ_1 of the triangulation α' into the nerve of β;

(b) a simplicial mapping σ_2 of the nerve of β into the triangulation α.

Carrying out first σ_1 and then σ_2, we obtain a simplicial mapping $\sigma = \sigma_2\sigma_1$ of the complex α' into α. Under this mapping the star of the vertex $e' \in \alpha'$ is contained in the star of the vertex $\sigma e' \in \alpha$, which means that $\sigma e'$ is one of the vertices of the carrier of the point e' in the complex α. In other words, σ is a canonical shift of the triangulation α' into α. But then, as we have seen, $\sigma\alpha' = \alpha$, which means that the whole n-dimensional triangulation α is the image (under the simplicial mapping σ_2) of the subcomplex $\sigma_1(\alpha')$ of the nerve of β. Under a simplicial mapping the dimension of a simplex cannot rise, so that the dimension of the complex $\sigma_1(\alpha')$, and, what is more, the dimension of all of β, is larger than or equal to n, as it was required to prove.

§3. Combinatorially close simplicial mappings. Prisms

1. Two simplicial mappings ϕ_0 and ϕ_1 of the complex X into the complex Y will be said to be *combinatorially close* if for any simplex T_X of the complex X there is a simplex $T_Y \in Y$ such that the simplexes $\phi_0 T_X$ and $\phi_1 T_X$ are proper or improper faces of the simplex T_Y.

The importance of the concept of combinatorially close mappings, which will be set forth fully in the following subsection, is already evident from the fact that any two canonical shifts of a subdivision β of a triangulation α are combinatorially close mappings in that triangulation.

Our object consists in the proof of the following proposition:

Theorem. *Every two combinatorially close mappings ϕ_0 and ϕ_1 of a complex X into a complex Y generate the same homomorphism of the group $\Delta^r X$ into the group $\Delta^r Y$ and one and the same adjoint homomorphism of the group $\nabla^r Y$ into the group $\nabla^r X$.*

For the proof we shall need the so-called prisms, to whose definition we shall now turn.

2. **Prisms.** Let $|a_0 \cdots a_n|$ be a simplex with a definite order for its vertices. Let $|b_0 \cdots b_n|$ be a second copy of the simplex $|a_0 \cdots a_n|$.[*] From these objects there is defined an (abstract) complete complex Π, consisting of all the simplexes of the form

$$|b_0 \cdots b_i a_i \cdots a_n|, \quad i = 0, 1, \cdots, n,$$

and of all their faces. This complex will be called the *prism* Π with bases $|a_0 \cdots a_r|$ (lower) and $|b_0 \cdots b_n|$ (upper). For short it is called the *prism over the simplex $|a_0 \cdots a_n|$*.

In the prism Π there is defined an $(n + 1)$-dimensional chain known as *the prism over the oriented simplex $t^n = (a_0 \cdots a_n)$* and denoted by Πt^n:

$$\Pi t^n = \sum_{i=0}^{n} (-1)^i (b_0 \cdots b_i a_i \cdots a_n).$$

Suppose that we are given a complex X, all of whose vertices are written in a definite order: $a_1, \cdots, a_s$. Then the vertices of each simplex of the complex X are also written in a definite order. Taking a second copy X_1 of the complex X with vertices $b_1, \cdots, b_s$, we may define a *prism over the complex X*, or, more precisely, *a prism Π with lower base $X_0 = X$ and upper base X_1*, as the complex which is the sum of the prisms constructed on all the simplexes of the complex X. In the complex Π there is defined a prism over each oriented simplex t^n, and also a prism Πx^n over each chain $x^n = \sum c_k t_k^n$:

$$\Pi x^n = \sum_{k} c_k \Pi t_k^n.$$

It is easy to verify (see [1], page 285, §9.3) that under these definitions

---

[*] If the simplex $|a_0 \cdots a_n|$ lies in the space R^n, then, imbedding R^n in R^{n+1} in the form of the plane $x_{n+1} = 0$, we may choose the points $b_0, \cdots, b_n$ in the plane $x_{n+1} = 1$ above the points $a_0, \cdots, a_n$, so that the simplex $|b_0 \cdots b_n|$ is congruent to the simplex $|a_0 \cdots a_n|$ and is obtained from it by a translation with the vector $\{0, 0, \cdots, 0, 1\}$.

$$\Delta\Pi x^n = x^n - x_1^n - \Pi\Delta x^n,$$

where x_1^n is the same chain x^n, but lying in the upper base X_1 of the prism Π. In particular, if z is a Δ-cycle of the complex X, then

$$\Delta_\pi z = z - z_1, \tag{1}$$

where again z_1 is the same chain z but lying in the upper base X_1 of the prism Π.

3. We turn to the proof of the theorem formulated at the beginning of this section.

The first of the assertions may obviously be expressed as follows:

Theorem IIΔ. *If ϕ_0 and ϕ_1 are two combinatorially close mappings of the complex X into the complex Y, and z is any Δ-cycle of the complex X, then the cycles $\phi_0 z$ and $\phi_1 z$ are homologous to one another in Y.*

In order to prove this, we choose two copies X_0 and X_1 of the complex X with corresponding copies z_0 and z_1 of the cycle z on them, and we map X_0 by the map ϕ_0, and X_1 by the map ϕ_1, into the complex Y. We relabel in some way all the vertices in the complex X, and thus also in the complexes X_0 and X_1, and we consider the prism Π with bases X_0 and X_1. In view of the combinatorial closeness of the mappings ϕ_0 and ϕ_1, these mappings, given respectively on X_0 and X_1, generate a mapping ϕ of the prism Π into the complex Y. The image of the chain Πz_0 under the mapping ϕ is some chain y of the complex Y_1. Since $\phi z_0 = \phi_0 z$, $\phi z_1 = \phi_1 z$ and $\Delta\Pi z_0 = z_0 - z_1$ it follows that

$$\Delta y = \Delta\phi\,\Pi z_0 = \phi\,\Delta\Pi z_0 = \phi z_0 - \phi z_1 = \phi_0 z - \phi_1 z,$$

i.e., $\phi_0 z \sim \phi_1 z$ in Y, as we were required to prove.

4. We shall prove the second assertion of our theorem, which may be formulated as follows:

Theorem II∇. *If ϕ_0 and ϕ_1 are combinatorially close simplicial mappings of the complex X into the complex Y, and z is some ∇-cycle of the complex Y, then for the adjoint homomorphisms $\overline{\phi}_0$ and $\overline{\phi}_1$ (of the group $Z_\nabla Y$ into $Z_\nabla X$) we have that $\overline{\phi}_0 z \sim \overline{\phi}_1 z$ in X.*

Proof. We consider again the mapping ϕ of the prism Π with bases X_0 and X_1 into the complex Y.

We denote by r the dimension of the ∇-cycle z. Now we shall construct an $(r-1)$-dimensional chain x, of the complex X for which

$$\nabla x = \overline{\phi}_0 z - \overline{\phi}_1 z, \tag{2}$$

as follows. Let t_0^{r-1} be some oriented $(r-1)$-dimensional simplex of the triangulation X. The orientation t_0^{r-1} generates a definite orientation of all the r-dimensional simplexes of the prism $\Pi \, | t_0^{r-1} | \subset \Pi$, constructed over that simplex. The sum of the values of the ∇-cycle $\bar{\phi} z$ on all these r-dimensional simplexes will be taken to be the value of the chain x on t_0^{r-1}. For the proof of formula (2) we have to show that on every r-dimensional simplex t_0^r of the triangulation X the chain

$$\bar{\phi}_0 z - \bar{\phi}_1 z - \nabla x$$

is equal to zero, i.e., that

$$(\phi_0 z \cdot t_0^r) - (\phi_1 z \cdot t_0^r) - \sum_{t^{r-1} < t_0^r} (x \cdot t^{r-1}) = 0,$$

the last sum being taken over all suitably oriented faces t^{r-1} of the simplex t_0^r.

To see this, we consider all the oriented $(r+1)$-dimensional simplexes t^{r+1} of the prism $\Pi \, | t_0^r | \subset \Pi$, and we sum over them the expressions

$$(\nabla \bar{\phi} z \cdot t^{r+1}) = \sum_{t^r < t^{r+1}} (\bar{\phi} z \cdot t^r),$$

each of which is equal to zero since $\bar{\phi} z$ is a ∇-cycle. In summing over these t^{r+1} each interior r-dimensional simplex t^r of the prism Πt_0^r appears twice with opposite signs, so that the whole sum is equal to zero and will consist of (i) two terms which correspond to the upper and lower bases of the prism $\Pi \, | t_0^r |$, i.e., (taking into account the orientation of the simplexes of the prism) the terms $(\phi_0 z t_0^r)$ and $- (\phi_1 z t_0^r)$, and (ii) the terms $\bar{\phi} z t^r$ which correspond to the simplexes t^r lying on the lateral surface of the prism $\Pi \, | t_0^r |$, i.e., the terms $(\bar{\phi} z t^r)$, where $t^r \in \Pi \, | t^{r-1} |$ and t^{r-1} is a face of the simplex t_0^r. But the terms $(\bar{\phi} z^r t^r)$, where t^r enters into the given prism Πt^{r-1}, form in their sum exactly $(x t^{r-1})$, so that in the result we obtain indeed the desired proof of equation (2).

§4. Definition of the Δ- and ∇-groups of compacta [*]

We consider the set Ω of all the coverings [**] of a given compactum Φ. This set turns out to be partially ordered in a natural way, and thus directed (we say that the covering β follows the covering α and write $\beta > \alpha$ if β is subordinate to α, i.e., if each element of the covering β is contained in at least one element of the covering α). If we put into correspondence with each element of the covering β one of the elements of the covering α which contains it, we obtain a simplicial mapping ω_α^β of the nerve of β into the nerve of α, called a projection. Since an element of the covering β may be contained in several elements of the covering α, there may be a number of projections of the covering β into the covering α. However any two projections of the nerve of β into the nerve of α

---

[*] The considerations of this section also apply, without serious changes, to the case of arbitrary bicompacta.

[**] All of the coverings considered in this book are finite open coverings unless the opposite is specified.

are combinatorially close mappings. Therefore all of them define the same homomorphism ω^{β}_{α} of the group $\Delta^r\beta$ into $\Delta^r\alpha$ and the same homomorphism π^{α}_{β} of the group $\nabla^r\alpha$ into the group $\nabla^r\beta$. Accordingly we have a direct spectrum *

$$\{\nabla^r\alpha,\ \pi^{\alpha}_{\beta}\} \tag{1}$$

and the inverse spectrum

$$\{\Delta^r\alpha,\ \omega^{\beta}_{\alpha}\}. \tag{2}$$

The limit group of the direct spectrum (1) *is denoted by* $\nabla^r\Phi$ *and is called the r-dimensional* ∇*-group of the compactum* Φ.

The limit group of the inverse spectrum (2) *is denoted by* $\delta^r\Phi$ *and is called the r-dimensional* Δ*-group of the compactum* Φ.

The following is quite essential:

Remark 1. In the directed set of all coverings α of the compactum Φ there is a countable cofinal section. For this we may choose any sequence

$$\alpha_0,\ \alpha_1,\ \cdots,\ \alpha_k,\ \cdots \tag{3}$$

of coverings which satisfy the condition that α_k is an ϵ_k-covering as $\epsilon_k \to 0$, while α_{k+1} is subordinate to α_k.

It follows from the lemma of Lebesgue that the sequence (3) forms a cofinal section of the whole directed set of coverings. If we retain in the spectra (1) and (2) only those α lying in the sequence (3), we obtain in each of these spectra a countable cofinal section.

Remark 2. The definition of the group $\delta^r\Phi$ may also be given in the following way, which is evidently equivalent to that given first.

We shall call a *projective cycle* a system of cycles

$$z^r = \{z^r_{\alpha}\},$$

with one cycle z^r_{α} to each nerve α, satisfying the condition: *if* $\beta > \alpha$, *then* $\omega^{\beta}_{\alpha}z^r_{\beta} \sim z^r_{\alpha}$ *in* α.

Projective cycles are added term by term: if

$$z^r = \{z^r_{\alpha}\},\ z'^r = \{z'^r_{\alpha}\},\ \text{then}\ z^r + z'^r = \{z^r_{\alpha} + z'^r_{\alpha}\},$$

with this addition they obviously form a group.

The projective cycle $z^r = \{z^r_{\alpha}\}$ is said to be *bounding* or *homologous to zero* in Φ if $z^r_{\alpha} \sim 0$ in α for any α.

The factor-group $Z^r\Phi - H^r\Phi$ of the group of all r-dimensjonal projective cycles of the compactum Φ modulo the subgroup of all bounding cycles is found to have a natural isomorphism with the group $\delta^r\Phi$ and may be identified with it.

---

* See Chapter II.

One may give a somewhat different and simpler form to the definition of the group $\nabla^r \Phi$. We shall call any ∇-cycle z_α^r of the nerve of any covering α of the compactum Φ an r-dimensional ∇-cycle of that compactum. Two ∇-cycles z_α^r and z_β^r of the compactum Φ will be said to be homologous to each other in Φ if there is a covering γ of that compactum following both α and β such that $\pi_\gamma^\alpha z_\alpha^r \sim \pi_\gamma^\beta z_\alpha^r$ in γ. This definition of homology leads to a decomposition of the set of all r-dimensional ∇-cycles of the compactum Φ into *classes* or *cosets* of mutually homologous cycles. These cosets are the elements of the group $\nabla^r \Phi$. If we are given two cosets ζ'^r and ζ''^r, then we can always find a covering α such that on α there lies some ∇-cycles $z_\alpha'^r \in \zeta'^r$ and some $z_\alpha''^r \in \zeta''^r$. Then $\zeta' = \zeta'^r + \zeta''^r$ is defined as the coset containing the cycle $z_\alpha'^r + z_\alpha''^r$. This definition of the addition of cosets, which does not depend, as one easily sees, on the choice of elements in them, completes the definition of the group $\nabla^r \Phi$.

§5. Δ-groups of polyhedra

1. A consequence of Theorem A of §1. We consider the case when the compactum Φ is a polyhedron.

We begin by turning to the statement (still not proved by us) in §1, on page 37, Theorem A:

(A) *If the triangulation β is a subdivision of the triangulation α, then each cycle z_β of the complex β is homologous in β to the subdivision $s_\beta^\alpha z_\alpha$ of some cycle z_α of the complex α.*

From Theorem A we derive the following proposition: [*]

Theorem 1. *The canonical shift σ_α^β of the subdivision β of the triangulation α generates an isomorphism with the same name of the group $\Delta^r \beta$ onto the group $\Delta^r \alpha$ (inverse isomorphism s_β^α).*

Proof. As does every simplicial mapping, the shift σ_α^β generates a homomorphism with the same name of the group $\Delta^r \beta$ into the group $\Delta^r \alpha$. From the Sperner-Alexander formula (§1 of this chapter, formula (4) on page 38, it follows immediately that σ_α^β is a homomorphism onto the whole group $\Delta^r \alpha$. *This homomorphism is an isomorphism*, as follows from the following assertion: *if $\sigma_\alpha^\beta \sim 0$ in α, then $z_\beta \sim 0$ in β.*

We shall prove this fact. In view of Theorem A (supposed proved), the cycle z_β is homologous in β to a subdivision $s_\beta^\alpha z_\alpha$ of some cycle z_α of the complex α. Since under the simplicial mapping σ_α^β cycles which are mutually homologous (with respect to β) go into cycles which are homologous in α, we have

---

[*] This theorem and its proof presented below are valid not only for triangulations, but also for their open subcomplexes.

$$z_\alpha = \sigma_\alpha^\beta s_\beta^\alpha z_\alpha \sim \sigma_\alpha^\beta z_\beta \sim 0 \text{ in } \alpha.$$

Thus, $z_\alpha \sim 0$ in α. But then also $s_\beta^\alpha z_\alpha \sim 0$ in β, and therefore also $z_\beta \sim 0$ in β. Our assertion, and with it, Theorem 1, is proved (under the assumption that Theorem A is proved).

Remark. We saw in §1 that it follows from Theorem A that the subdivision operator s_β^α generates an isomorphism with the same name of the group $\Delta^r \alpha$ into $\Delta^r \beta$. Since $\sigma_\alpha^\beta s_\beta^\alpha z_\alpha^r = z_\alpha^r$, the isomorphisms σ_α^β and s_β^α of the Betti groups are mutually inverse.

Before proving Theorem A, we shall deduce from it still another very important consequence, assuming only that Theorem A is proved in the case when β is a barycentric subdivision of the triangulation α.

We consider the sequence

$$\alpha_0, \; \alpha_1, \; \cdots, \; \alpha_k, \; \cdots \tag{1}$$

of triangulations of the polyhedron Φ, in which each triangulation is a barycentric subdivision of the one preceding it, [*] and the initial triangulation α_0 is arbitrary. Each of these triangulations α_k is the nerve of a covering consisting of the principal open stars of the triangulation α_k; this covering will be denoted also by α_k.[**] Since the diameters of the elements of the covering α_k tend to zero with increasing k, the sequence (1) of coverings forms a cofinal section of the set of all coverings of the polyhedron Φ. If the vertex e_{k+1} of the complex α_{k+1} has as its carrier the simplex T_k in α_k, then the star Oe_{k+1} is contained in the star of each vertex T_k. Therefore the canonical shift σ_k of the triangulation α_{k+1} into α_k realizes a projection of the nerve α_{k+1} into the nerve α_k, and these projections are isomorphic mappings of the group $\Delta^r \alpha_{k+1}$ onto $\Delta^r \alpha_k$. To the sequence (1) corresponds the cofinal section

$$\{\Delta^r \alpha_k, \; \sigma_k\} \tag{2}$$

in the spectrum $\{\Delta^r \alpha, \; \omega_\alpha^\beta\}$, while the projections realized by the canonical shifts are isomorphisms of all the groups of the spectrum (2) onto one another. Therefore also the limit group of the spectrum (2), i.e., the group $\Delta^r \Phi$, is isomorphic to all the groups $\Delta^r \alpha_k$, and in particular to the group $\Delta^r \alpha_0$, where α_0 was any triangulation of the polyhedron Φ.

---

[*] We leave to the reader the proof of the following elementary geometric fact (it is proved for example in the book [1], Supplement 2, §4, page 655): if the diameter of an n-dimensional simplex T^n is d, then the diameter of a barycentric subdivision of the simplex T^n does not exceed $\frac{n}{n+1} d$. It follows that the simplexes of our complexes α_k become arbitrarily small in diameter as $k \longrightarrow \infty$.

[**] Thus, we shall use the same letter for a covering and its nerve; this will be always done in what follows.

We have derived from Theorem A *the theorem on the invariance of the Betti groups of polyhedra:*

Theorem 2. *The group $\Delta^r \alpha$ of any triangulation α of the polyhedron Φ is isomorphic to the group $\delta^r \Phi$ and therefore is a topological invariant of the polyhedron Φ.*

Accordingly,

any two triangulations of two homeomorphic polyhedra have isomorphic Δ-groups.

Theorem 2 and its corollary have been proved under the assumption that Theorem A is true for a barycentric subdivision β of the triangulation α.

2. **Proof of Theorem A for barycentric subdivisions.** We shall employ induction on the dimension number n of the triangulation α. For $n = 0$ the triangulation α consists of a finite number of vertices, and $\beta = \alpha$, which means that our theorem is valid. Suppose that Theorem A has been proved for any triangulation α of dimension $< n$. From this hypothesis it follows that for barycentric subdivisions of triangulations of dimension $< n$ Theorem 1 is also valid. In particular, it therefore follows that for the barycentric subdivision θ^{n-1} of the boundary of an n-dimensional simplex, the groups $\Delta^r \theta^{n-1} = 0$ for $r < n - 1$.[*]

We shall now prove an auxiliary proposition.

Lemma. *In the barycentric subdivision $\tau^p = s_\beta^\alpha T^p$ of the simplex T^p of dimension $p \leq n$, every r-dimensional cycle, $0 \leq r < p$, is homologous to zero.*

For zero-dimensional cycles this is obvious, since the only vertex contained in the complex τ^p is the center o of the simplex T^p, and bounds in τ^p any segment issuing from that vertex. So let $r > 0$. Since the barycentric subdivision of the simplex T^p is a pyramid over the barycentric subdivision θ^{p-1} of the boundary S^{p-1} of the simplex T^p, every oriented r-dimensional simplex of the complex τ^p is a pyramid over some oriented simplex of the complex θ^{p-1}. If we put into correspondence with the oriented simplex $t^r = (ot^{r-1})$ of the complex τ^p the oriented simplex $t^{r-1} = ft^r$ of the complex θ^{p-1}, we readily see that f is an isomorphism of the group $L^r \tau^p$ onto the group $L^{r-1} \theta^{p-1}$, satisfying the condition

$$f \Delta x^r = - \Delta f x^r,\text{[**]}$$

---

[*] For $r = 0$ we choose here the group Δ^{00}, based on zero-dimensional cycles with the sum of their coefficients equal to zero.

[**] Indeed, if $t^{r-1} = (e_1 e_2 e_3 \cdots e_r) = ft^r$, $t^r = (o\, e_1 e_2 e_3 \cdots e_r)$, then, taking Δt^r in τ^p, we have

$$\Delta t^r = - (oe_2 e_3 \cdots e_r) + (oe_1 e_3 \cdots e_r) \text{ etc.,}$$

so that

$$f \Delta t^r = - (e_2 e_3 \cdots e_r) + (e_1 e_3 \cdots e_r) \text{ etc. } = - \Delta f t^r.$$

from which it follows that f generates an isomorphism of the groups $Z^r \tau^p$ and $H^r \tau^p$ respectively onto $Z^{r-1} \theta^{p-1}$ and $H^{r-1} \theta^{p-1}$. This means that the groups $\Delta^{r-1} \theta^{p-1}$ for $r < p$ are null groups, so that $\Delta^r \tau^p = 0$, as it was required to prove.

Now we prove Theorem A. *Suppose that β is a subdivision of an n-dimensional triangulation α, while it is known that for any simplex $T^p \in \alpha$ in its subdivision $s_\beta^\alpha T^p$ all the cycles of dimension $< p$ are homologous to zero* (for a barycentric subdivision we have proved this).

Suppose we are given any r-dimensional cycle z_β of the complex β.

Each simplex t_β^r of that cycle, i.e., a simplex entering into it with a coefficient differing from zero, has in the complex α its carrier t_α^p. The difference $p - r$ may be called the *depth* (of the "stratum") of the simplex t_β^r in the complex α; the maximum of these depths (taken over all simplexes t_β^r of the cycle z_β) is called the *depth* of the cycle; if this depth is equal to zero, i.e., if the given r-dimensional cycle of the complex β lies on the r-dimensional skeleton α^r of the complex α, then z_β is a subdivision of some cycle z_α of the complex α. Indeed, the subdivision $s_\beta^\alpha T_\alpha^r$ of any simplex T_α^r is an oriented r-dimensional pseudomanifold;[*] the r-dimensional cycles lying on it have the form $c s_\beta^\alpha t_\alpha$, where t_α^r is some orientation of the simplex T_α^r. The part of the cycle z_β lying on $s_\beta^\alpha T_\alpha^r$ is an r-dimensional cycle on the pseudomanifold $s_\beta^\alpha T_\alpha^r$, which means that it has the form $c s_\beta^\alpha t_\alpha^r$, so that the whole cycle z_β^r has the form $\sum_i c_i s_\beta^\alpha t_{\alpha_i}^r$, i.e., is a subdivision of some chain $\sum c_i t_{\alpha_i}^r$ (obviously a cycle) of the complex α. Thus Theorem A is proved as soon as we construct for each cycle z_β of positive depth q a cycle $z_\beta''^r$ homologous to it in β of smaller depth; for in a finite number of steps we shall thus reach a cycle of zero depth, i.e., a cycle of the form $s_\beta^\alpha z_\alpha$, which is homologous in β to the given cycle z_β.

Thus, suppose that the cycle z_β has depth $q > 0$. We choose a simplex T_α^p of maximal dimension $r + q = p$, in which the simplexes of the cycle z_β are contained. We shall denote by x the part of the cycle z_β lying in T_α^p (i.e., on the complex $s_\beta^\alpha T_\alpha^p$).

This part x is a cycle in $s_\beta^\alpha T_\alpha^p$. Indeed, the cycle z_β does not contain any simplexes whose carriers in α have dimension $> p$. Therefore z_β lies on the

---

[*] In fact, to every $(r-1)$-dimensional simplex of the complex $\tau = s_\beta^\alpha T_\alpha^r$ there evidently adjoin two r-dimensional simplexes of that complex; if now the complex τ is not strongly connected, it may be represented as the sum of two subcomplexes A and B in τ with an intersection $A \cap B$ of dimension $\leq r - 2$; but this is not possible, since any two points $a \in \tilde{A}$, $b \in \tilde{B}$ not belonging to the set $\tilde{A} \cap \tilde{B}$, may be joined in T^r by a line segment not intersecting $\tilde{A} \cap \tilde{B}$ (because of the fact that the dimension of $\tilde{A} \cap \tilde{B} \leq r - 2$). Thus τ is a pseudomanifold. It is oriented, since all the r-dimensional simplexes T_β^r may be given an orientation identical to the chosen orientation of T_α^r.

subdivision $s_\beta^\alpha \alpha^p$ of the p-dimensional skeleton α^p of the complex α. In $s_\beta^\alpha \alpha^p$ the complex $s_\beta^\alpha T_\alpha^p$ is an open subcomplex, so that the assertion follows.[*]

From the lemma, the cycle x of the complex $s_\beta^\alpha T^p$ bounds in $s_\beta^\alpha T_\alpha^p$ some chain x' whose boundary in all of β is $\Delta x' = x + y$, where y lies on the boundary of the simplex T_α^p (subdivided in β); since

$$\Delta y = - \Delta x = \Delta (z_\beta - x),$$

by replacing in z_β the piece x by the chain $- y$ we obtain the cycle $z_\beta' = z_\beta - x - y$, which no longer has any simplexes lying in T_α^p and is homologous to the cycle z_β in β inasmuch as $z_\beta - z_\beta' = x - y = \Delta x'$. The cycle z_β', which is homologous to the cycle z_β in β, has the property that the number of simplexes T_α^p of highest dimension p into which the cycle z_β' passes is one less than the number of the same simplexes for the cycle z_β.

Repeating this process a finite number of times, we obtain a cycle homologous in β to the cycle z_β and having a depth $< q$, which is what we were required to prove.

Thus Theorem 2 is proved completely and Theorem A and Theorem 1 for barycentric subdivisions.

3. **Proof of Theorem A and Theorem 1 for arbitrary subdivisions.**

For the proof of Theorem A (from which Theorem 1 will follow), it will suffice to prove that in any subdivision τ of the n-dimensional simplex T every cycle u of dimension $r < n$ is homologous to zero.

For the proof of the last assertion we take the combinatorial closure $[\tau]$ of the complex τ; it is a subdivision of the triangulation $[T^n]$. In the complete complex $[\tau]$ the chain u has the boundary $\Delta u = z$, lying on $[\tau] - \tau = \theta^{n-1}$, i.e., on the subdivided boundary S^{n-1} of the simplex T^n. The cycle z bounds in θ the chain y, since $\Delta^{r-1}\theta = 0$. The chain $z' = u - y$ is a cycle in $[\tau]$, bounding in $[\tau]$ some chain x'. The part of the chain x' lying in τ is a chain in τ, bounded by the cycle u, as was to be proved.

Remark. Let β be a subdivision of the triangulation α (it is sufficient to consider only barycentric subdivisions). One may prove that the homomorphism π_β^α of the group $\nabla^r\alpha$ into the group $\nabla^r\beta$, adjoint to the canonical shift σ_α^β, is an isomorphism of the group $\nabla^r\alpha$ onto the group $\nabla^r\beta$ (the proof was given, for example, in Chapter 4 of the book [1]). It therefore follows that *for any polyhedron Φ the group $\nabla^r\Phi$ is isomorphic to the group $\nabla^r\alpha$, where α is any triangulation of the*

---

[*] In view of the fact that the operator Δ commutes with the operator of excision by an open subcomplex (see Chapter I, §2.9, page 20) and $s_\beta^\alpha T_\alpha^r$ is an open subcomplex of the complex $s_\beta^\alpha \alpha^r$.

polyhedron Φ.

§6. Enlargement of a given triangulation

The isomorphism of the Δ-groups of the triangulation α onto the Δ-groups of one of its subdivisions β, generated by the subdivision operator s_β^α, naturally leads to the following more general construction which we shall need in the next chapter.

By an enlargement of an n-dimensional triangulation β we shall mean a system α of subcomplexes $U_i^r \subset \beta$ of various dimensions r (from 0 to n inclusive), satisfying the following two conditions: *

$1°$. All the U_i^r are simple pairwise disjoint combinatorial pseudomanifolds;** their sum is the whole complex β.

$2°$. For every U_i^r the complex $B_i^{r-1} = [U_i^r] - U_i^r$ (consisting of all the faces of the simplexes $T \in U_i^r$ not belonging to U_i^r) is an $(r-1)$-dimensional complex which is the sum of some of the pseudomanifolds U_j^s, $0 \leq s \leq r-1$.

We shall say that U_j^s is a *proper face* of the pseudomanifold U_i^r if $U_j^s \subseteq B_i^{r-1}$. In this case we have $U_j^s < U_i^r$.

We observe first of all:

if $T_1 < T_2 \in \beta$ and $T_1 \in U_1$, $T_2 \in U_2$, then certainly $U_1 \leq U_2$.

Indeed, from our hypotheses it follows that either $T_1 \in U_2$ (in which case $U_1 = U_2$), or $T_1 \in B_2$ (whereupon, from condition $2°$, $U_1 \subseteq B_2$), which means that $U_1 < U_2$.

From what has just been proved it easily follows that:

(a) if $U_k^r < U_i^{r+1}$, then U_k^r is an open subcomplex of the complex B_i^r;

(b) every U_k^r is open in the complex $\alpha^r \subseteq \beta^r$, the sum of all the U_j^s of dimension $s \leq r$.

The proof in both cases is the same. Suppose that $T \in U_k^r$, $T < T' \in B_i^r$ (respectively $T' \in \alpha^r$). It is required to prove that $T' \in U_k^r$. Suppose $T' \in U_h^p$; since $T' \in B_i^r$ (respectively $T' \in \alpha^r$) we also have $U_h^p \subseteq B_i^r$ (respectively $U_h^p \subseteq \alpha^r$), so that $p \leq r$.

On the other hand, it follows from the inequality $T < T'$ that $U_k^r \leq U_i^p$,

---

*In the particular case when β is a subdivision of the triangulation α, for the complex U_i^r we take the subdivisions $s_\beta^\alpha T_i^r$ of the simplexes $T_i^r \in \alpha$. Then the complexes U_i^r correspond in a one-to-one way to the simplexes $T_i^r \in \alpha$. Therefore it is natural to denote the system of all the U_i^r by α.

**An r-dimensional pseudomanifold U^r is said to be *simple* if it is oriented and if for $0 < p < r$ all its Δ^p-groups are null groups. We speak of a *combinatorial* pseudomanifold, since we have in mind the complex U_i^r and not its body the polyhedron $\widetilde{U}_i^r$.

i.e., $r \le p$, which means that $r = p$ and $U_k^r = U_h^p$, $T' \in U_k^r$, as was required to prove.

We now denote for each U_i^r one of the orientations of the pseudomanifold U_i^r by u_i^r, and the other by $- u_i^r$.

The chains u_i^r are evidently linearly independent (since the complexes U_i^r have no common elements).

We shall prove that Δu_i^r is a linear combination of chains u_j^{r-1}:

$$\Delta u_i^r = \sum_j c_{ij} u_j^{r-1}. \tag{1}$$

Since u_i^r is a cycle on the oriented pseudomanifold U_i^r, Δu_i^r can differ from zero only on the $(r - 1)$-diemnsional simplexes of the complex B_i^{r-1}.

But B_i^{r-1} is made up of U_j^s of dimensions $s \le r - 1$. Therefore, if Δu_i^r is distinct from zero on some simplex t_k^{r-1}, then $|t_k^{r-1}| \in U_j^{r-1} \subseteq B_i^{r-1}$. Since U_j^{r-1} is an open subcomplex of the complex B_i^{r-1} (on which lies the cycle Δ_i^r), the portion of the cycle Δ_i^r lying on it is a cycle of the pseudomanifold U_j^{r-1}, i.e., it has the form $c u_j^{r-1}$. This means that Δu_i^r, as the sum of its pieces lying on the various U_j^{r-1}, may be represented in the form (1), as was required to be proved.

The coefficient c_{ij} in the representation (1) is called the *incidence coefficient* $(u_i^r : u_j^{r-1})$. Naturally, $(- u_i^r : u_j^{r-1}) = (u_i^r : - u_j^{r-1}) = - (u_i^r : u_j^{r-1})$. We shall prove that *the set of all chains* $\pm u_i^r$, *with the incidence coefficients just defined, forms a cell complex which we shall also denote by* α *and call an* (*algebraic*) *enlargement of the complex* β.

To this end it will suffice to prove the identity

$$\sum_j (u_i^r : u_j^{r-1}) (u_j^{r-1} : u_k^{r-2}) = 0 \tag{2}$$

for any two u_i^r and u_k^{r-2}. As a result of formula (1), the chain $\Delta \Delta u_i^r$ may be represented as a linear combination of chains u_k^{r-2}, and from the definition of incidence coefficients it follows that the coefficient on u_k^{r-2} in this linear combination is exactly the left side of the equality (2).

Since $\Delta \Delta_i^r = 0$ and all the chains u_k^{r-2} are linearly independent, the coefficient in question is equal to zero, and thus equation (2) is proved.

We shall prove that *the Δ-groups of a cell complex are isomorphic to the corresponding Δ-groups of the simplicial complex* β. To this end we assign to each cell $\pm u_i^r$ (as an element of the cell complex α) the same cell as a chain of the complex β. Because of this correspondence, there corresponds to each chain x of the cell complex α a chain of the complex β, and we obtain an isomorphism s_β^α of the group $L^r \alpha$ into the group $L^r \beta$. It is easy to see that this isomorphism (completely analogous to the earlier subdivision isomorphism which we also called s_β^α) commutes with the operator Δ and accordingly generates an isomorphism with the same name of the group $\Delta^r\alpha$ into the group $\Delta^r\beta$. To prove that this homomorphism is an isomorphism of the group $\Delta^r\alpha$ onto the group $\Delta^r\beta$ it is sufficient to prove a proposition formulated in the same terms as Theorem A and B in §1, to wit:

(A) *Each cycle* z_β^r *of the complex* β *is homologous in* β *to some cycle of the form* $\Sigma c_i u_i^r$.

(B) *Each cycle of the form* $z_\beta^r = \Sigma c_i u_i^r$, *homologous to zero in* β, *bounds a chain of the form* $x_\beta^{r+1} = \Sigma a_h u_h^{r+1}$.

Lemma. *Suppose as before that* $\alpha^r \subseteq \beta$ *is the sum of all the* U_j^s, $s \leq r$. *If the chain* x^r *lies on* α^r *and has a boundary lying on* α^{r-1}, *then* x^r *is a linear combination of the chains* u_i^r.

It suffices to show that the piece $U_i^r x^r$ of the chain x^r lying on U_i^r has the form $c u_i^r$. But for this it suffices to show that $U_i^r x^r$ is a cycle on the pseudomanifold U_i^r, which in its turn immediately follows from the fact that U_i^r is an open subcomplex of the triangulation α^r, and Δx^r lies on α^{r-1}.

In particular, it follows from the lemma that every r-dimensional cycle on α^r has the form $\Sigma c_i u_i^r$.

The proof of Proposition A repeats the proof of the analogous proposition for subdivisions.

Indeed, we select as the carrier of the simplex T^r of the complex β that unique U_i^p which contains this simplex. The difference $p - r = q$ we again call the depth of the simplex T^r. After this, one defines in the natural way the depth of any chain of the complex β. We have proved: if the depth of the cycle z_β^r of the complex β is equal to zero, then the cycle has the form $z_\beta^r = \Sigma c_i u_i^r$.

This means that again everything reduces to the construction, for a given cycle z_β^r of positive depth, of a homologous (in β) cycle of smaller depth.

This is achieved, as in the case of subdivisions, by reducing the number of those $U_i^p \subseteq \alpha$ of highest dimension $p = r + q$ which are carriers of the simplexes of the cycle z_β^r. We choose one such U_i^p. Since all of the cycle z_β^r, having the depth q, lies on α^p and U_i^p is open in α^p, the piece $x^r = U_i^p z_\beta^r$ of the cycle z_β^r lying on U_i^p is a cycle on U_i^p, bounding in U_i^p a chain x^{r+1}. Then the boundary Δx^{r+1} of the chain x^{r+1} in β has the form

$$\Delta x^{r+1} = x^r + y^r,$$

where y^r lies on B_i^{p-1}. Replacing in z_β^r the piece x^r by $-y^r$, we obtain the cycle

$$z''^r_\beta = z^r_\beta - x^r - y^r,$$

homologous in β to the cycle z^r_β (since $z^r_\beta - z''^r_\beta = x^r + y^r = \Delta x^{r+1}$) and no longer
containing any simplexes having U^p_i as their carrier. Applying this device a
finite number of times, we obtain a cycle homologous in β to the cycle z^r_β and
lying on α^{p-1}, i.e., having depth $< q$.

Quite analogously one proves Proposition 2: if the chain x^{r+1}_β, bounded by
the cycle $z^r_\beta = \Sigma c_j u^r_j$, has a null depth, then it has the form $x^{r+1}_\beta = \Sigma a_i u^{r+1}_i$
(this is the assertion of the lemma). If now the chain x^{r+1}_β has the depth $q > 0$,
it lies on α^{q+r+1}; we choose one of the U^{q+r+1}_i containing the simplexes of the
chain x^{r+1}_β. It follows from our hypotheses that $U^{q+r+1}_i x^r_\beta$ is a cycle on
U^{q+r+1}_i, and that the same considerations as above lead to the possibility of replacing this piece by a chain lying on B^{q+r} and having the same boundary. Repeating this scheme, we may reduce the depth of the chain x^{r+1}, without changing its boundary z^r_β, until the depth becomes equal to zero.

The proposition just proved may be formulated as follows:

Theorem 3. *The Δ-groups of any enlargement of a given triangulation are
isomorphic to the corresponding groups of that triangulation.*

§7. The groups $\Delta^r_c \Phi$

1. **Definition of the groups $\Delta^r_c \Phi$ and the isomorphisms $\Delta^r_c \Phi = \delta^r \Phi$.** In many
questions in topology it turns out to be convenient to study not the groups $\delta^r \Phi$,
defined above by means of coverings, but rather their metric equivalent, the so-
called groups $\Delta^r_c \Phi$. They are defined as follows. We shall say that an *r-dimensional ϵ-simplex of the compactum* Φ is a set consisting of $r + 1$ points of that
compactum and having diameter $< \epsilon$. It is then natural to call an *ϵ-complex of the
compactum* Φ any (finite) set of ϵ-simplexes in it (satisfying, if necessary, the
condition of completeness). An *ϵ-chain of the compactum* Φ is any finite chain
lying on a ϵ-complex of the compactum Φ. In other words, an *r*-dimensional
ϵ-chain of the compactum Φ is a finite linear form of the type $x^r = \Sigma c_i t^r_i$, where
the t^r_i are oriented ϵ-simplexes. The *modulus* $|x^r|$ of the chain x^r is the com-
plete ϵ-complex consisting of all those simplexes entering into the chain x^r with
nonzero coefficients *and* of all the faces of these simplexes. The boundary of an
ϵ-chain $x^r = \Sigma c_i t^r_i$ is defined as usual by the formula

$$\Delta x^r = \Sigma c_i \Delta t^r_i.$$

If $\Delta x^r = 0$, then the ϵ-chain x^r is said to be an ϵ-cycle; a given ϵ-cycle of the
compactum Φ is said to be ϵ'-*homologous to zero* (ϵ'-*bounding*) in Φ if it is the
boundary of some ϵ'-chain of that compactum. Two ϵ-cycles are said to be
ϵ'-homologous to each other if their difference is ϵ'-homologous to zero in Φ.

After these auxiliary constructions,[*] we turn to the fundamental concept of a *true* or Δ_c-*cycle*[**] and Δ_c-*homology*. The sequence

$$z^r = (z^r_1, z^r_2, \cdots, z^r_k, \cdots) \tag{1}$$

of r-dimensional ε_k-cycles, with $\epsilon_k \to 0$, of the compactum Φ is said to be a Δ_c-cycle of the compactum Φ if for any ϵ all the cycles of this sequence, beginning with a certain one, are ϵ-homologous to each other in Φ.

A Δ_c-cycle (1) is *bounding* or *homologous to zero* in Φ, if for any ϵ all the z^r_k beginning at some point are ϵ-homologous to zero in Φ.

Δ_c-cycles are added termwise: if $z^r = \{z^r_k\}$, $z'^r = \{z'^r_k\}$, then $z^r + z'^r = \{z^r_k + z'^r_k\}$. Here the addition of all r-dimensional Δ_c-cycles of the compactum Φ form a group $Z^r\Phi$, which contains a subgroup $H^r\Phi$ of bounding Δ_c-cycles. The factor-group $Z^r\Phi - H^r\Phi$ is denoted by $\Delta^r_c\Phi$ and is called the r-dimensional Δ_c-group of the compactum Φ.

The fundamental result of this section is the following:

Theorem. *The groups $\Delta^r_c\Phi$ are isomorphic to the groups $\delta^r\Phi$ defined above.*

This isomorphism is realized by means of the so-called canonical shifts, a special case of which was considered earlier for polyhedra.

Let $\alpha = \{o_1, \cdots, o_s\}$ be a covering of the compactum Φ and let η_α be a Lebesgue number of this covering. Take $\eta < \frac{1}{2}\eta_\alpha$ and consider any η-complex β of the compactum Φ.[***] The stars of the complex β are in diameter less than $2\eta < \eta\alpha$, so that each of them is contained in at least one element o_i of the covering α. Now we put into correspondence with each vertex e_β of the complex β some element of the covering α containing the star of the vertex e_β. Thus, each vertex e_β of the complex β has been put into correspondence with some vertex $e_\alpha = \sigma_\alpha e_\beta$ of the nerve α, and this mapping of vertices accomplishes a simplicial mapping σ_α of the complex β into the nerve α, called the *canonical shift*[****] of the complex β into the nerve α.

Obviously one may speak of a canonical shift also for any η-chain of the compactum Φ into the nerve α, understanding by this a shift of modulus $|x^r|$ of that

---

chain. Two different canonical shifts of a given η-complex into the nerve α are combinatorially close; *therefore the homology class of those cycles of the nerve* α *into which a given* η-*cycle goes (or* η-*homologous* η-*cycles go with* $\eta < \frac{1}{2}\eta_\alpha$), *under the various canonical shifts, is uniquely determined and does not depend on the particular choice of the canonical shift* (relative to the given covering α).

Keeping from now on to the same covering α and to the number $\eta < \frac{1}{2}\eta_\alpha$, we choose any Δ_c-cycle

$$z^r = (z_1^r, z_2^r, \cdots, z_k^r, \cdots);$$

All of its elements z_k^r beginning at some point are η-cycles, and all of them are η-homologous to one another, and therefore, under a canonical shift into the nerve α, all the z_k^r with sufficiently large k define the same homology class, an element of the group $\Delta^r\alpha$, which we denote by $\zeta_\alpha^r = \sigma_\alpha z^r$. Two Δ_c-cycles homologous to each other evidently define the same element of the group Δ_α^r. Therefore *the canonical shift* σ_α *into the nerve* α *defines a homomorphism, which we shall denote by* σ_α, *of the group* $\Delta_c^r\Phi$ *into the group* Δ_α^r.

If the covering β follows the covering α and σ_β is a canonical shift into the nerve β, then the mapping $\omega_\alpha^\beta \sigma_\beta$ is a canonical shift into α. It follows therefore that the set of canonical shifts, taken over all coverings α of the compactum Φ, puts into correspondence, with each Δ_c-cycle z^r, the homology classes $\zeta_\alpha^r \in \Delta^r\alpha$ forming a thread of the spectrum $\{\Delta^r\alpha, \omega_\alpha^\beta\}$, while to mutually homologous Δ_c-cycles there corresponds one and the same thread, i.e., one and the same element of the group $\delta^r\Phi$. Thus, *we have a homomorphism, called the homomorphism of the canonical shift, or simply the canonical shift, denoted by* σ, *and mapping the group* $\Delta_c^r\Phi$ *into the group* $\Delta^r\Phi$.

We shall show that *this homomorphism is in fact an isomorphism.*

We shall preface this proof by two remarks. The first is *on the geometrical realization in* Φ *of the nerve of the given covering* $\alpha = \{o_1, \cdots, o_s\}$. By this we understand the choice, for the role of the vertex e_i of the nerve α, of points of the same sets o_i of elements of the covering α to which these vertices correspond. If the given covering is an ϵ-covering (for some given ϵ), then its nerve, realized in Φ in the way we have just stated, is a 2ϵ-complex[*] of the compactum

---

[*] Sometimes the realization of the nerve is understood in a somewhat wider and indeed more geometrical sense. If the compactum Φ is a set lying in n-dimensional (for sufficiently large n) space R^n (or S^n), and one is given a covering $\alpha = \{o_1, \cdots, o_s\}$, then the vertices e_i of the nerve α corresponding to the sets o_i are taken not directly in these sets, but in a given neighborhood of them in such a way that they are in general position. On the skeletons of the nerve one stretches regular geometrical simplexes of the space R^n. Since this enveloping space R^n has a sufficiently high dimension (namely, $n \geq 2r + 1$, where r is the dimension of the nerve α), it turns out that the nerve α is realized in the form of a triangulation lying in any given neighborhood of the original set Φ ("realization close to the set $\Phi \subset R^n$").

Φ and the canonical shift of any sufficiently fine complex into the nerve α turns out to be an ϵ-shift, i.e., under this shift each vertex of the complex is moved a distance less than ϵ.

Our second remark is related to these ϵ-shifts and consists in the following assertion:

In the compactum Φ (more generally in any metric space) any ϵ-cycle under any ϵ'-shift goes into an $(\epsilon + 2\epsilon')$-cycle homologous to it.

We obtain this result as a special case of an important formula. Let us consider any ϵ-chain x^r in Φ and subject it to an ϵ'-shift in Φ (here $\epsilon > 0$ and $\epsilon' > 0$ are quite arbitrary).

The ϵ'-shift under consideration consists in putting into correspondence with each vertex e of the modulus $K = |x^r|$ of the chain x^r a point $e' = Se \in \Phi$ at a distance less than ϵ'. Each simplex of the complex K, being an ϵ-simplex, goes into some $(\epsilon + 2\epsilon')$-simplex in Φ, so that we have a simplicial mapping S of the complex K onto some $(\epsilon + 2\epsilon')$-complex K' of the compactum Φ. Under this mapping the chain x^r of the complex K goes into some chain Sx^r of the complex K'.

Labelling the vertices of the chain x^r in some way once and for all, we consider the prism Π over the complex K. We shall define as follows a mapping S' of the set of all vertices of the prism Π into the space Φ. If e_i is a vertex of the lower base of the prism Π, (i.e., a vertex of the chain x^r), then we put $S'e_i = e_i$; if e_i' is a vertex of the upper base of the prism Π, and if e_i is the vertex of the lower base corresponding to it, then we put $S'e_i' = Se_i$.

If $|e_0' \cdots e_k' e_k \cdots e_r|$ is any simplex of the prism Π, then its image under the mapping S' is a skeleton $|Se_0 \cdots Se_k e_k \cdots e_r|$, whose diameter is equal to the largest of the numbers $\rho(e_i, e_j)$, $\rho(Se_i, Se_j)$, $\rho(e_i, Se_j)$, where e_i, e_j are any two vertices lying in the same skeleton of the complex K.

But each of these numbers is less than $\epsilon + 2\epsilon'$, so that under the mapping S' there corresponds to each simplex of the prism Π some $(\epsilon + 2\epsilon')$-simplex of the space Φ.

Under the simplicial mapping S' of the complex Π the prism Πx^r over the chain x^r goes into some $(\epsilon + 2\epsilon')$-chain, which we shall denote by $\Pi_S x^r$ and call the prism stretched over the chain x^r and on its ϵ'-displacement Sx^r, or simply the *prism of the shift of the chain x^r*.

In just the same way the prism $\Pi \Delta x^r$ over Δx^r (in the complex Π) under S' goes into the prism $\Pi_S \Delta x^r$ stretched over Δx^r and $S\Delta x^r$.

Here the simplicial mapping S' carries relation (1) of §3 into the relation

$$\Delta\Pi_S x^r = x^r - Sx^r - \pi_S \Delta x^r. \tag{2}$$

This is the important formula mentioned above. In particular, if x^r is an ϵ-cycle, then $\text{II}_S \Delta x^r = 0$, so that (2) goes into

$$\Delta \text{II}_S x^r = x^r - S x^r. \tag{3}$$

Since $\text{II}_S x^r$ is an $(\epsilon + 2\epsilon')$-chain, we obtain an $(\epsilon + 2\epsilon')$-homology

$$x^r \underset{\epsilon + 2\epsilon'}{\sim} S x^r,$$

as was to be proved.

Now we turn to the proof of the fact that the canonical shift σ or the group $\Delta_c^r \Phi$ into $\delta^r \Phi$ is an isomorphism of the first group onto the second.

To this end we select a sequence of ϵ_h-coverings of the compactum Φ as $\epsilon_h \rightarrow 0$:

$$\alpha_1, \alpha_2, \cdots, \alpha_h, \cdots, \tag{4}$$

where α_{h+1} follows α_h. This sequence is a cofinal section of the set of all coverings of the compactum Φ, so that we may retain, in the calculation of the spectrum which defines the group $\delta^r \Phi$, only the coverings of (4). We suppose that their nerves are realized in Φ.

We shall first of all prove that the homomorphism is an isomorphism, i.e., that only the null element of the group $\Delta_c^r \Phi$ maps into the null element of the group $\delta^r \Phi$.

Suppose that the image of the Δ_c-cycle $z^r = \{z_k^r\}$ is the null element of the group $\delta^r \Phi$. We shall prove that $z^r \sim 0$ in Φ. From our hypothesis it follows that under the canonical shifts into $\alpha_1, \alpha_2, \cdots, \alpha_h$, all the z_k^r, beginning in each case at some stage, go into cycles homologous to zero in the corresponding nerves. Suppose $\epsilon > 0$ is given arbitrarily. We take an h so large that α_h is an ϵ-covering. Then we take a k_0 so large that if $k > k_0$ all the z_k^r are η_h-cycles, where η_h is a Lebesgue number of the covering α_h, which means that *a fortiori* $\eta_h < \epsilon$. All the cycles z_k^r, $k \geq k_0$, under the canonical shift relative to α_h, go into cycles of the nerve α_h which are 3ϵ-homologous to them, and which in that nerve are homologous to zero, i.e., in every case 2ϵ-homologous to zero. In other words, all the z_k^r, beginning at some stage, are 3ϵ-homologous to zero in Φ. This proves that the homomorphism σ is a homomorphism into the group $\delta^r \Phi$. We shall prove that this is an isomorphism onto the whole group $\delta^r \Phi$. We select any element of the group $\delta^r \Phi$, i.e., a thread $\zeta^r = \{\zeta_h^r\}$, $\zeta_h^r \in \Delta^r \alpha_h$. We take in each ζ_h^r a cycle z_h^r. Then

$$\omega_h^{h+1} z_{h+1}^r \sim z_h^r \text{ in } \alpha_h \tag{5}$$

(we write ω_h^{h+1} instead of $\omega_{\alpha_h}^{\alpha_{h+1}}$) and the nerves α_h are realized in Φ. The projection ω_h^{h+1} is an ϵ_h-shift and the homology (5) is a $3\epsilon_h$-homology, so that

$z^r = \{z^r_h\}$ is a Δ_c-cycle of the compactum Φ which goes into itself under the canonical shift. Thus it follows that σ realizes a mapping of the group $\Delta^r_c \Phi$ into $\delta^r \Phi$.

The isomorphism of the groups $\Delta^r_c \Phi$ and $\delta^r \Phi$ is proved.

2. **The case when Φ is a polyhedron.** In the case when the compactum Φ is a polyhedron, we may immediately prove that the group $\Delta^r_c \Phi$ is isomorphic to the Δ-groups of any triangulation α of the polyhedron Φ, which gives a new proof of the theorem on the invariance of the Betti groups.

Thus, let α be a triangulation of the polyhedron Φ. We denote by α a covering formed by the principal stars of that triangulation. We denote by η a Lebesgue number of the covering α. We select any Δ_c-cycle

$$z^r = (z^r_1, z^r_2, \cdots, z^r_k, \cdots) \tag{6}$$

of the polyhedron Φ. All the z^r_k, beginning with some one, are η-cycles of the polyhedron Φ, η-homologous to each other in Φ. Under the canonical shift σ_α into the nerve α they go into cycles of that nerve which are homologous to one another in it. Thus, to each Δ_c-cycle z^r in Φ there corresponds a definite element $\zeta^r_\alpha = \sigma_\alpha z^r$ of the group $\Delta^r \alpha$, while to mutually homologous Δ_c-cycles there obviously corresponds one and the same element $\zeta^r_\alpha \in \Delta^r \alpha$. Thus we obtain *a homomorphism of the canonical shift σ_α*, mapping the group $\Delta^r \Phi$ into $\Delta^r \alpha$.

To show that this homomorphism is a mapping onto the whole group $\Delta^r \alpha$, we denote the kth barycentric subdivision of the triangulation α by α_k and consider any cycle z^r_0 of the complex α and its successive subdivisions z^r_k in the complexes α_k. We obtain a sequence

$$(z^r_0, z^r_1, \cdots, z^r_k, \cdots); \tag{7}$$

since under the canonical shift of α_{k+1} into α_k the cycle z^r_{k+1} goes into z^r_k and since that shift brings about an ϵ_k-homology with $\epsilon_k \to 0$ it follows that (7) is a Δ_c-cycle of the compactum Φ. Under the canonical shift σ_α into the nerve α^* there corresponds to the homology class of the Δ_c-cycle (7) the homology class of the cycle z^r_0, which proves that σ_α maps the group Δ^r_c onto $\Delta^r \alpha$.

We shall show that the homomorphism σ_α of the group $\Delta^r \Phi$ onto $\Delta^r \alpha$ is an isomorphism. Suppose we know about the Δ_c-cycle

$$z^r = (z^r_1, z^r_2, \cdots, z^r_k, \cdots), \tag{6}$$

that $\sigma_\alpha z^r = 0$. It is required to deduce from this that $z^r \sim 0$ in Φ, i.e., that for any ϵ all the z^r_k, beginning at some point, are ϵ-homologous to zero in Φ.

---

* We are consistently using one and the same symbol α to denote both the triangulation and the covering formed by the principal stars of that triangulation (having the triangulation α as its nerve).

Choose such an ϵ. Take a Lebesgue number $3\eta < \epsilon$ of the covering α and let h be so large that the diameters of all the stars of the triangulation α_h are less than η. Then we take a Lebesgue number $\eta_h < \eta$ for the covering α_h and retain in (6) only those z_k^r which are η_h-cycles, η_h-homologous to one other in Φ. Among these we select one or another z_k^r and denote by σ_h the canonical shift into the nerve α_h. This shift is an η-shift, realizing a 3η-homology in Φ. By Theorem A of §1.5, the cycle $\sigma_h z_k^r$ is homologous in α_h to the subdivision $s_h z_\alpha^r$ of some cycle z_α^r of the complex α and accordingly

$$z_k^r \underset{3\eta}{\sim} s_h z_\alpha^r \text{ in } \Phi.$$

Under the canonical shift [*] the right side of this homology goes into z_α^r and accordingly $z_\alpha^r \sim 0$ in α, and so $s_h z_\alpha^r \sim 0$ in α_h. Since α_h is an ϵ-complex, $s_h z_\alpha^r \underset{\epsilon}{\sim} 0$, which means that also $z_k^r \underset{\epsilon}{\sim} 0$ in Φ, as was required to be proved.

§8. Relative cycles and homologies (cycles and homologies modulo F)

Let K be a complete simplicial complex, G an open subcomplex of the complex K and $F = K - G$ the complementary closed subcomplex. We have already seen that one may consider cycles and homologies in G and thus obtain a group $\Delta^r G$. However, one may obtain groups isomorphic to the groups $\Delta^r G$ in a rather different way. We shall call a *cycle of the complex K modulo F* or a *cycle relative to F* any chain of the complex K whose boundary lies on F.

Remark. Here, two cycles modulo F are considered to be equal if they coincide on any simplex of G.[**]

We say that the cycle z (modulo F) is homologous to zero in K (modulo F) if there exists a chain x of the complex K, whose boundary in K has the form

$$\Delta x = z + y.$$

where y is a chain lying on F.

The factor-group of all r-dimensional cycles of the complex K (modulo F) with respect to the subgroup of cycles homologous to zero (modulo F) is called the r-dimensional Δ-group of the complex K (modulo F) and is denoted by $\Delta^r(K, \bmod F)$. It is easy to see that the groups $\Delta^r(K, \bmod F)$ and $\Delta^r G$ are isomorphic.[***]

---

[*] The displacement is carried out on a chain bounded by the cycle $z_k^r - s_h z_\alpha^r$.

[**] Thus, a cycle, or more generally, *a chain modulo F*, is the class of all possible chains of the complex K which coincide with one another on G. So in all computations with these chains we neglect simplexes lying on F.

[***] We have here two different points of view: first the one we have adopted up to now, of considering G as an independent complex. The other point of view uses, in studying the homology properties of an open subcomplex $G \subset K$, only chains of the complex K itself. This point of view leads to cycles and homologies modulo F.

These concepts carry over also to the case of open sets in the compactum Φ.

Let Γ be an open set of the compactum Φ. Let us denote by Ψ the complementary closed set $\Psi = \Phi - \Gamma$.

We shall say that *a given ϵ-complex of the compactum Φ lies in Ψ if all of its vertices are points of the set Ψ.* A chain of the compactum Φ lies in Φ if all its simplexes lie in Ψ. A chain of the compactum Φ is called an ϵ-cycle (modulo Ψ) if it is an ϵ-chain whose boundary lies in Ψ. The ϵ-cycle z (modulo Ψ) is ϵ'-homologous to zero (modulo Ψ) if there exists an ϵ'-chain in Ψ whose boundary has the form

$$\Delta x = z + y,$$

where y lies on Ψ.

A Δ_c-cycle on Φ (modulo Ψ) is by definition a sequence

$$z^r = (z_1^r,\ z_2^r,\ \cdots,\ z_k^r,\ \cdots), \tag{1}$$

where the z_k^r are ϵ_k-cycles (modulo Ψ), $\epsilon_k \to 0$, while for any $\epsilon > 0$ all the z_k^r beginning at some point are ϵ-homologous to each other in Φ (modulo Ψ).

A Δ_c-cycle (1) (modulo Ψ) is said to be homologous to zero in Φ (modulo Ψ) if, for any $\epsilon > 0$, all the z_k^r, beginning at some point, are ϵ-homologous to zero in Φ (modulo Ψ).

The addition of Δ_c-cycles (modulo Ψ) is carried out in the same way as the addition of ordinary Δ_c-cycles, i.e., term by term. Here the addition of r-dimensional Δ_c-cycles (modulo Ψ) forms a group $Z^r(\Phi, \bmod \Psi)$, in which there is contained a subgroup $H^r(\Phi, \bmod \Psi)$ of cycles homologous to zero in Φ (modulo Ψ); the factor group $Z^r(\Phi, \bmod \Psi) - H^r(\Phi, \bmod \Psi)$ is called the r-dimensional Δ-group of the compactum Φ (modulo Ψ). *

We shall have need of the special case when Φ is a polyhedron, α a triangulation of it, and Ψ the body of some closed subcomplex α' of the triangulation α. In this case the group $\Delta^r(\Phi, \bmod \Psi)$ is isomorphic to the group $\Delta^r(\alpha, \bmod \alpha')$ (or isomorphic to the group $\Delta^r(\alpha - \alpha')$ isomorphic to it). The proof is quite analogous to the one presented at the end of the preceding section on the isomorphism of the groups $\Delta^r\Phi$ and $\Delta^r\alpha$. Let us outline it.

Let α be any triangulation of the polyhedron Φ, and also the covering consisting of the principal stars of this triangulation. This covering cuts out of the

---

* This group also could have been defined by means of coverings as the limit group of the spectrum

$$(\Delta^r(\alpha, \bmod \Psi_\alpha),\ \omega_\alpha^\beta),$$

where Ψ_α is a subcomplex of the nerve α consisting of all the simplexes each of whose vertices corresponds to an element of the covering α intersecting with Ψ.

The proof may be left to the reader.

polyhedron $\Psi = \widetilde{\alpha}'$ a covering α' consisting of the principal stars of the triangulation α' of the polyhedron Ψ. We take an $\eta > 0$ so small that it is a Lebesgue number of the coverings α and α', and we consider only those k which are so large that the z_k^r are η-cycles (modulo Ψ), η-homologous to each other (modulo Ψ). Under the canonical shift of any η-complex into the nerve α, to the vertices lying in Ψ correspond vertices of the nerve α'. Therefore, all the z_k^r go into cycles of the nerve α (modulo α'), homologous to one another in α (modulo α'). Thus, the canonical shift into the nerve α defines a homomorphism of the group $\Delta^r (\Phi, \bmod \Psi)$ into the group $\Delta^r(\alpha, \bmod \alpha')$. In the same way as at the end of the preceding section, we show that this isomorphism is an isomorphism of the group $\Delta^r (\Phi, \bmod \Psi)$ onto the group $\Delta^r (\alpha, \bmod \alpha')$. Here we need only prove Theorem A of $\S 2$ for relative cycles and homologies. But this proof requires only quite automatic changes in our preceding considerations and may be left to the reader.[*]

§9. Local Δ-groups

Suppose that there are given in the compactum Φ two open sets Γ_α and Γ_β with $\Gamma_\beta \subset \Gamma_\alpha$. Then for the complements $\Psi_\alpha = \Phi - \Gamma_\alpha$, $\Psi_\beta = \Phi - \Gamma_\beta$ we have the inclusion $\Psi_\beta \supset \Psi_\alpha$, from which it follows that every Δ_c-cycle of the compactum Φ (modulo Ψ_α) is *a fortiori* a Δ_c-cycle (modulo Ψ_β), and that each such cycle homologous to zero (modulo Ψ_α) is homologous to zero also modulo Ψ_β. Therefore, putting into correspondence with each Δ_c-cycle modulo Ψ_α the same cycle taken modulo Ψ_β, we obtain a homomorphism, which we shall denote by J_β^α of the group $\Delta_c^r (\Phi, \bmod \Psi_\alpha)$ into the group $\Delta_c^r (\Phi, \bmod \Psi_\beta)$.

Now let $\{\Gamma_\alpha\}$ be the directed, by inclusion, set of all the neighborhoods Γ_α of some point $\xi \in \Phi$. It follows from what has been said that we have a direct spectrum

$$\{\Delta_c^r (\Phi, \bmod \Psi_\alpha), J_\beta^\alpha\}, \tag{1}$$

whose limit group is denoted by $\Delta_c^r (\xi, \Phi)$ and called *the r-dimensional Δ_c-group of the compactum Φ at the point ξ*. Here the coefficient domain is any group $\mathfrak{A}$.

In the spectrum (1) there is a countable cofinal section; to obtain it we may choose any countable sequence of neighborhoods of the point ξ closing down on that point.

In particular, if Φ is a polyhedron and K is any of its triangulations, then, taking the sequence of subdivisions

$$K_1, K_2, \cdots, K_\nu, \cdots$$

of the triangulation K and passing to arbitrarily fine meshes as ν increases, we

---

[*] Since the cycles and homologies in question, in α modulo α', are equivalent to those entering in the open subcomplex $\alpha - \alpha'$, one may refer simply to the footnote on page 46.

may choose as neighborhoods of the point open stars $\tilde{O}_\nu$ of the carriers $T_\nu \in K$ of the point ξ. The triangulations K_ν may, as one easily sees, be chosen so that all the combinatorial stars $O_\nu = O_{K_\nu} T_\nu$ are isomorphic to each other. Therefore, retaining in the spectrum (1) only a countable final section corresponding to the stars O_ν, and observing that all the groups constituting this section are isomorphic to one another[*] and that the corresponding projections are isomorphisms of these groups onto one another, we conclude that the limit group $\Delta_c^r (\xi, \Phi)$ is also isomorphic to all of the groups $\Delta^r O_\nu$, i.e., in particular, to the group $\Delta^r O_K T$, where $\xi \in T \in K$. Thus we have proved the following proposition:

Theorem. *Let Φ be a polyhedron, K any of its triangulations, T the carrier of the point ξ in the triangulation K. Then the group $\Delta_c^r (\xi, \Phi)$ is isomorphic to the group $\Delta^r O_K T$.*

Hence it follows in particular that *the group $\Delta^r O_K T$ (where T is the carrier of the point ξ in the triangulation K of the polyhedron Φ) does not depend on the choice of the triangulation K (theorem of invariance for local Betti groups).* This theorem is fundamental in the study of homology manifolds and the duality theorem for them will be applied in the following chapter.

Corollary. *At any interior point ξ of the n-dimensional closed simplex $\bar{T}^n$, the group $\Delta^n(\xi, \bar{T}^n)$ over the group of coefficients $\mathfrak{A}$ is isomorphic to the group $\mathfrak{A}$, and at each boundary point $\xi \in \bar{T}^n - T^n$ the group $\Delta^n(\xi, \bar{T}^n)$ is a null-group. The groups $\Delta^r(\xi, \bar{T}^n)$ for $r < n$ are null groups for all $\xi \in \bar{T}^n$.*

Indeed, for the interior point $\xi \in T^n$ the star of its carrier in the complex $[T^n]$ consists of only the one simplex T^n and $\Delta^n T^n = \mathfrak{A}$, $\Delta^r T^n = 0$ for $r < n$; if now the point ξ has as its carrier the face T^p of the complex T^n, then for $r < p$ there are no r-dimensional simplexes in the star $O_{[T^n]} T^p$, so that there are no r-dimensional cycles different from zero. One may easily prove that also all the remaining Δ-groups of the simplex $O_{[T^n]} T^p$ are null. However, one may avoid this in noting that the topological image of a closed simplex onto itself can always carry the point $\xi \in \bar{T}^n - T^n$ into a point having as its carrier an $(n-1)$-dimensional face of the simplex T^n. In the star $O_{[T^n]} T^{n-1}$ itself there are no simplexes of dimension $< n - 1$, and all the $(n-1)$-dimensional cycles have the form ct^{n-1} and are homologous to zero, since they bound a chain ct^n, where t^n is the orientation assigned to the simplex T^n. For the same reason, there is no n-dimensional nonzero chain in $O_{[T^n]} T^{n-1}$ which is not a cycle.

From what has just been proved, there follows further the theorem on the invariance of interior points of n-dimensional Euclidean space. Under a topological

---

[*] We use here the isomorphism proved in the last section between the groups $\Delta^r (\Phi, \mathrm{mod}\,(\Phi - \tilde{O}_\nu))$ and $\Delta^r (K_\nu, \mathrm{mod}\,(K_\nu - O_\nu))$.

mapping of any set lying in n-dimensional Euclidean space onto a set lying in the same space, any interior point of one set goes into an interior point of the other.

CHAPTER IV

Homology manifolds, Poincaré duality.
Combinatorial case of Pontrjagin duality (in ∇-form)

§1. Definition and elementary properties of homology manifolds. Star complex of an h-manifold

1. A *homology manifold* or an *h-manifold* of dimension n is a connected polyhedron M^n whose local Δ-groups (with integer coefficients) are the same as the groups of a (interior) point of an n-dimensional simplex, i.e., for any point $\xi \in M^n$ the group $\Delta^r(\xi, M^n)$ is the null group for $r \neq n$, and the group $\Delta^n(\xi, M^n)$ is an infinite cyclic group.* It follows from the invariance theorem proved at the end of the last chapter that the same definition may be expressed in the following form:

A connected polyhedron M^n is said to be an n-dimensional h-manifold if in any of its triangulations K the stars $O_K T$ of all the simplexes have the following Betti groups: $\Delta^n O_K T$ is infinite cyclic, $\Delta^r O_K T = 0$ for $r \neq n$.

From what was stated at the end of the preceding section it follows that every n-dimensional *closed manifold* (i.e., a polyhedron each point of which has a neighborhood homeomorphic to the n-dimensional open simplex) is also an n-dimensional h-manifold.

Along with the arbitrarily chosen triangulation K of the given h-manfiold M^n we shall consider its *barycentric subdivision* K_1 *and star complex* K^*, i.e., the set of all of its barycentric (open) stars. As is well known (see Ch. 1, §1.7), the barycentric star U_i^{n-p} is said to be adjoint to the simplex $T_i^p \in K$ if the center of the simplex T_i^p is the lowest vertex of the star U_i^{n-p}. Here the boundary $B_i^* = [U_i^{n-p}] - U_i^{n-p}$ of the star U_i^{n-p} is composed of the barycentric stars of lowest dimension, called the *"faces" of the star* U_i^{n-p}: the star U_j is a face of the star U_i (written $U_j < U_i$) if for the adjoint simplexes T_j, T_i we have $T_i < T_j$. The basic problem of this section is the proof of the following proposition:

The star complex K^* is an enlargement of the triangulation K_1 (in the sense of §6 of Ch. III). Therefore the cell complex, consisting of the oriented barycentric

---

*Therefore it already follows that $\dim M^n = n$. Indeed, if $k = \dim M^n > n$, we would have a simples $T^k \subset M^n$ and for its points ξ, we would have $\Delta^k(\xi, M^n) = \Delta^k(\xi, T^k) \neq 0$. If $\dim M^n < n$, we would have $\Delta^n(\xi, M^n) = 0$ for all $\xi \in M^n$.

The last remark shows, moreover, that for any triangulation of our h-manifold M^n, every simplex $T^r \in K$ of dimension $r < n$ is the face of some n-dimensional simplex (i.e., a triangulation of an n-dimensional h-manifold is a *dimensionally homogeneous* n-dimensional complex).

stars (we denote this complex also by K^*), has Betti groups isomorphic to the corresponding groups of the triangulations K_1 and K.

In order to prove the proposition just stated, we have to establish several simple properties of triangulations of h-manifolds (these triangulations are sometimes called *combinatorial h-manifolds*).

2. Stars in h-manifolds.(1). If T^{n-1} is some $(n-1)$-dimensional simplex of the n-dimensional combinatorial h-manifold K, then adjoining T^{n-1} there are exactly two n-dimensional simplexes (this easily follows from the fact that $\Delta^n O_K T^{n-1}$ is an infinite cyclic group).

(2). The star $O_K e$ of any vertex is an n-dimensional orientable pseudomanifold.

Indeed, we saw [*] that K, and therefore also $O_K e$, is a dimensionally homogeneous complex. Suppose that $T^{n-1} \in O_K e$. Then both the n-dimensional simplexes of the complex K adjoining T^{n-1} belong to the star $O_K e$. It remains to be proved that O_K is a strongly connected complex. This will prove that $O_K e$ is a pseudomanifold, and hence, since $\Delta^n O_K e$ is different from zero, that the pseudomanifold $O_K e$ is oriented.

To prove the strong connectivity of the complex $O_K e$ we prove first that in $O_K e$ there can exist at most one n-dimensional cycle modulo 2 which is different from zero.

Let z^n be any cycle modulo 2 in $O_K e$. Taking the n-dimensional simplexes of $O_K e$ with given orientations t_i^n, we consider the coefficients 0 or 1, with which these simplexes enter into the cycle z^n, as integers. We obtain an integer-valued chain x^n, in whose boundary all the coefficients are divisible by 2, so that we may write

$$\Delta x^n = 2 z^{n-1},$$

where z^{n-1}, evidently, is a (integer-valued) cycle. Since $\Delta^{n-1} O_K e$ is a null group, we have $z^{n-1} \sim 0$ in $O_K e$, i.e., in that star there exists an integer-valued chain y^n bounded by the cycle z^{n-1}. But then $\Delta 2 y^n = 2 z^{n-1} = \Delta x^n$, so that $x^n - 2 y^n$ is an integer-valued cycle and the cycle z^n is obtained from this cycle by reducing its coefficients modulo 2.

Thus, each n-dimensional cycle modulo 2 of the complex $O_K e$ is obtained from some integer-valued cycle of the same complex, reduced modulo 2. But the group $\Delta^n O_K e = Z^n O_K e$ (with integer coefficients) is by hypothesis an infinite cyclic group, which means that all the n-dimensional integer-valued cycles in $O_K e$ have the form $c z_0^n$, where z_0^n is some definite n-dimensional cycle and c

---

[*] See the note on page 64.

is an integer. On reducing modulo 2, all the cycles cz_0^n with even c give zero, and, the cycles with odd c give one unique cycle modulo 2, namely the one obtained by reducing the cycle z_0^n. Thus, in $O_K e$ there cannot be more than one n-dimensional cycle modulo 2 which differs from zero.

Now we suppose that $O_K e$ is not a strongly connected complex. Then $O_K e = K' \cup K''$, where K' and K'' are two nonempty n-dimensional closed subcomplexes of the complex $O_K e$, yielding in their intersection a complex of dimension $\leq n - 2$. Hence it follows that if T^{n-1} lies in one or another of the complexes K' or K'', then both of the n-dimensional simplexes adjoining T^{n-1} belong to the same complex. Therefore, taking all the n-dimensional simplexes of the complex K' with coefficient 1, we obtain one cycle z^n distinct from zero modulo 2, and taking all the n-dimensional simplexes of the complex K'', we obtain a further cycle distinct from zero and from z^n. The contradiction thus obtained proves the strong connectivity of the star $O_K e$, and, at the same time, the fact that this star is an oriented pseudomanifold.

(3). Every h-manifold is a closed pseudomanifold.

Indeed, to each $(n-1)$-dimensional simplex of the complex K there adjoin exactly two n-dimensional simplexes, so that we need only prove that K is a strongly connected complex. Suppose this were not the case. Then K may be represented as the sum of two nonempty dimensionally homogeneous n-dimensional closed subcomplexes K' and K'', whose intersection is a complex of diemsnion $\leq n - 2$.

Let e be some vertex of $K' \cap K''$. We have

$$O_K e = (O_K e \cap K') \cup (O_K e \cap K''),$$

while from the fact that $e \in K' \cap K''$ and that the n-dimensional complexes K' and K'' are dimensionally homogeneous it follows that $O_K e \cap K'$ and $O_K e \cap K''$ are nonempty n-dimensional (still dimensionally homogeneous) subcomplexes. These complexes are closed in $O_K e$ since K' and K'' are complete complexes. Finally, the dimension of the intersection $(O_K e \cap K') \cap (O_K e \cap K'')$ does not exceed $n - 2$. But this contradicts the fact that $O_K e$ is a strongly connected complex.

The proposition just proved evidently makes it possible to speak of *orientable* and *nonorientable* h-manifolds.

(4). For $n \geq 2$ the edge $B_K e = [O_K e] - O_K e$ of the star $O_K e$ is an $(n-1)$-dimensional simple* h-manifold.

---

*A pseudomanifold (in particular, an h-manifold) of dimension n is said to be *simple* if it is orientable and if its Δ^r-groups for $0 \leq r < n$ are null-groups (here for $r = 0$ we choose the group Δ^∞).

Indeed, $O_K e$ is a pyramid over $B = B_K e$, so that simplexes $T^{p+1} \in O_K e$ of dimension > 0 and having the form $T^{p+1} = |e T^p|$, where $T^p \in B$, correspond in a one-to-one way to the simplexes $T^p \in B$. Here to the star $O_B T^p$ of the simplex T^p in B there corresponds the star of the simplex T^{p+1} in $O_K e$, i.e., simply the star $O_K T^{p+1}$.

From the strong connectivity of the star $O_K e$ and the inequality $n \geq 2$ follows the strong connectivity of B. To the chains x^r in B correspond pyramids over them, which are $(r + 1)$-dimensional chains in $O_K e$, while (in particular) to the chains of the star $O_B T^p$ correspond the chains of the star $O_K T^{p+1}$. Therefore we obtain an isomorphism between the groups $L^r B$ and $L^{r+1} O_K e$, in which the groups $L^r O_B T^p$ correspond to the groups $L^{r+1} O_K T^p$.

This isomorphism generates an isomorphism between $\Delta^{r+1} O_K e$ and $\Delta^r B$, and also between $\Delta^{r+1} O_K T^{p+1}$ and $\Delta^r O_B T^p$, from which it follows that (i) the groups $\Delta^r B$ are isomorphic to the groups $\Delta^{r+1} O_K e$ and (ii) the groups $\Delta^r O_B T^p$ are isomorphic to the groups $\Delta^{r+1} O_K T^{p+1}$.

From the second assertion it follows that $\Delta^{n-1} O_B T^p$ is an infinite cyclic group, while $\Delta^r O_B T^p = 0$ for $r < n - 1$. Moreover, since B is connected it is an $(n-1)$-dimensional orientable h-manifold. It is a simple h-manifold since $\Delta^r B = 0$ for all $r < n - 1$.

3. **Barycentric stars in h-manifolds.** Now we can prove the fundamental proposition formulated at the end of §1.1 of this chapter.

(5). *All the barycentric stars of a combinatorial h-manifold K are simple pseudomanifolds.*

Proof. Let the dimension of the complex K be n. For $n = 1$ the theorem is evident. We shall prove that if it is true for $n - 1$ it is true for n.

To this end we choose any vertex e of the complex K. Now to the center of each simplex $T^{p+1} \in O_K e$, $p > 0$, different from the vertex e we make correspond the center of the face T^p of that simplex which is opposite the vertex e.

This, as is easily seen, establishes a one-to-one simplicial mapping (isomorphism) of the edge B^* of the barycentric star* U^n adjoint to the vertex e onto the barycentric subdivision B_1 of the complex $B = [O_K e] - O_K e$, i.e., the edge of the star $O_K e$. Here to each simplex of the complex B^* having as its lowest vertex the center of some $T^{p+1} = |e T^p| \in O_K e$ corresponds the simplex of the complex B_1 having as its lowest vertex the center of the simplex $T^B \in B$. From this it follows that every barycentric star lying on B^* of the complex K is

---

* By a barycentric star of the complex K we shall always mean a subcomplex of the complex K_1.

isomorphic to some barycentric star of the complex B.

But B, in view of Proposition 4, is an $(n-1)$-dimensional h-manifold. This means that from the induction hypothesis all the barycentric stars of the complex B are simple pseudomanifolds. Accordingly, all the barycentric stars of the complex K lying on B^* are simple h-manifolds. Since each barycentric star of the complex K having dimension $< n$ lies on the edge B of some n-dimensional barycentric star, we have proved that all the barycentric stars of the complex K having dimension $< n$ are simple pseudomanifolds.

It remains only to prove that every n-dimensional barycentric star U^n, i.e., a star adjoint to some vertex e of the complex K, is a simple pseudomanifold. But this follows from the fact that the barycentric star U^n is a pyramid (with vertex e) over its edge B^*, which is isomorphic to the barycentric subdivision of the edge B of the star $O_K e$, and is therefore, from Theorem 4, a simple n-dimensional h-manifold, as we were required to prove.

From Theorem (5) it follows that the star complex K^* of the complex K is an enlargement of the complex K_1. Accordingly, the Δ- and ∇-groups of the complex K^* are isomorphic to the respective groups of the complexes K_1 and K. The goal set by us at the end of §1.1 of this chapter has been reached.

We shall now strengthen somewhat the result obtained.

We denote by A^* a closed subcomplex of the complex K^*. By $A_1^* \subset K_1$ we denote a barycentric subdivision of the complex A^*, i.e., the subcomplex of the complex K_1 defined as the sum of the barycentric stars which are elements of the complex A^*. We know that A_1^* is a triangulation.

The complex A^* is evidently an enlargement of the complex A_1^*.

Therefore, on the basis of the results of §6 of the preceding chapter, we have:

1°. Each cycle z of the complex A_1^* is homologous in A_1^* to the barycentric subdivision $s_1 z^*$ of some cycle $z^* = \Sigma a_i u_i{}^*$ of the complex A^*.

2°. If a cycle of the form $z^r = s_1 z^*$, where $z^* = \Sigma a_i u_i^r$, is homologous to zero in A_1^*, then that cycle is the boundary of the subdivision in A_1^* of some chain of the form $x^{r+1} = \Sigma_i c_i u_i^{r+1}$.

§2. Poincaré duality

1. Among the elements of the complexes K and K^* there is, as we have seen, a one-to-one relation such that to each simplex $T_i^\mu \in K$ there corresponds the adjoint barycentric star $U_i^\nu \in K^*$ (here and from now on $\mu + \nu = n$, where n

---

*By u_i we mean the orientation of the barycentric star U_i.

is the dimension of the h-manifold $M^n = \tilde{K}$). Our immediate problem consists in extending this relation to oriented elements of both complexes, i.e., in putting into correspondence with each oriented simplex t_i^μ the completely defined orientation u_i^ν of the barycentric star U_i^ν adjoint to this simplex (of course, in such a way that to different orientations of the simplex there correspond different orientations of the barycentric star adjoint to it). If this problem is solved, then to each μ-dimensional chain x^μ of the complex K there will be assigned a ν-dimensional chain in the complex K^*, taking on each oriented star u_i^ν the value which the chain x^μ takes on on the oriented simplex t_i^μ of the complex K adjoint to that star. Thus we will have proved an isomorphism, which will be denoted by D and will be called the *Poincaré isomorphism*, between the groups $L^\mu K$ and $L^\nu K^*$. It turns out that under this isomorphism the ∇-boundary of the chain x^μ corresponds to the Δ-boundary of the chain $y^\nu = Dx^\nu$ multiplied by $(-1)^\mu$ (and conversely), from which it follows that under the Poincaré isomorphism μ-dimensional ∇-cycles of the complex K correspond to ν-dimensional Δ-cycles of the complex K^*, while cycles homologous to zero in K correspond to cycles homologous to zero in K^*, so that we obtain an *isomorphism* (also denoted by D) between the groups $\nabla^\mu K$ and $\Delta^\nu K^*$. Since the Δ-groups of the complex K^* are isomorphic to the same groups of the complex K, we obtain the proposition:

For $\mu + \nu = n$ the groups ∇^μ and Δ^ν of an oriented h-manifold M^n are isomorphic to each other.

This proposition is called the *duality law of Poincaré* (in the form given by A. N. Kolmogorov).*

We proceed to the completion of our plan.

2. The intersection index $(t^\mu \times u^\mu)$. Let t^μ be an oriented simplex of the complex K, and u^ν an oriented barycentric star. If the simplex $T^\mu = |t^\mu|$ and the star $U^\nu = |u^\nu|$ are not adjoint, then the intersection index $(t^\mu \times u^\nu)$ is set equal to zero by definition. If they are adjoint to each other, then the intersection index will be set equal to ± 1, where the sign is determined as follows.

We choose any simplex T_1^μ of the barycentric subdivision K_1 which lies on T^μ, oriented in the same way as t^μ. Let

$$t_1^\mu = \epsilon(a_0 \cdots a_\mu), \quad \epsilon = \pm 1,$$

where the vertices are written in increasing order of the dimension of the simplexes

---

*Since the group ∇^μ is the direct sum of the $(\mu - 1)$-dimensional torsion group $\Theta^{\mu-1}$ and of as many infinite cyclic groups as the size of the μ-dimensional Betti number π^μ (Ch. I, §2.10), it follows from the isomorphism $\nabla^\mu = \Delta^\nu$ that the Betti numbers π^μ and π^ν are equal: $\pi^\mu = \pi^\nu$: and hence the torsion groups $\Theta^{\mu-1}$ and Θ^ν are isomorphic: $\Theta^{\mu-1} = \Theta^\nu$. This is the original form of the Poincaré duality law.

of the complex K whose centers they are, so that a_ν is the leading vertex, i.e., the center of the simplex T^μ.

In the oriented star u^ν, adjoint to the simplex $|t^\mu|$, we also choose any simplex $t_1^\nu \in K_1$, oriented in the same way as the star u^ν, i.e., entering in the chain u^ν with coefficient $+1$. Let

$$t_1^\nu = \eta(a_\mu \cdots a_n), \quad \eta = \pm 1,$$

where the vertices are again written in the order of increase, so that in particular a_n is the center of the n-dimensional simplex $T^n \in K$ carrying the simplex $|t_1^\nu|$. Suppose that we have chosen a definite orientation of the whole h-manifold K and let t^n be the corresponding orientation of the simplex T^n. Then

$$(a_0 \cdots a_\mu \cdots a_n) = \gamma t^n, \text{ where } \gamma = \pm 1.$$

By definition

$$(t^\mu \times u^\nu) = \epsilon\eta\gamma.$$

We have to prove that the intersection index so defined for the oriented simplex t^μ of the complex K and the oriented barycentric star u^ν depends only on t^μ and u^ν and on the choice of the orientation of all of K, and does not depend on the choice of the simplexes $|t_1^\mu|$ and $|t_1^\nu|$. To this end let us choose another simplex $t_1'^\mu$, lying in $|t^\mu|$ and oriented in the same way as t^μ. In view of the strong connectivity of the complex, which is a barycentric subdivision of the simplex $|t^\mu|$, we may restrict ourselves to the case when the simplexes $|t_1^\mu|$ and $|t_1'^\mu|$ have a common $(\mu - 1)$-dimensional face $T_1^{\mu-1}$. Let this face be

$$T_1^{\mu-1} = |a_0 \cdots a_{k-1}a_{k+1} \cdots a_\mu|,$$

and the vertices opposite it in $|t_1^\mu|$ and $|t_1'^\mu|$ respectively a_k and a_k'. Then the simplexes

$$T_1^n = |a_0 \cdots a_k \cdots a_\mu \cdots a_n|$$

and

$$T'^n = |a_0 \cdots a_k' \cdots a_\mu \cdots a_n|$$

have a common $(n - 1)$-dimensional face

$$|a_0 \cdots a_{k-1}a_{k+1} \cdots a_\mu \cdots a_n|.$$

If $t_1^\mu = \epsilon(a_0 \cdots a_k \cdots a_\mu)$, $t_1'^\mu = \epsilon'(a_0 \cdots a_k' \cdots a_\mu)$ and $t_1^{\mu-1}$ is any orientation of the simplex $T_1^{\mu-1}$, for example $t_1^{\mu-1} = (a_0 \cdots a_{k-1} a_{k+1} \cdots a_\mu)$, then $(t_1^\mu : t_1^{\mu-1}) = -(t_1'^\mu : t_1^{\mu-1})$, since the simplexes t_1^μ and $t_1'^\mu$ are oriented the same way in $|t^\mu|$ and situated on opposite sides of their common face.

But $(t_1^\mu : t_1^{\mu-1}) = (-1)^k \epsilon$, $(t_1'^\mu ; t_1^{\mu-1}) = (-1)^k \epsilon'$, so that $\epsilon' = -\epsilon$.

Analogously, if the orientations $t_1^n = \gamma(a_0 \cdots a_k \cdots a_\mu \cdots a_n)$ and $t_1'^n = \gamma'(a_0 \cdots a_k \cdots a_\mu \cdots a_n)$ are the same as the orientation t^n, and consequently the same as each other, and if $t_1^{n-1} = (a_0 \cdots a_{k-1} a_{k+1} \cdots a_n)$, then $(t_1^n : t_1^{n-1}) = -(t_1'^n : t_1^{n-1})$, $(t_1''^n : t_1^{n-1}) = (-1)^k \gamma$, $(t_1''^n : t_1^{n-1}) = (-1)^k \gamma'$. This means that $\gamma' = -\gamma$, and therefore $\epsilon' \eta \gamma' = (-\epsilon)\eta(-\gamma) = \epsilon \eta \gamma$, so that $\epsilon \eta \gamma$ does not change sign when t_1^μ is replaced by $t_1'^\mu$. In exactly the same way, using the strong connectivity of the complex U^ν, we prove that the number $\epsilon \eta \gamma$ does not depend on the choice of the simplex $|t_1^\nu|$.

Thus the correctness of our definition of the intersection index $(t^\mu \times u^\nu)$ is proved.

Evidently, if one replaces t^μ by $-t^\mu$, or u^ν by $-u^\nu$, or, finally, if one replaces the given orientation of the whole h-manifold K by the opposite orientation, the intersection index $(t^\mu \times u^\nu)$ changes sign.

Remark. Let X^μ, Y^ν, and Z^n be the planes carrying respectively the simplexes t^μ, t_1^ν, t^n, oriented in accordance with the orientation of these simplexes. Then $(t^\mu \times u^\nu)$ is simply the intersection index $(X^\mu \times Y^\nu)$ in Z^n (see Chapter VI, §5).

Now we shall once and for all make the following convention:

By t_i we shall denote an arbitrary, but definitely chosen orientation of all the simplexes of the complex K. For the barycentric stars we choose orientations u_i such that, denoting by the same index i a simplex and its adjoint star, we shall always have $(t_i^\mu \times u_i^\nu) = +1$.

3. **Connection between the incidence coefficients** $(u_j^{\nu+1} : u_i^\nu)$ **and** $(t_i^\mu : t_j^{\mu-1})$. The goal set in §1.1 will be reached if we prove the following equality:

$$(u_j^{\nu+1} : u_i^\nu) = (-1)^\mu (t_i^\mu : t_j^{\mu-1}), \tag{1}$$

where the stars $u_j^{\nu+1}$ and u_i^ν are adjoint to the simplexes $t_j^{\mu-1}$ and t_i^μ. From equation (1) and the definition of the operators ∇ and Δ it immediately follows that for any chain $x^{\mu-1}$ of the complex K we have *

$$D\nabla x^{\mu-1} = (-1)^\mu \Delta D x^{\mu-1}, \tag{A}$$

from which follow all of the consequences enumerated in §1.1.

Now we shall prove formula (1); we shall drop the indices i and j.

On the simplex $|t^\mu|$ we choose a simplex $|t_1^\mu| = |a_0 \cdots a_\mu| \in K_1$, while, as always, the vertices go in increasing order, so that $a_{\mu-1}$ is the center of the simplex $|t^{\mu-1}|$ and a_μ is the center of the simplex $|t^\mu|$. In the star $u^{\nu+1}$ choose the simplex $|t_1^{\nu+1}| = |a_{\mu-1} a_\mu \cdots a_n| \in K_1$. We choose the orientations

---

* Indeed, it suffices to prove equation (A) for the case $x^{\mu-1} = t_j^{\mu-1}$, but

$$(D\nabla t_j^{\mu-1} \cdot u_i^\nu) = (\nabla t_j^{\mu-1} \cdot t_i^\mu) = (t_i^\mu : t_j^{\mu-1});$$

$$(\Delta D t_j^{\mu-1} \cdot u_i^\nu) = (\Delta u_j^{\nu+1} \cdot u_i^\nu) = (u_j^{\nu+1} : u_i^\nu).$$

t_1^μ and $t_1^{\nu+1}$ in accordance with the orientations t^μ and $u^{\nu+1}$, so that, denoting by s_1 the operation of barycentric subdivision, we have

$$s_1 t^\mu = t_1^\mu + \cdots ; \tag{2}$$

$$u^{\nu+1} = t_1^{\nu+1} + \cdots . \tag{2*}$$

Put

$$t_1^\mu = \eta(a_0 \cdots a_\mu); \; t_1^{\nu+1} = \eta^*(a_{\mu-1} a_\mu \cdots a_n), \tag{3}$$

$$\eta = \pm 1, \; \eta^* = \pm 1.$$

Choose

$$t_1^{\mu-1} = \eta(a_0 \cdots a_{\mu-1}).$$

Then $(t_1^\mu : t_1^{\mu-1}) = (-1)^\mu$ and the simplex $|t_1^{\mu-1}|$ lies on $|t^{\mu-1}|$. Accordingly,

$$\Delta s_1 t^\mu = \Delta t_1^\mu + \cdots = (-1)^\mu t_1^{\mu-1} + \cdots .$$

Put $\epsilon = (t^\mu : t^{\mu-1})$, $\epsilon^* = (u^{\nu+1} : u^\nu)$. Then

$$\Delta s_1 t^\mu = s_1 \Delta t^\mu = \epsilon s_1 t^{\mu-1} + \cdots ,$$

so that $(-1)^\mu t_1^{\mu-1}$ is oriented in the same way as $\epsilon t^{\mu-1}$, i.e., $(-1)^\mu \epsilon t_1^{\mu-1}$ has the same orientation as $t_1^{\mu-1}$.

Now put

$$t_1^\nu = \eta^*(a_\mu a_{\mu+1} \cdots a_n).$$

Then

$$(t_1^{\nu+1} : t_1^\nu) = 1 \text{ and } \Delta u^{\nu+1} = \Delta t_1^{\nu+1} + \cdots = t_1^\nu + \cdots .$$

On the other hand, if the coefficient with which t_1^ν appears in u^ν is denoted by c, then

$$\Delta u^{\nu+1} = \epsilon^* u^\nu + \cdots = \epsilon^* c t_1^\nu + \cdots ,$$

so that $\epsilon^* c = 1$, $c = \epsilon^*$.

Suppose, finally, that the orientation chosen for the h-manifold M^n is given by the orientation

$$t_1^n = \gamma(a_0 \cdots a_{\mu-1} a_\mu, \cdots a_n), \gamma = \pm 1$$

of the simplex $|a_0 \cdots a_n| \in K_1$.

Bringing everything together we have:

the orientation $t_1^\mu = \eta(a_0 \cdots a_{\mu-1} a_\mu)$ coincides with the orientation t^μ,

the orientation $\epsilon^* t_1^\nu = \epsilon^* \eta^*(a_\mu \cdots a_n)$ coincides with u^ν,

the orientation $t_1^n = \gamma(a_0 \cdots a_\mu, a_\mu \cdots a_n)$ coincides with the orientation of M^n,

which means

$$\eta \epsilon^* \eta^* \gamma = (t^\mu \times u^\nu) = 1.$$

At the same time,

the orientation $(-1)^{\mu} \epsilon t^{\mu-1} = (-1)^{\mu} \epsilon \eta (a_0 \cdots a_{\mu-1})$ coincides with $t^{\mu-1}$,

the orientation $t_1^{\nu+1} = \eta^* (a_{\mu-1} \cdots a_n)$ coincides with $u^{\nu+1}$,

i.e.,

$$(-1)^{\mu} \epsilon \eta \, \eta^* \gamma = (t^{\mu-1} \times u^{\nu+1}) = 1.$$

Thus, $\eta \epsilon^* \eta^* \gamma = (-1)^{\mu} \epsilon \eta \eta^* \gamma$, which means that $\epsilon^* = (-1)^{\mu} \epsilon$, as was to be proved.

§3. Combinatorial case of the Pontrjagin duality law

Let M^n be a homology manifold of dimension n.

By K we denote any triangulation of the manifold M^n. By α we denote a closed subcomplex of the triangulation K, by K_1 a barycentric subdivision, and by K^* the star complex of the triangulation K. We denote by α^* the (open) subcomplex of the star complex K^* consisting of all the barycentric stars adjoint to simplexes of the complex α. Then $\beta = K^* - \alpha^*$ is a closed subcomplex of the complex K^*.

This section is devoted to a proof of the following theorem, which is the combinatorial case of the Pontrjagin duality law (in ∇-form).

Theorem 1. *Let the h-manifold M^n be acyclic in the dimensions p and $p+1$, where $0 \leq p \leq n-1$; Put $q = n - p - 1$.*[*] *Then the group $\nabla^p \alpha$ is isomorphic to the group $\Delta^q \beta$. Here, if $q = 0$, we must take the group $\Delta^{00} \beta$, and if $p = 0$, the group $\nabla^{00} \alpha$ (i.e., the direct sum of infinite cyclic groups, taken in number one higher than the number of components of the complex α).*

The proof will be prepared by a number of auxiliary remarks.

We observe first of all that every ∇-cycle z_α of the complex α, homologous to zero in α, is a cycle, *which can be extended to the whole complex K*, in the sense that there exists a ∇-cycle z of the complex K coinciding on α with z_α. Indeed, let $z_\alpha = \nabla x_\alpha$, where x_α lies on α. We consider the chain $E x_\alpha$ of the complex K coinciding with x_α on α and equal to zero off α, and we put $z = \nabla E x_\alpha$. Denoting by αz the part of the chain z lying on α, we have (since α is a closed manifold)

$$\alpha z = \alpha \nabla E x_\alpha = \nabla \alpha E x_\alpha = \nabla x_\alpha = z_\alpha,$$

as was to be proved.

Now we shall prove that *if $\nabla^p K = 0$, then the inverse is true: every p-dimensional ∇-cycle z_α^p of the complex α, which is extensible onto the whole complex K, is homologous to zero in α.*

Indeed, suppose we have a ∇-cycle z^p of the complex K such that $\alpha z^p = z_\alpha^p$.

---

[*] I.e., $\nabla^p K = \nabla^{p+1} K = 0$, so that by the Poincaré duality $\Delta^{q+1} K = 0$, $\Delta^q K = 0$.

From the hypothesis $\nabla^p K = 0$ follows the existence of a chain x of the complex K, bounding the ∇-cycle z^p:

$$\nabla x = z^p.$$

We operate on both sides of this equation with the excision operator. We obtain

$$a\nabla x = \nabla a x = a z^p = z^p_a,$$

$$z^p_a \sim 0 \text{ in } a,$$

as was to be proved.

Thus, under the assumptions of our theorem the group ∇^p_a coincides with the factor-group* $\nabla^p(a:K) = Z^p_\nabla a - C^p a$ of the group of all p-dimensional ∇-cycles of the complex a modulo the subgroup $C^p a$ of all those ∇-cycles which are extensible to the whole complex K.

Now we consider the subgroup $Z^q(\beta:K^*)$ of the group $Z^q\beta$, consisting of the cycles which are homologous to zero in K^*. In it is contained, evidently, the group $H^q\beta$ of cycles homologous to zero in β. Therefore the factor-group $Z^q(\beta:K^*) - H^q\beta$ is isomorphic in a natural way to some subgroup $\Delta^q(\beta:K)$ of the group $\Delta^q\beta$. Under the hypothesis $\Delta^{q+1}K = 0$, laid down in the conditions of Theorem 1, the groups $Z^q\beta$ and $Z^q(\beta:K^*)$ evidently coincide with each other. Therefore Theorem 1 may be generalized to the following proposition, which contains no hypothesis of any kind concerning the acyclicity of the h-manifold M^n.

Theorem 2. *In an h-manifold M^n the groups $\nabla^p(a:K)$ and $\Delta^q(\beta:K^*)$ are isomorphic to each other.*

We shall prove this theorem by directly constructing an isomorphism of the group $\nabla^p(a:K)$ onto the group $\Delta^q(\beta:K)$.

To this end we consider the operator ΔD, which puts into correspondence with each ∇-cycle z^p_a of the complex a the cycle $\Delta D z^p_a \in H^q K^*$. Evidently, for any $u^q_i \in a^*$ we have $(\Delta D z^p_a \cdot u^q_i) = \pm (D\nabla z^p_a \cdot u^q_i) = \pm (\nabla z^p_a \cdot t^{p+1}_i) = 0$ (since $t^{p+1}_i \in a$), i.e., $\Delta D z^p_a \in Z^q\beta$, and also, since $\Delta D z^p_a \in H^q K^*$, we have $\Delta D z^p_a \in Z^q(\beta:K^*)$. Thus, *the operator ΔD realizes an (obviously) homomorphic mapping of the group $Z^p_\nabla a$ into the group $Z^q(\beta:K^*)$.*

---

*We observe that if $p = 0$ the group $\nabla^0(a:K)$ is exactly the group $\nabla^{00}a$, formally defined above as the direct sum of $k-1$ infinite cyclic groups, where k is the number of components of the complex a. Indeed, the zero-dimensional chain of any complete complex is a ∇-cycle if and only if it is constant on every component of that complex. Since the number of components of the complex a is equal to k, the group $\Delta^0(a:K)$ is the factor-group of the group of all zero-dimensional chains of the complex a, constant on each component of the complex a, by the subgroup of all the chains which are constant on all of K. It is easy to see that this factor-group is isomorphic to the direct sum of $k-1$ infinite cyclic groups.

An analogous result is evidently true for any coefficient domain.

We shall show:

1°. $\Delta D z_\alpha^p \sim 0$ in β if and only if z_α^p is an extensible ∇-cycle. This will show that the operator ΔD generates an *isomorphic* mapping of the group $\nabla^p(\alpha:K)$ into the group $\Delta^q(\beta:K^*)$.

2°. For each Δ-cycle z^q of the complex β, homologous to zero in K^*, there exists a ∇-cycle z_α^p of the complex α such that $\Delta D z_\alpha^p \sim z^q$ in β. This will prove that the operator ΔD generates a mapping of the group $\nabla^p(\alpha:K)$ onto the whole group $\Delta^q(\beta:K^*)$.

The proof of Theorem 2 will then be complete.

Proof of assertion 1°.

(a) Suppose that z_α^p is an extensible ∇-cycle of the complex α. We shall show that $\Delta D z_\alpha^p \sim 0$ in β.

Since z_α^p is an extensible cycle, there exists a ∇-cycle z^p of the complex α such that $\alpha z^p = z_\alpha^p$. Then $D z^p$ is a Δ-cycle on K^*. We break it into two pieces: $D z^p = x^{q+1} + y^{q+1}$, where x^{q+1} lies on α^*, and y^{q+1} on β. Evidently,

$$\Delta(-y^{q+1}) = \Delta x^{q+1} = \Delta \alpha^* D z^p.$$

But $\alpha^* D z^p = D \alpha z^p$, so that

$$\Delta(-y^{q+1}) = \Delta D \alpha z^p = \Delta D z_\alpha^p.$$

Since $-y^{q+1}$ lies on β, we have $\Delta D z_\alpha^p \sim 0$ in β, and assertion (a) is proved.

(b) Let z_α^p be a ∇-cycle on α relative to which we know that $\Delta D z_\alpha^p \sim 0$ in β. We shall show that then z_α^p is an extensible cycle.

To this end we choose first of all a chain y^{q+1} on β, bounded by a cycle $\Delta D z_\alpha^p$. Then $D z_\alpha^p - y^{q+1} = z^{q+1}$ is a cycle of the complex K^*. We define a chain z^p on K, putting $z^p = D^{-1} z^{q+1}$, i.e., giving it on each t_i^p the value equal to the value of the Δ-cycle z^{q+1} on the star u_i^{q+1} adjoint to the simplex t_i^p. Then z^p is a ∇-cycle on K, and it remains for us to prove that $\alpha z^p = z_\alpha^p$. But this is done without difficulty, since obviously $z^p = z_\alpha^p - D^{-1} y^{q+1}$, where $D^{-1} y^{q+1}$ lies on $K - \alpha$.

Proof of assertion 2°. Let z^q be a cycle on β, bounding the chain x^{q+1} of the complex K^*. We represent x^{q+1} as the sum of two pieces

$$x^{q+1} = x_\alpha^{q+1} + y^{q+1},$$

of which x_α^{q+1} lies on α^*, and y^{q+1} on β. Then

$$z^q = \Delta x^{q+1} = \Delta x_\alpha^{q+1} + \Delta y^{q+1}.$$

But the piece x_α^{q+1} may be represented in the form

$$x_\alpha^{q+1} = \alpha^* x^{q+1} = \alpha^* D D^{-1} x^{q+1} = D \alpha D^{-1} x^{q+1},$$

so that

$$\Delta x_{\alpha}^{q+1} = \Delta D \alpha D^{-1} x^{q+1}.$$

Denoting the chain $\alpha D^{-1} x^{q+1}$ of the complex α by z_{α}^{p}, we have

$$z^{q} - \Delta D z_{\alpha}^{p} = \Delta y^{q+1},$$

so that

$$z^{q} \sim \Delta D z_{\alpha}^{p} \text{ in } \beta.$$

It remains to prove that z_{α}^{p} is a ∇-cycle of the complex α, i.e., that for any $t_{\alpha}^{p+1} \in \alpha$,

$$(\nabla z_{\alpha}^{p} \cdot t_{\alpha}^{p+1}) = 0.$$

But

$$\nabla z_{\alpha}^{p} = \nabla \alpha D^{-1} x^{q+1} = \alpha \nabla D^{-1} x$$

$$= \pm \alpha D^{-1} \Delta x^{q+1} = \pm \alpha D^{-1} z^{q}.$$

Since z^{q} lies on β it follows that $\alpha D^{-1} z^{q} = 0$, i.e., also ∇z_{α}^{p} on α is equal to zero, with which the proof is complete.

CHAPTER V

General duality law of Pontrjagin (in ∇-form)

§1. Formulation of the theorem and reduction to a homological lemma

1. **Definition of the groups Δ_{c}^{r} for non-closed, in particular for open, sets.**
First definition: [*] By a Δ_{c}-cycle of an arbitrary set B (lying in a given h-manifold M^{n}) we mean a Δ_{c}-cycle lying on any compactum $\psi \subseteq B$. A given Δ_{c}-cycle is said to be *homologous* in B if it is homologous to zero on some compactum ψ', $\psi \subseteq \psi' \subseteq B$. The factor group of the group Z_{c}^{r} of all r-dimensional Δ_{c}-cycles of the set B, modulo the subgroup H_{c}^{r} of cycles homologous to zero in B, is called the group $\Delta_{c}^{r} B$.

In the case when B is open in M^{n}, we may employ simpler methods; for example, one of the following:

Second definition: A cycle of the set B may be defined as an arbitrary polyhedral cycle, i.e., any cycle of any triangulation K lying in B, and we say that this cycle is homologous to zero in B if it is homologous to zero in some triangulation K' lying in B.

Third definition: We may understand by a chain x^{r} lying in B a linear form (as always, finite), $x^{r} = \Sigma c_{i} t_{i}^{r}$, where the simplexes $|t_{i}^{r}|$ (which may be degenerate, i.e., with skeletons lying in dimensions $< r$) lie in B. This definition of chains leads as usual to the definition of a cycle and of homology, and conse-

---

[*] The definition serves for any metric space.

sequently to the definition of Δ-groups. It is easy to prove that all three of these definitions of the Betti groups of an open set B are equivalent, i.e., lead to isomorphic groups. We shall use the first definition as being formally the simplest. In my book [1] the third definition was applied and was proved equivalent to the first.

We observe that the first definition may be formulated in the following way, which we will use extensively in the second part of this monograph. We consider the set of all compacta ψ lying in an (arbitrary) B, partially ordered by inclusion. If $\psi \subset \psi'$, then every Δ_c-cycle lying on ψ lies also on ψ', and a Δ_c-cycle homologous to zero in ψ is homologous to zero also in ψ'. Therefore we have defined a homomorphism $E_{\psi'}^{\psi}$ (the "inclusion homomorphism") of the group $\Delta_c^r \psi$ into the group $\Delta_c^r \psi'$, and we have the direct spectrum

$$\{\Delta_c^r \psi, \, E_{\psi'}^{\psi}\}. \tag{1}$$

The limit group of this spectrum is evidently the group $\Delta_c^r B$.

In this chapter we are interested only in the case when B is an open set in M^n. In this case each compactum $\psi \subset B$ lies inside some polyhedron $\psi' \subset B$ and we obtain a cofinal section of the spectrum (1) if among all the compacta we retain only those which are polyhedra and we regard ψ' as following ψ when ψ lies *inside* ψ'. Namely, we replace the whole spectrum (1) by this cofinal section, which we shall again denote by (1). Thus *for a set B open in M^n the group $\Delta_c^r B$ may be defined as the limit of the direct spectrum* (1), *where the ψ are polyhedra lying in B, and ψ' follows ψ if ψ lies inside ψ'.*

In the spectrum (1) defined in this way one may obtain a countable cofinal section

$$\{\Delta_c^r \psi_\nu, \, E_{\nu+1}^{\nu}\}, \tag{1'}$$

if one chooses only a sequence of polyhedra ψ_ν having the property that ψ_ν lies inside $\psi_{\nu+1}$ and that $B = \bigcup_{\nu=1}^{\infty} \psi_\nu$. We shall use this circumstance also in the sequel.

2. The Pontrjagin duality law consists in the following:

Let M^n be an n-dimensional h-manifold, acyclic over any given coefficient domain $\mathfrak{A}$ in the sense that $\nabla^p M^n = \nabla^{p+1} M^n = 0$, which means that if $q = n - p - 1$, then also $\Delta^{q+1} M^n = \Delta^q M^n = 0$. Let A be a set closed in M^n, and $B = M^n - A$ the complementary open set. Then the groups $\nabla^p A$ and $\Delta_0^q B$ (taken over the same coefficient domain $\mathfrak{A}$) are isomorphic to each other. [*]

---

[*] In such a form the Pontrjagin duality law was first formulated and proved in my book [1], page 14. The initial formulation of the theorem of Pontrjagin, the so-called Δ-form of this theorem, is given in Chapter VI of this monograph.

3. **Plan of the proof of the Pontrjagin theorem. Reduction to the combinatorial case and to a homological lemma.** If K is any triangulation of the manifold M^n, then by α we denote the closed subcomplex of the complex K consisting of those simplexes which contain points of the set A and of all the faces of these simplexes (so that $A \subseteq \widetilde{\alpha}$).

The complex α is the nerve of the covering of the set A whose elements are the intersection $A \cap Oe$ of the set A with the principal stars of the triangulation K; [*] *this covering will also be denoted by α.*

By K_1 we shall always mean a barycentric subdivision of the complex K, and by K^* the corresponding star complex (complex of barycentric stars); by α^* we denote the open subcomplex of the complex K^* consisting of all the barycentric stars adjoint to the simplexes of the complex α. Finally, by β we shall denote the complementary closed subcomplex $\beta = K^* - \alpha^*$ of the complex K^*.

Now we select a sequence of triangulations

$$K^{(1)}, K^{(2)}, \ldots, K^{(\nu)}, \ldots$$

of the manifold M^n having the following properties:

$1°$. $K^{(\nu+1)}$ is a multiple (not less than twofold) barycentric subdivision of the triangulation $K^{(\nu)}$. [**]

$2°$. The polyhedron $\widetilde{\alpha}^{(\nu+1)}$ lies inside $\widetilde{\alpha}_\nu$, and $\widetilde{\beta}^{(\nu)}$ inside $\widetilde{\beta}^{(\nu+1)}$.

Then the coverings $\alpha^{(1)}, \alpha^{(2)}, \ldots, \alpha^{(\nu)}, \ldots$ form a cofinal section in the directed set of all coverings of the set A, and the polyhedra $\widetilde{\beta}^{(1)}, \widetilde{\beta}^{(2)}, \ldots$ $\ldots, \widetilde{\beta}^{(\nu)}, \ldots$ form a cofinal section in the directed set of all $\psi \subset B$. Denoting by $\overline{\sigma}^\nu_{\nu+1}$ the homomorphism of the group $\nabla^p \alpha^{(\nu)}$ into the group $\nabla^p \alpha^{(\nu+1)}$ adjoint to the canonical shift $\sigma^{(\nu+1)}_\nu$ of the triangulation $\alpha^{(\nu+1)}$ into $\alpha^{(\nu)}$, we see that

$$\nabla^p A = \varinjlim (\nabla^p \alpha^{(\nu)}, \overline{\sigma}^\nu_{\nu+1}). \tag{2}$$

At the same time,

$$\Delta^q B = \varinjlim_c (\Delta^q_c \widetilde{\beta}^{(\nu)}, E^\nu_{\nu+1}). \tag{3}$$

Now we shall consider our spectra (2) and (3). We have proved in the preceding chapter that there exists an isomorphic mapping of the group $\nabla^p \alpha^{(\nu)}$ onto $\Delta^q \beta^{(\nu)} = \nabla^q \widetilde{\beta}^{(\nu)}$, realized by the operator $\Delta D = \pm DV E^{\alpha^{(\nu)}}_{K^{(\nu)}}$. Thus, for the proof

---

[*] In fact, since all the principal simplexes of the complex α intersect A, the sets $A \cap Oe_0, \ldots, A \cap Oe_r$, where $T = |e_0 \cdots e_r|$ is a simplex of the complex α, intersect in a nonempty set $A \cap OT$, where OT is the star of the simplex T in the complex α.

[**] Thus, $K^{(\nu+1)}$ is a subdivision of the complex $K_1^{(\nu)}$ (the subscript 1 always denotes the passage to a barycentric subdivision).

of the theorem of Pontrjagin we need only to prove that the operator $D\nabla E_{K(\nu)}^{\alpha(\nu)}$ commutes (in the sense of subsection 4 of the second chapter) with the projections $\bar{\sigma}_{\nu+1}^{\nu}$ and $E_{\nu+1}^{\nu}$. In order to simplify the notation we shall drop the index ν and replace the index $\nu+1$ by a stroke. Then we need only prove, for any p-dimensional ∇-cycle z on α, the formula *

$$D\nabla E_{K'}^{\alpha'}\,\bar{\sigma}_{\alpha'}^{\alpha}z \sim E_{\tilde{\beta}'}^{\tilde{\beta}}\,D\nabla E_{K}^{\alpha}z \text{ in } \tilde{\beta}'. \tag{4'}$$

First we shall transform it slightly. To this end we observe that $\bar{\sigma}$ and ΔE commute with each other in the sense that

$$\nabla E_{K'}^{\alpha'}\,\bar{\sigma}_{\alpha'}^{\alpha}\,z \sim \bar{\sigma}_{K'}^{K}\,\nabla E_{K}^{\alpha}z \text{ in } K'-\alpha'. \tag{$4_0'$}$$

Indeed, since the operators $\bar{\sigma}_{K}^{K}$ and ∇ are commutative, the formula $(4_0')$ which we need may be written in the form

$$\nabla E_{K'}^{\alpha'}\,\bar{\sigma}_{\alpha'}^{\alpha}\,z \sim \nabla \sigma_{K'}^{K}\,E_{K}^{\alpha}z \text{ in } K'-\alpha',$$

i.e., in the form

$$\nabla(E_{K'}^{\alpha'}\bar{\sigma}_{\alpha'}^{\alpha}\,z - \bar{\sigma}_{K'}^{K}\,E_{K}^{\alpha}z) \sim 0 \text{ in } K'-\alpha'.$$

But this last homology follows easily from the fact that the chain in parenthesis obviously lies on $K'-\alpha'$.

Thus the homology $(4_0')$ is proved. The operator D carries it into the homology

$$D\nabla E_{K'}^{\alpha'}\bar{\sigma}_{\alpha'}^{\alpha}\,z \sim D\bar{\sigma}_{K'}^{K}\,\nabla E_{K}^{\alpha}z \text{ in } \tilde{\beta}',$$

which we apply to the left side of formula $(4')$, at the same time dropping the symbol of the operator $E_{\tilde{\beta}'}^{\tilde{\beta}}$ on the right side of $(4')$. In this way we rewrite the homology $(4')$ in the following way:

$$D\bar{\sigma}_{K'}^{K}\,\nabla E_{K}^{\alpha}\,z \sim D\nabla E_{K}^{\alpha}z \text{ in } \tilde{\beta}'.$$

Now we observe that $\nabla E_{K}^{\alpha}z$ is a ∇-cycle of the triangulation K, equal to zero on α. Therefore the proof is complete if we prove the following proposition:

Homology lemma. *Let z be a p-dimensional ∇-cycle of the triangulation K, equal to zero on α. Then*

$$D\bar{\sigma}\,z \sim Dz \text{ in } \tilde{\beta}' \text{ (where } \bar{\sigma} = \bar{\sigma}_{K'}^{K}). \tag{4}$$

§2. Reduction of the homology lemma to a combinatorial lemma

1. **An elementary lemma.** We begin with the proof of the following elementary

---

*Here every chain of the star complex K^* (of any triangulation K) is considered as a chain of the barycentric subdivision K_1 common for K and K^*.

proposition.[*]

Elementary lemma. *There exists a continuous mapping ϕ of the set $\tilde{K} - \tilde{\alpha}$, open in $\tilde{K}$, onto the polyhedron $\tilde{\beta} \subset K - \tilde{\alpha}$ that leaves all the points of $\tilde{\beta}$ invariant.*

Proof. The set $G = \tilde{K} - \tilde{\alpha}$ is the sum of simplexes of the barycentric subdivision of the open complex $K - \alpha$. Let T_1^r be one of these simplexes, not contained in $\tilde{\beta}$. Then the leading vertex of T_1^r is the center of some simplex $T^{n_0} \in K - \alpha$, which is the carrier of the simplex T_1^r in K, while the lowest vertex of the simplex T_1^r is the center of some simplex $T^{n_r} \in \alpha$. In other words, T_1^r has the form

$$T_1^r = |T^{n_0} > T^{n_1} > \cdots > T^{n_r}|, \; T^{n_0} \in K - \alpha, \; T^{n_r} \in \alpha$$

(here and in the sequel, in similar cases, we give instead of the vertices of the simplex $T_1^r \in K_1$ those simplexes $T^{n_i} \in K$ whose centers are these vertices). Let $T^{n_{i+1}}$ be the first in the series of simplexes $T^{n_0}, \cdots, T^{n_r}$, which belongs to the complex α. Then in the simplex T_1^r we have two opposite faces

$$T_1' = |T^{n_0} > \cdots > T^{n_i}| \;\text{ and }\; T_1'' = |T^{n_{i+1}} > \cdots > T^{n_r}|,$$

of which the first lies on $\tilde{\beta}$ and the second on $\tilde{\alpha}$. Each point ξ of the simplex T_1^r uniquely defines an open segment $(\xi''\xi')$ containing it with the endpoints

$$\xi'' \in T_1'' \subset \tilde{\alpha}, \; \xi' \in T_1' \subset \tilde{\beta}.$$

If we let each point $\xi \in G - \tilde{\beta}$ move rectilinearly and uniformly, during a unit time interval, along the segment $\xi''\xi'$ containing it into the point ξ', and leave invariant all the points of $\tilde{\beta}$, we obtain a retraction carrying G into $\tilde{\beta}$ as we were required to prove.

2. **The mapping σ_1.** We shall denote by σ_1 the simplicial mapping of the triangulation K_1' into K_1,[**] obtained as a result of the canonical shift σ of the triangulation K' into K and consisting in putting into correspondence with the center of any simplex of the triangulation K' the corresponding center of its image under the mapping σ.

Lemma (A). *Suppose that on the simplex t_1^{q+1} of the complex K_1' the cycle Dsz is different from zero. Then both of the closed simplexes $\overline{t_1'}^{q+1}$ and $\sigma_1 \overline{t_1'}^{q+1}$ are contained in some convex subset $\tilde{O}$ of the set $G = \tilde{K} - \tilde{\alpha}$.*

---

[*] The notations introduced in the beginning of §1.3 will be preserved throughout this chapter.

[**] We retain the notation of the end of the preceding subsection ($K' = K^{(\nu+1)}, K = K^{(\nu)}$, the subscript always indicating the passage to a barycentric subdivision.

Proof of Lemma (A). From the hypothesis of Lemma (A) it follows that $|t_1^{q+1}|$ is the principal simplex of the barycentric star adjoint to some simplex $T'^p \in K'$ on which $\bar{\sigma} z \neq 0$. This means that under the canonical shift σ we have

$$\sigma T'^p = T^p \in K - a.$$

Let

$$|t_1'^{q+1}| = |T'^n > T'^{n-1} > \cdots > T'^p|$$

(where $T'^n, \cdots, T'^p$ are simplexes of K' whose centers are vertices of $|t_1'^{q+1}|$).

The simplex $\sigma_1 |t_1'^{q+1}|$ has the form

$$\sigma_1 |t_1'^{q+1}| = |\sigma T'^n \geq \sigma T'^{n-1} \geq \cdots \geq \sigma T'^p|,$$

where, as we know, $\sigma T'^p = T^p$.

Let T^n be the carrier of the simplex T'^n in K. We denote by $\tilde{O}$ the set which is the sum of the simplex T^n and of all of its faces $T^k \geq T^p$. In other words, $\tilde{O}$ is the body of the star $O_{[T^n]} T^p$ of the simplex T^p in the complex $[T^n]$. We shall show that $\tilde{O}$ is the desired convex set, lying in G and containing both simplexes $|t_1'^{q+1}|$ and $\sigma_1 |t_1'^{q+1}|$.

First of all, the set $\tilde{O}$ is convex (since it consists of those points T^n whose barycentric coordinates taken relative to the vertices of T^p are positive).

Further, $\tilde{O} \subset G$. This follows from the fact that $K - a$ is an open subcomplex, and $T^p \in K - a$. Hence the star $O_K T^p$, and a fortiori the star $O_{[T^n]} T^p$, lie in $K - a$.

Since σ is a canonical shift, the simplex $\sigma T'^n$ is a proper or improper face of the simplex T^n, and since under these circumstances $\sigma T'^p = T^p$, all the simplexes $\sigma T'^n, \sigma T'^{n-1}, \cdots, \sigma T'^p$, and consequently all their centers, lie in $\tilde{O}$; since $\tilde{O}$ is convex, the whole simplex $|\sigma_1 |t_1'^{q+1}| \subset O$.

It remains to prove that $|t'^{q+1}|$ lies in $\tilde{O}$. But this immediately follows from the fact that

$$|t_1'^{q+1}| = |T'^n > T'^{n-1} > \cdots > T'^p|$$

and that the carriers in K of the simplexes T'^n and T'^p are respectively T^n and T^p.

Thus Lemma (A) is proved.

3. Appearance of the combinatorial lemma. From what has been proved, if we make each vertex of the cycle $D\bar{\sigma} z$ move uniformly and rectilinearly into its image under the mapping σ_1, we carry the chain $D\bar{\sigma} z$ into $\sigma_1 D\bar{\sigma} z$ by means of a displacement taking place in G. Accordingly,

$$D\bar{\sigma} z \sim \sigma_1 D\bar{\sigma} z \text{ in } G$$

(the homology is understood in the sense of homology of Δ_c-cycles), where this homology is realized on some polyhedron $\Pi \subset G$.

In view of the elementary lemma proved above, there is a continuous mapping ϕ ("retraction"), carrying $G' \supset G$ into $\tilde{\beta}'$ and leaving invariant all the points of $\tilde{\beta}'$.

Under the mapping ϕ the cycles $D\bar{\sigma}z$ and $\sigma_1 D\bar{\sigma}z$ (lying in $\tilde{\beta}'$) remain in place, and the film bounded by them (in $G \subset G'$) goes into a film lying in $\tilde{\beta}'$. Accordingly,

$$D\bar{\sigma}z \sim \sigma_1 D\bar{\sigma}z \text{ in } \tilde{\beta}'. \tag{1}$$

But it turns out that the following fundamental identity holds, known as the *combinatorial lemma*:

$$\sigma_1 D\bar{\sigma}z = Dz.$$

As a consequence of this identity the homology (1) goes into the homology

$$D\bar{\sigma}z \sim Dz \text{ in } \tilde{\beta}',$$

constituting the content of the homology lemma.

Thus the homology lemma is completely reduced to the combinatorial lemma, and it remains only to prove the latter.

§3. Proof of the combinatorial lemma

I present this proof in the form given by K. Sitnikov,[*] which is somewhat of a simplification of my original proof (see [1], pp. 528–531).

The manifold M^n is supposed oriented. The oriented n-dimensional simplexes of the triangulation K will be denoted by t^n, and of the triangulation K' by t'^n.

We begin with the following auxiliary assertion. Suppose that under the canonical shift σ of the triangulation K' into K the given simplex t'^n goes into the simplex ϵt^n, $\epsilon = \pm 1$. Then any oriented face t'^p of the simplex t'^n goes into some oriented face t^p of the simplex t^n. Our assertion is that for the pieces v'^{q+1}, v^{q+1} of the oriented barycentric stars adjoint respectively to t'^p and t^p, lying in $|t'^n|$ and $|t^n|$ respectively, the following equality holds:

$$\sigma_1 v'^{q+1} = \epsilon v^{q+1}.$$

Indeed, since σ on $|t'^n|$ is a nondegenerate affine mapping, then $\sigma_1 v'^{q+1} = \sigma v'^{q+1} = \pm v^{q+1}$.

---

[*] Mat. Sb. 34(76) (1954), 3–54, pp. 18 ff.; ibid. 31(73) (1952), 441–458.

We choose in v'^{q+1} a simplex $t_1'^{q+1}$. Then $(t'^P \times t_1'^{q+1}) = 1$ [*] and $(\sigma t'^P \times \sigma_1 t'^{q+1}) = \epsilon$, i.e., $(t^P \times \sigma_1 t'^{q+1}) = \epsilon$, whence (since $(t^P \times v^{q+1}) = 1$) it follows that $\sigma_1 v'^{q+1} = \epsilon v^{q+1}$, i.e., assertion (1).

We shall now investigate what happens to the cycle $D\bar\sigma z$ under the mapping σ_1. That cycle, like every cycle of the complex K_1, decomposes into pieces lying in various simplexes t'^n. We select some one of these simplexes t'^n. Two cases are possible: either the simplex t'^n under the mapping σ degenerates, or it does not. We shall prove that in the first case the image of the piece lying in t'^n of the chain $D\bar\sigma z$ is equal to zero. Indeed, if the piece of the chain $D\bar\sigma z$ in question does not equal zero, then on the face t'^P of the simplex t'^n adjoint to the star carrying the piece in question the chain $\bar\sigma z$ is also different from zero, which means that the mapping σ of the simplex $|t'^P|$ is nondegenerate. But if the simplex $|t'^n| = T'^n$ degenerates under the mapping σ_1, and its face $|t'^P| = T'^P$ does not degenerate, then under the mapping σ_1 all the principal (i.e., $(q+1)$-dimensional) simplexes $|t'^{q+1}|$ of the barycentric star adjoint to this simplex degenerate, [**] which proves our assertion.

Now suppose that the second case holds, i.e., $\sigma t'^n = \epsilon t^n$. Then equation (1) comes into play. Since the algebraic number of simplexes covering the given simplex t^n under the mapping σ is equal to 1 it follows from equation (1) that the sum of the pieces of the chain $D\bar\sigma z$ lying in all possible t'^n and mapping onto the given simplex t^n goes under the mapping σ_1 into a piece of the cycle Dz^P lying in t^n. Summing this result over all t^n, we obtain the identity $\sigma_1 D\bar\sigma z = Dz$.

This completes the proof of the combinatorial lemma and therefore of the Pontrjagin duality law.

---

[*] We understand $(t'^P \times t_1'^{q+1})$ as defined in §2 of Chapter IV to be the intersection of the simplex t'^P with the barycentric star u'^{q+1} adjoint to this simplex (and carrying the piece v'^{q+1} and the simplex $t_1'^{q+1}$). This definition was made by bringing in the simplexes t'^P and $t_1'^{q+1}$, and is furthermore a definition of the intersection of the planes X'^P, X'^{q+1} carrying the simplexes t'^P, $t_1'^{q+1}$ and oriented by them in the oriented space R^n carrying the simplex t'^n (see Chapter VI, §5). If we are given an affine mapping σ of the space R'^n onto the oriented space $R^n = \sigma R'^n$ (we have in mind that orientation into which the given orientation of the space R'^n goes under the mapping σ), then, evidently,
$$(\sigma X'^P \times \sigma X'^{q+1})_{\text{in } R^n} = (X'^P \times X'^{q+1})_{\text{in } R'^n}$$
Therefore we have the formula $(\sigma t'^P \times \sigma t'^{q+1}) = \epsilon$, which we need at present (a similar derivation is in §2 of Chapter 7 of the book [1]).

[**] Indeed, suppose that $|t_1'^{q+1}| = |T'^n > T'^{n-1} > \cdots > T'^P|$. Then $\sigma_1 |t_1'^{q+1}| = |\sigma T'^n \geq \sigma T'^{n-1} \geq \cdots \geq \sigma T'^P|$. Since $\sigma T'^P$ is by hypothesis a p-dimensional simplex, while $\sigma T'^n$ has dimension n, it follows that at least one of the signs $\geq$ is in fact an equality sign, which means that the simplex $|t_1^{q+1}|$ degenerates under the mapping σ_1.

§4. Elementary special cases of the Pontrjagin duality law

Since the rank of the group $\overset{\circ}{\Delta} B$ defines the number of components of the open set B, it follows from the Pontrjagin duality law that homeomorphic closed sets of the space S^n divide the space into the same number of components.

Further, a compactum A which divides the space S^n is the common boundary of all the regions into which it divides S^n if and only if no proper closed subset $A' \subset A$ divides the space S^n (see, for example, [1], Chapter 2, §1:2). Therefore, if one of two mutually homeomorphic closed sets in S^n is a common boundary for all the components of the complementary open sets, then the other is also.

Now let A be a curved polyhedron. Then, as follows from the fundamental theorem of §2.11 of Chapter I, and from the remark at the end of §5 of Chapter III, the group $\nabla^p A$ is isomorphic to the direct sum of the $(p-1)$-dimensional torsion group $\Theta^{p-1} A$ and as many infinite cyclic groups as the pth Betti number of the polyhedron A.

Accordingly, the group $\Delta^q B$ also decomposes into the direct sum of a finite group (which it is natural to call the q-dimensional torsion group of the open set B) and of a free abelian group whose rank is equal to the rank of the group $\Delta^p B$, called the q-dimensional Betti number of the open set B. Therefore *for the curved polyhedron A the duality theorem of Pontrjagin may be formulated in the following way:*

Let $A \subset S^n$ be a curved polyhedron. Then the group $\Delta^q B$ is a group with a finite number of generators, while its rank, the qth Betti number of the open set B, is equal to the pth Betti number of the polyhedron A, and the q-dimensional torsion group $\Theta^q B$ is isomorphic to the $(p-1)$-dimensional torsion group $\Theta^{p-1} A$.

CHAPTER VI

The Pontrjagin duality law in Δ-form

§1. Introduction. Topology in the groups $\Delta^r \Phi$.* Formulation of the Pontrjagin duality law

In this chapter we shall assume known the fundamental concepts and results of the theory of characters to the extent indicated in subsection 5 of Chapter II.

In order to formulate the Pontrjagin duality law in Δ-form, we need first of all to introduce a topology into the groups $\Delta^r \Phi$, where Φ is a compactum, under the assumption that the coefficient domain is a bicompact topological group $\mathfrak{B}$.

---

*Editor's remark: In Part II, Chapter I, §3.2 the author remarks in a footnote (see p. 123 of the present translation) that these groups should have been denoted by $\mathscr{S}^r \Phi$ but that, as explained in the footnote, this error is immaterial.

To this end we introduce a topology into the group $\Delta^r \alpha$, where α is a finite complete simplicial complex, and the coefficient domain is always a bicompact group $\mathfrak{B}$. This is done in a simple and natural way. First of all, it is clear that the group $L^r \alpha$ is the direct sum of as many groups isomorphic to the group $\mathfrak{B}$ as there are r-dimensional simplexes in the complex α. Accordingly, the group $L^r \alpha$ is bicompact. Further, it is easy to verify that the homomorphism Δ of the group $L^r \alpha$ into $L^{r-1} \alpha$ is a continuous homomorphism. Therefore the group $H^r \alpha$, as the image of the bicompact group $L^{r+1} \alpha$ under the homomorphism Δ, is a bicompact group; the group $Z^r \alpha$, as the kernel of the homomorphism Δ of the group $L^r \alpha$ onto $H^{r-1} \alpha$, is closed, and therefore is a bicompact subgroup of the bicompact group $L^r \alpha$. Accordingly, also the group $\Delta^r \alpha = Z^r \alpha - H^r \alpha$ is bicompact. The group $\Delta^r \Phi$ was defined in Chapter III as the limit group of the inverse spectrum of the groups $\Delta^r \alpha$, where α ran over all coverings of the compactum Φ. Since all the groups $\Delta^r \alpha$ are compact, and the homomorphisms ω_α^β are continuous, the group $\Delta^r \Phi$, by subsection 4 of Chapter II, also acquires a topology and turns out to be bicompact in this topology. Thus, in the case of a bicompact coefficient domain $\mathfrak{B}$ we shall speak of the bicompact group $\Delta^r \Phi$. As always, $\mathfrak{A}$ will denote a discrete group dual to the group $\mathfrak{B}$.

Now we may formulate the fundamental result of this chapter in the following way:

Theorem 1 (The Pontrjagin duality law in Δ-form). *Let M^n be an oriented homology manifold of dimension n. Let p and q be nonnegative integers with the sum p + q = n − 1. Suppose that M^n is acyclic in dimensions q and q + 1, which means also in dimension p.* [*] *Let A be a compactum lying in M^n, and $B = M^n - A$. Then we have the duality*

$$\Delta^p (A, \mathfrak{B}) \,|\, \Delta_c^q (B, \mathfrak{A}).$$

This theorem is easily derived from the Pontrjagin duality theorem in ∇-form proved in the preceding chapter. To this end it suffices to prove only the duality[**] and to observe that the homomorphisms ω_α^β and π_β^α in the definition of the groups

---

[*] Indeed, if $\Delta^q M^n = 0$, then $\nabla^{p+1} M^n = 0$, so that $\theta^p M^n = 0$; from $\Delta^{q+1} M^n = 0$ it follows that $\pi^p M^n = 0$. Therefore $\Delta^p M^n = 0$.

[**] Let us prove this duality. To this end we recall some further fundamental propositions of the theory of characters. First of all the *theorem on annihilators: If the groups X and Y are dual to each other and the subgroup $A \subseteq X$ is the annihilator of the subgroup $B \subseteq Y$ (i.e., A consists of all those elements $x \in X$ whose scalar products with an arbitrary element of the group B is equal to zero), then B is the annihilator of the group A.* If A is the annihilator of the group B (and thus B is the annihilator of the group A) in the duality $X \,|\, Y$, then we write

$$X \supseteq A \perp B \subseteq Y. \tag{1}$$

Such a relation is called an *annihilation*.

One of the fundamental propositions of the theory of characters runs as follows:

$\Delta^p A$ and $\nabla^p A$ for any compactum Z are adjoint. Therefore, on the basis of Theorem 4 of Chapter II, it follows that the groups $\Delta^p(A, \mathfrak{B})$ and $\nabla^p(A, \mathfrak{A})$ are adjoint, and since $\nabla^p(A, \mathfrak{A})$ and $\Delta^q_c(B, \mathfrak{A})$ are isomorphic, $\Delta^p(A, \mathfrak{B})$ and $\Delta^q_c(B, \mathfrak{A})$ are dual.

In this chapter we shall give one further proof of Theorem 1 which does not depend on the considerations of Chapter V and uses only §§1 and 2 of Chapter IV.

If two groups $A \subseteq X$ and $B \subseteq Y$ satisfy the annihilation (1), *then the following dualities also hold:*

$$(X - A)\,|\,B;\ (Y - B)\,|\,A. \tag{2}$$

Suppose the pair of dualities

$$X\,|\,X^*,\ Y\,|\,Y^*$$

are given along with adjoint homomorphisms: a homomorphism ϕ of the group X into the group Y and a homomorphism ϕ^* of the group Y^* into the group X^*. If the kernels of the homomorphisms ϕ and ϕ^* are respectively the groups $X_0 \subseteq X$ and $Y^*_0 \subseteq Y$, and the images of the groups X and Y^* are respectively $Y_1 = \phi X \subseteq Y$ and $X^* = \phi^* Y^* \subseteq X^*$, then we write this in the form of a "diagram":

$$\phi \left| \begin{array}{cc} X \supseteq X_0 & X^*_1 \subseteq X^* \\[4pt] Y \supseteq Y_1 & Y^*_0 \subseteq Y^* \end{array} \right| \phi^* . \tag{3_0}$$

Under these conditions we have the following lemma.

Lemma. *From* (3_0) *follows*

$$\phi \left| \begin{array}{cc} X \supseteq X_0 \perp & X^*_1 \subseteq X^* \\[4pt] Y \supseteq Y_1 \perp & Y^*_0 \subseteq Y^* \end{array} \right| \phi^* \tag{3}$$

(where the diagonal stroke denotes the duality $X^*_1\,|\,Y_1$) *and*

$$X_0\,|\,X^* - X^*_1);\ (Y - Y_1)\,|\,Y^*_0 \tag{3$'$}$$

Proof. The dualities (3) follow from the theorem on annihilators. The duality (3$'$) follows from the same theorem and from the isomorphism $Y_1 = X - X_0$. It suffices to prove the annihilations in (3), and for that matter the first of them. Let $x \in X_0$, $y^* \in Y^*$. We have $(x \cdot \phi^* y^*) = (\phi x \cdot y^*) = 0$. On the other hand, if $\phi x \neq 0$, then there exists a $y^* \in Y^*$ such that $(x \cdot \phi^* y^*) = (\phi x \cdot y^*) \neq 0$, as was required to prove.

We turn now to the proof of the duality $\Delta^p(\alpha, \mathfrak{B})\,|\,\nabla^q(\alpha, \mathfrak{B})$. By definition we have for the operators Δ and ∇,

$$(\Delta t^{r+1} \cdot t^r) = (\nabla t^r \cdot t^{r+1}) = (t^{r+1} : t^r),$$

from which the operators Δ and ∇ turn out to be adjoint. Therefore, from the lemma, we have the following diagram 4, in which Z^* and H^* are the groups of Δ-cycles of the complex α and Δ-cycles homologous to zero in α, and Z, H are the groups of ∇-cycles and ∇-cycles homologous to zero in α, while in L^* the coefficient domain is $\mathfrak{B}$, and in L the coefficient domain is $\mathfrak{A}$:

$$\nabla \left| \begin{array}{cc} L^{p-1} \supseteq Z^{p-1} \perp & H^{*p-1} \subseteq L^{*p-1} \\[4pt] L^p \supseteq H^p & \perp\ Z^{*p} \subseteq L^{*p} \end{array} \right| \Delta; \tag{4}$$

$$(L^p - H^p) \perp Z^{*p}. \tag{4$'$}$$

These two sections are closely related to the theory of linking cycles (§§2 and 5) (with which we shall meet also in the second part of this monograph). We shall discuss this theory before proving Theorem 1.

§2. Linking cycles

1. **Intersection of chains.** We shall preserve the notation introduced in the beginning of §3 of Chapter IV. By K we mean some triangulation of the orientable h-manifold M^n. By K_1, as always, we mean a barycentric subdivision of the complex K, and by K^* the complex of barycentric stars of the triangulation K. The orientations t_i of the simplexes of the complex K will be taken arbitrarily, the orientation of the barycentric stars being chosen so that for each simplex t_i^μ and the barycentric star u_i^ν adjoint to it we have $(t_i^\mu \times u_i^\nu) = 1$ (always $\mu + \nu = n$).

For any two chains, $x^\mu = \sum_i a_i t_i^\mu$ of the complex K and $y^\nu = \sum_j b_j u_j^\nu$ of the complex K^*, we define the intersection $(x^\mu \times y^\nu)$ by the formula

$$(x^\mu \times y^\nu) = \sum_{i,j} a_i b_j (t_i^\mu \times u_j^\nu).$$

Here we have to suppose, for example, that the role of coefficient domains in K and K^* is played by one and the same ring (for example, the ring of integers or rational numbers), or that as the coefficient domains in K and K^* one has two arbitrary, mutually dual groups $\mathfrak{A} \,|\, \mathfrak{B}$; [*] we shall be occupied in this chapter principally with this second case; in this case $(x^\mu \times y^\nu) \in K$. *Obviously, the intersection operation is distributive with respect to addition.*

For us the following property of the intersection index has fundamental importance:

$$(x^\mu \times \Delta y^{\nu+1}) = (-1)^\mu (\Delta x^\mu \times y^{\nu+1}) \tag{1}$$

or (reading from right to left)

$$(\Delta x^\mu \times y^{\nu+1}) = (-1)^\mu (x^\mu \times \Delta y^{\nu+1}).$$

In view of the distributivity of the intersection operator, formula (1) will be proved if we prove it for the case

$$x^\mu = t_i^\mu, \quad y^\nu = u_j^{\nu+1}.$$

Putting $(t_i^\mu : t_k^{\mu-1}) = \epsilon_{ik}$, we have, from the results of Chapter IV, §2.3:

We recall a theorem from the theory of characters: if $X \,|\, X^*$, $Y \,|\, Y^*$, $X \supseteq Y$, $(X - Y) = Z \,|\, Z^*$, then $Y \,|\, (X^* - Z^*)$. We apply this to the isomorphism

$$(L^p - H^p) - (Z^p - H^p) = L^p - Z^p.$$

Substituting from (4'), (4) the relations $(L^p - H^p) \,|\, Z^{*p}$, $L^p - Z^p = H^{p+1} \,|\, H^{*p}$, we obtain the desired duality $(Z^p - H^p) \,|\, (Z^{*p} - H^{*p})$.

   [*] Or, finally, that one of the two coefficient domains is any group, and the second is the group of integers.

$$\Delta t_i^\mu = \sum_k \epsilon_{ik} t_k^{\mu-1}, \quad \Delta u_j^{\nu+1} = (-1)^\mu \sum_h \epsilon_{hj} u_h^\nu.$$

Therefore, recalling that $(t_i^\mu \times u_h^\nu) = \delta_{ih}$, $(t_k^{\mu-1} \times u_j^{\nu+1}) = \delta_{kj}$,

we shall have: $(t_i^\mu \times \Delta u_j^{\nu+1}) = (-1)^\mu \sum_h \epsilon_{hj} (t_i^\mu \times u_h^\nu) = (-1)^\mu \epsilon_{ij}$;

$$(\Delta t_i^\mu \times u_j^{\nu+1}) = \sum_k \epsilon_{ik} (t_k^{\mu-1} \times u_j^{\nu+1}) = \epsilon_{ij},$$

as we were required to prove.

Corollary to formula (1). *If of the two cycles $z \in Z^\mu K$, $z_* \in Z^\nu K^*$ at least one is homologous to zero* (in K or in K^*), *then $(z \times z_*) = 0$.*

Indeed, if for example $z = \Delta x$, then

$$(z \times z_*) = (\Delta x \times z_*) = \pm (x \times \Delta z_*) = 0.$$

2. **Linking coefficient.** Formula (1) and its corollary allow us to define the *linking coefficient* $\flat(z^p, z_*^q)$, $p + q = n - 1$, *of two cycles, homologous to zero in K and K^* respectively*. First of all:

(A) *If x_1^{p+1} and x_2^{p+2} are two chains, bounded in K by the cycle z^p, then $(x_1^{p+1} \times z_*^q) = (x_2^{p+1} \times z_*^q)$.*

Indeed, $x_1^{p+1} - x_2^{p+2}$ is a cycle, so that its intersection with the cycle z_*^q, homologous to zero in K^*, is equal to zero.

This makes it possible to introduce:

Definition: *Let $z^p \sim 0$ in K, $z_*^q \sim 0$ in K^*. The linking coefficient $\flat(z^p, z_*^q)$ is the intersection index $(x^{p+1} \times z_*^q)$, where x^{p+1} is any chain of the complex K bounded by the cycle z^p.*

Analogously one also defines $\flat(z_*^q, z^p)$, which we do not need, however.

The linking coefficient has the distributive property (both left and right).

It is also easy to verify the following formula, in which y_*^{q+1} is any chain in K^* bounded by the cycle z_*^q:

$$\flat(z^p, z_*^q) = (-1)^{p+1} (z^p \times y_*^{q+1}). \tag{2}$$

Indeed, if $\Delta x^{p+1} = z^p$, then

$$\flat(z^p, z_*^q) = (x^{p+1} \times z_*^q) = (x^{p+1} \times \Delta y_*^{q+1}) = (-1)^{p+1}(\Delta x^{p+1}, y_*^{q+1}) = (-1)^{p+1}(z^p \times y_*^{q+1}).$$

It follows from formula (2) that $\flat(z_*^q, z^p)\flat = \pm(z^p, z_*^q)$.

Now suppose that we are given a closed subcomplex α of the complex K.

As always, we denote by α^* the open subcomplex of the complex K^* consisting of those barycentric stars adjoint to the elements of α, and by $\beta = K^* - \alpha^*$ the complementary open subcomplex. We shall suppose that the cycles z^p, z_*^q (homologous to zero in K and K^* respectively) lie on α and β respectively. If here $z^p \sim 0$ in α, then taking the chain x^{p+1} bounded by the cycle z^p in α, we see that

$$\flat(z^p, z_*^q) = 0.$$

In just the same way, if $z_*^q \sim 0$ in β, it may be shown from formula (2) that $\flat(z^p, z_*^q) = 0$.

Hence it follows that cycles z^p, which are homologous to one another in α, have on any cycle z_*^q lying on β (and homologous to zero in K^*) the same linking coefficient. Analogously, cycles homologous to one another in β have the same linking coefficients as any (homologous to zero in K) cycle z^p of the complex α. Hence it follows that under the assumption of acyclicity of the manifold M^n in dimensions q, $q+1$ (and thus p) one may speak of the linking coefficient $\flat(\zeta^p, \zeta_*^q)$ of any two homology classes $\zeta^p \in \Delta^p(\alpha, \mathfrak{B})$ and $\zeta_*^q \in \Delta^q(\beta, \mathfrak{A})$. In other words, we may speak of the *scalar product*

$$(\zeta^p, \zeta_*^q) = \flat(\zeta^p, \zeta_*^q)$$

of any two elements $\zeta^p \in \Delta^p(\alpha, \mathfrak{B})$, $\zeta_*^q \in \Delta^q(\beta, \mathfrak{A})$.

In the following section we prove that this scalar product[*] defines a duality

$$\Delta^p(\alpha, \mathfrak{B}) \,|\, \Delta^q(\beta, \mathfrak{A}), \tag{3}$$

which constitutes the content of the combinatorial case of the Pontrjagin duality law.

§3. Combinatorial case of the Pontrjagin duality law in Δ-form

1. We shall say that the two cycles z^p, z_*^q are *linked* if their linking coefficient differs from zero. The proof of the combinatorial case of the duality law of Pontrjagin consists in the proof of the following two theorems:

A. *To each cycle z^p not homologous to zero in α there corresponds a cycle z_*^q of β linked with it.*

B. *To each cycle z_*^q not homologous to zero in β there corresponds a cycle z^p of the complex α linked with it.*

Coefficient domain: $\mathfrak{B}$ on α and $\mathfrak{A}$ on β.

It suffices to prove either one of these theorems. The proof of the other is quite analogous. We shall prove Theorem A.

Lemma 1. *Suppose that x_*^{q+1} is a chain of the complex K^*. If $(bt_i^p \times x_*^{q+1}) = 0$ (for any $b \in \mathfrak{B}$), then the star u_i^{q+1} does not enter in the chain x_*^{q+1}.*

Indeed, if $x_*^{q+1} = \sum_j a_j u_j^{q+1}$, then

$$0 = (bt_i^p \times x_*^{q+1}) = \sum_j ba_j(t_i^p \times u_j^{q+1}) = ba_i$$

---

[*] It is bilinear and continuous.

for any $b \in \mathfrak{B}$, which means that $a_i = 0$.

Lemma 2. *The groups* $L = L^p(\alpha, \mathfrak{B})$ *and* $L^* = L^{q+1}(\alpha^*, \mathfrak{A})$ *are dual to one another in the sense of the theory of characters if one takes as the scalar product of two chains* $x^p \in L$ *and* $x_*^{q+1} \in L^*$ *their intersection* $(x^p \times x_*^{q+1})$.

For the proof we observe that the group L is the direct sum of groups, isomorphic to the group $\mathfrak{B}$ and taken in the number $\rho^{(p)}$ equal to the number of p-dimensional simplexes of the complex α. The group L^* is the direct sum of the same number $\rho^{(p)}$ of groups isomorphic to the group $\mathfrak{A}$. Since there exists a natural duality between the groups L and $L^p(\alpha, \mathfrak{A})$, realized by the usual scalar multiplication, for $x^p = \Sigma\, b_i t_i^p \in L$ and $x'^p = \Sigma\, a_i t_i^p \in L^p(\alpha, \mathfrak{A})$ we have $(x \cdot x'^p) = \underset{ij}{\Sigma}\, b_i a_j$. But the operator D realizes a natural isomorphism of the group $L(\alpha, \mathfrak{A})$ onto the group L^*; to the chain x'^p corresponds the chain $x_*^{q+1} = \Sigma\, a_i u_i^{q+1}$, while $(x^p \times x_*^{q+1}) = \Sigma\, b_i a_i = (x^p \cdot x'^p)$, with which Lemma 2 is proved.

Now we shall prove the fundamental lemma:

Lemma 3. *By* z_*^q *we denote a cycle of the complex* K^*, *and by* x_*^{q+1} *a chain of the complex* K^* *bounded by the cycle* z_*^q.

In order that the cycle z_*^q *should lie on* β *it is sufficient*[*] *that*

$$(\Delta x^{p+1} \times x_*^{q+1}) = 0 \tag{1}$$

for any choice of the chain x^{p+1} *of the complex* α.[**]

Proof. In view of Lemma 1 it is sufficient to prove that from (1) follows

$$(b t_i^{p+1} \times z_*^q) = 0$$

for any $|t_i^{p+1}| \in \alpha$ and $b \in \mathfrak{B}$. But

$$(b t_i^{p+1} \times z_*^q) = (b t_i^{p+1} \times \Delta x_*^{q+1}) = \pm\, (\Delta b t_i^{p+1} \times x_*^{q+1})$$

and under the hypothesis of Lemma 3 the last parenthesis is equal to zero, as we were required to prove.

2. **Proof of Theorem A.** This theorem is deduced from Lemmas 2 and 3 in the following way. Let z^p be a cycle not bounding in A. I assert that it is sufficient to find a character $\phi \equiv x_*^{q+1} \in L^*$ of the group L satisfying the following conditions:

1°. $(h^p \times x_*^{q+1}) = 0$ for all cycles h^p bounding on α.

2°. $(z^p \times x_*^{q+1}) \neq 0$.

---

[*] And evidently necessary! (But we shall not need the necessity.)

[**] Thus, the geometrical condition that the body of the cycle z_*^q lie on the polyhedron β reduces to the algebraic condition (1).

Indeed, suppose that such a character $\phi \equiv x_*^{q+1}$ has been found, and put $z_*^q = \Delta x_*^{q+1}$. Then condition $1°$ may be rewritten in the form $(\Delta x^{p+1} \times x_*^{q+1}) = 0$ for all chains x^{p+1} of the complex α, which from Lemma 3 means that the cycle z_*^q lies in β. But in view of condition $2°$ we have

$$\mathfrak{b}(z^p,\, z_*^q) = \pm\,(z^p \times x_*^{q+1}) \neq 0,$$

with which Theorem A is proved.

Thus it remains to construct the character $\phi \equiv x_*^{q+1}$ satisfying the conditions $1°$ and $2°$. But the possibility of such a construction is given by a strengthening of the theorem on the existence of characters. Observe that in the fundamental case of the simplest coefficient domains J, K, and also J_m we use the theorem of existence only in the application to "elementary groups".*

§4. Reduction of the general case of the Pontrjagin theorem to the combinatorial case and to the linking of true cycles

The reduction of the general case of the Pontrjagin duality theorem to the combinatorial case does not present difficulties in principle. But several tedious points arise from the necessity, as we shall now see, of defining the linking coefficient (while preserving its basic properties) for a wider class of cycles than those considered above. Since such an extension of the concept of linking is necessary, the easiest thing to do is to define the linking coefficient directly for two true cycles z^p, z^q lying on intersecting compacta in M^n. Furthermore, such a definition will permit a certain degree of strengthening of the formulation of the Pontrjagin theorem itself (see §5.8).

This definition must evidently be a generalization of the one given above, i.e., reduce to that one when our true cycles are the cycles considered above of the complexes α and β.**

Suppose that for any two true cycles z_1^p, z_2^q, homologous to zero in M^n and lying on intersecting carrier-compacta Φ_1 and Φ_2 in M^n and taken respectively with respect to coefficient domains $\mathfrak{A}$ and $\mathfrak{B}$, there is defined a linking coefficient $\mathfrak{b}(z_1^p,\, z_2^q)$ which does not change when either of the two cycles z_i, $i = 1,\, 2$, is replaced by a cycle homologous to it in Φ_i. We shall show that in only a few words that in only a few words we may bring the proof of the duality law of Pontrjagin to a conclusion.

---

* This is the expression used for the direct sums of finitely many groups J_m and J in the discrete case, and of the groups J_m and K in the bicompact case.

** Recall once again that every cycle of the complex $\beta \subset K^*$ is a cycle of a barycentric subdivision K_1 of the complex K and that any cycle of any triangulation may be considered as a true cycle.

Consider the sequence of triangulations

$$K^{(1)}, K^{(2)}, \ldots, K^{(\nu)}, \ldots$$

of the manifold M^n, each of which is a single or multiple barycentric subdivision of the preceding. The $K_*^{(\nu)}$ are the corresponding complexes of barycentric stars. By $\alpha^{(\nu)}$ we denote the closed subcomplex of the triangulation $K^{(\nu)}$ consisting of those simplexes which contain the points of A and of all the faces of these simplexes. Then the subcomplexes $\alpha_*^{(\nu)}$ and $\beta^{(\nu)} = K^* - \alpha_*^{(\nu)}$ are defined as usual, and writing $\sigma_\nu^{\nu+1}$ in place of $\sigma_{\alpha(\nu)}^{\alpha(\nu+1)}$, and so forth, we have

$$\Delta^p(A, \mathfrak{B}) = \varprojlim \{\Delta^p(\alpha^{(\nu)}, \mathfrak{B}), \sigma_\nu^{\nu+1}\};$$

$$\Delta_c^q(B, \mathfrak{A}) = \varinjlim \{\Delta^q(\widetilde{\beta}^{(\nu)}, \mathfrak{A}), E_{\nu+1}^\nu\},$$

where $E_{\nu+1}^\nu$ is the inclusion homomorphism of the group $\Delta_c^q \widetilde{\beta}^{(\nu)}$ into $\Delta_c^q \widetilde{\beta}^{(\nu+1)}$.

Identifying the groups $\Delta^p \alpha^{(\nu)}$ and $\Delta^p \widetilde{\alpha}^{(\nu)}$ and remarking that the canonical shift $\sigma_\nu^{\nu+1}$ of a cycle of the complex $\alpha^{\nu+1}$ does not take that cycle out of its homology class in the polyhedron $\widetilde{\alpha}^\nu$, we may write

$$\Delta^p(A, \mathfrak{B}) = \varprojlim \{\Lambda_c^p(\widetilde{\alpha}^\nu, \mathfrak{B}), \bar{E}_\nu^{\nu+1}\}, \tag{1}$$

$$\Delta^q(B, \mathfrak{A}) = \varinjlim \{\Delta_c^q(\widetilde{\beta}^\nu, \mathfrak{A}), E_{\nu+1}^\nu\}, \tag{2}$$

where $\bar{E}_\nu^{\nu+1}$ is the inclusion homomorphism of the group $\Delta_c^p \widetilde{\alpha}^{(\nu+1)}$ into $\Delta^p \widetilde{\alpha}^{(\nu)}$ But in view of the combinatorial case of the theorem of Pontrjagin we have the duality

$$\Delta_c^p(\widetilde{\alpha}^{(\nu)}, \mathfrak{B}) \mid \Delta_c^q(\widetilde{\beta}^{(\nu)}, \mathfrak{A}),$$

while the scalar product of two elements $\zeta_\nu^p \in \Delta_c^p(\widetilde{\alpha}^\nu, \mathfrak{B})$ and $\zeta_\nu^q \in \Delta_c^q(\widetilde{\beta}^{(\nu)}, \mathfrak{A})$ is equal to the linking coefficient $\mathfrak{v}(z_\nu^p, z_\nu^q)$ of any cycles[*] $z_\nu^p \in \zeta_\nu^p$ and $z_\nu^q \in \zeta_\nu^q$.

Here, evidently, the adjointness condition is satisfied:[**]

$$(\bar{E}_\nu^{\nu+1} \zeta_{\nu+1}^p \cdot \zeta_\nu^q) = (\zeta_{\nu+1}^p \cdot E_{\nu+1}^\nu \zeta^q),$$

---

[*] These cycles may be considered as true cycles, and *we shall consider it to be proved* that the coefficient for them coincides with that defined in the preceding section.

[**] Hence it is clear that, properly, we need only the following: if under the canonical shift of the complex $\alpha^{(\nu+1)}$ into the complex $\alpha^{(\nu)}$ the cycle $z_{\nu+1}^p$ goes into z_ν^p and z_ν^q is a cycle of the complex $\beta^{(\nu)}$, then the linking coefficient $\mathfrak{v}(z_{\nu+1}^p, z_\nu^q)$ is defined and is equal to the linking coefficient $\mathfrak{v}(z_\nu^p, z_\nu^q)$. Therefore it suffices to define the linking coefficient for polyhedral cycles at a positive distance from one another (which are cycles $z_{\nu+1}^p$, z_ν^q on the one hand, and cycles z_ν^p, z_ν^q on the other), and prove that $\mathfrak{v}(z_\nu^p, z_\nu^q)$ does not change when we replace these cycles by cycles homologous to them in the polyhedron $\widetilde{\alpha}^{(\nu)}$ (i.e., a cycle $z_{\nu+1}^p$ outside of z_ν^q).

All of this will be proved, with considerable room to spare, in §5.

so that the spectra (1) and (2) are adjoint, which means that their limit groups $\Delta^p(A, \mathfrak{B})$ and $\Delta^q_c(B, \mathfrak{A})$ are dual to each other.

Thus, it remains only to extend the concept of linking coefficients to true cycles. This will be done in the following section.

§5. Linking coefficients for true cycles

1. **The index of intersection for planes and simplexes.** In order to simplify the technical situation (connected with bringing complexes into general position and so forth), we restrict ourselves to the case when the manifold in question is a simple sphere S^n, which we consider as Euclidean space R^n completed by a point at infinity. The space R^n will be assumed oriented once and for all. Then the *intersection index* $(X^\mu \times Y^\nu)$ of two of its oriented planes X^μ and Y^ν, $\mu + \nu = n$, intersecting in the point o, is defined. Namely, taking oriented simplexes $x^\mu = (oa_1 \cdots a_\mu)$, $y^\nu = (ob_1 \cdots b_\nu)$ which lie respectively in X^μ and Y^ν and are oriented in accordance with the orientations chosen for X^μ and Y^ν, we find that the orientation of the simplex $(oa_1 \cdots a_\mu b_1 \cdots b_\nu)$ either coincides with the orientation chosen for the space R^n, or else is opposite to it. In the first case we put $(X^\mu \times Y^\nu) = +1$, in the second $(X^\mu \times Y^\nu) = -1$.

It is easy to verify that this definition does not depend on the choice of the simplexes $x^\mu = (oa_1 \cdots a_\mu)$, $y^\nu = (ob_1 \cdots b_\nu)$ (i.e., of their vertices $a_1, \cdots, a_\mu$; $b_1, \cdots, b_\nu$).

Now we define the intersection index of any two oriented simplexes x^μ and y^ν of the oriented space R^n. Here we shall suppose that the simplexes x^μ and y^ν *lie in relatively general position*, i.e., that they either lie at a positive distance [*] from one another, or else the simplexes x^μ and y^ν are nondegenerate, so that their skeletons form a system of $(\mu + 1) + (\nu + 1)$ points in general position and that the planes X^μ and Y^ν carrying them intersect in a point belonging to both of these simplexes.[**]

In the first case we put $(x^\mu \times y^\nu) = 0$, in the second we define $(x^\mu \times y^\nu)$ to be the intersection index of the planes carrying the simplexes x^μ and y^ν and oriented in accordance with them.

2. **The intersection index of chains lying in relatively general position.** For the chains $x^\mu = \sum_i a_i x_i^\mu$, $y^\nu = \sum_j b_j y_j^\nu$ (where x_i^μ and y_j^ν denote oriented simplexes), one says that *they lie in relatively general position* if each pair of simplexes x_i^μ, y_j^ν, appearing in the respective chains with nonzero coefficients, lie in relatively

---

[*] In this case we admit degeneracy. The skeleton of a simplex may lie in a plane of a smaller number of dimensions than the number of its vertices less one.

[**] The reader will recall that by a simplex we always understand an *open* simplex.

general position. We define the intersection index by the formula [*]

$$(x^\mu \times y^\nu) = \sum_{i,j} a_i b_j (x_i^\mu \times y_j^\nu)$$

and directly prove the following important proposition:

Theorem A. *Let x^μ and y^ν be two chains in relatively general position. There exists an $\epsilon > 0$ so small that every pair of chains x'^μ, y'^ν obtained from x^μ and y^ν respectively by an ϵ-shift also lies in relatively general position, while*

$$(x'^\mu \times y'^\nu) = (x^\mu \times y^\nu).$$

Indeed, it is sufficient to require that ϵ be smaller than half the smallest positive number which is a distance between any simplexes belonging to the chains x^μ and y^ν.

3. **Formulas for calculating with intersection indices.** Obviously the intersection index $(x^\mu \times y^\nu)$ depends linearly on both chains x^μ and y^ν. Furthermore, it is easy to prove the formula

$$(x^\mu \times y^\nu) = (-1)^{\mu\nu} (y^\nu \times x^\mu), \tag{1}$$

which one need only verify for the case of two simplexes.

The following theorem has a basic significance:

Theorem B. *If the chains x^r and y^s, $r + s = n + 1$, lie relatively general position, then*

$$(x^r \times \Delta y^s) = (-1)^r (\Delta x^r \times y^s). \tag{2}$$

This also need only be proved for the case when each of the two chains consists only of one simplex.

It suffices to consider the case when the simplexes x^r and y^s intersect, which is evidently along an (open) segment $o_1 o_2$.

Each of the points o_1, o_2 lies on the boundary of one of our simplexes and is an (interior) point of the other. Therefore two cases are possible:

$1°$. Both the points o_1 and o_2 lie on the boundary, say of the simplex x^r (and inside y^s).

$2°$. The point o_1 lies on the boundary x^r, and the point o_2 on the boundary y^s.

Proof in Case $1°$. Because of the general position, the points o_1 and o_2 are interior points of two different faces $|x_1| = |x_1^{r-1}|$ and $|x_2| = |x_2^{r-1}|$. Let us denote by $|x^{r-2}|$ the common face of the simplexes $|x_1|$ and $|x_2|$, and by a_1, a_2 vertices of the simplexes $|x_1|$, $|x_2|$ not belonging to the face $|x^{r-2}|$. No matter how $|x^{r-2}|$ may be oriented, the orientations $x_1 = (a_1 x^{r-2})$ and $x_2 = (a_2 x^{r-2})$ enter into Δx^r with opposite signs. Since the left side of equation (1) in our case becomes zero, we need to prove that also the right

---

[*] The coefficient domains are as in $\S 2.1$ of this chapter.

side is zero, i.e., that

$$(x_1 \times y^s) = (x_2 \times y^s).$$

It remains to compute both sides and thereby to verify that they are equal. To save space we shall speak of the product of the orientations of two simplexes of the same dimension d lying in the same space (of the same dimension d) as a number equal to $+1$ or -1 according as these orientations are the same or opposed.

Now we take in the plane Y^s some $(s-1)$-dimensional simplex y^{s-1} such that its plane does not intersect with the segment $o_1 o_2$ and its vertices lie in general position in R^n with the vertices of the simplex x^{r-2}.

Now we shall compute $(x_1 \times y^s)$. The point o_1 is the intersection point of the simplexes x_1 and y^s. The simplex $(o_1 x^{r-2})$ has an orientation the same as $x_1 = (a_1 x^{r-2})$, since the points a_1 and o_1 lie in the plane of the simplex x_1 on one side of the plane of the face x^{r-2}.

Therefore $(x_1 \times y^s)$ is equal to the product of the orientations of y^s and $(o_1 y^{s-1})$, multiplied by the product of the orientations of R^n and $(o_1 x^{r-2} y^{s-1})$, which we write as follows:

$$(x_1 \times y^s) = y^s (o_1 y^{s-1}) R^n (o_1 x^{r-2} y^{s-1}).$$

Analogously

$$(x_2 \times y^s) = y^s (o_2 y^{s-1}) \cdot R^n \cdot (o_2 x^{r-2} y^{s-1}).$$

Now we observe that

$$y^s \cdot (o_1 y^{s-1}) = y^s (o_2 y^{-1})$$

since the points c_1 and o_2 lie on one side of the plane y^{s-1}.

In the same way

$$R^n (o_1 x^{r-2} y^{s-1}) = R^n (o_2 x^{r-2} y^{s-1}),$$

since the points o_1 and o_2 lie in R^n on one side of the plane of the simplex $|x^{r-2} y^{s-1}|$ (the intersection of this plane with the plane Y^s is the plane of the simplex y^{s-1}, and the segment $o_1 o_2$ does not intersect the latter).

The equation $(x_1 \times y^s) = (x_2 \times y^s)$ is thus proved.

Proof in Case 2°. Suppose that the point o_1 lies on the face x^{r-1} of the simplex x^r (in view of the general position, the points o_1 and o_2 cannot lie on a face of lower dimension). We shall denote the vertices of the simplexes $|x^r|$ and $|y^s|$ opposite the faces $|x^{r-1}|$ and $|y^{s-1}|$ respectively by a and b. We choose the orientations x^{r-1} and y^{s-1} so that

$$x^r = (ax^{r-1}), \ y^s = (by^{s-1}).$$

Then the formula (2) which we are trying to prove takes the form

$$(x^r \times y^{s-1}) = (-1)^r (x^{r-1} \times y^s).$$

Let a_1, b_1 be any vertices of the simplexes x^{r-1}, y^{s-1}. We choose the orientations x^{r-2}, y^{s-2} of the faces opposite to these vertices so that

$$x^{r-1} = (a_1 x^{r-2}), \ y^{s-1} = (by^{s-2}).$$

Since their segment $o_1 a_1$ does not intersect the plane x^{r-2}, the points o_1 and a_1 lie in the plane of the face x^{r-1} on one side of the plane x^{r-2}. Therefore the simplex $(o_1 x^{r-2})$ has the same orientation as x^{r-1}.

For the same reason $(o_2 y^{s-2})$ has the same orientation as y^{s-1}.

We now show that $(o_2 o_1 x^{r-2})$ has the same orientation as $x^r = (ax^{r-1})$. This follows from the fact that the points o_2 and a lie in the plane X^r on one side of the plane of the face x^{r-1} (this is the same as the plane of the simplex $|o_1 x^{r-2}|$).

Analogously, the orientations $(o_1 o_2 y^{s-2})$ and y^s are the same. Therefore

$$(x^r \times y^{s-1}) = R^n (o_2 o_1 x^{r-2} y^{s-2});$$

$$(x^{r-1} \times y^s) = R^n (o_1 x^{r-2} o_2 y^{s-2}).$$

But evidently

$$(o_2 o_1 x^{r-2} y^{s-2})(o_1 x^{r-2} o_2 y^{s-2}) = (-1)^r,$$

with which the proof is complete.

4. Generalized conditions under which the intersection index is defined. The *body* of a chain x is a polyhedron $\bar{x}$ which is the sum of the simplexes entering into x with non-null coefficients and all the faces of these simplexes.

The following proposition is of basic importance:

Theorem C. *Let there be given in* R^n *(in* S^n*) a cycle* z^μ *and a chain* x^ν, $\mu + \nu = n$, *lying in relatively general position. If* $z^\mu \sim 0$ *in* $R^n - \overline{\Delta x^\nu}$, *then* $z^\mu \times x^\nu = 0$.

Proof. Let $x^{\mu+1}$ be a chain lying in $R^n - \overline{\Delta x^\nu}$ and bounded by the cycle z^μ.

We take for the chains z^μ and x^ν the number $\epsilon < \frac{1}{2}(\bar{x}^{\mu+1}, \overline{\Delta x^\nu})$ in accordance with Theorem A, and by means of an ϵ-shift we carry the chains $x^{\mu+1}$ and x^ν into $x'^{\mu+1}$ and x'^ν in such a way that the following conditions are satisfied:

1°. The set of all vertices[*] of both chains lies in general position.

Let $z'^\mu = \Delta x'^{\mu+1}$. Then, from Theorem A, the following condition will also be satisfied:

$$(z'^\mu \times x'^\nu) = (z^\mu \times x^\nu), \tag{2°}$$

and therefore it is sufficient to prove that $(z'^\mu \times x'^\nu) = 0$. But the chains $x'^{\mu+1}$ and x'^ν meet the conditions of Theorem B, so that

$$(z'^\mu \times x'^\nu) = (\Delta x'^{\mu+1} \times x'^\nu) = \pm (x'^{\mu+1} \times \Delta x'^\nu) = 0,$$

as was to be proved.

Corollary. If z_1^μ and z_2^ν are cycles in R in relatively general position,[**] then $(z_1^\mu \times z_2^\nu) = 0$.

Indeed, also $\Delta z_2^\nu = 0$ and $z_1^\mu \sim 0$ in R^n, so that our assertion is contained in Theorem C.

---

[*] A vertex of the chain is a vertex of any simplex entering into that chain with a non-zero coefficient.

[**] For the null-dimensional cycles z^0 we have to require that $z^0 \sim 0$ in R^n, i.e., that the sum of the coefficients in z^0 is equal to zero. If instead of R^n one considers S^n, then for the n-dimensional cycles z^n we have to make the additional requirement that $z^n \sim 0$ in S^n.

Now we may significantly generalize the conditions under which the intersection index of two chains is defined. Namely, *we may define the intersection index for any two chains x^μ and y^ν of the space R^n satisfying the condition that the sets $\overline{x}^\mu \cap \overline{\Delta y^\nu}$ and $\overline{y}^\nu \cap \overline{\Delta x^\mu}$ are empty.*

This condition may be formulated more concisely as follows: *neither of the two chains intersects the boundary of the other one.*

To do this we suppose that 3ϵ is less than $\rho(\overline{x}^\mu, \overline{\Delta y^\nu})$, $\rho(\overline{y}^\nu, \overline{\Delta x^\mu})$ and less than the least of the distances between any two distinct points which are vertices of either of the chains x^μ and y^ν.

We subject the vertices of the chain x^μ to an ϵ-shift f' such that the chain $x'^\mu = f'x^\mu$ obtained as a result lies in relatively general position with the chain y^ν. The intersection index $(x'^\mu \times y^\nu)$ is defined, and we put

$$(x^\mu \times y^\nu) = (x'^\mu \times y^\nu).$$

We shall show that this definition of the intersection index does not depend on the special choice of the ϵ-shift f'. Suppose that for another ϵ-shift f'' we obtain a chain x''^μ. We shall show that

$$(x''^\mu \times y^\nu) = (x'^\mu \times y^\nu).$$

Each of the shifts f' and f'', because of the smallness of ϵ, has the property that no two distinct vertices of the chain x^μ go into the same vertex of the chain x'^μ, or of the chain x''^μ. Therefore the displacement f'^{-1} is defined, and accordingly, also the 2ϵ-shift $f = f''f'^{-1}$, which carries the chain x'^μ into the chain x''^μ.

Labelling all the vertices of the chain x^μ in a definite order, and carrying over this order to the vertices of the chains x'^μ and x''^μ, we may speak of the prism $\Pi x'^\mu$ spanning the chain x'^μ and its image x''^μ under the mapping f. All of this prism lies in a 3ϵ-neighborhood of the polyhedron $\overline{x}^\mu$, i.e., in any case in $R^n - \overline{\Delta y^\nu}$.

We have (see Chapter III, §3)

$$\Delta \Pi x'^\mu = x'^\mu - x''^\mu - \Pi \Delta x'^\mu,$$

so that

$$x'^\mu - x''^\mu - \Pi \Delta x'^\mu \sim 0 \text{ in } R^n - \overline{\Delta y^\nu},$$

and accordingly

$$((x'^\mu - x''^\mu - \Pi \Delta x'^\mu) \times y^\nu) = 0.$$

But since $\Pi \Delta x'^\mu$ lies in a 3ϵ-neighborhood of the polyhedron $\overline{\Delta x^\mu}$, that is to say in $R^n - \overline{y}^\nu$, we have $(\Pi \Delta x'^\mu \times y^\nu) = 0$, i.e.,

$$((x'^\mu - x''^\mu) \times y^\nu) = 0, \quad (x'^\mu \times y^\nu) = (x''^\mu \times y^\nu),$$

as we were required to prove.

Remark. We have defined the intersection index $(x^\mu \times y^\nu)$ by means of the equation $(x^\mu \times y^\nu) = (x'^\mu \times y^\nu)$, where x'^μ is a chain which is the result of a sufficiently small shift of the chain x^μ and which lies in relatively general position with the chain y^ν. But it would be just as natural to define the index $(x^\mu \times y^\nu)$ by means of the equation

$$(x^\mu \times y^\nu) = (x^\mu \times y'^\nu),$$

where y'^ν lies in relatively general position with x^μ and is gotten from y^ν by a sufficiently small shift. Thus there arises the question as to the equality of the intersection indices defined in these two fashions. As one might expect, this question has a positive answer, as we may see in the following way.

† We choose the chain x'^μ, into which x^μ goes under the ϵ-shift, in such a way that it is in relatively general position not only with respect to y^ν but also with respect to the (already chosen) chain y'^ν. Then the intersection indices $(x'^\mu \times y^\nu)$ and $(x'^\mu \times y'^\nu)$ are immediately defined (in the sense of subsection 2 of this section), while considerations quite analogous to the preceding show that

$$(x'^\mu \times y'^\nu) = (x'^\mu \times y^\nu);$$
$$(x^\mu \times y'^\nu) = (x'^\mu \times y'^\nu)$$

and so

$$(x^\mu \times y'^\nu) = (x'^\mu \times y^\nu),$$

as it was required to prove.

Thus we obtain the following fundamental result:

Theorem D. *If neither of the chains x^μ, y^ν intersects the boundary of the other one, then the intersection index $(x^\mu \times y^\nu)$ may be defined by means of one of the formulas*

$$(x^\mu \times y^\nu) = (x'^\mu \times y^\nu); \; (x^\mu \times y^\nu) = (x^\mu \times y'^\nu);$$
$$(x^\mu \times y^\nu) = (x'^\mu \times y'^\nu), \tag{D}$$

where x'^μ and y'^ν are chains arising from x^μ, y^ν by means of a sufficiently small shift, subject to the sole additional requirement that the chains appearing in each parenthesis on the right side of formulas (D) stand in relatively general position.

5. The case when x^μ is a chain of a triangulation K, and y^ν is a chain of the corresponding star complex K^*. We shall prove that in this case the intersection index just defined coincides with the intersection index defined in Chapter IV, §§2.2 and 2.6.

We return to the notations of these sections. Let t_i^μ be arbitrarily chosen orientations of the simplexes of the complex K. We denote by u_i^ν that orientation of the barycentric star adjoint to the simplex t_i^ν which gives for the intersection index, defined in Chapter IV, §2.2, the value $(t_i^\mu \times u_i^\nu) = +1$.

It suffices to prove that for the intersection index defined in the present paragraph, *which will be denoted throughout this discussion by* $(t_i^\mu \otimes u_i^\nu)$, *we also have* $(t_i^\mu \otimes u_i^\nu) = 1$ (for $j \neq i$, evidently, $(t_i^\mu \times u_j^\nu) = (t_i^\mu \otimes u_j^\nu) = 0$).

From the definition of the number $(t_i^\mu \times u_i^\nu)$ in Chapter IV it follows that if $\mu = n$, $\nu = 0$, the star u_i^0 is the center of the simplex t_i^n taken with the coefficient $+1$. But the definition of the current paragraph gives in this case also $(t_i^n \otimes u_i^0) = +1$, so that the formula

$$(t_i^\mu \otimes u_i^\nu) = 1$$

is proved for $\nu = 0$. Now we shall apply induction. Supposing that we have proved the equality $(t_i^\mu \otimes u_i^\nu) = 1$, we shall prove that also

$$(t_j^{\mu-1} \otimes u_j^{\nu+1}) = 1.$$

To this end we choose any simplex $|t_i^\mu| \in K$ belonging to the star of the simplex $|t_j^{\mu-1}|$. Then by formula (1) of Chapter 4, §2.3, we have

$$(u_j^{\nu+1} : u_i^\nu) = (-1)^\mu (t_i^\mu : t_j^{\mu-1}).$$

Therefore, applying formula (2) of this paragraph, we obtain

$$(t_j^{\mu-1} \otimes u_j^{\nu+1}) = (t_i^\mu : t_j^{\mu-1})(\Delta\, t_i^\mu \otimes u_j^{\nu+1}) =$$
$$= (-1)^\mu (t_i^\mu : t_j^{\mu-1})(t_i^\mu \otimes \Delta\, u_j^{\nu+1}) =$$
$$= (-1)^\mu (t_i^\mu : t_j^{\mu-1})(u_j^{\nu+1} : u_i^\nu)(t_i^\mu \otimes u_i^\nu) =$$
$$= (-1)^{2\mu}(t_i^\mu : t_j^{\mu-1})^2 (t_i^\mu \otimes u_i^\nu) = 1,$$

as was to be proved.

6. **Linking coefficient for cycles in R^n (in S^n).** Let z_1^p and z_2^q, $p + q = n - 1$, be two cycles in R^n (in S^n). We suppose these cycles are homologous to zero (which in the case of R^n is a restriction only in the case of the zero-dimensional cycle, when it means that the sum of the coefficients is equal to zero.[*] We assume that the bodies of these cycles are nonoverlapping polyhedra. We choose any chain x^{p+1} bounded by the cycle z^p. The condition of Theorem (D) is satisfied, and the intersection index $(x^{p+1} \times z^q)$ is defined. We put

$$\flat(Z_1^p, Z_2^q) = (X^{p+1} \times Z_2^q).$$

This definition does not depend on the choice of the particular chain X^{p+1} bounded by the cycle Z_1^p: if X'^{p+1} is another such chain, then $X^{p+1} - X'^{p+1}$

---

[*] See the footnote on page 96.

is a cycle homologous in R^n to zero, and therefore [*]

$$((x^{p+1} - x'^{p+1}) \times z_2^q) = 0, \quad \text{i.e.,} \quad (x^{p+1} \times z_2^q) = (x'^{p+1} \times z_2^q).$$

The linking coefficient thus defined, $\mathfrak{v}(z_1^p, z_2^q)$, coincides with the one defined earlier in the case that z_1^p is taken in the complex $\alpha \subset K$, and z_2^q in the complex $\beta = K^* - \alpha^*$. This follows immediately from the fact that the definitions of the intersection index of the chains x^{p+1} and z_2^q of the complexes K and K^*, given in this subsection and in §2, coincide with each other.

The linking coefficient $\mathfrak{v}(z_1^p, z_2^q)$ has a number of important properties.

The first of them is that it is linear with respect to each of its two arguments z_1^p and z_2^q. The second is that it is commutative up to sign:

$$\mathfrak{v}(z_1^p, z_2^q) = \pm \, \mathfrak{v}(z_2^q, z_1^p),$$

or more precisely,

$$\mathfrak{v}(z_1^p, z_2^q) = (-1)^{pq+1} \, \mathfrak{v}(z_2^q, z_1^p).$$

For the proof we select chains x^{p+1} and y^{q+1}, bounded by the cycles z_1^p and z_2^q and in relatively general position. Then from formulas (2) and (1) we have

$$\mathfrak{v}(z_1^p, z_2^q) = (x^{p+1} \times \Delta y^{q+1}) = (-1)^{p+1}(\Delta x^{p+1} \times y^{q+1}) =$$

$$= (-1)^{p+1}(-1)^{p(q+1)}(y^{q+1} \times \Delta x^{p+1}) =$$

$$= (-1)^{pq+1} \, \mathfrak{v}(z_2^q, z_1^p).$$

Finally, we have the fundamental theorem:

$$\textit{If } z_1^p \sim 0 \textit{ in } R^n - \overline{z}_2^q, \textit{ then} \qquad\qquad \text{(E)}$$
$$\mathfrak{v}(z_1^p, z_2^q) = \mathfrak{v}(z_2^q, z_1^p) = 0.$$

For the proof it suffices to take a chain x^{p+1}, bounded by the cycle z_1^p, in $R^n - \overline{z}_2^q$.

What has been proved is a sufficient justification (in the case $M^n = S^n$) of the arguments of the preceding section, thus completing the proof of the theorem of Pontrjagin. However, for completeness, and for interesting applications in the following part of this monograph, we will give a further definition of the linking

---

[*] In the case S^n and $p = n - 1$ the discussion loses its force (since $x^{p+1} - x'^{p+1}$ may be non-homologous to zero in S^n), but the derivation still goes through: if $x^{p+1} - x'^{p+1}$ is non-homologous to zero in S^n, then the body of this cycle is all of S^n, and the equality

$$((x^{p+1} - x'^{p+1}) \times z^q) = 0$$

follows from the fact that the null-dimensional cycle has the sum of its coefficients equal to zero. Without this last condition, the assertion $(x^{p+1} \times z_2^q) = (x'^{p+1} \times z_2^q)$ for $p = n - 1$ in S^n is in general no longer true.

coefficient, this time for true cycles.

7. **Linking coefficient for true cycles.** This now may be defined without difficulty. Suppose that the true cycles

$$z^p = (z_1^p, z_2^p, \cdots, z_k^p, \cdots) \tag{3}$$

and

$$z'^q = (z_1^q, z_2^q, \cdots, z_k^q, \cdots) \tag{3'}$$

have as their carriers nonintersecting compacta Φ and Φ' in S^n. All of the cycles z_k^p and z_k^q will be conceived as cycles in S^n, i.e., we shall suppose that the simplexes of these cycles are given as the usual (possibly degenerate) geometrical simplexes in S^n. Without loss of generality we may assume that the cycles z_k^p, $k = 1, 2, 3, \cdots$, are at a distance $> \delta = \frac{1}{3}\rho(\Phi, \Phi')$ from the cycles * $z_{k'}^{(q)}$, $k' = 1, 2, 3, \cdots$. This may be accomplished by dropping a number of terms in the sequences (3), (3'). Moreover, we may assume that the ϵ_k-chains x_k^{p+1} in Φ and $x_{k'}'^{q+1}$ in Φ', satisfying the conditions $\Delta x_k^{p+1} = z_{k+1}^p - z_k^p$ and $\Delta x_{k'}'^{q+1} = z_{k+1}'^q - z_k'^p$, are chosen so that $\rho(\overline{x}_k^{p+1}, \overline{x}_{k'}'^{q+1}) > \delta$ (for any k, k'). Under these conditions the linking coefficients $\mathfrak{v}(z_k^p, z_k'^q)$ are defined, and all of them are equal to one another (moreover, all the $\mathfrak{v}(z_k^p, z_{k'}'^q)$ are equal to one another for any k, k'). The common value of all of these linking coefficients will be called the linking coefficient $\mathfrak{v}(z^p, z'^q)$ of the true cycles z^p and z'^q.

In particular, if z^p is a true cycle of the compactum $A \subset S^n$, and z'^q is a true cycle of the open set $B = S^n - A$ (i.e., a true cycle lying on some compactum $\Phi' \subset B$), then the linking coefficient is $\mathfrak{v}(z^p, z'^q)$.

8. **Pontrjagin duality as linking duality.** As we know, the groups $\delta^p A$ and $\Delta_c^p A$ (where A is compact) are connected to each other by a certain natural isomorphism, and therefore may be identified with each other. Thus, we may speak of the duality

$$\Delta_c^p(A, \mathfrak{B}) \mid \Delta_c^q(B, \mathfrak{A}),$$

which is usually called *Pontrjagin duality.*

From the considerations of the preceding section it follows that *this duality is the so-called linking duality*, in the sense that the scalar product $(\zeta_A^p \cdot \zeta_B^q)$ of two arbitrary elements $\zeta_A^p \in \Delta_c^p A$, $\zeta_B^q \in \Delta_c^q B$ is equal to the linking coefficient $\mathfrak{v}(\zeta_A^p, \zeta_B^q)$ of the homology classes ζ_A^p and ζ_B^q, i.e., to the linking coefficient $\mathfrak{v}(z_A^p, z_B^q)$ of arbitrary cycles $z_A^p \in \zeta_A^p$, $z_B^q \in \zeta_B^q$.

---

*By the distance of chains we mean the distance of their bodies.

BIBLIOGRAPHY

[1] P. S. Aleksandrov, *Combinatorial topology*, OGIZ, Moscow, 1947 (Russian); English transl., Vols. I and II, Graylock Press, Rochester, New York, 1956, 1957.

[2] ——, *Introduction to the general theory of sets and functions*, GITTL, Moscow, 1948. (Russian)

[3] L. S. Pontrjagin, *Continuous groups*, 2nd ed., GITTL, Moscow, 1954. (Russian)

Translated by:

J. M. Danskin, Jr.

TOPOLOGICAL DUALITY THEOREMS. II
NON-CLOSED SETS

P. S. ALEKSANDROV

CONTENTS

Introduction [*]

When, in a report to the Moscow International Topological Conference in 1935, I gave a formulation of fundamental problems of the combinatorial (homological) theory of non-closed sets, it was not at all clear that there could exist such a theory in general, i.e., that such a vast class of mathematical objects as the set of all point sets of Euclidean space could admit investigations by the methods of combinatorial topology. Let me recall the relevant passage in that report. [**]

"The significance and the prospects of the homological theory of compacta are based on the concrete geometrical knowledge to which we have been led by the concept of homology.

"The combinatorial topology of noncompact spaces in this sense does not exist, and the first question which arises in this connection is the following:

"Do generalized non-compact spaces admit algebraic-combinatorial investigations in the sense of homology theory?

"A positive answer to this question appears to me to be probable but by no means certain. The hypothesized positive answer would mean first of all the construction of a homology theory of finite-dimensional spaces, i.e., practically speaking, of point sets of Euclidean spaces, that is to say:

"1. A theory of dimension of these sets along the lines [***] on which it is now constructed for compacta.

"2. Theorems on duality, or at least theorems of existence for linking cycles.

"3. A theory of continuous (to all appearances, closed) mappings.

---

[*] This introduction is a reproduction, with a few changes, of the introductory remarks of the report which I gave to the Congress of Polish Mathematicians in 1952 and published in the Journal Fundamenta Mathematicae 41 (1954). In the introduction we give a short historical summary of the development of the combinatorial topology of non-closed sets. The rest of the exposition is independent of the introduction.

[**] Mat. Sb. 1 (43) (1936), p. 603.

[***] I.e., homological.

"Duality theorems are especially decisive for the fruitfulness of the theory: a theory without a duality law would hardly lead to important new knowledge. Does there exist for any set $M \subseteq R^n$ a duality relation of homological character between M and $R^n - M$? That is the question of decisive significance for the whole further development of the set-theoretic topology. I am abstaining for the present from any speculation on that subject. It may be that it would be better to put the question first in less than full generality. Either the extension of the duality law to the case of F_σ-and F_δ-sets, or the establishment of the impossibility of such an extension, would be extremely important."

Thus in my 1935 report a quite definite program was laid down. Now one may say that the first two points of this program have been in their main lines completed, essentially during the last decade. The doubts expressed in 1935, fortunately, did not turn out to be justified; the combinatorial topology of non-closed sets is at the present time the fastest growing new part of topology. As always happens in the fruitful development of a new direction of scientific activity the solution of the first problems, basic for that direction, naturally calls into being new questions, opening a fascinating prospect for future investigations.

This state of affairs was reached in several stages. First of all one should recall that the fundamental theorem on the nerve of a covering was freed from the assumption of compactness of the space by Kuratowski in 1933 [1] and proved by a method which turned out to be extremely fruitful. At about the same time E. Čech [2], on exactly the same basis of the concept of the nerve of a covering, laid down the beginnings of a homology theory of arbitrary (not only compact) topological spaces. It is true that the Čech homology theory, based entirely on finite coverings, did not turn out to be fruitful in the study of non-compact sets, nor in fact in the study of non-closed sets of Euclidean space, with which we are solely concerned in this paper.

Further, Čogošvili [3] gave in 1945 a generalization of the Pontrjagin duality law for, it is true, a very specialized class of non-closed sets, containing in particular all retracts, and then he greatly advanced [4] the systematic investigation of various types of homology groups of topological spaces, and developed in this connection a very useful algebraic apparatus (direct spectra of bicompact groups, which we shall discuss later). In the formulation and proof established by myself of the first general Čogošvili duality theorem, invariance theorems alone did not suffice.

At about the same time (1946–1947) the theorem on equivalence of the Uryson definition (by means of coverings) and the homology definition of dimension was proved by me [5] for all normal topological spaces (an analog of the theorem on ϵ-shifts and a theorem on essential coverings had been proved by me [6] already in 1940). At the same time Dowker [7] extended to non-closed sets the Hopf classification theorem and deduced from it new consequences for the theory of dimension, and in 1948 gave [8] a complete analysis of the possible generalizations of my theorem on ϵ-shifts. Dowker [7] and Hemmingsen [9] proved the validity of the utilization of infinite coverings in the theory of dimension of non-closed sets.

In 1947 I proved, in the paper [10], the first general duality theorem for arbitrary sets of Euclidean spaces, and in the proof of this theorem applied the algebraic constructions of Čogošvili. In the same paper I investigated a wide class of sets for which the duality law preserves its original form as given by L. S. Pontrjagin for closed sets.* These investigations were continued by E. F. Miščenko [13], but they can in no way be regarded as finished.

A further fundamental step in the whole theory of combinatorial topology of non-closed sets was taken in 1951, when K. Sitnikov [14a] proved his duality law, just as general, but stronger, than the theorem proved by me.

At the same time Sitnikov also solved [15] the fundamental problem of the theory of dimension for non-closed sets, giving a characterization of the dimension of a non-closed set by means of local-homology properties of the complementary set, i.e., proving for non-closed sets the "obstruction theorem."

But these last investigations, as well as the theory of dimension in general, remain at the outskirts of the present work, which is devoted exclusively to the duality theorems for non-closed sets proved by me in the papers [10] and [16], and by Sitnikov in his papers [14], [17].

Only in the last chapter do I overstep the limits of the duality theory proper and prove my theorem on homeomorphism of point sets [18], [19] in its final form, which was given to it by I. Švedov [20].

The distribution of this material in the separate chapters is clear from the table of contents.

In conclusion I want to express my deep thanks to N. A. Berikašvili for his very essential assistance in editing and proofreading this paper, and also to my students A. Arhangel'skiĭ, B. Pasynkov, and V. Ponomarev, who aided me in putting the text of the paper into final form, and to A. P. Leonova and E. V. Šurova for the great amount of work they have done in getting out both parts of this paper.

---

*In connection with my paper [10] I should further mention the paper of Kaplan [11], in which he proved some interesting results concerning the homology properties of non-closed sets. However, in the duality relation he proved only the equality of the Betti numbers for mutually complementary sets, i.e., he did what had already been done for closed sets in 1927 (see, for example, [12]). Kaplan did not prove any duality theorem in the current sense of the word, i.e., a theorem concerning the Betti groups and not their ranks.

NOTATION

In this work, by a space we mean a metrizable space with a countable basis, and by a point set we always mean a set lying in some n-dimensional spherical space S^n. Frequently we shall say simply "set" instead of "point set".

If we are simultaneously considering two sets A and B lying in the same S^n, then we always suppose that they are mutually complementary: $B = S^n - A$.

By Λ we denote the empty set. By n, p, q we denote non-negative integers, and by n we always mean the dimension of the spherical space in which the given point sets are being considered. If n, p, q are being considered simultaneously, then we suppose that

$$p + q = n - 1.$$

All of the groups considered are commutative. By $\mathfrak{A}$ we denote any discrete group, and by $\mathfrak{B}$ any bicompact group; if $\mathfrak{A}$ and $\mathfrak{B}$ are being considered simultaneously, then we suppose that between them Pontrjagin duality holds, which we write as follows:

$$\mathfrak{A} \mid \mathfrak{B}.$$

By a covering of the space (point set) we always understand an open covering (for more in connection with this see Chapter I, §1).

The closure of any set A is denoted by $[A]$.

In the whole of the following exposition it is supposed that the reader is acquainted with all the fundamental concepts and facts expounded in the first part of this paper. We shall use the Roman numeral I in referring to it.

CHAPTER I

Betti groups of non-closed sets

Introductory remarks

By "Betti groups" in the sequel we shall understand both Δ- and ∇-groups, i.e., homology and cohomology groups. There exist two fundamental approaches to the definition of these groups for an arbitrary space. The first consists in defining these groups by a limiting process applied to the corresponding groups of nerves of the coverings of the space in question. This approach is a carry-over to the case of arbitrary spaces of the definition of Betti groups for compacta which was given in the first part of this book, Chapter III, §4. The second approach consists in considering the directed (in the sense of ordinary inclusion) set of all compacta Φ lying in some space X. To this directed set (as we shall analyze in detail in what follows) there correspond direct and inverse spectra, respectively, of the ∇- and Δ-groups of these compacta. The Betti groups of the space X itself, called "groups with compact carriers," are defined as limit groups of these spectra. Both approaches are essential. They lead to different results; the duality theorems arise from comparison of the groups defined by these two methods, the one for the set $A \subseteq S^n$, and the other for its complement B.

In this chapter we shall consider in detail both methods of defining the Betti groups of arbitrary compacta.

§1. Betti groups defined by means of coverings
("Projective groups")

1. **Preparatory definitions.** For any two coverings α and β of the space X, we say that the covering β follows the covering α (or $\beta > \alpha$), if the covering β is subordinated to the covering α, i.e., if each element of the covering β is contained in at least one element of the covering α. Thus there is established an order in the set of all coverings of the set X.

We shall consider only directed subsets in the set of all coverings of the given space X.

First of all we observe: *a star-finite* * *covering of the space X forms a cofinal section in the directed set of all coverings of that set.* **

This allows us to restrict ourselves in what follows to star-finite coverings only. For each such covering, its nerve is defined in just the same way as for a finite covering; the nerve of a covering will be denoted by the same symbol as the covering itself.

If the covering β follows the covering α, then, putting into correspondence

---

* It is easy to see that every star-finite (even locally-finite) covering of a space with a countable basis is no more than countable.

** This proposition was proved in the paper [10] for all point sets. Simultaneously, Kaplan in his paper [11] proved the same proposition for all metric spaces with countable bases. We shall present Kaplan's proof.

Let X be a metric space with a countable basis, and let $\omega = \{o_i\}$ be any covering of the space X. By Uryson's theorem one may regard the space X as a set lying in the Hilbert cube H. We choose a system Ω of sets open in H, satisfying the condition $o_i = X \cap O_i$.

It is sufficient to find a star-finite covering $\Omega' = \{o'\}$ of the set $\Gamma = \bigcup_i O_i$, subordinate to the covering Ω of that set. Since Γ is open in the compactum H, and therefore locally compact, there exists a representation $\Gamma = \bigcup_\nu \Gamma_\nu$, where Γ_ν is open and the compactum $[\Gamma_\nu]$ is contained in $\Gamma_{\nu+1}$; here $\nu = 0, 1, 2, \cdots, \Gamma_0 = \Lambda$. Choose a finite number of elements O_i, covering $[\Gamma_\nu]$; suppose that this is

$$O_{\nu 1}, \cdots, O_{\nu s_\nu}.$$

Put

$$O'_{\nu j} = O_{\nu j} \cap (\Gamma_{\nu+1} / [\Gamma_{\nu-1}]).$$

The sets $O'_{\nu j}$, where $\nu = 1, 2, 3, \cdots$ and $j = 1, 2, \cdots, s_\nu$, form the desired star-finite covering of the set Γ, subordinate to the covering Ω.

with each element of the covering β any one of the elements of the covering α containing it, we obtain a simplicial mapping ("projection") ω_α^β of the nerve β into the nerve α. Any two projections of the nerve β into the nerve α are combinatorially close mappings (see I, page 41). Moreover, if $\gamma > \beta > \alpha$, then, whatever the projections ω_β^γ and ω_α^β are, the mapping $\omega_\alpha^\beta \omega_\beta^\gamma = \omega_\alpha^\gamma$ also is a projection (transitivity law).

Now we suppose that for all coverings α of the set X, entering into a given directed family Σ, there are defined groups $\mathfrak{G}_\alpha$, which under the projections ω_α^β of the nerve β into the nerve α are connected by homomorphisms ω_α^β, satisfying the condition of transitivity, of the groups $\mathfrak{G}_\beta$ into $\mathfrak{G}_\alpha$ (homomorphisms π_β^α of the groups $\mathfrak{G}_\alpha$ into $\mathfrak{G}_\beta$). Here we suppose that the identity homomorphism ω_α^β generates the identity homomorphism and that the *correctness condition* is satisfied: namely, to combinatorially close mappings there corresponds one and the same homomorphism ω_α^β (respectively π_β^α). In this case one says that the groups $\mathfrak{G}_\alpha$ are groups of Δ-type (∇-type). In view of these conditions, groups of Δ-type form inverse spectra and groups of ∇-type form direct spectra. The limit group of this spectrum is called a "$\mathfrak{G}$-group" of the set X, respectively of Δ- or ∇-type. *This group is evidently defined by the following set of data:*

1. *The directed family Σ of coverings of the set X.*

2. *The groups $\mathfrak{G}_\alpha$ for each $\alpha \in \Sigma$ (along with their homomorphisms ω_α^β, respectively π_β^α).*

2. The groups $\Delta_f^r X$; $\nabla_f^r X$; the groups $\delta^r X$ and $\nabla^r X$. We present the most important special cases of the definition just given.

1°. The directed family of coverings Σ is the family Σ_f of all finite coverings α, with $\mathfrak{G}_\alpha = \Delta^r \alpha$, respectively $\mathfrak{G}_\alpha = \nabla^r \alpha$. This group is of Δ-type, respectively of ∇-type. The condition of correctness is satisfied (for the proof, see Part I, Chapter III, §3).

The corresponding limit groups are called Betti groups of the set X, *based on finite coverings* (or Čech groups); they are denoted by $\Delta_f^r X$ or $\nabla_f^r X$. We see that for the construction of a duality theory they are not suitable, since they are not dualizable (to two homeomorphic sets A, A' there may correspond complements B, B' in S^n with nonisomorphic Čech groups).

2°. *The most important case is the following.* We consider the directed *family Σ of all star-finite coverings α of the set X.* As the coefficient domain we take any group $\mathfrak{A}$ (without topology). As the groups of Δ-type we take the groups $\Delta^r(\alpha, \mathfrak{A})$ of the nerve α, based on considering only finite chains. *

As groups of ∇-type we take the groups $\nabla^r(\alpha, \mathfrak{A})$, based on infinite chains. **
The projection ω_α^β of the nerve β into the nerve α defines, according to the
known (Part I, Chapter III, §1) rule, a homomorphism π_β^α of the group $\nabla^r \alpha$ into
the group $\nabla^r \beta$; first we define a homomorphism π_β^α of the group $L^r \alpha$ of all infinite chains of the complex α into the group $L^r \beta$ by putting *** for each
$x_\alpha^r \in L^r \alpha$ and for each oriented simplex t_β^r of the complex β,

$$(\pi_\beta^\alpha x_\alpha^r \cdot t_\beta^r) = (x_\alpha^r \cdot \omega_\alpha^\beta t_\beta^r).$$

The correctness condition is satisfied in both cases (for a proof see Part I,
Chapter III, §3; it does not depend on the assumption of finiteness of the complexes α, β).

We denote the corresponding limit groups by

$$\delta^r(X, \mathfrak{A}) = \varprojlim \{\Delta^r(\alpha, \mathfrak{A}), \ \omega_\alpha^\beta\},$$

$$\nabla^r(X, \mathfrak{A}) = \varinjlim \{\nabla^r(\alpha, \mathfrak{A}), \ \pi_\beta^\alpha\}.$$

3. **Projective cycles and ∇-cosets.** The same as in the case of compacta
(Part I, page 45), one may also for arbitrary spaces X give a definition of the
groups $\delta^r X$ and $\nabla^r X$ in a rather different form, immediately equivalent to the
preceding, but not explicitly using the concept of spectrum. In many cases this
form turns out to be convenient. Moreover, it is very easy to visualize. For the
convenience of the reader we present it here (at the cost of some repetition of
the exposition on page 45 of the first part).

A projective cycle on the set X over the coefficient domain $\mathfrak{A}$ is a system
of cycles

$$z^r = \{z_\alpha^r\},$$

with one finite cycle z_α^r on each nerve **** α, satisfying the condition: if $\beta > \alpha$,
then *****

---

    * This means that by *a chain of the (infinite) complex* α we mean only finite linear forms
$x^r = \Sigma c_i t_i^r$ (where the t_i^r are oriented simplexes of the given complex α), that is,
functions $x^r(t_i^r)$ with values from the group $\mathfrak{A}$, satisfying the evenness condition
$x^r(t_i^r) = - x^r(-t_i^r)$, *and taking on values different from zero only on a finite number of
simplexes of the given complex.* The boundary Δx^r, cycles, and the groups $\Delta^r \alpha$ are then
defined exactly as in Part I, Chapter I, §2, the projections ω_α^β as in Part I, Chapter III,
§§1.1 and 1.2; if $x_\beta^r = \Sigma c_i t_{\beta i}^r$, then $\omega_\alpha^\beta x_\beta^r = \Sigma c_i \omega_\alpha^\beta t_{\beta i}^r$, and so forth.

    ** This means that by a chain one understands any linear form $x^r = \Sigma c_i t_i^r$, without
the requirement that it be finite; the ∇-boundary Δx^r is defined as in Part I, Chapter I,
§2.

    *** We recall that the scalar product of a chain with an oriented simplex is the value
of that chain on that simplex Part I, Chapter I, §2.5.

    **** Recall that α runs through the directed set of all star-finite coverings of the
set.

    ***** The homology (1) means the existence of a finite chain on the nerve α,
bounded by the cycle $z_\alpha^r - \omega_\alpha^\beta z_\beta^r$.

$$z^r_\alpha \sim \omega^\beta_\alpha z^r_\beta \text{ on } \alpha. \tag{1}$$

Projective cycles are added termwise: if

$$z' = \{z'_\alpha\}, \ z'' = \{z''_\alpha\}, \text{ then } z' + z'' = \{z'_\alpha + z''_\alpha\}.$$

Under this addition they evidently form a group which we shall denote by $Z^r(X, \mathfrak{A})$.

The projective cycle $z^r\{= z^r_\alpha\}$ is said to be bounding or homologous to zero in X, if $z^r_\alpha \sim 0$ for each α (i.e., each z^r_α bounds a *finite* chain in α). The bounding projective cycles form a subgroup $H^r(X, \mathfrak{A})$ of the group $Z^r(X, \mathfrak{A})$. The factor-group $Z^r(X, \mathfrak{A}) - H^r(X, \mathfrak{A})$ stands, as is easily seen, in a natural relation of isomorphism with the group $\delta^r(X, \mathfrak{A})$ and may be identified with this last, which makes it possible for us to consider the equation

$$\delta^r X = Z^r X - H^r X$$

as the definition of the group $\delta^r X$. Sometimes the group $\delta^r X$ is called a projective Δ-group (over a discrete coefficient domain) of the set X.

In order to obtain an analogous definition of the group $\nabla^r X$, we shall call a ∇-cycle of the set X any (generally speaking infinite) ∇-cycle z^r_α of the nerve of any star-finite covering α of the set X. Two ∇-cycles z^r_α and z^r_β of the set X are said to be homologous to each other in X if there exists a γ following both α and β such that

$$\pi^\alpha_\gamma z^r_\alpha \sim \pi^\beta_\gamma z^r_\beta \text{ in } \gamma,$$

i.e., in the nerve α there exists a chain x^{r-1}_γ, generally speaking infinite, such that $\nabla x^{r-1}_\gamma = \pi^\alpha_\gamma z^r_\alpha - \pi^\beta_\gamma z^r_\beta$.

This definition of homology leads to the decomposition of the set of all r-dimensional ∇-cycles of the set X into classes (or "cosets") of cycles homologous to one another. These cosets form the elements of the group $\nabla^r(X, \mathfrak{A})$. If we are given two cosets ζ'^r and ζ''^r, then in order to define their sum we first find a covering α such that on the nerve α there lies some ∇-cycle $z'^r_\alpha \in \zeta'^r$ and some ∇-cycle $z''^r_\alpha \in \zeta'^r$ (it is easy to see that such an α always exists). Then $\zeta^r = \zeta'^r + \zeta''^r$ is defined as the coset containing the ∇-cycle $z'^r_\alpha + z''^r_\alpha$. It is easy to verify that this definition of addition of cosets does not depend on the choise of the elements entering into it. This completes the definition of the group $\nabla^r X$.

Remark 1. It is no accident that finite chains on the nerves α form the basis of the definition of the group $\delta^r X$, while arbitrary chains form the basis of the definition of the group $\nabla^r X$. If we were to try to define the group $\delta^r X$ in the above fashion but using infinite chains, we would not obtain any results, since the operator ω^β_α is not defined for infinite chains: onto one and the same simplex t^r_α of the nerve, there might under ω^β_α be mapped infinitely many simplexes of the nerve β, entering into a given chain x^r_β with non-zero coefficients. In order to define the value of the chain $\partial^\beta_\alpha x^r_\beta$ on the simplex t^r_α, we would have to take an (infinite!) sum of these coefficients, but such a sum is not defined.

On the other hand, as the basis of the definition of the groups $\nabla^r X$ one could never take finite chains on the nerves α, since under the mapping π^α_β a finite chain (and even one oriented simplex) of the nerve β might go into an infinite chain of the nerve α.

4. **Non-dualizability of groups based on finite coverings.** We shall construct two sets A and A' in S^2, whose complements B and B' are homeomorphic and for which at the same time $\nabla^2_f A' = 0$, and $\nabla^2_f A \neq 0$ (even consisting of an infinite number of elements). Of course analogous examples may be constructed in any S^n.

We represent the sphere as the usual plane R^2 completed by one point at infinity. By A we denote a closed circle in this plane, from whose circumference one point has been deleted. By A' we denote some half-segment in the plane R^2. It is easy to see that the complementary sets B and B' are homeomorphic. Obviously, the two-dimensional Čech groups for the set A' are null-groups. We shall show that the two-dimensional Čech group for the set A is different from zero. This may be most easily proved for the two-dimensional ∇-group (with integer coefficients) by using the theorem of Dowker [7], which asserts that this group is always different from zero given that there exists an essential (in the sense of *uniform* homotopy) mapping of the set A onto the two-dimensional sphere. Such a mapping f may be constructed as follows. We represent the set A as a half-plane $y \geq 0$ (without the point at infinity). We map all the lines $y = n$ (where n is a non-negative integer) into one point (the "pole"), and topologically map each strip $n < y < n+1$ onto the whole sphere S^2 without the pole P, so that as a result one obtains a continuous mapping f of the whole half-plane $y \geq 0$ onto S^2. It is easy to verify that the mapping f thus obtained is essential in the sense of uniform homotopy.

§2. Betti groups with compact carriers

1. **The groups $\Delta^r_c X$.** We shall consider the set, directed by inclusion, of all compacta lying in a given space X. For each compactum $\phi \subseteq X$ we choose the group $\Delta^r_c \phi$ (defined in Part I, Chapter III, §7), i.e., the so-called Vietoris group of this compactum. If $\phi \subseteq \phi'$, then the identity mapping of the compactum ϕ into the compactum ϕ' generates a homomorphism $E^\phi_{\phi'}$, — the so-called inclusion homomorphism — of the group $\Delta^r_c \phi$ into the group $\Delta^r_c \phi'$. Indeed, the group $\Delta^r_c \phi$ is the group of classes of mutually homologous Δ_c-cycles (or "Vietoris" cycles) of the compactum ϕ. But each Δ_c-cycle of the compactum ϕ is at the same time a Δ_c-cycle of the compactum $\phi' \supseteq \phi$, and two cycles homologous to each other in

ϕ are *a fortiori* homologous to one another in ϕ'. Therefore each homology class of the compactum ϕ lies in some (clearly unique) homology class of the compactum ϕ', and this defines the homomorphism $E^{\phi}_{\phi'}$. The groups $\Delta^r_c\phi$, along with the homomorphisms $E^{\phi}_{\phi'}$ connecting them, form a direct spectrum

$$\{\Delta^r_c\phi,\ E^{\phi}_{\phi'}\},$$

whose limit group is called the group $\Delta^r_c X$ (or Vietoris group of the space X). The coefficient group here is arbitrary and it is taken without any topology.

One may reformulate this definition so that there does not appear in it any explicit concept of group spectra. We shall call a Vietoris or Δ_c-*cycle of the set* X any Δ_c-cycle of any compactum $\phi \subseteq X$. If we are given two Δ_c-cycles z^r, z'^r, lying respectively on compacta $\phi \subseteq X$ and $\phi' \subseteq X$, then both of them lie on the compactum $\phi \cup \phi' \subseteq X$ and have the sum $z^r + z'^r$, which is accordingly also a Δ_c-cycle of the set X. Therefore all the r-dimensional Δ_c-cycles of the set X form a group $Z^r_c X$. In this group there is a subgroup $H^r_x X$ consisting of all the Δ_c-cycles which are homologous to zero in X. Here, a Δ_c-cycle z^r, lying on some $\phi \subseteq X$, is called a bounding cycle (or is said to be homologous to zero) in X if there exists a compactum ϕ' on which z^r is homologous to zero. Evidently the group $\Delta^r_c X$ is the factor-group of the group $Z^r_c X$ by the subgroup $H^r_c X$:

$$\Delta^r_c X = Z^r_c X - H^r_c X.$$

The group $\Delta^r_c X$ is historically the first "group with compact carriers" defined for any space X. But at the present time another group with compact carriers is known, namely the Sitnikov group $\Delta^r X$. It is richer than the group $\Delta^r_c X$, and to all appearances ought to be considered the fundamental Δ-group with compact carriers. The group $\Delta^r X$ is obtained if one modifies, as K. Sitnikov did, the concept of true cycle and homology.

2. Sitnikov cycles. The group $\Delta^r\phi$. We shall call a Δ-cycle or a Sitnikov cycle of the compactum ϕ the two-rowed matrix (with an infinite number of columns):

$$Z^r = \begin{pmatrix} z^r_1, & z^r_2, & \cdots, & z^r_k, & \cdots \\ x^{r+1}_1, & x^{r+1}_2, & \cdots, & x^{r+1}_k, & \cdots \end{pmatrix}, \quad k = 1, 2, 3, \cdots \tag{1}$$

consisting of ϵ_k-cycles z^r_k and ϵ_k-chains x^{r+1}_k of the compactum ϕ, $\epsilon_k \to 0$, connected to one another by the relations $\Delta x^{r+1}_k = z^r_{k+1} - z^r_k$, $k = 1, 2, 3, \cdots$. Sitnikov cycles of a given dimension r, lying on the compactum ϕ, form a group $Z^r\phi$ under termwise addition.

We shall say that the cycle (1) bounds, or is homologous to zero, on the compactum ϕ, if there exist ϵ'_k-chains x^{r+2}_k and y^{r+1}_k such that $\epsilon'_k \to 0$ and $\Delta y^{r+1}_k = z^r_k$, $\Delta x^{r+2}_k = y^{r+1}_{k+1} - y^{r+1}_k - x^{r+1}_k$. Bounding cycles form a subgroup

$H^r\phi$ of the group $Z^r\phi$. The factor-group $\Delta^r\phi = Z^r\phi - H^r\phi$ is called the *Sitnikov group of the compactum* ϕ.

Remark 1. Taking in the matrix (1) only the first row, we obtain a Vietoris cycle, uniquely determined by the given Sitnikov cycle, while two Sitnikov cycles, evidently, determine the same Vietoris cycle if and only if their first rows coincide.

It is also evident that a bounding Sitnikov cycle determines a bounding Vietoris cycle. On the other hand, since for each Vietoris cycle z^r_k, by its very definition, there exists a sequence of ϵ'_k-chains x^{r+1}_k, $\epsilon'_k \to 0$, for which $\Delta x^{r+1}_k = z^r_{k+1} - z^r_k$, each Vietoris cycle may be complemented by adjoining to it a suitably chosen second row to a Sitnikov cycle. Thus *there exists a natural homomorphism of the group $Z^r\phi$ of all Sitnikov cycles onto the group $Z^r_c\phi$ of Vietoris cycles. Under this homomorphism every bounding cycle goes into a bounding cycle, so that we have a natural homomorphism of the Sitnikov group $\Delta^r\phi$ onto the Vietoris group $\Delta^r_c\phi$.*

3. **The Sitnikov group $\Delta^r X$.** If the compactum ϕ is contained in the compactum ϕ', then again we have a natural inclusion homomorphism $E^\phi_{\phi'}$ of the group $\Delta^r\phi$ into the group $\Delta^r\phi'$, and therefore the set of all compacta lying in the given space X, directed by inclusion, generates a direct spectrum

$$\{\Delta^r\phi, \, E^\phi_{\phi'}\},$$

whose limit group is, by definition, the Sitnikov group $\Delta^r X$.

It may also be defined as follows. We shall call a Sitnikov, or a Δ-cycle, of the set X any cycle lying on some compactum $\phi \subseteq X$. We shall say that that cycle bounds in X, if it bounds on some compactum $\phi' \subseteq X$. The group $\Delta^r X$ may be defined as the factor-group of the group $Z^r X$ of all r-dimensional Sitnikov cycles of the set X modulo the subgroup $H^r X$ of all bounding cycles. The coefficient group is again arbitrary and is considered without topology.

Remark 2. The Sitnikov cycle

$$\begin{bmatrix} z^r_1, \; z^r_2, & \cdots, & z^r_k, & \cdots \\ x^{r+1}_1, \; x^{r+1}_2, & \cdots, & x^{r+1}_k & \cdots \end{bmatrix} \tag{1}$$

is called weakly homologous to zero, if the Vietoris cycle $(z^r_1, \, z^r_2, \, \cdots, \, z^r_k, \, \cdots)$ constituting its first row is homologous to zero. The Sitnikov cycles weakly homologous to zero define in $\Delta^r(X, \mathfrak{A})$ a subgroup $H^r_c(X, \mathfrak{A})$. This subgroup is the kernel of that natural homomorphism of the group $\Delta^r(X, \mathfrak{A})$ onto the group $\Delta^r_c(X, \mathfrak{A})$ which we obtain if we put into correspondence with each Sitnikov cycle its first row. Therefore we have an isomorphism

$$\Delta^r_c(X, \mathfrak{A}) = \Delta^r(X, \mathfrak{A}) - H^r_c(X, \mathfrak{A}).$$

Remark 3. The groups $\Delta^r X$ and $\Delta^r_c X$ may be nonisomorphic to each other, even if X is a compactum and the group of coefficients is the group of all integers.

Indeed, let X be the one-dimensional continuum called a solenoid. [*] We take in the solenoid X two points a and b, which can not be joined in X by any simple arc. This pair of points forms a zero-dimensional cycle, homologous to zero in X in the sense of Vietrois (since X is a continuum), but not homologous to zero in the sense of Sitnikov. This last fact follows from the fact that ϵ-chains, joining the points a and b as $\epsilon \rightarrow 0$, run through the solenoid a larger number of times than it is possible to construct the chains x_k^r entering into the definition of Sitnikov homology. Thus, in the case of the solenoid, $\Delta_c^0 X = 0$ and $\Delta^0 X \neq 0$. If X is the topological product of a solenoid and a circumference, then $\Delta_c^1 X = 0$ and $\Delta^1 X \neq 0$. In this connection one should note that the kernel $H_c^r(X, \mathfrak{A})$ of the natural homomorphism of the group $\Delta^r X$ onto the group $\Delta_c^r X$ has been studied very little. In particular, very little study has been given to the group $N^r X$ lying in this kernel generated by the "null-cycles", i.e., those r-dimensional Sitnikov cycles whose first row consists of zeros only and whose second row accordingly consists of $(r + 1)$-dimensional cycles. The null-cycles form a subgroup of the group $Z^r X$ of Sitnikov cycles, which is the kernel of the natural homomorphism of this group onto the group $Z_c^r X$.

4. The following theorem holds.

Theorem. [**] *If the coefficient group admits a bicompact topology, then the groups $\Delta_c^r X$ and $\Delta^r X$ over this coefficient domain are isomorphic to each other (i.e., $H_c^r(X, \mathfrak{B}) = 0$).*

Proof. We may assume that the coefficient domain is provided with a bicompact topology, and accordingly is a bicompact group $\mathfrak{B}$.

The theorem obviously follows from the following lemma:

If

$$(z_1^r, z_2^r, \cdots, z_k^r, \cdots) \tag{1}$$

is a Vietoris cycle, homologous to zero in the compactum Φ over the bicompact

---

[*] We shall restrict ourselves here to the simplest so-called dyadic solenoids, which are defined as follows. We take a sequence $Q_1 \supset Q_2 \supset \cdots \supset Q_\nu \supset \cdots$ of decreasing ring-like bodies, each of which is homeomorphic to a solid torus (equal to a torus along with the region interior to it), while each of these bodies is the trace left by the motion in three-dimensional space of some circle, whose plane, while remaining vertical, moves in such a way that the center describes a simple closed curve and two arbitrary positions of the circle do not have common points. The radius of the moving circle is the "thickness" of the ring-like body. We suppose that the ring-like bodies Q_ν are so chosen that their thickness converges to zero as $\nu \rightarrow \infty$ and that the trajectory of the center of the circle describing the body $Q_{\nu + 1}$ runs twice around the body Q_ν. The intersection of all the Q_ν. The intersection of all the Q_ν, $\nu = 1, 2, 3, \cdots$, is a dyadic solenoid. Solenoids are indecomposable continua; on each solenoid there is a pair of points which can not be joined by any simple arc on it (there is in fact an uncountable set of such pairs of points).

[**] This theorem is contained in my paper, *Zur Homologie-Theorie der Kompakten*, Compositio Math. 4 (1937), pp. 256–271 (as regards details, see the paper of Sitnikov [17a], footnote on page 281 of the English translation). In order not to have to present here auxiliary concepts superfluous to our present objectives, introduced in my paper cited above, I present below the proof of Sitnikov.

coefficient domain $\mathfrak{B}$, then, supplementing it arbitrarily to a Sitnikov cycle

$$\begin{bmatrix} z_1^r, & z_2^r, & \cdots, & z_k^r, & \cdots \\ x_1^{r+1}, & x_2^{r+1}, & \cdots, & x_k^{r+1}, & \cdots \end{bmatrix}, \tag{2}$$

we obtain a cycle (2) which is also homologous to zero in Φ, but this time understood in the Sitnikov sense.

Since the Vietoris cycle (1) bounds in the compactum Φ, there exists an ϵ_k-chain y_k^{r+1} of that compactum, $\epsilon_k \to 0$, satisfying the conditions

$$\Delta y_k^{r+1} = z_k^r. \tag{3}$$

Our problem consists in finding an ϵ_k'-chain x_k^{r+2} under the conditions

$$\Delta x_k^{r+2} = y_{k+1}^{r+1} - y_k^{r+1} - x_k^{r+1}. \tag{4}$$

First let us transform our data in such a way that, without making any essential change, we give them a more convenient form.

To this end we choose a sequence of finite coverings

$$\alpha_1, \alpha_2, \cdots, \alpha_k, \cdots.$$

of the compactum Φ, where the diameters of the elements of α_k tend to zero as $k \to \infty$ and the diameter of each star[*] of the covering α_{k+1} is less than the Lebesgue number of the covering α_k.

Without loss of generality (compare Part I, Chapter III, §7.1), we may assume that z_k^r and y_k^{r+1} are chains of the geometric realization of the nerve α_k.

We shall call a projective chain a sequence of chains of a given dimension lying respectively on the nerves α_k. First of all we shall construct a certain special sequence

$$\eta_1^{r+1}, \eta_2^{r+1}, \cdots, \eta_\nu^{r+1}, \cdots$$

of $(r+1)$-dimensional projective chains, namely

$$\eta_\nu^{r+1} = (y_{\nu,1}^{r+1}, y_{\nu,2}^{r+1}, \cdots, y_{\nu,\nu-1}^{r+1}, y_{\nu,\nu}^{r+1}, \cdots, y_{\nu,k}^{r+1}, \cdots),$$

where $y_{\nu,k}^{r+1} = y_k^{r+1}$ for all $k \geq \nu$, and $y_{\nu,\nu-1}^{r+1}, \cdots, y_{\nu,1}^{r+1}$ are defined step by step in the following way.

We denote by $\phi_{\nu-1}$ a definite "canonical shift" of the chains $y_\nu^{r+1} = y_{\nu,\nu}^{r+1}$ and $x_{\nu-1}^{r+1}$ into the nerve $\alpha_{\nu-1}$ determined in the following way: on the vertices of these chains which are vertices of the nerve α_ν, the shift $\phi_{\nu-1}$ is defined as some projection $\omega_\nu^{\nu+1}$, and on the vertices which are vertices of $\alpha_{\nu-1}$, the shift $\phi_{\nu-1}$ is defined as the identity mapping.

---

[*] A star of the covering α is the sum of all the elements of the covering intersecting with any one of the elements of the same covering.

Now put

$$y_{\nu,\nu-1}^{r+1} = \phi_{\nu-1} y_{\nu,\nu}^{r+1} - \phi_{\nu-1} x_{\nu-1}^{r+1}.$$

Then

$$\Delta y_{\nu,\nu-1}^{r+1} = \Delta\phi_{\nu-1} y_{\nu,\nu}^{r+1} - \Delta\phi_{\nu-1} x_{\nu-1}^{r+1} = \phi_{\nu-1} z_\nu^r - \Delta\phi_{\nu-1} x_{\nu-1}^{r+1}.$$

But

$$\Delta\phi_{\nu-1} x_{\nu-1}^{r+1} = \phi_{\nu-1} \Delta x_{\nu-1}^{r+1} = \phi_{\nu-1}(z_\nu^r - z_{\nu-1}^r) = \phi_{\nu-1} z_\nu^r - z_{\nu-1}^r,$$

so that

$$\Delta y_{\nu,\nu-1}^{r+1} = z_{\nu-1}^r.$$

Analogously we obtain step by step $y_{\nu,\nu-2}^{r+1}, \cdots, y_{\nu,1}^{r+1}$ by the formula

$$y_{\nu,k}^{r+1} = \phi_k y_{\nu,k+1}^{r+1} - \phi_k x_k^{r+1}, \text{ for } k \leq \nu - 1. \tag{5}$$

Here, for all k,

$$\Delta y_{\nu,k}^{r+1} = z_k^r, \text{ for any } \nu. \tag{6}$$

Since the coefficient domain $\mathfrak{B}$ is bicompact, the group $L^{r+1} a_k$ of all $(r+1)$-dimensional chains of the nerve α_k is also bicompact (that is, the group $\mathfrak{B}$, raised to a degree equal to the number of $(r+1)$-dimensional simplexes of the nerve α_k). Therefore the product of all the groups $L_{\alpha_k}^{r+1}$, $k = 1, 2, \cdots$, is also compact; its elements are evidently all the $(r+1)$-dimensional projective chains.

This means that there exists a projective chain

$$\eta^{r+1} = (Y_1^{r+1}, Y_2^{r+1}, \cdots, Y_k^{r+1}, \cdots)$$

which is the limit for the sequence

$$\eta_1^{r+1}, \eta_2^{r+1}, \cdots, \eta_\nu^{r+1}, \cdots.$$

We shall consider this projective chain η^{r+1}. Since the form (6) is valid — for any given k — for all ν, therefore, passing to the limit with respect to ν, we have

$$\Delta Y_k^{r+1} = z_k^r, \text{ for any } k. \tag{7}$$

Further, from formula (5) we obtain (passing to the limit on ν) for any k,

$$Y_k^{r+1} = \phi_k Y_{k+1}^{r+1} - \phi_k x_k^{r+1} \tag{8}$$

The desired chain x_k^{r+2}, satisfying the condition

$$\Delta x_k^{r+2} = Y_{k+1}^{r+1} - Y_k^{r+1} - x_k^{r+1}, \tag{9}$$

may now be constructed by the formula

$$x_k^{r+2} = \Pi_k Y_{k+1}^{r+1} - \Pi_k x_k^{r+1}, \tag{10}$$

where Π_k denotes the prism constructed over the corresponding chain under the shift ϕ_k into the nerve α_k.

The condition (9) is indeed fulfilled, since under the shift ϕ_k of the cycle z_k^r remains in place, and accordingly

$$\Delta\Pi_k Y_{k+1}^{r+1} = Y_{k+1}^{r+1} - \phi_k Y_{k+1}^{r+1} - \Pi_k z_{k+1}^r,$$

$$\Delta\Pi_k x_k^{r+1} = x_k^{r+1} - \phi_k x_k^{r+1} - \Pi_k z_{k+1}^r.$$

Subtracting the last equality from the next-to-last, we find, keeping (10) and (8) in mind, that

$$\Delta x_k^{r+2} = Y_{k+1}^{r+1} - (\phi_k Y_{k+1}^{r+1} - \phi_k x_k^{r+1}) - x_k^{r+1} =$$
$$= Y_{k+1}^{r+1} - Y_k^{r+1} - x_k^{r+1},$$

as was to be proved.

5. **Relations among the groups $\Delta_c^r X$, $\Delta^r X$, and the group $\delta^r X$.** If a Δ_c- or Δ-cycle z^r of the space X lies on the compactum $\Phi \subseteq X$, then it may be carried by means of a canonical shift into a projective cycle $\{z_\alpha^r\}$ defined, it is true, not uniquely, but up to a "projective" homology; this was proved in detail in Part I, Chapter III, §7.1 for Δ_c-cycles. Since to each Δ-cycle there uniquely corresponds a certain Δ_c-cycle (the first row of the Sitnikov matrix), the same is valid also for Δ-cycles. Here, to a bounding Δ_c-cycle (Δ-cycle) there corresponds a bounding projective cycle, so that we have a completely defined "natural" homomorphism h of the group $\Delta_c^r X$ and the group $\Delta^r X$ into the group $\delta^r X$.

Frequently it is convenient to replace the group $\Delta_c^r X$ itself (over any coefficient domain $\mathfrak{A}$ or $\mathfrak{B}$) by the group $\delta_c^r X$ isomorphic to it, for which the homomorphism h becomes particularly clear. This is done as follows. In each (star-finite) covering α there are only a finite number of elements intersecting with a given compactum $\Phi \subseteq X$. Let $O_{\alpha_1}, \cdots, O_{\alpha_s}$ be these elements. Put $O'_{\alpha_i} = \Phi \cap O_{\alpha_i}$. Without loss of generality we may assume that the sets $O_{\alpha_1}, \cdots, O_{\alpha_s}$ are numbered so that O'_{α_i} is nonempty for all $i \leq s'_\alpha$ and only for these. We obtain a covering $\Phi\alpha = \{O'_{\alpha_1}, \cdots, O'_{\alpha_{s'_\alpha}}\}$ of the compactum. The elements O'_{α_i} and O'_{α_j} of the covering $\Phi\alpha$ will be considered different for $i \neq j$ even if they coincide geometrically. The nerve of the covering $\Phi\alpha$ is a finite subcomplex of the nerve α. We immediately observe that if $\Phi' \supset \Phi$, then the nerve $\Phi\alpha$ is a subcomplex of the nerve $\Phi'\alpha$. We shall now say that *the projective cycle $z^r = \{z_\alpha^r\}$ lies on the compactum $\Phi \subseteq X$* if the cycle z_α^r for any α lies on the subcomplex $\Phi\alpha$ of the nerve α and if for any $\beta > \alpha$ we have a homology $\omega_\alpha^\beta z_\beta^r \sim z_\alpha^r$ on $\Phi\alpha$. We shall further say that the projective cycle $z^r = \{z_\alpha^r\}$, lying on the compactum Φ, bounds on the compactum $\Phi' \subseteq X$ if $z_\alpha^r \sim 0$ on $\Phi'\alpha$ for any α.

Finally, we shall say that *a projective cycle of the set X is compact* if it lies on some compactum Φ, and that it *bounds compactly* (or is *compactly homo-*

logous to zero) if it is homologous to zero on some compactum $\Phi' \subset X$. The factor-group of all r-dimensional compact cycles of the set X modulo the subgroup of compactly bounding cycles will then be called the group $\delta_c^r X$. It is easy to prove (repeating essentially the discussion of Part I, Chapter III, §7) that there exists an isomorphism, realized by a canonical shift, between the groups $\Delta_c^r X$ and $\delta_c^r X$. Since all compact cycles are projective, and each compactly bounding cycle is a bounding projective cycle, the natural homomorphism h of the group $\delta_c^r X$ into the group $\delta^r X$ is defined automatically.

6. ∇-groups with compact carriers: the groups $\nabla_c^r (X, \mathfrak{A})$. In order to define them, we need first of all to define the excision homomorphism $J_{\phi'}^{\phi}$, for the group $\nabla^r \phi$, where ϕ is a compactum.*

Let $\phi' \subseteq \phi$, where ϕ and ϕ' are compacta. A ∇-cycle of the compactum ϕ is, as we know, any cycle z_α^r of the nerve α of any (finite) covering of that compactum. Two ∇-cycles z_α^r and z_β^r of the compactum ϕ are said to be homologous to each other in ϕ if there exists a covering γ following both α and β and such that $\pi_\gamma^\alpha z_\alpha^r \sim \pi_\gamma^\beta z_\beta^r$ in γ. In order to define the homomorphism $J_{\phi'}^{\phi}$, of the group $\nabla^r \phi'$ into the group $\nabla^r \phi$, called the excision homomorphism, we need for each ∇-cycle z_α^r of the compactum ϕ to define a ∇-cycle

$$z_{\alpha'}^r = J_{\phi'}^{\phi}, z_\alpha^r$$

of the compactum ϕ', and to prove that, under this mapping, to homologous ∇-cycles of the compactum ϕ correspond homologous, in ϕ', ∇-cycles of that compactum.

We consider a covering of the compactum ϕ

$$\alpha = \{O_1, \cdots, O_s\},$$

on whose nerve lies the ∇-cycle z_α^r. The covering α excises from the compactum ϕ' the covering

$$\alpha' = \{O_1', \cdots, O_{s'}'\},$$

whose elements are the non-empty sets among the sets $\phi' \cap O_i$. Here we are supposing that the numeration of the sets $O_1, \cdots, O_s$ is chosen so that among the sets $\phi' \cap O_1, \cdots, \phi' \cap O_s$ the non-empty ones are precisely the first s'. Then the sets $O_i' = \phi' \cap O_i$, $i = 1, 2, \cdots, s'$ (among which some may geometrically coincide) are regarded as distinct elements of the covering α'.

Putting into correspondence with each O_i' the corresponding (i.e., furnished with the same index) set O_i, we obtain an isomorphic simplicial mapping $E_\phi^{\phi'}$ of the nerve α' onto some subcomplex of the nerve α, so that to each simplex

---

§4.   * The definition of the groups $\nabla^r \phi$ for compacta ϕ was given in Part I, Chapter III,

$t'_{\alpha'}$ of the nerve α' there corresponds a simplex t_α (of the same dimension number) of the nerve α. In view of this isomorphism, to each chain x^r_α of the nerve α there corresponds a chain $x'^r_{\alpha'} = J^\phi_{\phi'} x^r_\alpha$ of the nerve α', according to the formula

$$(x'^r_{\alpha'} \cdot t'^r_{\alpha'}) = (x^r_\alpha \cdot t^r_\alpha),$$

while the operator $J^\phi_{\phi'}$ commutes with the operator ∇, so that to the ∇-cycle x^r_α corresponds the ∇-cycle $x'^r_{\alpha'}$ (see Part I, Chapter III, §1). Here, to two ∇-cycles z^r_α and z^r_β which are homologous to one another in ϕ' there correspond ∇-cycles $z'^r_{\alpha'}$ and $z'^r_{\beta'}$, homologous to each other in ϕ', since if

$$\nabla x^{r-1}_\gamma = \pi^\alpha_\gamma z^r_\alpha - \pi^\beta_\gamma z^r_\beta,$$

then

$$\nabla x'^{r-1}_{\gamma'} = \pi^{\alpha'}_{\gamma'} z'^r_{\alpha'} - \pi^{\beta'}_{\gamma'} z'^r_{\beta'}.$$

Thus, we have defined a homomorphism $J^\phi_{\phi'}$ of the group $\nabla^r \phi$ into the group $\nabla^r \phi'$, called the excision homomorphism; accordingly, if one considers the set of all compacta ϕ, lying in the given space X and ordered by inclusion, we have an inverse spectrum of discrete groups:

$$\{\nabla^r(\phi, \mathfrak{A}), J^\phi_{\phi'}\},$$

whose limit group is called *the r-dimensional ∇-group with compact carriers of the set X*, or the group $\nabla^r_c(X, \mathfrak{A})$. This group is defined for any discrete coefficient domain $\mathfrak{A}$ and is itself a discrete group.

A natural question arises: do the character groups of the groups $\delta^r(X, \mathfrak{A})$, $\nabla^r(X, \mathfrak{A})$; $\Delta^r(X, \mathfrak{A})$, $\Delta^r_c(X, \mathfrak{A})$; $\nabla^r_c(X, \mathfrak{A})$ introduced up to now have an independent geometric meaning? Unfortunately, for all the enumerated groups other than the first and last, this question remains open. As to the groups $\delta^r(X, \mathfrak{A})$ and $\nabla^r_c(X, \mathfrak{A})$, their character groups are, for the former, a singular bicompact ∇-group over a bicompact coefficient domain $\mathfrak{B} | \mathfrak{A}$, and for the latter a (naturally, also bicompact) Δ-group of singular kind. Both of these groups are defined by means of some algebraic constructions due to Čogošvili, to whose exposition we shall now turn.

§3. A theorem of Čogošvili.

The groups $\overline{\Delta}^r(X, \mathfrak{B})$, $N^r_\Delta(X, \mathfrak{B})$ and $\overline{\delta}^r(X, \mathfrak{B})$

1. **The bicompact limit of the direct spectrum of bicompact groups. Čogošvili's theorem.** We begin by putting into correspondence with each direct spectrum of bicompact topological groups

$$\{Y_\alpha, \pi^\alpha_\beta\}, \tag{1}$$

in which all the projections π^α_β are continuous homomorphisms, a certain uniquely defined, bicompact group $\overline{Y}$, which we shall call the *bicompact limit of the direct spectrum* (1). Generally speaking, this limit is, of course, different from the usual "algebraic" limit, which is a discrete group, and is obtained if one considers the groups Y_α to be without topology (see Part I, Chapter II, subsection 3). The theorem of Čogošvili, which will be formulated and proved below, gives us a better understanding of the relationships existing between the groups Y_α constituting the spectrum (1) and the bicompact limit $\overline{Y}$ of that spectrum.

Thus, suppose that we are given the spectrum (1). Denote by X_α the discrete character group of the bicompact group Y_α, and by ω_α^β the homomorphism of the group X_β into the group X_α, adjoint to the continuous homomorphism π_β^α of the group Y_α into the group Y_β (see Part I, Chapter II). We obtain the inverse spectrum

$$\{X_\alpha, \omega_\alpha^\beta\}$$

of discrete groups, whose limit group (see Part I, Chapter II, subsection 4) is a discrete group X.

Definition. *The bicompact limit of the direct spectrum* (1) *of bicompact groups is called the bicompact character group of the group X, and is denoted by $\overline{Y}$:*

$$\overline{Y} = lb\{Y_\alpha, \pi_\beta^\alpha\}. \tag{2}$$

We shall denote by Y the algebraic limit of the direct spectrum (1). The Čogošvili theorem establishes a connection between the groups Y and $\overline{Y}$; in this theorem it is proved that a certain factor group Y' of the group Y is a subgroup, and indeed an everywhere dense subgroup of the group $\overline{Y}$. In order to define the subgroup $Y_0 \subset Y$ over which Y' is a factor-group, we observe that between any two elements

$$x \in X = \varprojlim \{X_\alpha, \omega_\alpha^\beta\}$$

and

$$y \in Y = \varinjlim \{Y_\alpha, \pi_\beta^\alpha\}$$

there is naturally defined a scalar product satisfying the usual conditions. Indeed, we recall that y is some coset of elements y_α from the various groups Y_α, while x is a thread

$$x = \{x_\alpha\},$$

passing through all the groups X and containing one element from each group X_α and satisfying the condition $x_\alpha = \omega_\alpha^\beta x_\beta$ for $\beta > \alpha$. Let us choose some $y_\alpha \in y$, $y \in Y$, and choose for that α the unique element $x_\alpha \in x$. Since $X_\alpha | Y_\alpha$, the scalar product $(x_\alpha \cdot y_\alpha)$ is defined. We put

$$(x \cdot y) = (x_\alpha \cdot y_\alpha).$$

We shall show that this definition does not depend on the particular choice of the arbitrary element contained in it — i.e., of the choice of $y_\alpha \in y$. Indeed, suppose

$y_\beta \in y$. Then there exists a γ, following both α and β, such that

$$\pi_\gamma^\alpha y_\alpha = \pi_\gamma^\beta y_\beta = y_\gamma \in y$$

and

$$(x_\gamma, y_\gamma) = (x_\gamma \cdot \pi_\gamma^\alpha y_\alpha) = (\varpi_\alpha^\gamma x_\gamma \cdot y_\alpha) = (x_\alpha \cdot y_\alpha).$$

In the same way

$$(x_\gamma \cdot y_\gamma) = (x_\beta \cdot y_\beta),$$

so that

$$(x_\alpha \cdot y_\alpha) = (x_\beta \cdot y_\beta)$$

We now denote by Y_0 the subgroup of the group Y consisting of all those elements $y \in Y$ which have a null scalar product with each element $x \in X$. We define Y' as a factor-group modulo the group Y_0:

$$Y' = Y - Y_0.$$

Obviously, the scalar product defined for any $y \in Y$ and $x \in X$ automatically generates a scalar product for any $y' \in Y'$ and $x \in X$, since each element $y' \in Y'$ is a character of the group X. It is clear also that two different elements y_1' and y_2' of the group Y' are different characters of the group X (since in the contrary case their difference $y_1' - y_2'$ would have a null scalar product with all the elements of the group X, i.e., it would be the null element Y_0 of the group Y').

Thus, the group Y' is a subgroup of the group $\overline{Y}$.

Now we may formulate the Čogošvili theorem and prove it.

Čogošvili's theorem. *The group Y' is an everywhere dense subgroup of the bicompact group $\overline{Y}$.*

Proof. Suppose that the subgroup Y' of the group $\overline{Y} \mid X$ is not everywhere dense in $\overline{Y}$. Then the closure $[Y'] \subset \overline{Y}$ of the group Y' in $\overline{Y}$ does not coincide with $\overline{Y}$, so that the annihilator in X of the group $[Y']$ is a subgroup $H \subseteq X$ distinct from zero. Choose an element $x = \{x_\alpha\} \in H$ distinct from zero. Then $(x \cdot y') = 0$ for all $y' \in Y'$. But, taking in the thread $x \neq 0$ an element $x_\alpha \neq 0$, we may fit to it some element $y_\alpha \in Y_\alpha$ for which $(x_\alpha \cdot y_\alpha) \neq 0$, and taking the coset $y \in Y$ containing the element y_α, and the class $y' \in Y'$ containing y, we obtain in turn $(x \cdot y') = (x \cdot y) = x_\alpha \cdot y_\alpha \neq 0$, which contradicts the fact that $x \in H$.

This contradiction proves the Čogošvili theorem.

2. **The group $\overline{\Delta}^r(X, \mathfrak{B})$. The first non-linkability group $N_\Delta^r(X, \mathfrak{B})$.** Let X be any space. Consider the set of all compacta Φ lying in X, directed by inclusion. For each compactum $\Phi \subset X$ we choose the group $\delta^r(\Phi, \mathfrak{B}) = \Delta_c^r(\Phi, \mathfrak{B}) =$

$\Delta^r(\Phi, \mathfrak{B})$ with the topology defined in Part I, Chapter VI, $\S 1.$ [*] We obtain the direct spectrum of bicompact groups

$$\{\delta^r(\Phi, \mathfrak{B}), E_{\Phi}^{\Phi'}\}, \quad \Phi' \subset \Phi$$

whose bicompact limit is by definition the group

$$\overline{\Delta}^r(X, \mathfrak{B}) = lb\{\delta^r\Phi, E_{\Phi}^{\Phi'}\}.$$

In order to understand its geometric meaning, we recall first of all that the groups $\delta^r(\Phi, \mathfrak{B})$ and $\nabla_c^r(\Phi, \mathfrak{A})$ are dual to each other (see Part I, Chapter VI, $\S 1$), and that this duality is given by a scalar product, defined on the nerves [**] α and then for $x_{\alpha_0} \in x \in \nabla^r(\Phi, \mathfrak{A}); \ y_{\alpha_0} \in \{y_\alpha\} \in y \in \delta^r(\Phi, \mathfrak{B})$ by the formula

$$(x \cdot y) = (x_{\alpha_0} \cdot y_{\alpha_0}).$$

Now we observe that if $\Phi' \subset \Phi$ the homomorphism $E_{\Phi}^{\Phi'}$ of the group $\delta^r\Phi'$ into the group $\delta^r\Phi$ and the homomorphism $J_{\Phi'}^{\Phi}$ of the group $\nabla^r\Phi$ into the group $\nabla^r\Phi'$ are adjoint to each other. Indeed, we need to show that for any $x \in \nabla^r\Phi, \ y' \in \delta^r\Phi'$, we have

$$(x \cdot E_{\Phi}^{\Phi'} y') = (J_{\Phi'}^{\Phi} x \cdot y').$$

To this end we choose $x_\alpha \in x$ and $y'_\alpha \in \{y'_\alpha\} \in y' \in \delta^r\Phi'$. Then (using the notations of $\S 2.6$) we have $(x \cdot E_{\Phi}^{\Phi'} y') = (x_\alpha \cdot y'_\alpha) = (x'_{\alpha'} \cdot y'_{\alpha'}) = (J_{\Phi'}^{\Phi} x \cdot y')$, from which the assertion follows.

Now we see that the inverse spectrum adjoint to the spectrum (1) is nothing other than the spectrum

$$\{\nabla^r(\Phi, \mathfrak{A}), J_{\Phi'}^{\Phi}\} \tag{2}$$

whose limit group is the group $\nabla_c^r(X, \mathfrak{A})$. Therefore the bicompact limit of the spectrum (1), i.e., the group $\overline{\Delta}^r(X, \mathfrak{B})$, is the character group of the group $\nabla_c^r(X, \mathfrak{A})$, so that we have the duality

$$\nabla_c^r(X, \mathfrak{A}) \,|\, \overline{\Delta}^r(X, \mathfrak{B}).$$

In order to get a deeper understanding of the meaning of this duality, we apply the

---

[*] In the place noted the groups $\delta^r\Phi$ were mistakenly denoted by $\Delta^r\Phi$. However this error, as we now see, is immaterial: in Chapter VI of Part I we consider only the group $\Delta^r\Phi$ over the bicompact coefficient domain $\mathfrak{B}$, and in this case $\Delta^r(\Phi, \mathfrak{B}) = \delta^r(\Phi, \mathfrak{B})$. Moreover, for any coefficient domain $\delta^r\Phi = \Delta_c^r\Phi$, so that in our case $\Delta^r\Phi = \delta^r\Phi = \Delta_c^r\Phi$.

[**] Indeed, on each finite complex α we have a duality $L^r(\alpha, \mathfrak{A}) \,|\, L^r(\alpha, \mathfrak{B})$ of the groups of chains, given by the scalar product. This duality (as was proved in Part I, Chapter VI, $\S 1$, footnote on pages 85–87) generates a duality $\nabla^r(\alpha, \mathfrak{A}) \,|\, \Delta^r(\alpha, \mathfrak{B})$, from which it follows in particular that, taking arbitrarily $x \in \xi^r \in \nabla^r(\alpha, \mathfrak{A})$ and $y \in \eta^r \in \Delta^r(\alpha, \mathfrak{B})$, we obtain always one and the same $(x \cdot y) = (\xi \cdot \eta)$.

Now suppose that $y_{\alpha_0} \in \{y_\alpha\} \in \delta^r\Phi$, $x_\alpha \in x \in \nabla^r\Phi$. In order to prove the legality of the definition $(x \cdot y) = (x_\alpha \cdot y_\alpha)$, i.e., the independence of this product from the choice $x_\alpha \in \nabla^r\Phi$, it is sufficient, as always in such cases, to show that for $x_\beta = \pi_\beta^\alpha x_\alpha$ we have $(x_\beta \cdot y_\beta) = (x_\alpha \cdot y_\alpha)$. But $(x_\beta \cdot y_\beta) = (\pi_\beta^\alpha x_\alpha \cdot y_\beta) = (x_\alpha \cdot \omega_\alpha^\beta y_\beta) = (x_\alpha \cdot y_\alpha)$.

Čogošvili theorem. The algebraic limit of the spectrum (1) is, evidently, the group

$$\Delta^r_c(X, \mathfrak{B}) = \Delta^r(X, \mathfrak{B}).$$

We denote by $N^r_\Delta(X, \mathfrak{B})$ the subgroup of this group consisting of those elements which have a null scalar product with all the elements of the group $\nabla^r_c(X, \mathfrak{A})$. This scalar product has a very simple meaning: any element y of the group $\Delta^r_c X$ is defined by some element y_Φ of the group $\delta^r \Phi$ for some $\Phi \subset X$, while the element $x \in \nabla^r_c X$ is a thread $x = \{x_\Phi\}$, $x_\Phi \in \nabla^r \Phi$. Then, as we have seen, $(x \cdot y) = (x_\Phi \cdot y_\Phi)$.

The group $N^r_\Delta(X, \mathfrak{B})$ has a very considerable significance in the combinatorial topology of non-closed sets; for reasons that will become clear in this same section, it is called *the nerve of the (r-dimensional) nonlinkability group of the set*.

In view of the Čogošvili theorem the factor-group

$$\Delta'^r(X, \mathfrak{B}) = \Delta^r_c(X, \mathfrak{B}) - N^r_\Delta(X, \mathfrak{B}) \tag{3}$$

is an everywhere dense subgroup of the group $\overline{\Delta}^r(X, \mathfrak{B})$. As is known, if a given topological group G is an everywhere dense subgroup of the bicompact group $\overline{G}$, then the group $\overline{G}$ is the completion of the group G and is therefore uniquely defined. Therefore, up to an isomorphism, we may realize not more than one bicompact group having the given topological group G as an everywhere dense subgroup.

Thus, the bicompact group $\overline{\Delta}^r(X, \mathfrak{B})$ may be defined as the completion of the group (3); here the group (3) is taken with the topology which arises from the fact that its elements are (in view of the scalar product) characters of the group $\nabla^r_c(X, \mathfrak{A})$.

The group $\overline{\Delta}^r(X, \mathfrak{B})$ may be defined also as the unique (up to an isomorphism) bicompact group containing the group $\Delta'^r(X, \mathfrak{B})$ as an everywhere dense subgroup.

In order to clear up completely the geometric meaning of the group $\overline{\Delta}^r(X, \mathfrak{B})$, it remains to deal with the group $N^r_\Delta(X, \mathfrak{B})$. We now turn to this task.

3. **The geometrical meaning of the nonlinkability group** $N^r_\Delta(X, \mathfrak{B})$. **Sliding cycles.** Up to this point we have been considering an arbitrary space X. Here we shall suppose that the dimension of this space is finite, which is equivalent to the assumption that X is a set lying in some S^n.

Now suppose we are given $A \subset S^n$, $B = S^n - A$. An arbitrary compactum, lying in A, will be denoted by ϕ, and we shall denote by μ its complement in S^n:

$$\mu = S^n - \phi.$$

Evidently, if ϕ runs through the set of all compacta lying in A, then μ runs through the set of all neighborhoods of the set B. Then, by the Pontrjagin duality law

$$\Delta_c^p(\phi, \mathfrak{B}) = \delta^p(\phi, \mathfrak{B}) \mid \Delta_c^q(\mu, \mathfrak{A}), \tag{1}$$

while this duality is a linking duality: for $x_\phi \in \Delta_c^p(\phi, \mathfrak{B})$, $y_\mu \in \Delta_c^q(\mu, \mathfrak{A})$, and any cycles $z_\phi^p \in x_\phi$, $z_\mu^q \in y_\mu$, we have

$$(x_\phi \cdot y_\mu) = \mathfrak{v}(z_\phi^p, z_\mu^q), \tag{1'}$$

where $\mathfrak{v}$ is the linking coefficient (see Part I, Chapter VI, §5). If $\phi' \subset \phi$, then $\mu' \supset \mu$ and the inclusion homomorphisms $E_\phi^{\phi'}$ of the group $\Delta_c^p\phi'$ into $\Delta_c^p\phi$ and $E_{\mu'}^\mu$ of the group $\Delta_c^q\mu$ into the group $\Delta_c^q\mu'$ are adjoint. Therefore the inverse spectrum

$$\{\Delta_c^q\mu, E_{\mu'}^\mu\}, \tag{2}$$

where $\mu > \mu'$ if $\mu \subset \mu'$, is a spectrum adjoint to the direct spectrum

$$\{\Delta_c^p\phi, E_\phi^{\phi'}\}, \tag{3}$$

where $\phi > \phi'$ if $\phi \supset \phi'$. Therefore putting for the coefficient domain $\mathfrak{A}$

$$D^q B = \varprojlim (\Delta_c^q\mu, E_{\mu'}^\mu) \tag{4}$$

we have by the definition of the group $\overline{\Delta}^p(A, \mathfrak{B})$ a duality

$$D^q(B, \mathfrak{A}) \mid \overline{\Delta}^p(A, \mathfrak{B}), \tag{5}$$

which means also an isomorphism

$$D^q(B, \mathfrak{A}) = \nabla_c^p(A, \mathfrak{A}). \tag{6}$$

Now we return to our purpose: namely, to clarify the geometrical significance of the group $N_\Delta^p(A, \mathfrak{B})$. This group is defined as a subgroup of the group

$$\Delta_c^p(A, \mathfrak{A}) = \varinjlim(\Delta_c^p\phi, E_\phi^{\phi'}),$$

consisting of those elements whose scalar product with any element of the group $D^q(B, \mathfrak{A})$ is equal to zero. In the case at hand this scalar product has a particularly simple meaning. Indeed, let us give the name "sliding cycle" to any system ("thread") of finite, polyhedral cycles z_μ^q, taken one in each neighborhood μ of the set B and satisfying the condition: if $\mu \subset \mu'$, then $z_\mu^q \sim z_{\mu'}^q$ in μ'. Under termwise addition

$$\{z_\mu^{(1)q}\} + \{z_\mu^{(2)q}\} = \{z_\mu^{(1)q} + z_\mu^{(2)q}\}$$

the sliding cycles form a group $Z_{sl}^q B$. In this group there is contained a subgroup $H_{sl}^q B$, consisting of bounding sliding cycles $\{z_\mu^q\}$ (satisfying the condition: $z_\mu^q \sim 0$ in μ for any μ).

The group $D^q B$ is (up to a natural isomorphism) nothing other than the factor group

$$Z_{sl}^q B - H_{sl}^q B.$$

For any sliding cycle $z_{sl}^q = \{z_\mu^q\}$ of the set B and any Δ_c-cycle z^p of the set A, we may define as follows a linking coefficient $\mathfrak{o}(z^P, z^q_{\mathrm{sl}})$: the cycle z^P lies on some compactum ϕ_0; then for all $\mu > \mu_0$ (i.e., $\mu \subset \mu_0$) the linking coefficient $\mathfrak{o}(z^P, z^q_\mu)$ is defined and (in view of the homology $z^P_\mu \sim z^q_{\mu_0}$ in μ_0) has one and the same value, equal to $\mathfrak{o}(z^P, z^q_{\mu_0})$. This common value will be taken as the linking coefficient $\mathfrak{o}(z^P, z^q_{\mathrm{sl}})$. On the other hand, if $z^q_{\mathrm{sl}} = \{z^q_\mu\} \in \eta \in D^q(B, \mathfrak{A})$ and $z^P_{\phi_0} \in \xi \in \Delta^P_c(A, \mathfrak{B})$, then by the definition of the groups $\Delta^P_c A$, $D^q B$ and of the scalar product of their elements, we have

$$(\xi \cdot \eta) = (x_{\phi_0} \cdot y_{\mu_0}) = \mathfrak{o}(z^P_{\phi_0}, z^q_{\mu_0}) = \mathfrak{o}(z^P, z^q_{\mathrm{sl}}).$$

In other words: *the scalar product of two elements* $\xi \in \Delta^P_c(A, \mathfrak{B})$ *and* $\eta \in D^q_{\mathrm{sl}}(B, \mathfrak{A})$ *is equal to the linking coefficient of any* Δ_c-*cycle* $z^P \in \xi$ *with any sliding cycle* $z^q_{\mathrm{sl}} \in \eta$.

Therefore the nonlinkability group $N^P_\Delta(A, \mathfrak{B})$ *is a subgroup of the group* $\Delta^P_c(A, \mathfrak{B})$, *whose elements are defined by the* Δ_c-*cycles of the set* A *which have a zero linking coefficient with every sliding cycle of the set* B. This proposition, which justifies calling the group a nonlinkability group, may be taken as its definition. The second definition of the nonlinkability group obtained in this way is very easy to visualize but is non-invariant in form. Thus, its actually expressing a topological invariant is a significant geometrical fact, which results from the equivalence of the second definition of the group $N^P_\Delta(A, \mathfrak{B})$ to the first. In what follows we shall see that the nonlinkability group $N^P_\Delta(A, \mathfrak{B})$ is not only invariant but is dualizable (i.e., may be expressed through topological invariants of the set B).

Remark 1. We shall call a Δ_c-cycle of the set A (over the coefficient domain $\mathfrak{B}$) *nonlinkable* if its linking coefficient with every sliding cycle of the set B (over the coefficient domain $\mathfrak{A}$) is equal to zero. The nonlinkable cycles of a given dimension p form a group $Z^P_0(A, \mathfrak{B})$, containing the subgroup of all the bounding cycles. From the general theorem of E. Noether "on isomorphisms" it follows that the group $N^P_\Delta(A, \mathfrak{B})$ is isomorphic to the factor-group $Z^P_0(A, \mathfrak{B}) - H^P_c(A, \mathfrak{B})$:

$$N^P_\Delta(A, \mathfrak{B}) = Z^P_0(A, \mathfrak{B}) - H^P_c(A, \mathfrak{B}).$$

Remark 2. In view of the Pontrjagin duality law, every cycle (over the coefficient domain $\mathfrak{B}$) lying on the compactum $A \subset S^n$ and not homologous to zero on A links with some cycle of the open set B, so that, in the case of compacta A, we have $Z^P_0(A, \mathfrak{B}) = H^P_c(A, \mathfrak{B})$,

$$N^P_\Delta(A, \mathfrak{B}) = 0.$$

At the same time, for non-closed point sets, the appearance of nonlinkability is encountered quite frequently. Choose, for example, a set A (on the plane completed by the point at infinity) consisting of the graph of the curve $y = \sin\dfrac{1}{x}$ for

$0 < |x| < \dfrac{1}{\pi}$ and of all the rational points of the interval $-1 < y < 1$ of the ordinate axis. Any pair of these last points defines in A a zero-dimensional cycle, non-homologous to zero in A and at the same time not linking. In the same way, any pair of irrational points of the interval $-1 < y < 1$ of the ordinate axis forms a nonlinkable cycle of the set B. It is necessary to point out that both of the groups $D^1 A$ and $D^1 B$ in our case are null-groups; thus, the zero-dimensional Δ_c-groups of both sets A and B coincide with the nonlinkability groups (which are not only different from zero but are even noncountable). The presence of non-linkability groups represents one of the most significant phenomena in the combinatorial topology of non-closed sets. It has no analogue in the topology of compacta.

4. **Formulation of the invariance theorem and the first general duality law.** In the second chapter we shall prove that *the groups $D^p(A, \mathfrak{A})$ and $\delta^p(A, \mathfrak{A})$ are isomorphic to each other*. From this *"invariance theorem"* it follows, in view of the isomorphism $\nabla_c^q(B, \mathfrak{A}) = D^p(A, \mathfrak{A})$, already proved by us in subsections 2 and 3, and the duality $\nabla_c^q(B, \mathfrak{A}) \mid \overline{\Delta}^q(B, \mathfrak{B})$, that the following law holds.

First general duality law:

$$
\begin{cases}
\text{in } \nabla\text{-form: the isomorphism } \delta^p(A, \mathfrak{A}) = \nabla_c^q(B, \mathfrak{A}), \\[2mm]
\text{in } \Delta\text{-form: The duality } \delta^p(A, \mathfrak{A}) \mid \overline{\Delta}^q(B, \mathfrak{B}).
\end{cases}
$$

The reader may now immediately proceed to the proof of the theorem of invariance in the second chapter, setting aside the reading of the following subsection (on the groups $\overline{\delta}^r(X, \mathfrak{B})$) until these groups are used (in Chapter IV).

5. **The bicompact projection group $\overline{\delta}^r(X, \mathfrak{B})$.** This was defined by K. Sitnikov. It allowed him to establish a duality of the groups Δ_c^p and a series of other groups of point sets. The group $\overline{\delta}^r(X, \mathfrak{B})$ is defined as follows. For the nerve of any star-finite covering α of the set X we define the group $\Delta^r(\alpha, \mathfrak{B})$, *based on finite chains*; we consider it without topology. In the group $\nabla^r(\alpha, \mathfrak{A})$, *based on infinite chains*, we consider the subgroup $N_{\nabla}^r(\alpha, \mathfrak{A})$, consisting of all the elements which have a zero scalar product with all the elements of the group $\Delta^r(\alpha, \mathfrak{B})$. In an analogous way we choose in the group $\Delta^r(\alpha, \mathfrak{B})$ a subgroup $N_{\Delta}^r(\alpha, \mathfrak{B})$, consisting of all the elements having a zero scalar product with every element of the group $\nabla^r(\alpha, \mathfrak{A})$. Each element of the group $\Delta'^r(\alpha, \mathfrak{B}) = \Delta^r(\alpha, \mathfrak{B}) - N_{\Delta}^r(\alpha, \mathfrak{B})$ is a character of the group $\nabla'^r(\alpha, \mathfrak{A}) = \nabla^r(\alpha, \mathfrak{A}) - N_{\nabla}^r(\alpha, \mathfrak{A})$, while different elements are different characters. Therefore the group $\Delta'^r(\alpha, \mathfrak{B})$ is a subgroup of the bicompact character group $\overline{\delta}^r(\alpha, \mathfrak{B})$ of the group $\nabla'^r(\alpha, \mathfrak{A})$ and receives its topology from it. We shall show that $\Delta'^r(\alpha, \mathfrak{B})$ is an everywhere dense subgroup of the group $\overline{\delta}^r(\alpha, \mathfrak{B})$. To this end it is sufficient to prove that the annihilator in $\nabla'^r(\alpha, \mathfrak{B})$ of the closure of the groups $\Delta'^r(\alpha, \mathfrak{B})$ in $\overline{\delta}^r(\alpha, \mathfrak{B})$ is the null group.

But we shall prove still more, namely that for any element $\xi \in \nabla'^r(\alpha, \mathfrak{A})$ distinct from zero there is an element $\eta \in \Delta'^r(\alpha, \mathfrak{B})$ for which $(\xi \cdot \eta) \neq 0$. This last assertion results from the fact that if $\xi \neq 0$, then for any element x of the equivalence class ξ there exists an element $y \in \Delta^r(\alpha, \mathfrak{B})$ such that $(x \cdot y) \neq 0$. Denoting the equivalence class of the element y by $\eta \in \Delta'^r(\alpha, \mathfrak{B})$, we obtain $(\xi \cdot \eta) \neq 0$, as we were required to prove.

We consider the direct spectrum

$$\{\nabla^r(\alpha, \mathfrak{A}), \pi_\beta^\alpha\}. \tag{1}$$

It is easy to see that

$$\pi_\beta^\alpha N_\nabla^r(\alpha, \mathfrak{A}) \subseteq N_\nabla^r(\beta, \mathfrak{A}). \tag{1'}$$

Indeed, if $x_\beta \in \pi_\beta^\alpha N_\nabla^r(\alpha, \mathfrak{A})$, then for any $y_\beta \in \Delta^r(\beta, \mathfrak{B})$ we have (taking $x_\alpha \in N_\nabla^r(\alpha, \mathfrak{A})$ so that $x_\beta = \pi_\beta^\alpha x_\alpha$)

$$(x_\beta \cdot y_\beta) = (\pi_\beta^\alpha x_\alpha \cdot y_\beta) = (x_\alpha \cdot \omega_\alpha^\beta y_\beta) = 0,$$

from which it follows that $x_\beta \in N_\nabla^r(\beta, \mathfrak{A})$.

It follows from (1') that the spectrum (1) defines a direct spectrum

$$\{N_\nabla^r(\alpha, \mathfrak{A}), \pi_\beta^\alpha\}, \tag{2}$$

whose limit group we shall denote by

$$N_\nabla^r(A, \mathfrak{A}) = \varinjlim \{N_\nabla^r(\alpha, \mathfrak{A}), \pi_\beta^\alpha\}. \tag{3}$$

One may also speak of the spectrum formed by the factor-groups

$$\nabla'^r(\alpha, \mathfrak{A}) = \nabla^r(\alpha, \mathfrak{A}) - N_\nabla^r(\alpha, \mathfrak{A}),$$

with projections which are naturally defined by the projections in (1) and therefore also denoted by π_β^α. It is easy to see that (up to a naturally defined isomorphism) the limit group of the spectrum

$$\{\nabla'^r(\alpha, \mathfrak{A}), \pi_\beta^\alpha\} \tag{4}$$

is

$$\nabla'^r(A, \mathfrak{A}) = \nabla^r(A, \mathfrak{A}) - N_\nabla^r(A, \mathfrak{A}). \tag{4'}$$

Now we shall prove that the groups $\overline{\delta}^r(\alpha, \mathfrak{B})$ join together into an inverse spectrum. Indeed, if $\beta > \alpha$, the projection ω_α^β generates a homomorphism, of the same name, of the group $\Delta^r(\beta, \mathfrak{B})$ into the group $\Delta^r(\alpha, \mathfrak{B})$, while $\omega_\alpha^\beta N_\Delta^r(\beta, \mathfrak{B}) \subseteq N_\Delta^r(\alpha, \mathfrak{B})$ (which is proved in a way quite analogous to the way in which the inclusion (1') was proved). Therefore we have defined a continuous homomorphism of the group

$$\Delta'^r(\beta, \mathfrak{B}) \text{ into } \Delta'^r(\alpha, \mathfrak{B}),$$

denoted, naturally, also by ω_α^β. This homomorphism, extended by continuity to the whole group $\overline{\delta}^r(\beta, \mathfrak{B})$, gives us the desired projection ω_α^β of the group

$\overline{\delta}^{\,r}(\beta, \mathfrak{B})$ into $\overline{\delta}^{\,r}(\alpha, \mathfrak{B})$, so that we have the inverse spectrum

$$\{\overline{\delta}^{\,r}(\alpha, \mathfrak{B}), \, \varpi^{\beta}_{\alpha}\}, \tag{5}$$

whose limit group is then called the group $\overline{\delta}^{\,r}(X, \mathfrak{B})$

$$\overline{\delta}^{\,r}(X, \mathfrak{B}) = \varinjlim \{\overline{\delta}^{\,r}(\alpha, \mathfrak{B}), \, \varpi^{\beta}_{\alpha}\}.$$

From duality

$$\nabla'^{r}(\alpha, \mathfrak{A}) \,|\, \overline{\delta}^{\,r}(\alpha, \mathfrak{B})$$

and from the fact that the homomorphisms π^{α}_{β} and ϖ^{β}_{α} in the spectra (4) and (5) are adjoint, it follows that the limit groups of these spectra are also dual to one another. In other words, we have the duality

$$\nabla'^{r}(X, \mathfrak{A}) \,|\, \overline{\delta}^{\,r}(X, \mathfrak{B}). \tag{6}$$

6. **Concluding remarks. Formulation of the central duality law of Sitnikov and his duality theorems for the groups** Δ^{r}_{c}. By now we have become accustomed to the following situation: a group dual to some group Δ-type turns out to be a ∇-group and conversely. Thus, for example, we have seen that

$$\nabla^{r}_{c}(X, \mathfrak{A}) \,|\, \overline{\Delta}^{r}(X, \mathfrak{B}).$$

Now we shall see that a ∇-group dual to the group $\overline{\delta}^{\,r}(X, \mathfrak{B})$ is not the whole group $\nabla^{r}(X, \mathfrak{A})$ but only its factor-group $\nabla'^{r}(X, \mathfrak{A})$. This forces us to regard the group $\overline{\delta}^{\,r}(X, \mathfrak{B})$ as less rich than the group $\nabla^{r}(X, \mathfrak{A})$. One should however remark that this statement "to every Δ-group there corresponds some dual ∇-group and conversely" is not corroborated in two very important cases: we do not know what geometrical meaning the group dual to the group $\nabla^{r}(X, \mathfrak{A})$, or the group dual to the group $\Delta^{r}(X, \mathfrak{A})$, may have. For a number of reasons, which will become clearer in the following exposition, we are inclined to regard the groups $\nabla^{r}X$ and $\Delta^{r}X$ as the richest among groups of ∇- and Δ-type respectively. *The central duality law of Sitnikov*, whose proof constitutes one of the fundamental purposes of this article, asserts that *for mutually complementary sets A and B in S^{n} we have the isomorphism*

$$\nabla^{p}(A, \mathfrak{A}) = \Delta^{q}(B, \mathfrak{A}).$$

However, no connection between the groups $\nabla^{r}X$ and $\Delta^{r}X$ for one and the same set is at present known. The answer to the following natural question is also unknown. Do the groups $\Delta^{r}(X, \mathfrak{A})$, taken in all dimensions r, and if necessary over all coefficient domains $\mathfrak{A}$, define the groups $\nabla^{r}X$?

We remark in closing, that, as we shall prove in Chapter IV, *the groups $\Delta^{p}_{c}(A, \mathfrak{A})$ and $\overline{\delta}^{\,q}(B, \mathfrak{B})$ are dual to each other, so that each of these groups is dualizable.*

CHAPTER II

First general duality law

Introductory remarks

In Chapter I, §3.4, the first general duality law was formulated in two equivalent forms: in the duality form (Δ-form) and in the isomorphism form (∇-form), namely,

$$\Delta\text{-form: } \delta^p(A, \mathfrak{A}) \mid \overline{\Delta}^q(B, \mathfrak{B}),$$

$$\nabla\text{-form: } \delta^p(A, \mathfrak{A}) = \nabla^q_c(B, \mathfrak{A}),$$

while the proof of the theorem in these two forms was reduced to the proof of the "invariance theorem," expressed in the form of an isomorphism

$$\delta^p(A, \mathfrak{A}) = D^p(A, \mathfrak{A}). \tag{1}$$

The proof of the invariance theorem constitutes the content of the first section of this chapter. The second and third sections are devoted to special cases of the duality law (in particular, to the proof of the dualizability of the number of components of an arbitrary point set, and also to the so-called second duality law for closed sets in S^n) and to various remarks connected with these.

§1. Proof of the invariance theorem

1. **Triangulations and coverings adjoint to them.** In the first part we considered exclusively finite complexes, in particular, triangulations. Now we shall assume that the complexes considered by us (i.e., the sets of open simplexes lying in the given S^n) satisfy the following significantly weaker requirement:

Condition of local finiteness. [*] *Each point contained in any simplex of the given complex K (lying in S^n) has a neighborhood, relative to all of S^n, which intersects with only finitely many simplexes of the complex K.*

A triangulation is a complex τ, consisting of pairwise nonintersecting open simplexes of various dimensions lying in the given S^n and satisfying, besides the condition of local finiteness, also the condition of completeness: each face of a simplex which is an element of the triangulation τ is also an element of that triangulation.

From these conditions it immediately follows that each triangulation consists of not more than a countable number of simplexes. As we know from the first part, the set-theoretical sum of all the simplexes which are the elements of a given

---

[*] The concept of local finiteness of a system of sets is one of the most important concepts of topology. It was first introduced by me in 1924 in the paper *"Les ensembles de la première classe et les espaces abstraits,"* C. R. Acad. Sci., Paris 178(1924), pp. 185–187.

complex K form the *body* of that complex and is denoted by $\widetilde{K}$.

Each set X which is the body of some triangulation τ is said to be a *polyhedron*, while the triangulation τ is said to be the triangulation of the polyhedron $\widetilde{\tau} = X$. Since they form a special case of point sets, it goes without saying that the polyhedra get their topologies from the enveloping space S^n. An evident consequence of the local finiteness of the given complex K is that it is *star finite*: the star $O_K e$ of any vertex e of the complex K, and consequently the star $O_K T$ of each simplex* of the complex K, is a finite subcomplex of the complex K. As in the finite case, one defines open and closed subcomplexes of triangulations (see Part I, Chapter I, §1.3), while the closed subcomplexes of a triangulation coincide with the complete subcomplexes. Therefore, and from the condition of local finiteness, it easily follows that *the body* $\widetilde{K}_0$ *of the subcomplex* K_0 *of the triangulation* τ *is a closed subset of the polyhedron* $\widetilde{\tau}$ *if and only if the subcomplex* K_0 *is a closed (i.e., complete) subcomplex of the triangulation* τ, *or in other words, if and only if it is itself a triangulation.* The passage to the complement gives the analogous condition for open subcomplexes. In particular, *the body of the star of a given triangulation* τ *is a set open in the polyhedron* $\widetilde{\tau}$.

As in the finite case, we shall principally have need not of the stars of the simplexes $T \in \tau$ but rather of their bodies. These we shall call *open stars*. The open star of the simplex $T \in \tau$ will be denoted by $o_\tau T$. Particularly important are the open stars $o_\tau e$ of the vertices of the triangulation τ; we shall call them the principal stars of the triangulation τ. The system of all principal stars of the triangulation τ forms a covering of the polyhedron $\widetilde{\tau}$, whose nerve is the triangulation τ.

We consider any sequence

$$\tau_1, \tau_2, \cdots, \tau_\nu, \cdots$$

of triangulations of the polyhedron X that becomes arbitrarily fine in the sense that the sequence

$$\epsilon_1, \epsilon_2, \cdots, \epsilon_\nu, \cdots,$$

where ϵ_ν is the upper bound of the diameters of the simplexes $T \in \tau_\nu$, converges to zero.

It is not hard to see that by taking the principal stars of all the triangulations τ_ν we obtain a basis for the polyhedron X, considered as a topological space.**

---

* By a star of the simplex T of the complex K we understand the set of all simplexes of that complex which have T as a proper or improper face.

** From this it immediately follows that *all polyhedra are locally compact spaces*; here compactness of the polyhedron is equivalent to the finiteness of some one (and therefore all) of its triangulations.

This result may be strengthened as follows.

Into each covering ω of the polyhedron X we may insert a subordinated covering formed by the principal stars of some triangulation of that polyhedron.

We shall present a proof of this proposition.[*] Let σ be some triangulation of the polyhedron X. If it is finite, i.e., if the polyhedron X is a compactum, then the proposition is proved; it suffices to take a subdivision τ of the triangulation σ, whose fineness (i.e., upper bound of the diameters of its principal stars) is a Lebesgue number for the covering ω. Hence, let σ be an infinite triangulation. Then it may be easily represented as a sum of finite triangulations

$$\sigma = \bigcup_{\nu = 0}^{\infty} \sigma_\nu,$$

where σ_0 consists, for example, of one simplex and all of its faces, and σ_ν may have common and moreover non-principal elements only with $\sigma_{\nu-1}$ and $\sigma_{\nu+1}$. We choose a subdivision τ_0 of the triangulation σ_0, sufficiently fine that its principal stars are subordinate to ω. We suppose that we have a triangulation τ_i which is a subdivision of the triangulation $\bigcup_{0 \le \nu \le i} \sigma_\nu$ and such that its principal stars are subordinated to ω. We subdivide the triangulation σ_{i+1} so that its stars are subordinated to ω and we carry out the corresponding central subdivision of the triangulation τ_i. This subdivision affects only those simplexes of τ_i which lie on certain simplexes of σ_i. As a result we obtain a triangulation τ_{i+1}, which is a subdivision of the complex $\bigcup_{0 \le \nu \le i+1} \sigma_\nu$, while the principal stars of this triangulation are subordinated to ω. Under our process simplexes entering into σ_i after the $(i+1)$st subdivision already remain invariant, so that as a result we obtain a triangulation $\tau = \bigcup_{i=0}^{\infty}$ of the whole polyhedron X, whose principal stars are subordinated to ω.

We observe that one can also prove in an analogous way the so-called Theorem of Runge:

Every set Γ open in S^n is a polyhedron.

For the proof we choose a sequence of triangulations of the sphere S^n, $\nu = 0, 1, 2, \cdots$, τ_ν of unboundedly increasing refinement, each of which is a subdivision of the preceding, while the initial triangulation τ_0 is itself taken so fine that its subcomplex σ_0, consisting of all the simplexes whose closures lie in Γ, is nonempty. After this, we define by induction the triangulation σ_ν consisting of all the simplexes of the triangulation τ_ν not contained in $\tilde{\sigma}_{\nu-1}$ but lying

---

[*] In my paper [10] it was proved for the only case needed by us in the sequel, namely when the polyhedron X is an open set in S^n. The general formulation and proof presented below are due to S. Kaplan. The two proofs of this (essentially quite elementary) theorem differ rather little from each other and appeared approximately at the same time.

along with their faces in Γ. Evidently $\bigcup_{\nu=0}^{\infty} \tilde{\sigma}_\nu = \Gamma$. Then by subdividing the "stratum" σ_ν, as in the proof of the preceding theorem, we obtain the desired triangulation τ of the whole set Γ.

For a detailed construction of the proof, the reader should look in [10], Chapter 1, §3.

From what has been proved follows

Lemma 1 (fundamental elementary lemma).[*] *Suppose that in S^n we are given an arbitrary system $\mathfrak{G}$ of open sets G_α. There exists a triangulation τ of the set $\Gamma = \bigcup_{G_\alpha \in \mathfrak{G}} G_\alpha$ such that the covering consisting of the principal stars of this triangulation is subordinated to the system $\mathfrak{G}$.*

We shall need some further propositions.

Fundamental definition. *We say that the triangulation τ' follows the triangulation τ if each simplex $T' \in \tau'$ lies in some* (evidently, unique) *simplex $T \in \tau$* (its *carrier* in τ).

Remark. We shall also need a certain strengthening of the following concept: we say that *the triangulation τ' strongly follows* the triangulation τ if τ' follows a double barycentric subdivision of the triangulation τ.

Lemma 2. *Let P, P', P'' be polyhedra in S^n, while $P'' \subseteq P \cap P'$. For any choice of the triangulations τ, τ' of the polyhedra P and P', one may find a triangulation τ'' of the polyhedron P'' that follows both τ and τ'.*

For the proof we introduce the definition of a generalized triangulation, differing from the definition of a triangulation only in that the elements of a generalized triangulation may be any convex polyhedra (and not just complexes). Here, as before, the polyhedra are taken to be open (i.e., their faces are not counted with them). Now let σ be any triangulation of the polyhedron P''. Since the closure of any simplex $T_0'' \in \sigma$ is a compactum lying in the polyhedron P, it — and *a fortiori* the simplex T_0' itself — intersects with only a finite number of simplexes[**] of the triangulation τ. For the same reason the simplex T_0'' intersects with only a finite number of simplexes of the triangulation τ'. Therefore there exists only a finite number of convex polyhedra which can be represented in the form

$$T_0'' \cap T \cap T',$$

where T_0'' is a fixed simplex of the triangulation σ, and T and T' are arbitrary

---

simplexes, belonging respectively to the triangulations τ and τ'. These polyhedra, constructed for all T''_0, are elements of a generalized triangulation of the polyhedron P'', any simplicial subdivision of which may be taken as the desired triangulation τ''.

2. **Covering triangulations. Triangulations canonical with respect to a given set.** In the remainder of this section, let X be a given set lying in a given S^n. We shall say that *the triangulation τ covers the set X* if $X \subseteq \tilde{\tau} \subseteq S^n$; if here each principal simplex of the triangulation τ contains at least one point of the set X, then we say that *the triangulation τ strongly covers the set X*. Each triangulation τ covering a set X contains a unique triangulation τ' strongly covering that set: the triangulation τ' consists of all those simplexes $T \in \tau$ containing points of the set X, and of all the faces of these simplexes. The unique triangulation τ', contained (as a subcomplex) in the triangulation τ and strongly covering the set X, is called the *derived triangulation* of the triangulation τ (with respect to the set X).

We shall now call a triangulation *canonical if it is derived from some triangulation of some neighborhood of the set X.* The canonical triangulations form a directed set in view of the definition of following given above for triangulations. Indeed, let τ'_1 and τ'_2 be two canonical triangulations, corresponding to the triangulations τ_1 and τ_2 of open sets. We choose some triangulation τ of the open set $\tilde{\tau}_1 \cap \tilde{\tau}_2$ following both of the triangulations τ_1 and τ_2 (such a triangulation exists because of Lemma 2). The derived triangulation τ' of the triangulation τ is the desired canonical triangulation following both τ'_1 and τ'_2.

3. **The group $d^r X$. The plan of proof of the invariance theorem.** If the triangulation θ follows the triangulation τ, then, putting into correspondence with each vertex e_θ of the triangulation θ some vertex of the carrier of the point e_θ in τ, we obtain, as we know, a simplicial mapping, the "canonical shift" σ_τ^θ of the triangulation θ into the triangulation τ; there may be many canonical shifts of the triangulation θ into the triangulation τ, but any two of them are combinatorially close simplicial mappings, so that all the canonical shifts generate one and the same *homomorphism σ_τ^θ* of the group $\Delta^r \theta$ into the group $\Delta^r \tau$ — *the homomorphism of the canonical shift* (see Part I, Chapter III, §§1 and 3).

Now we consider the directed set of all the canonical triangulations τ' with the corresponding canonical shifts; we obtain an inverse spectrum

$$\{\Delta^r \tau', \sigma_{\tau'}^{\theta'}\}, \tag{1}$$

whose limit group we denote by $d^r X$. The invariance theorem will be proved if we show that each of the groups $\delta^r X$ and $D^r X$ is isomorphic to the group $d^r X$.

4. **Proof of the isomorphism $\delta^r X = d^r X$.** If the triangulation τ covers the set

X, then the intersections $X \cap o_e$ of the principal stars o_e of that triangulation with the set X form a covering ω_τ of that set, which we call *adjoint to the given triangulation* τ. Regarding the elements $X \cap o_e$ and $X \cap o_{e'}$ as different, if o_e and $o_{e'}$ are different (i.e., the vertices e and e'), we see that the nerve of the covering ω_τ of the set X, adjoint to the given triangulation τ covering that set is a proper or improper subcomplex of the triangulation τ. The fundamental fact for what follows is: *If the triangulation τ strongly covers the set X, then the nerve of the covering ω_τ adjoint to it is the whole triangulation τ.*

For the proof we need to show that if the principal stars $o_{e_0}, \cdots, o_{e_r}$ of the triangulation τ have a nonempty intersection, then the set

$$(X \cap o_{e_0}) \cap \cdots \cap (X \cap o_{e_r}) = X \cap o_{e_0} \cap \cdots \cap o_{e_r} \qquad (2)$$

is also nonempty; but

$$o_{e_0} \cap \cdots \cap o_{e_r} = o_T,$$

where $T = |e_0 \cdots e_r|$, and the star o_T, as does any star of the triangulation, contains at least one principal simplex of that triangulation, so that (since each principal simplex of the triangulation τ contains points of the set X) it also contains points of the set X; therefore the set (2) coincides with the nonempty set $X \cap o_T$, with which our assertion is proved.

We shall now consider only canonical triangulations τ. We shall call the coverings of the set X adjoint to them *canonical coverings*. We have just shown that *any canonical triangulation is the nerve of its adjoint canonical covering of the set X.*

Now we shall prove that *any covering α of the set X is followed* (in the directed set of all coverings of X) *by some canonical covering.* Indeed, let $\alpha = \{\Gamma_i\}$ be the given covering of the set X. For each $\Gamma_i \in \alpha$ we choose a set G_i open in S^n and excising the set Γ_i from the set X:

$$X \cap G_i = \Gamma_i.$$

From Lemma 1 of subsection 1, we can find a triangulation τ of the set $G = \bigcup_i G_i$, which is open in S^n, such that the set of all principal stars of the triangulation τ, and therefore *a fortiori* the covering ω_τ, adjoint to this triangulation, are subordinated to the covering α. Moreover, this is valid for the canonical covering $\omega_{\tau'}$, adjoint to the triangulation τ' derived from τ. With this our assertion is proved.

Therefore, if in the spectrum

$$\{\Delta^r \alpha, \, \mathfrak{S}_\alpha^\beta\} \qquad (3)$$

with the limit group $\delta^r X$ we retain only the canonical coverings, we obtain a spectrum

$$\{\Delta^r \alpha', \, \mathfrak{S}_{\alpha'}^{\beta'}\}, \qquad (4)$$

where the α' are canonical coverings. But the nerve of each such covering is a canonical triangulation τ' for which the given covering is adjoint. There may be many such triangulations but, since they are nerves of one and the same covering α', they are isomorphic to one another. Therefore we may replace each α' in the spectrum (4) (as we are interested in this case only up to an isomorphism) by a canonical triangulation τ' (generating the covering α'). In the spectrum obtained in this way,

$$\{\Delta^r\tau', \, \eth_{\tau'}^{\theta'}\},\tag{5}$$

both the order and the projection are the same as in (4) (or in (3)). We obtain a multiplication of the spectrum (5)* if we consider all the canonical triangulations τ', generating one and the same canonical covering, as geometrical data (accordingly, coinciding with each other only if they coincide geometrically). The order and the projections remain the same as before (i.e., we consider that $\theta' > \tau'$ if the covering β' adjoint to the triangulation θ' is subordinate to the covering α' adjoint to the triangulation τ' with the projections defined in (3)). This new agreement does not change the limit group of the spectrum (5), which remains the group $\delta^r X$.

Finally, we carry out a last step: in the multiplication of the spectrum (5) thus obtained, in which τ' already runs through all the canonical triangulations (as geometrical data), we weaken the order** by setting up the order which was earlier defined for the triangulations. Under this the homomorphism $\eth_{\tau'}^{\theta'}$ for the group $\Delta^r\tau'$ is nothing other than the homomorphism of the canonical shift. This last

---

* The concept of multiplication of a directed set (in particular of a spectrum) is easily reduced to the concept of a cofinal section. It consists in the following.

We obtain a multiplication of a given directed set (respectively, of a given inverse or direct spectrum) Σ if we replace each element $\alpha \in \Sigma$ by a family σ_α of new elements, which we shall denote by $\alpha\lambda$, $\alpha\mu$, $\cdots$, while any two elements $\alpha\lambda$ and $\alpha\mu$ (and also $\alpha\lambda$ and $\beta\mu$) are considered to be different if either $\lambda \neq \mu$ or $\alpha \neq \beta$. Into the set Σ^* of elements thus obtained we introduce an order, putting $\beta\mu > \alpha\lambda$ in Σ^* if and only if $\beta > \alpha$ in Σ. If Σ is a spectrum, then we assume that the groups α and $\alpha\lambda$ are not only isomorphic to each other, but connected by a natural isomorphism chosen once and for all, as a consequence of which also $\alpha\lambda$ and $\alpha\mu$ lie in relation to one another by a natural isomorphism. Under this isomorphism, if $\beta\mu > \alpha\lambda$, we may carry the projection $\eth_\alpha^\beta$ (respectively π_β^α) in a natural way onto the groups $\beta\mu$ and $\alpha\lambda$, so that we obtain a projection $\eth_{\alpha\lambda}^{\beta\mu}$ (respectively $\pi_{\beta\mu}^{\alpha\lambda}$) for any λ and μ. The spectrum $\Sigma^* = \{\alpha\lambda, \, \eth_{\alpha\lambda}^{\beta\mu}\}$, respectively $\Sigma^* = \{\alpha\lambda, \, \pi_{\beta\mu}^{\alpha\lambda}\}$, is called a multiplication of the spectrum Σ.

Identifying the element $\alpha \in \Sigma$ with some definite element $\alpha\lambda \in \Sigma^*$, we may regard the directed set (spectrum) Σ as a cofinal section of the directed set (spectrum) Σ^*.

** We say that the directed set Σ' is obtained from the directed set Σ by weakening the order, if both sets consist of the same elements, while from $\beta > \alpha$ in Σ follows $\beta > \alpha$ in Σ' (but, generally speaking, not the reverse). From this definition it follows (see Part I, Chapter II, subsection 1) that the directed set Σ' is a cofinal section of the directed set Σ.

Remark. It is essential that the set Σ', obtained from Σ by weakening the order, remains, in the weakened order, a *directed* set. Without this we of course can not speak of cofinality.

transformation turns the spectrum (5), without changing its limit group $\delta^r X$, into the spectrum (1), whose limit group is the group $d^r X$. Thus, the isomorphism $\delta^r X = d^r X$ is proved.

5. Proof of the isomorphism $d^r X = D^r X$. Taking each canonical triangulation τ' as many times as there are triangulations τ of all possible neighborhoods $\widetilde{\tau}$ of the set X for which τ' is the derived triangulation, we obtain the spectrum[*]

$$\{\Delta^r \tau', \sigma_\tau^\theta\}, \tag{6}$$

which is a multiplication of the spectrum

$$\{\Delta^r \tau', \sigma_{\tau'}^{\theta'}\} \tag{1}$$

and therefore has the same limit group

$$d^r X = \varprojlim \{\Delta^r \tau', \sigma_\tau^\theta\}.$$

Now we put into correspondence with each triangulation τ of some one of the neighborhoods $\widetilde{\tau}$ of the set X a completely defined neighborhood λ'_τ of the set X, the *"canonical neighborhood"* (corresponding to the triangulation τ), in such a way that the following conditions are satisfied:

(a) The polyhedron $\widetilde{\tau}'$ (i.e., the body of the derived triangulation τ' of the triangulation τ) is contained in λ'_τ and is a retract of the set λ'_τ.

(b) If $\theta > \tau$, then $\lambda'_\theta \subseteq \lambda'_\tau$.

(c) The canonical neighborhoods form, under inclusion (independently of the triangulations τ defining them) a directed set.

(d) The directed set of all canonical neighborhoods of the set X forms a cofinal section in the directed set of all general neighborhoods of the set X.

Before turning to the construction of canonical neighborhoods, we observe that from their existence there already follows the isomorphism $d^r X = D^r X$. In fact:

(1) If the triangulation θ follows the triangulation τ, then the "canonical shift homomorphism" σ_τ^θ of the group $\Delta^r \theta$ into $\Delta^r \tau$ coincides with the inclusion homomorphism $E_{\widetilde{\tau}}^{\widetilde{\theta}}$ of the group $\Delta^r \widetilde{\theta} = \Delta^r \theta$ into the group $\Delta^r \widetilde{\tau} = \Delta^r \tau$, defined by the identity mapping of the polyhedron $\widetilde{\theta}$ into $\widetilde{\tau}$.

(2) Since, from condition (a) on canonical neighborhoods, the polyhedron $\widetilde{\tau}$ is a retract of the polyhedron λ'_τ, therefore the identity mapping of $\widetilde{\tau}'$ into λ'_τ defines an isomorphism $E_{\lambda'_\tau}^{\widetilde{\tau}'}$ of the group $\Delta^r \tau'$ onto the group $\Delta^r \lambda'_\tau$. From these

---

[*] Evidently, under each canonical shift σ_τ^θ the derived triangulation θ' goes into τ' (so that the shift σ_τ^θ uniquely generates also the shift $\sigma_{\tau'}^{\theta'}$).

two remarks it follows, keeping condition (b) in mind, that

$$d^r X = \varprojlim \{ \Delta^r \lambda'_\tau, \; E^{\lambda'_\theta}_{\lambda'_\tau} \}.$$

But in view of property (a) we may consider also the spectrum

$$\{ \Delta^r \lambda', \; E^{\mu'}_{\lambda'} \}, \tag{7}$$

where the canonical neighborhoods λ', μ', $\cdots$, considered independently from the triangulations τ, θ, $\cdots$ to which they correspond, are ordered by inclusion. Since the spectrum $\{ \Delta^r \lambda_\tau, \; E^{\lambda'_\theta}_{\lambda'_\tau} \}$ is evidently a multiplication of the spectrum (7), therefore

$$d^r X = \varprojlim \{ \Delta^r \lambda', \; E^{\mu'}_{\lambda'} \}.$$

But in view of property (c) the canonical neighborhoods form a cofinal section in the set of all neighborhoods of the set X, so that the limit group of the spectrum (7) coincides with the group $D^r X$, from which the isomorphism $d^r X = D^r X$ follows.

Thus, the question comes down to the construction of canonical neighborhoods satisfying the conditions (a)–(d). Suppose that we are given a triangulation τ whose body is an open set containing the set X. The derived triangulation will be denoted, as always, by τ'. For the construction of the set λ'_τ we shall for the time being say that the simplex $T' \in \tau'$ is exposed if the star of that simplex in τ is not contained in τ'. We denote by $O'T'$ the sum of all the simplexes of the double barycentric subdivision of the triangulation τ adjoining the exposed simplex T' (i.e., having a face lying in T'). The set λ'_τ is obtained from the polyhedron $\tilde{\tau}'$ by joining to it the sets $O'T'$ for all exposed simplexes $T' \in \tau'$.

It is easy to verify that λ'_τ is open in $\tilde{\tau}$ and hence open also in S^n, that the polyhedron $\tilde{\tau}'$ is a retract of the set λ'_τ, and that $\lambda'_\theta \subseteq \lambda'_\tau$ whenever $\theta > \tau$.

Now we shall prove property (c). Suppose that we are given two canonical neighborhoods λ'_{τ_1} and λ'_{τ_2}. We choose a triangulation τ of the open set $\lambda'_{\tau_1} \cap \lambda'_{\tau_2}$ that follows both τ_1 and τ_2 (such a triangulation exists because of Lemma 2 of subsection 1). But in view of property (b) we have $\lambda'_\tau \subseteq \lambda'_{\tau_1} \cap \lambda'_{\tau_2}$, so that property (c) is proved. For the proof of property (d) it is sufficient to take some neighborhood λ of the set X and a triangulation τ derived from it. Since $\lambda'_\tau \subseteq \tilde{\tau} = \lambda$, then all is proved.

Thus, the isomorphism $d^r X = D^r X$ and, along with it, the invariance theorem and the first duality law are completely proved.

$$\S 2. \ \text{Second duality law for compacta}$$

As a special case of the general duality law proved above, we consider the case when A is a compactum[*] lying in S^n. The Pontrjagin duality law asserts in this case the duality

$$\delta^p(A, \mathfrak{B}) \,|\, \Delta^q_c(B, \mathfrak{A})$$

in which it is essential that the group of the compactum A taken over a bicompact coefficient domain $\mathfrak{B}$ and topologized in a suitable way. The same question on the dualizability of the groups $\delta^r(A, \mathfrak{A}) = \Delta^r_c(A, \mathfrak{A})$ of a compactum A, taken over a discrete coefficient domain (even, for example, over the group of integers), remained open until the proof of the general duality law.

Now we see that

$$\delta^p(A, \mathfrak{A}) \,|\, \overline{\Delta}^q(B, \mathfrak{B}). \tag{1}$$

Here, for the compactum A, the group $\delta^p(A, \mathfrak{A})$ may be defined — as in the first part of this paper — by the use of finite coverings only (forming in this case a cofinal section of the set of all coverings), so that it is isomorphic to the group $\Delta^p_c(A, \mathfrak{A})$ (see Part I, Chapter III, §7), and therefore the duality (1) may be re-written in the form

$$\Delta^p_c(A, \mathfrak{A}) \,|\, \overline{\Delta}^q(B, \mathfrak{B}). \tag{2}$$

This relation may be considered as a "second duality law" for compacta $A \subset S^n$ over discrete coefficient domains. From this results the dualizability of the group $\Delta^r_c(A, \mathfrak{A})$ for closed $A \subset S^n$. In Chapter IV we shall see that $\Delta^r_c A$ is dualizable for arbitrary sets as well.

§3. Zero-dimensional case — components and quasicomponents
of a point set

1. We shall now consider the zero-dimensional case of the general duality law, which makes it possible, for example, to prove the dualizability of the number of components of a point set (Eilenberg's theorem). Here, as the sole coefficient group we choose a group of order 2, which reduces all the algebraic considerations to the simplest concepts of the elementary theory of sets. It will be convenient for us to use the following notations. Let C be any ("abstract") set. We shall denote by C^0, and call the *large group of the set C*, the group whose elements are all possible finite subsets of C, with addition defined as addition modulo 2 (the sum of two subsets of the set C is their symmetric difference, i.e., the set of those elements of C which lie only in one or the other of the two sets but not in both). In the group C^0 there is a subgroup C^{00}, whose elements are those and only those finite subsets of the set C which consist of an even number of elements. The group C^{00} will be called the *small group of the set C*.

---

[*] We have put this special case into a special section, since it is related to a group of questions which have long since become classical.

We shall define the *number* of elements of a set as being its cardinality, if the set is finite, and as the symbol ∞, if the set is infinite. Here we put, for any finite number a,

$$a < \infty,$$
$$\infty \pm a = \pm a + \infty = \infty.$$

If the set C contains the p elements

$$a_1, \cdots, a_p \tag{1}$$

(and perhaps contains other elements as well), then (1) gives us a system of linearly independent (modulo 2) elements[*] of the group C^0, and the (unordered) pairs

$$(a_1, a_2), (a_1, a_3), \cdots, (a_1, a_p) \tag{2}$$

are linearly independent elements of the group C^{00}. If (1) exhausts the whole set C, then, as is seen without difficulty, the corresponding elements of the group C^0 form a basis of that group and (2) form a basis of the group C^{00}. Hence it follows that:

I. *If C is a finite set consisting of p elements, then the rank of the group C^0 is equal to p, and the rank of the group C^{00} is equal to $p-1$; if the set C is infinite, then the ranks of both groups C^0 and C^{00} are ∞.*

In other words, *in all cases the rank of the group C^0 is equal to the number of elements of the set C, and the rank of the group C^{00} is equal to the number of elements in the group C less unity.*

Now suppose we are given a partially ordered system of sets $\{C_\alpha\}$. Suppose, moreover, that for each pair $\beta > \alpha$ we are given a mapping ("projection") ϖ_α^β of the set C_β into the set C_α, while the condition of transitivity is satisfied: if $\gamma > \beta > \alpha$, then $\varpi_\alpha^\gamma = \varpi_\alpha^\beta \varpi_\beta^\gamma$. The spectrum $\{C_\alpha, \varpi_\alpha^\beta\}$ of the sets C_α defined in this way has a limit set $C = \varprojlim (C_\alpha, \varpi_\alpha^\beta)$ whose elements are the "threads" $\{x_\alpha\}$, where x_α is an element of the set C_α (one in each C_α), and if $\beta > \alpha$ we have $\varpi_\alpha^\beta x_\beta = x_\alpha$. The mapping ϖ_α^β of the set C_β into the set C_α generates a mapping of the same name of each finite subset $c_\beta \subseteq C_\beta$ onto some finite subset $c_\alpha \subseteq C_\alpha$, while this mapping is defined also modulo 2: the image of the set $c_\beta \subseteq C_\beta$ is

---

[*] When we say that (1) is a system of elements of the group C^0, then we consider each a_i as a set consisting of the one element a_i. The elements $c_1, \cdots, c_s$ of the group C^0 are said to be linearly independent if, for any nonempty subset $\{i_1, \cdots, i_r\}$ of the set of numbers $1, 2, \cdots, s$, we have $c_{i_1} + \cdots + c_{i_s} \neq 0 \pmod 2$; evidently this definition reduces to the usual one if one takes linear combinations with coefficients from I_2. The definition of linear independence naturally leads also to the definition of the rank of the group (modulo 2).

We observe that the groups C^0 and C^{00} are (discrete) direct sums of some set of groups of the second order. The number of direct summands is equal to the rank of the group. The character groups of our groups are the already topological (bicompact) direct sums of groups of the second order. The number of these terms may again be defined as the rank of the corresponding groups.

considered to be the set of those elements of the set C_α into each of which there map an odd number of elements of the set C_β. Here it is evident that the set c_β, consisting of an even number of elements, maps onto a set c_α consisting of an even number of elements. It is clear that the mapping ϖ_α^β of the set C_β into the set C_α generates thus a homomorphic mapping (which we shall also denote by ϖ_α^β) of the group C_β^0 into the group C_α^0, under which the group C_β^{00} maps into the group C_α^{00}. Thus, we obtain the inverse spectra

$$\{C_\alpha^0, \varpi_\alpha^\beta\}, \{C_\alpha^{00}, \varpi_\alpha^\beta\}$$

whose limit groups will be denoted by (C^0) and (C^{00}). The elements of the group (C^0) are the "threads" $\{c_\alpha\}$ of finite sets $c_\alpha \subseteq C_\alpha$, with one subset c_α in each C_α, while it is required that if $\beta > \alpha$ then $\varpi_\alpha^\beta c_\beta = c_\alpha$.

Among the threads $\{c_\alpha\}$, those which are elements of the group (C^{00}) are distinguished by the condition that all of the sets c_α entering into their structure consist of an even number of elements.

Along with the limit groups (C^0) and (C^{00}) we may consider the large and small groups of the limit set C of our spectrum. These groups we denote respectively by C^0 and C^{00}.

If we put into correspondence with each thread $x = \{x_\alpha\} \in C$ its coordinate x_α, i.e., the element of the set C_α contained in it, understanding this mapping again modulo 2, we see that to each finite set of elements of the set C there corresponds in each set C_α some finite subset c_α, while these subsets (considered as elements of the corresponding groups C_α^0) form again a thread, i.e., an element of the group (C^0). Thus, we have set up a mapping f which, as is easily seen, is an isomorphism[*] of the group C^0 into the group (C^0), in view of which the group C^0 may be considered as a subgroup of the group (C^0) and the group C^{00} as a subgroup of the group (C^{00}). However, even in the simplest cases, the group C^0 does not coincide with the group (C^0) and the group C^{00} does not coincide with the group (C^{00}).

2. **Example. Remark.** We consider a countable sequence of sets C_k, $k = 1$, $2, 3, \cdots$, where C_k consists of the elements $x_1^k, x_2^k, \cdots, x_k^k$. The mapping ϖ_k^{k+1} is defined as follows: $\varpi_k^{k+1} x_1^{k+1} = \varpi_k^{k+1} x_2^{k+1} = x_1^k$, $\varpi_k^{k+1} x_3^{k+1} = x_2^k, \cdots, \varpi_k^{k+1} x_{k+1}^{k+1} = x_k^k$. It is easy to see that the set C is countable. Accordingly, the groups C^0 and C^{00} are also countable. Meanwhile, the groups (C^0)

---

[*] Indeed, let $c = \{x^1, \cdots, x^s\}$ be a nonempty finite subset of the set C (i.e., an element of the group C^0 which is different from zero). If $x^i = \{x_\alpha^i\}$, $i = 1, \cdots, s$, are elements of the set c, then for each pair x^i, x^j of these elements we can find an α_{ij} such that $x_{\alpha_{ij}}^i \neq x_{\alpha_{ij}}^j$. Choosing an α which follows all the α_{ij}, we see that $x_\alpha^1, \cdots, x_\alpha^s$ are all different from one another, so that c_α is not empty, which means that $f(c) \neq 0$.

and (C^{00}) each have the power of the continuum, as one sees from the following table

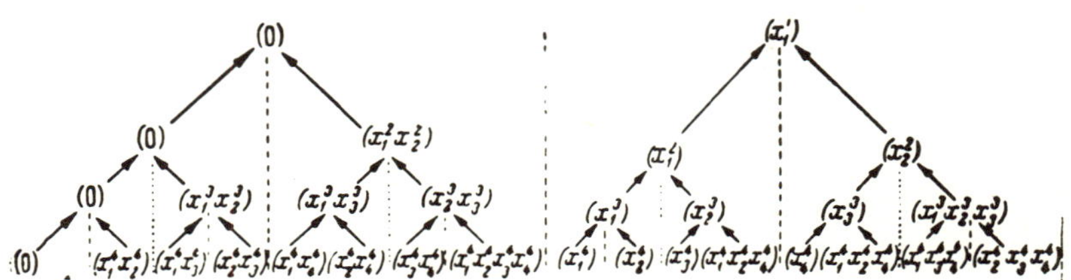

in which the rows indicate for $k = 1, 2, 3, 4$ all the subsets[*] respectively of the sets C_k, and the arrows indicate how these subsets map under the projections ω_k^{k+1}. Retaining in this table only the sets consisting of an even number of elements, we see that the group (C^{00}) also has the power of the continuum.

We indicate an essential though trivial case in which the groups C^0 and C^{00} are isomorphic to the groups (C^0) and (C^{00}) respectively.

II. *Suppose we are given a partially ordered system (of any cardinality) of sets C_α, each consisting of the same finite number p of elements. Suppose, moreover, that the projections ω_α^β are one-to-one mappings of C_β (following C_α) onto C_α. Then each of the groups C_α^0, C^0, (C^0) is the direct sum of p groups of second order, and each of the groups C_α^{00}, C^{00}, (C^{00}) is the direct sum of p − 1 groups of second order.*

3. **The groups $\delta^0 A$ and $\delta^{00} A$.** If K is a complex and Γ is a set open in S^n, then the groups $\Delta^0 K$ and $\Delta^{00} K$ (respectively $\delta^0 \Gamma$, $\delta^{00} \Gamma$) are the large and small groups of the set C of all the components of the complex K (respectively of the open set Γ).

Further, the groups $\delta^0 A$, $\delta^{00} A$ for any A, being isomorphic to the groups $D^0 A$, $D^{00} A$, may be defined in the following way. Suppose, as always, that λ is any set open in S^n and containing the set A. Denote by C_λ the set of all the components of the set λ. If $\lambda' > \lambda$, i.e., if $\lambda' \subset \lambda$, then each component of λ' is contained in a definite component of λ, so that the projection $\omega_\lambda^{\lambda'}$ of the set $C_{\lambda'}$ into C_λ is defined, and with it the spectrum $\{C_\lambda, \omega_\lambda^{\lambda'}\}$; the group (C^0), respectively (C^{00}), is thus the group $D^0 A = \delta^0 A$, respectively $D^{00} A = \delta^{00} A$.

Remark. Taking along with the components of the sets λ the components of the nerve α, we obtain an invariant definition of the groups $\delta^0 A = D^0 A$ and $\delta^{00} A = D^{00} A$.

---

[*] For example, (x_1^2) denotes the set consisting of the single element x_1^2, (x_1^2, x_2^2) denotes the set consisting of the two elements x_1^2 and x_2^2, and so forth. The empty set is denoted by (0).

For a countable compactum A consisting of a convergent sequence of points a_k, $k = 1, 2, \cdots$, and of the limit point $a = \lim\limits_{k \to \infty} a_k$, the groups $\delta^0 A$ and $\delta^{00} A$ have the power of the continuum. Indeed, for such a compactum, in the system of all λ one can find a cofinal subsequence λ_k, $k = 1, 2, 3, \cdots$, where the λ_k consist of pairwise nonintersecting intervals $x_1^k, \cdots, x_k^k$ in which x_1^k contains the point a and all of the points a_h for $h = k, k+1, \cdots$, while the remaining intervals $x_2^k, x_3^k, \cdots, x_k^k$ (each of length less than $1/k$) contain respectively the points $a_{k-1}, \cdots, a_1$. Therefore, it is seen immediately that the groups $D^0 A$, $D^{00} A$ are in our case exactly those groups having the power of the continuum which were considered in the example of the preceding subsection. Hence we easily deduce that, for each null-dimensional set A containing at least one non-isolated point, the groups $\delta^0 A$ and $\delta^{00} A$ have the power of the continuum. From the general duality law it further follows that two sets A and A', having homeomorphic complements, have isomorphic groups $\delta^{00} A = \delta^{00} A'$. For closed A this assertion follows, of course, from the Alexander-Pontrjagin duality law even in the strengthened form of a topological isomorphism of the groups $\Delta^{00} A$ and $\Delta^{00} A'$ (receiving the natural Pontrjagin topology); therefore, in particular, it easily follows that all infinite null-dimensional compacta A have a group $\Delta^{00} A$ topologically isomorphic to the "Cantor group" (i.e., the bicompactly-topologized direct sum of a countable set of groups of second order); for the proof it is sufficient to observe that every null-dimensional compactum consisting of an infinite set of points may be considered as a set in S^1 whose complement is the sum of a countable set of intervals. It is also easy to obtain a direct proof of this fact (see the example in subsection 2).

4. **Components and quasicomponents (real and ideal) of the set A.** We preserve the notation introduced in subsection 3; with this notation, as we have seen, the groups (C^0) and (C^{00}) are respectively isomorphic to the groups $\delta^0 A$ and $\delta^{00} A$.

Along with these groups it is natural to consider the large and small groups C^0 and C^{00} of the limit set C of the spectrum $\{C_\lambda, \varpi_\lambda^{\lambda'}\}$. The elements of that limit set are the "threads" of the form $\{\Gamma_\lambda\}$, where Γ_λ is some component of the open set λ, while if $\lambda' \subset \lambda$, we have $\Gamma_{\lambda'} \subseteq \Gamma_\lambda$. Such threads will be called (for reasons which are now becoming clear) *ideal quasicomponents of the set A*. The large and small groups of the set of all ideal quasicomponents of the set A will be denoted respectively by $C^0 A$ and $C^{00} A$.

Remark. We may show that the ideal quasicomponents of the set A stand in a one-to-one relation with the threads of the form $\{Q_\alpha\}$, where each Q_α is a component of the nerve of the covering α of the set A, and under the projection ϖ_α^β of the nerve β into α the component Q_β maps into Q_α (the proof of this assertion is easily obtained if one turns from the partially ordered system of all coverings α to a cofinal subsystem consisting of the canonical coverings). Hence it follows that the groups $C^0 A$ and $C^{00} A$ are topological invariants of the set A.

Now we recall the concept introduced by Hausdorff (see [21], p. 248) of a (real) quasicomponent A_x of the point x in the (topological) space A: the quasicomponent A_x is defined as the intersection of all the "open-closed" (i.e., simultaneously open and closed) sets in A which contain the point x. A quasicomponent A_x is closed in A; the component of the point x in A is contained in the quasicomponent A_x of this point and coincides with it if and only if that quasicomponent is connected. The quasicomponents of two distinct points either coincide or do not intersect, so that we may speak of the decomposition of the set A into quasicomponents, and this decomposition, generally speaking, has larger sets than the decomposition into components.

We turn to the general case of an (not always compact) $A \subset S^n$. We shall say that a given real quasicomponent A_x of A lies (or is contained) in an ideal quasicomponent $\{\Gamma_\lambda\}$ if A_x is contained in one of the sets Γ_λ which constitute that ideal quasicomponent.

III. *Each real quasicomponent A_x of the set A lies in one unique ideal quasicomponent $\{\Gamma_\lambda\}$ and is the intersection of all the sets Γ_λ which constitute that ideal quasicomponent. In each ideal quasicomponent there lies not more than one real quasicomponent, and if there lies none, then the intersection of all the Γ_λ which constitute the given ideal quasicomponent is empty.*

The proof is almost obvious. Take in some A_x a point x, and in each λ choose the component Γ_λ that contains x. Since Γ_λ is open in S^n and closed in λ it follows that $\Gamma_\lambda \cap A$ is open-closed in A, and therefore, since it contains x, it also contains the whole quasicomponent A_x. Thus, in each λ there is a completely defined component Γ_λ that contains the quasicomponent A_x. These Γ_λ in turn form an ideal quasicomponent $\{\Gamma_\lambda\}$ in which the given quasicomponent A_x lies. We shall prove that no point y which does not belong to the set A_x can be contained in all the elements Γ_λ of the ideal quasicomponent $\{\Gamma_\lambda\}$. Since the intersection of all the Γ_λ is contained in A^* it suffices to consider a point $y \in A$. Since y is not in A_x, there exists an open-closed set H in A which contains A_x but not y, while the set $A - H$ is open-closed in A and contains y. Therefore there exist in S nonintersecting open sets Γ' and Γ'' of which the first contains H and the second contains $A - H$ and hence the point y. The set

---

*This follows from $\bigcap \lambda = A$.

$\Gamma' \cup \Gamma'' = \lambda$ contains A; the corresponding Γ_λ, being subsets of Γ', do not contain y. From what has been proved already, it follows that in no ideal quasi-component can there lie more than one real quasicomponent. If now some point $x \in A$ is contained in all the elements of a given ideal quasicomponent $\{\Gamma_\lambda\}$, then it is evident that the quasicomponent A_x of the point x lies in $\{\Gamma_\lambda\}$, with which all the parts of Theorem III are proved.

5. **Null-dimensional case of the duality law. Eilenberg's theorem.** We shall now denote the number of components of the set A (in the sense established in subsection 1) by p_c, and the number of real quasicomponents by p_0, the number of ideal quasicomponents by p_i (that same number is the rank of the group $C^0 A$, so that the rank of the group $C^{00} A$ is equal to $p_i - 1$). Further the rank of the group $\delta^0 A = (C^0)$ is denoted by p^0 and the rank of the group $\delta^{00} A = (C^{00})$ by p^{00}. We shall first prove that for any A the following equations hold:

$$p_c = p_0 = p_i = p^0,$$
$$p^{00} = p^0 - 1. \tag{3}$$

The number p^{00} will be called the zero-dimensional Betti number of the set A.

We shall prove formula (3). First of all, we evidently have

$$p_0 \leq p_i \leq p^0,$$
$$p_0 \leq p_c,$$
$$p^{00} \leq p^0.$$

We prove: if p_0 is finite, then each quasicomponent of the set A is a component, and accordingly $p_c = p_0$. Indeed, let $A_1, \cdots, A_p$ be all the quasicomponents of the set A, and let their number be finite. We have to prove that each of them, for example A_1, is a connected set. If A_1 were not connected, then there would be a decomposition

$$A_1 = A_1' \cup A_1''$$

into two nonempty nonintersecting sets A_1' and A_1'', each of which is open-closed in A_1. However, A_1, being a quasicomponent is closed in A; on the other hand, since A_1 is the complement in A of the set $A_2 \cup \cdots \cup A_p$, which is closed in A, it is also open in A. Therefore the sets A_1' and A_1'', being open-closed in A_1, are open-closed in A. Hence it follows that a quasicomponent in A of any point $x \in A_1'$ has to be contained in A_1' and yet coincide with A_1. This contradiction proves the assertion.

We note: if p_c is finite, then *a fortiori* p_0 is finite, and accordingly $p_c = p_0$. In the case of finite p_c, we have from what has been proved

$$p_c = p_0 \leq p_i \leq p^0, \quad p^{00} \leq p^0.$$

Let

$$A_1, \cdots, A_p, \quad p = p_c = p_0$$

be all the components of the set A. Then the sets $A_1, \cdots, A_p$ may be included in pairwise nonintersecting, open in S^n sets

$$G_1, \cdots, G_p;$$

to this end it is sufficient for each pair A_i, A_j to choose sets H_i^j, H_j^i which are nonintersecting and open in S and to define G_i as the intersection of all the H_i^j, $j \neq i$. We evidently obtain a cofinal section of the partially ordered set of all λ if we restrict ourselves to only those λ's which are contained in

$$\lambda_0 = G_1 \cup \cdots \cup G_p.$$

Denoting by λ_i a component of the set A_i in $\lambda \subset \lambda_0$, we obtain $A_1 \subseteq \lambda_1 \subseteq G_1, \cdots, A_p \subseteq \lambda_p \subseteq G_p$. Put

$$\lambda' = \lambda_1 \cup \cdots \cup \lambda_p.$$

Evidently the sets λ' constructed in this way for each $\lambda \subseteq \lambda_0$ also form a cofinal section of the set of all λ. But the set C of components of the set λ' falls under the conditions of Theorem II, and therefore it follows that, in our case, $p^0 = p$, $p^{00} = p - 1$, which proves formula (3) in the case of finite p.

Suppose now that $p_c = \infty$. Then, by what has been proved, also $p_0 = \infty$, so that *a fortiori* $p_i = \infty$, $p^0 = \infty$. Since $p^0 = \infty$, the set C (the limit set for the spectrum $\{C_\alpha, \mathfrak{d}_\alpha^\beta\}$) is infinite, and therefore its small group C^{00} has infinite rank. Hence, the group (C^{00}), containing the group C^{00} as a subgroup, has infinite rank, i.e., $p^{00} = \infty$. Thus the proof of formula (3) is complete.

By the general duality law, $(C^{00}) = \delta^{00}A$ and $\Delta^{n-1}B$ are dual to each other, from which it follows that p^{00} is the rank of the group $\Delta^{n-1}B$.

Thus, the zero-dimensional case of the duality law may be formulated as follows:

Fundamental theorem. *The zero-dimensional Betti number of any set $A \subseteq S^n$, i.e., the number of its components* (equal to the number of its real, and also to the number of its idealized, quasicomponents), *is one larger than the rank of the group $\Delta^{n-1}B$.*

From this results the following.

Theorem of Eilenberg. *If B and B' are homeomorphic sets lying in S^n, then the complementary sets A and A' have one and the same number of components.*

Remark. However, neither the cardinality of the set of components of the set A, nor the cardinality of the set of real quasicomponents, nor the cardinality of the set of ideal components[*] of the set A remain invariant when one passes from the set A to the set A' whose complement $B' = S^n - A'$ is homeomorphic to the complement $B = S^n - A$ of the set A. It is sufficient to take as $A \subset S^1$ a zero-dimensional uncountable compactum, and as $A' \subset S^1$ a countable compactum. The sets B and B' are homeomorphic, while A consists of an uncountable, and A' of a countable, number of components (real quasicomponents), corresponding in a one-to-one way in each case to ideal components.

CHAPTER III

Sitnikov isomorphism and the central duality law[**]

$\S 1$. Auxiliary groups $\Delta^r_{\text{ext}} A$ and $\nabla^r_{\text{ext}} A$.

Construction of the Sitnikov isomorphism M and outline of the proof.

1. **Construction of the Sitnikov isomorphism M.** This isomorphism between the groups $\nabla^p A$ and $\Delta^q B$ requires the definition of some auxiliary groups, the so-called exterior groups $\Delta^r_{\text{ext}} A$ and $\nabla^r_{\text{ext}} A$, which are to a certain extent analogous to the groups $D^r A$ of the preceding chapter. However, in distinction from the group $D^r A$, not only the group $\nabla^r_{\text{ext}} A$ but also the group $\Delta^r_{\text{ext}} A$ will be based on infinite chains.

We now proceed to the definition of these groups. *Throughout this section we shall take triangulations to mean triangulations of sets open in S^n.*

Suppose that the triangulation τ' follows the triangulation τ and suppose that we are given an arbitrary (generally speaking, infinite) chain x^r of the triangulation τ. We define a chain $s^\tau_{\tau'} x^r$ of the triangulation τ', called the Δ-*extension of the chain x^r in the triangulation τ'*. To this end we choose any r-dimensional oriented simplex t' of the triangulation τ' and its carrier t in τ. If the dimension of the simplex t is larger than r, we put $(s^\tau_{\tau'} x^r \cdot t') = 0$; if the dimension of the simplex t is equal to r, then we put $(s^\tau_{\tau'} x^r \cdot t') = (x^r \cdot t)$. We shall show that the operator $s^\tau_{\tau'}$ defined in this way commutes with the operator Δ:

$$\Delta s^\tau_{\tau'} x^r = s^\tau_{\tau'} \Delta x^r. \tag{1}$$

In view of the linearity of both operators, it is sufficient to prove formula (1) for the chain $x^r = t^r$, which consists of one oriented simplex. On each simplex $t'^{r-1} \in \tau'$ lying inside t^r the chain $\Delta s^\tau_{\tau'} t^r$ takes on a null value since to each such simplex there adjoin exactly two r-dimensional simplexes of τ' lying in t^r, seeing that the body of the triangulation τ' is an open set. Since on the simplex

---

*Translator's note: Presumably these are quasicomponents.

**In the exposition of Sitnikov's results, we follow word for word his original paper [17] (cf. the introduction to the paper [17a]).

t'^{r-1} in question it is obvious that also $s^{\tau}_{\tau'}\Delta t^r$ takes on a zero value, then, for simplexes t'^{r-1} lying inside t^r, the equation (1) is proved for $x^r = t^r$. Now let t'^{r-1} lie on the boundary of the simplex t^r. Then to t'^{r-1} there adjoins a unique r-dimensional simplex of $s^{\tau}_{\tau'}t^r$, which is oriented in the same way as t^r. Therefore t'^{r-1} enters into $\Delta s^{\tau}_{\tau'}t^r$ with the same coefficient as in $s^{\tau}_{\tau'}\Delta t^r$. Formula (1) is thus proved.

From this it follows that the extension $s^{\tau}_{\tau'}z$ of any Δ-cycle z of the triangulation τ is a Δ-cycle of the triangulation τ', and that, if $z \sim 0$ in τ, then $s^{\tau}_{\tau'}z \sim 0$ in τ', so that $s^{\tau}_{\tau'}$ generates a homomorphism of the group[*] $\Delta^r\tau$ into $\Delta^r\tau'$. Therefore, considering the directed set of all triangulations τ of all possible neighborhoods of the set A, we obtain the direct spectrum

$$\{\Delta^r\tau,\ s^{\tau}_{\tau'}\},$$

whose limit group will be called the group $\Delta^r_{\text{ext}}A$. This definition may be re-written as follows. We shall call *an exterior cycle of the set A* any (generally speaking, infinite) Δ-cycle of any triangulation τ of any neighborhood $\tilde{\tau} = \lambda$ of the set A. Two exterior cycles z_1 and z_2, lying respectively on the triangulations τ_1 and τ_2, will be said to be *homologous to each other around the set A* if there exists a triangulation τ following both τ_1 and τ_2 such that

$$s^{\tau_1}_{\tau}z_1 \sim s^{\tau_2}_{\tau}z_2 \text{ on } \tau.$$

It is easy to verify that this definition, which obviously satisfies the reflexivity and symmetry conditions, also satisfies the transitivity condition and accordingly generates a decomposition of the set of all r-dimensional exterior Δ-cycles of the set A into classes of cycles that are homologous to one another around A. These classes are asserted to be the elements of the group $\Delta^r_{\text{ext}}A$. Addition in the group $\Delta^r_{\text{ext}}A$ is defined as follows. If $\zeta' \in \Delta^r_{\text{ext}}A$ and $\zeta'' \in \Delta^r_{\text{ext}}A$, and if $z^r_1 \in \zeta'$, $z^r_2 \in \zeta''$, are cycles lying respectively on the triangulations τ_1 and τ_2, then by taking a triangulation τ following both τ_1 and τ_2 we obtain the cycles $s^{\tau_1}_{\tau}z^r_1 = z' \in \zeta'$ and $s^{\tau_2}_{\tau}z^r_2 = z'' \in \zeta''$, which lie already on one and the same triangulation τ. The class ζ that contains the cycle $z' + z''$ is called the sum of the classes ζ' and ζ''. It is easy to verify that this definition is acceptable since it does not require a particular choice of elements.

Thus the group $\Delta^r_{\text{ext}}A$ is defined.

Now we turn to the definition of the group $\nabla^r_{\text{ext}}A$. Again let τ and τ' be tri-angulations of neighborhoods λ and λ' of the set A, where τ' follows τ. Then there are defined (non-uniquely!) canonical shifts $\sigma^{\tau'}_{\tau}$, of the triangulation τ' into

---

    * Based on infinite chains.

τ, as simplicial mappings which put in correspondence to each vertex of the triangulation τ' a vertex of its carrier in the triangulation τ. Each canonical shift defines, as is known (see Part I, Chapter III, §1), for each chain x^r of the triangulation τ its ∇-extension $\bar\sigma^{\tau}_{\tau'} x^r$ according to the formula

$$(\bar\sigma^{\tau}_{\tau'} x^r \cdot t'^r) = (x^r \cdot \sigma^{\tau'}_{\tau} t'^r)$$

for any oriented simplex $t'^r \in \tau'$. The homomorphism $\bar\sigma^{\tau}_{\tau'}$ commutes with the operator ∇

$$\nabla\bar\sigma^{\tau}_{\tau'} x^r = \bar\sigma^{\tau}_{\tau'} \nabla x^r,$$

which may be checked by a direct calculation (see Part I, page 36, subsection 4); it accordingly carries ∇-cycles of the triangulation τ into ∇-cycles of the triangulation τ', while cycles bounding in τ go into cycles bounding in τ'.

Two different canonical shifts of the triangulation τ' into τ, being combinatorially close simplicial mappings, carry any ∇-cycle of the triangulation τ into mutually homologous ∇-cycles of the triangulation τ' (see Part I, page 43, subsection 4). Thus *all the canonical shifts $\sigma^{\tau'}_{\tau}$ generate one and the same homomorphism $\bar\sigma^{\tau}_{\tau'}$ of the group $\nabla^r\tau$ into the group $\nabla^r\tau'$.*

Therefore if one considers the directed set of all possible triangulations τ of open sets covering the set A, *one obtains a direct spectrum*

$$\{\nabla^r\tau, \ \bar\sigma^{\tau}_{\tau'}\},$$

whose limit group is defined as the group $\nabla^r_{\mathrm{ext}} A$. As in the case of the group $\Delta^r_{\mathrm{ext}} A$, it is convenient to use the name, exterior ∇-cycle of the set A, for any ∇-cycle of any triangulation of any neighborhood of the set A, and to say that two ∇-cycles z^r_1 and z^r_2 that lie respectively on the triangulations τ_1 and τ_2 are *homologous to one another around A* if there is a triangulation τ following τ_1 and τ_2 such that $\bar\sigma^{\tau_1}_{\tau} z^r_1 \sim \bar\sigma^{\tau_2}_{\tau} z^r_2$ in τ. Then the elements of the group $\nabla^r_{\mathrm{ext}} A$ are homology classes of r-dimensional exterior ∇-cycles, and for the definition of the sum of two such classes $\mathfrak{z}$ and $\mathfrak{z}'$ we have to take the ∇-cycles $z\in\mathfrak{z}$ and $z'\in\mathfrak{z}$ which lie on one and the same triangulation τ; the sum $\mathfrak{z}+\mathfrak{z}'$ is the class containing the ∇-cycle $z+z'$.

In the succeeding sections of this chapter we shall successively construct completely defined, "natural" isomorphisms as follows:

an isomorphism J of the group $\nabla^p_{\mathrm{ext}} A$ onto the group $\nabla^p A$;

an isomorphism D of the group $\nabla^p_{\mathrm{ext}} A$ onto the group $\Delta^{n-p}_{\mathrm{ext}} A$;

an isomorphism Γ of the group $\Delta^{n-p}_{\mathrm{ext}} A$ onto the group $\Delta^q B$.

From these isomorphisms one can put together the Sitnikov isomorphism M of the group $\nabla^p A$ onto the group $\Delta^q B$ by means of the formula

$$M = \Gamma D J^{-1}.$$

§2. The isomorphism J of the group $\nabla^p_{\mathrm{ext}} A$ onto the group $\nabla^p A$

1. The operator J, which exhibits the isomorphism of the group $\nabla^p A$ onto the group $\nabla^p A$, is defined as follows. Let τ be a triangulation of some neighborhood of the set A. We put into correspondence with the triangulation τ the covering ω_τ of the set A which is adjoint to that triangulation.[*] The nerve ω_τ is a subcomplex of the triangulation τ, and to each chain u on τ there corresponds a chain Ju on ω_τ which assumes the same value as u, on each simplex of ω_τ. The "excision operator J" thus defined is well known (see [4]), and it commutes with the operator ∇, as one can easily show. Therefore if u is a ∇-cycle of the triangulation τ, then Ju is a ∇-cycle of the nerve ω_τ and hence (according to the definition given in Chapter I, §1.3) a ∇-cycle of the set A.

We shall show that if the exterior ∇-cycles u_1^p and u_2^p are homologous to one another around A, then the ∇-cycles Ju_1^p and Ju_2^p of the set A corresponding to them are mutually homologous in A. This will prove that the operator J defines a homomorphism of the group $\nabla^p_{\mathrm{ext}} A$ into the group $\nabla^p A$.

Thus, suppose that the ∇-cycles u_1^p and u_2^p, which are homologous to each other around A, lie respectively on triangulations τ_1 and τ_2. This means that there exists a triangulation τ following both τ_1 and τ_2 such that

$$\bar\sigma_\tau^{\tau_1} u_1 \sim \bar\sigma_\tau^{\tau_2} u_2 \quad \text{in } \tau, \tag{1}$$

where $\bar\sigma_\tau^{\tau_1}$ and $\bar\sigma_\tau^{\tau_2}$ are generated by arbitrary canonical shifts $\sigma_{\tau_1}^\tau$ and $\sigma_{\tau_2}^\tau$ of the triangulation τ, respectively, into τ_1 and τ_2. The covering ω_τ is subordinate to ω_{τ_1} and to ω_{τ_2}, and the nerve ω_{τ_2} goes under the shifts $\sigma_{\tau_1}^\tau$ and $\sigma_{\tau_2}^\tau$ into subcomplexes of the nerves ω_{τ_1} and ω_{τ_2}. Therefore the shifts $\sigma_{\tau_1}^\tau$ and $\sigma_{\tau_2}^\tau$ define projections $\varpi_{\tau_1}^\tau$ and $\varpi_{\tau_2}^\tau$, of the nerve ω_τ into the nerves ω_{τ_1} and ω_{τ_2}. For the "adjoint" homomorphisms $\pi_\tau^{\tau_1}$ and $\pi_\tau^{\tau_2}$ of the groups $L^p \omega_{\tau_1}$ and $L^p \omega_{\tau_2}$ into $L^p \omega_\tau$ corresponding to the projections $\varpi_{\tau_1}^\tau$ and $\varpi_{\tau_2}^\tau$, we have, as is easily verified,

$$\pi_\tau^{\tau_1} Ju_1 = J\bar\sigma_\tau^{\tau_1} u_1, \ \pi_\tau^{\tau_2} Ju_2 = J\bar\sigma_\tau^{\tau_2} u_2.$$

But the right sides of these equalities are homologous to each other in ω_τ (from (1) and the commutability of the operator J with the operator ∇). This means that the left sides are homologous to each other in ω_τ, with which we have proved that $Ju_1 \sim Ju_2$ in A. Accordingly, J is a homomorphism of the group $\nabla^p_{\mathrm{ext}} A$ into the group $\nabla^p A$.

2. We shall now prove that J is a homomorphism onto the whole group $\nabla^p A$.

The proof rests on a lemma which is useful in many other cases.

Let $\omega = \{o_k\}$ be a covering of the set $A \subseteq S^n$. Suppose that $O_1, \ldots, O_k, \cdots$

---

[*] In the sense of Chapter II, §1.4.

are sets open in S^n which excise the sets $o_1, \cdots, o_k, \cdots$ (i.e., $o_k = A \cap O_k$), and suppose that here we have satisfied the condition: if any collection $O_{k_1}, \cdots, O_{k_s}$ has a nonempty intersection, then the corresponding $o_{k_1}, \cdots, o_{k_s}$ must also have a nonempty intersection. Under these conditions the covering $\Omega = \{O_k\}$, which evidently has the same nerve as the covering ω, is called an extension of the covering ω to the neighborhood $\lambda = \bigcup_k O_k$ of the set A.

Lemma. *Every covering ω of the set A may be extended to some neighborhood of that set.*

This lemma was proved under very general assumptions by E. Čech. The elementary proof presented here (for sets lying in metric spaces) is due to Ju. M. Smirnov. It consists in the following. To each point $x \in A$ we put into correspondence a set Ux open in A, which is the intersection of all those sets $o_k \in \omega$ containing the point x (of which there are a finite number). We put $\epsilon_x = \frac{1}{2} \rho (x, A - Ux)$. We denote by Ox a spherical neighborhood of radius ϵ_x of this point x in S^n, and we put $O_k = \bigcup_{x \in o_k} Ox$. We shall show that $\Omega = \{O_k\}$ is the desired extension of the covering ω. It is easily seen that $A \cap O_k = o_k$. Indeed, if $y \in A - o_k$, then for each $x \in o_k$ we will have $y \in A - Ux$, which means that $\rho (x, y) > 2x$, so that y does not enter into O_k.

It remains to prove that if some O_k — say, for simplicity, $O_1, \cdots, O_r$ —, have a nonempty intersection, then the intersection of the corresponding o_k is also nonempty. Choose the point y in $O_1 \cap \cdots \cap O_r$. Then for $k = 1, \cdots, r$ there exist points $x_k \in o_k$ such that $y \in Ox_k$, i.e., that $\rho (y, x_k) < \epsilon_{x_k} = \epsilon_k$. Suppose that the enumeration of the elements of the covering ω is chosen so that $\epsilon_1 \leq \epsilon_2 \leq \cdots \leq \epsilon_r$, then we shall show that $x_1 \in o_1 \cap \cdots \cap o_k$, with which the lemma will be proved. By definition we have $x_1 \in o_1$. We choose some k such that $1 < k \leq r$. Then

$$\rho (x_k, x_1) \leq \rho (x_k, y) + \rho (y, x_1) < \epsilon_k + \epsilon_1 \leq 2\epsilon_k.$$

But $2\epsilon_k = \rho (x_k, A - Ux_k) \leq \rho (x_k, A - o_k)$ and therefore $x_1 \in o_k$, as was required to prove.

As a result we shall prove that the homomorphism J maps the group $\nabla^p_{\text{ext}} A$ onto the whole group $\nabla^p A$. Let z be any ∇-cycle of the set A. It lies on the nerve of some covering $\omega = \{o_i\}$ of the set A. Following the lemma, we choose a covering $\Omega = \{O_i\}$ that extends the covering ω to the neighborhood $\lambda = \bigcup_i O_i$. Now we take a triangulation τ of the neighborhood λ such that each star of the triangulation τ is contained in some element $O_i \in \Omega$ (such a triangulation will be called *subordinate* to the covering Ω). For the triangulation τ, the canonical shift (see Part I, page 55) σ^τ_ω into the nerve Ω (coinciding with the nerve ω) is defined; i.e., a simplicial mapping is defined by the fact that each vertex e of the triangulation τ is put into correspondence with the set $O_i \in \Omega$ containing the star of the vertex e. The simplicial mapping σ_ω^τ generates, as usual, an adjoint homomorphism $\bar{\sigma}_\tau^\omega$ of the group $L^p\omega$ into the group $L^p\tau$, under which to a ∇-cycle z there corresponds a ∇-cycle $u = \bar{\sigma}_\tau^\omega z$ of the triangulation τ, i.e., an exterior ∇-cycle of the set A. Our object will be achieved as soon as we prove that $Ju \sim z$ in A.

To this end we consider the covering ω_τ. Its nerve is a subcomplex of the triangulation τ. The canonical shift σ_ω^τ, considered on ω_τ, is a canonical shift $\eth_\omega^{\omega\tau}$ of the nerve * ω_τ into the nerve ω. From the definition itself of the homomorphism $\pi_{\omega_\tau}^\omega$ adjoint to the shift $\eth_\omega^{\omega\tau}$ it follows that on each simplex $t^p \in \omega_\tau$

$$(\pi_{\omega_\tau}^\omega z \cdot t^p) = (z \cdot \eth_\omega^{\omega\tau} t^p) = (z \cdot \sigma_\omega^\tau t^p) = (u \cdot t^p).$$

But also $(Ju \cdot t^p) = (u \cdot t^p)$. Thus $\pi_{\omega_\tau}^\omega z = Ju$, which means that *a fortiori* $z \sim Ju$ in A. The assertion is proved.

3. We shall show finally, that the homomorphism J of the group $\nabla_{ext}^p A$ onto the group $\nabla^p A$ is an isomorphism. In other words, we shall prove that from $Ju \sim 0$ in A it follows that $u \sim 0$ around A. Let τ be a triangulation on which the cycle u lies. By definition of the operator J, the cycle Ju lies on the nerve of the covering ω_τ, excised from the set A by the principal stars of the triangulation τ. Since $Ju \sim 0$ in A, there exists a covering $\omega = \{o_i\}$, subordinate to ω_τ, such that $\pi_\omega^{\omega\tau} Ju \sim 0$ in ω. Following the lemma, we construct, for the covering ω, a covering $\Omega = \{O_i\}$ of the neighborhood $\lambda = \bigcup_i O_i$ of the set A excising it, so that $A \cap O_i = o_i$ for all i and the covering Ω has the same nerve as ω. Moreover, we require that each element of the covering Ω be contained in some star of the triangulation τ (this requirement may be satisfied since ω is subordinate to ω_τ). We put into correspondence with each set $O_i \in \Omega$ some principal star containing it, i.e., the star of some vertex of the triangulation τ. In this way a canonical shift $\eth = \eth_\tau^\Omega$ of the nerve Ω into the complex τ is defined, and moreover, evidently a canonical shift into the subcomplex ω_τ of that complex, so that we may write $\eth_\tau^\Omega = \eth_{\omega_\tau}^\Omega = \eth$. The natural isomorphism between Ω and ω makes it possible to consider the canonical shift $\eth$ as a canonical shift of the covering ω into ω_τ, so that $\eth = \eth_{\omega_\tau}^\omega$. Since $\eth$, considered as a mapping of the nerve $\Omega = \omega$ into τ_1, is in reality a mapping into ω_τ, on each simplex of the nerve $\Omega = \omega$ the chains $\pi_\Omega^\tau u$ and $\pi_\omega^{\omega\tau} Ju$ have one and the same value. But by definition,

---

* Indeed, the shift σ_ω^τ is defined by the fact that for each principal star of the complex τ one has chosen an element O_i of the covering Ω containing it. The intersection of this star with A is an element of the covering ω_τ, to which we have put into correspondence the element $o_i = A \cap O_i$ of the covering ω.

$\pi_{\omega}^{\omega\tau} Ju \sim 0$ in ω, so that

$$\pi_{\Omega}^{\tau} u \sim 0 \text{ in } \Omega. \tag{2}$$

We choose a triangulation τ' of the neighborhood $\lambda = \bigcup_i O_i$ of the set A that is *subordinate to the covering* Ω (in the sense that each star of the triangulation τ is contained in at least one $O_i \in \Omega$). Then there exists a canonical shift $\mathfrak{D}_{\Omega}^{\tau'}$ of the triangulation τ' into the nerve $\Omega = \omega$, and the simplicial mapping $\mathfrak{D}_{\tau}^{\Omega}\mathfrak{D}_{\Omega = \omega}^{\tau'}$ represents a canonical shift $\mathfrak{D}_{\tau}^{\tau'}$ of the triangulation τ' into τ. Therefore the ∇-cycle

$$\pi_{\tau'}^{\tau} u = \pi_{\tau'}^{\Omega} \pi_{\Omega}^{\tau} u$$

is defined. It is a ∇-extension of the ∇-cycle u in the triangulation τ'. Since in view of (2) we have $\pi_{\Omega}^{\tau} u \sim 0$ in Ω, then

$$\pi_{\tau'}^{\tau} u = \pi_{\tau'}^{\Omega} \pi_{\Omega}^{\tau} u \sim 0 \text{ in } \tau',$$

i.e., $u \sim 0$ around A, as was required to be proved.

§3. The isomorphism D of the group $\nabla_{\text{ext}}^p A$ onto the group $\Delta_{\text{ext}}^{n-p} A$

1. Construction of the isomorphism D. Reduction to the fundamental lemma.

To each exterior p-dimensional ∇-cycle z^p of the set A that lies on some triangulation τ of some neighborhood λ of the set A there corresponds an $(n-p)$-dimensional star cycle $D^* z^p$ of dimension $n-p$, which on the oriented barycentric star $^* t^{n-p}$ adjoint to a given arbitrary oriented simplex t^p of the triangulation τ takes the same value as the cycle z^p takes on t^p (see Part I, Chapter IV, §2; the discussion presented there is not affected by the fact that τ, being the triangulation of the open manifold λ is an infinite complex).

Considering the star cycle $D^* z^p$ as a cycle of the barycentric subdivision $\tau_{(1)}$ of the triangulation τ, and subjecting the complex $\tau_{(1)}$ to a canonical shift into the triangulation τ, we carry the cycle $D^* z^p$ into some cycle Dz^p of the triangulation τ, whose homology class in τ does not depend on the particular choice of the canonical shift of the complex $\tau_{(1)}$ into τ. It is easy to see (see the place in Part I referred to above) that from $z^p \sim 0$ in τ it follows that $Dz^p \sim 0$ in τ. We shall prove that if

$$z_1^p \sim z_2^p \text{ around } A, \tag{1}$$

then

$$Dz_1^p \sim Dz_2^p \text{ around } A, \tag{1_D}$$

from which it follows that we have a homomorphism D of the group $\nabla_{\text{ext}}^p A$ into $\Delta_{\text{ext}}^{n-p} A$. In order to prove that indeed (1) implies (1_D), we need to prove the following proposition:

Fundamental lemma. *If τ' follows τ, then for any ∇-cycle z of the triangu-*

lation τ we have

$$Dsz \sim sDz \text{ in } \tau',$$

where s means: on the left, the operator of ∇-extension, and on the right the operator of Δ-extension from τ to τ'.

Suppose that the fundamental lemma has been proved, and suppose that

$$z_1^p \sim z_2^p \text{ around } A. \tag{1}$$

We denote the triangulations on which the ∇-cycles z_1 and z_2, respectively, lie, by τ_1 and τ_2. The homology (1) means that there exists a triangulation τ of some neighborhood of the set A, following both τ_1 and τ_2, on which $s_\tau^{\tau_1} z_1 \sim s_\tau^{\tau_2} z_2$. Then $D s_\tau^{\tau_1} z_1 \sim D s_\tau^{\tau_2} z_2$ in τ. But in view of the fundamental lemma, $D s_\tau^{\tau_1} z_1 \sim s_\tau^{\tau_1} D z_1$ and $D s_\tau^{\tau_2} z_2 \sim s_\tau^{\tau_2} D z_2$ in τ, so that $s_\tau^{\tau_1} D z_1 \sim s_\tau^{\tau_2} D z_2$ in τ, i.e., we have the homology (1_D).

From the fundamental lemma it follows further that the homomorphism D is an isomorphism of the group $\nabla_{\text{ext}}^p A$ into the group $\Delta_{\text{ext}}^{n-p} A$. To show this, we need to prove that from

$$Dz \sim O \text{ around } A \tag{2_D}$$

follows

$$z \sim O \text{ around } A. \tag{2}$$

But the homology (2_D) means that for an appropriate choice of the triangulation τ', following τ, we have $s_{\tau'}^\tau D z \sim O$ in τ'. But then, in view of the fundamental lemma we have $D s_{\tau'}^\tau z \sim O$ in τ', so that also $D^* s_{\tau'}^\tau z \sim O$ in the star complex $\overset{*}{\tau}{}'$ adjoint to the triangulation τ'. But then $s_{\tau'}^\tau z \sim O$ in τ', which means that (2) holds.

It is easy, finally, to see that the isomorphism D is a mapping onto the whole group $\Delta_{\text{ext}}^{n-p} A$. Indeed, taking arbitrarily $\eta \in \Delta_{\text{ext}}^{n-p} A$ and a cycle $y \in \eta$ that lies on the triangulation τ, we may find a star cycle y^* homologous to it in $\tau_{(1)}$ (see Part I, Chapter IV, end of §1). For the ∇-cycle $z = (D^*)^{-1} y^*$, we have $D^* z = y^*$, so that, denoting by $\zeta \in \nabla_{\text{ext}}^p A$ the homology class of the ∇-cycle z, we obtain $D\zeta = \eta$.

Thus, everything has been reduced to the proof of the fundamental lemma. This proof will take up the rest of the present section.

2. Auxiliary material and plan of proof of the fundamental lemma. First we consider the case when the triangulation τ' is a subdivision of the triangulation τ, i.e., when τ' follows τ and has the same body as the triangulation τ. In this case we have the lemma proved in Part I, Chapter V, §§2 and 3.

Combinatorial lemma. *Let τ' be a subdivision of the triangulation τ; denote by $\tau_{(1)}$ and $\tau'_{(1)}$ barycentric subdivisions of the triangulations τ and τ' respectively. Let σ be a canonical shift of the triangulation τ' into the triangulation τ.*

Denote by $\sigma_{(1)}$ *the simplicial mapping of the complex* $\tau'_{(1)}$ *into* $\tau_{(1)}$ *that puts into correspondence with each vertex of the triangulation* $\tau'_{(1)}$ *(i.e., the barycenter of some simplex* $t' \in \tau'$*) the center of the simplex* $\sigma t' = t \in \tau$. *Then*

$$\sigma_{(1)} D^* s^{\tau}_{\tau'} z = D^* z \tag{3}$$

(where the star cycles are considered as cycles of the triangulation $\tau'_{(1)}$*, respectively* $\tau_{(1)}$*).*

From this formula (proved, as already mentioned, in Part I, Chapter V, §3) it follows that the Δ-cycle $D^* sz$ (considered as a cycle of the complex $\tau'_{(1)}$) goes into $D^* z$ (considered as a cycle in the complex $\tau_{(1)}$) under the shift $\sigma_{(1)}$, and thus is homologous to the cycle $D^* z$ in the polyhedron $\tilde{\tau}'$; accordingly, $Dsz \sim Dz$ in $\tilde{\tau}'$, and therefore $Dsz \sim sDz$ in τ'.

Thus, the fundamental lemma for subdivisions is a consequence of the combinatorial lemma (3). This last formula is quite elementary and remains valid if one regards z as an arbitrary chain, not necessarily a ∇-cycle. One has to observe that in the general case (when τ' follows the triangulation τ but is not a triangulation of it) there is no kind of identity analogous to the identity (3), and therefore the proof of the fundamental lemma turns out to be complicated. It will be carried out according to the following plan.

In the formulation of the fundamental lemma we replace the triangulation τ' by one of its subdivisions τ'', and begin by showing that if the lemma holds for τ and τ'', then it holds in its original form, i.e., for τ and τ'. This is the first step of the proof.

The second step consists in proving the lemma for τ and for a particular subdivision τ'' of the triangulation τ' that satisfies certain special conditions which are formulated below in the "geometrical lemma". Finally, we prove that such a subdivision τ'', satisfying these special conditions, actually exists, i.e., we prove the "geometrical lemma". This is the third stage of the proof. With these, the fundamental lemma will be completely proved.

3. *The first step in the proof of the fundamental lemma.* Accordingly, let τ'' be a subdivision of the triangulation τ'. If $s^{\tau}_{\tau'}$ is the extension operator from τ to τ', and $s^{\tau'}_{\tau''}$ the extension operator (in the present case even simply the subdivision operator) from τ' to τ'', then $s^{\tau'}_{\tau''} s^{\tau}_{\tau'}$ is the extension operator from τ to τ''. We assume proved the homology

$$(s^{\tau'}_{\tau''} s^{\tau}_{\tau'}) Dz \sim D(s^{\tau'}_{\tau''} s^{\tau}_{\tau'}) z \quad \text{in } \tau'' \tag{4}$$

and have to derive from it the homology

$$s^{\tau}_{\tau'} Dz \sim D s^{\tau}_{\tau'} z \quad \text{in } \tau'. \tag{5}$$

The fundamental lemma has just been proved for subdivisions, so that

$$s^{\tau'}_{\tau''} D s^{\tau}_{\tau'} z \sim D s^{\tau'}_{\tau''} s^{\tau}_{\tau'} z \ \text{ in } \tau''.$$

Moreover, in view of (4),

$$D s^{\tau'}_{\tau''} s^{\tau}_{\tau'} z \sim s^{\tau'}_{\tau''} s^{\tau}_{\tau'} D z \ \text{ in } \tau''.$$

This means that

$$s^{\tau'}_{\tau''} s^{\tau}_{\tau'} D z \sim s^{\tau'}_{\tau''} D s^{\tau}_{\tau'} z \ \text{ in } \tau''.$$

But once the subdivisions $s^{\tau'}_{\tau''}(s^{\tau}_{\tau'} D z)$ and $s^{\tau'}_{\tau''}(D s^{\tau}_{\tau'} z)$ of the cycles $s^{\tau}_{\tau'} D z$ and $D s^{\tau}_{\tau'} z$ are homologous in τ'', then the cycles themselves are homologous in τ', i.e., the homology (5) holds.

4. **Second step of the proof of the fundamental lemma; auxiliary material.** We begin with the formulation of the "geometrical lemma."

Geometrical lemma. *Suppose that we are given a triangulation τ of the open set λ, and a triangulation τ' following it, of the open set λ'. Then there exists a subdivision τ'' of the triangulation τ' that satisfies the following condition: the triangulation τ'' may be represented as the sum of an increasing sequence of finite closed subcomplexes K_m of it, where the complex K_m is contained in the open kernel of the complex* K_{m+1} and is itself the closure of its open kernel**, and all the K_m, $m = 0, 1, 2, \cdots$, are respectively subcomplexes of triangulations τ_m, of which, beginning with $m = 1$, each is a subdivision of the preceding one and $\tau_0 = \tau$.*

As we have said, the proof of this lemma constitutes the concluding portion of the present section, and now, assuming that the geometrical lemma has been proved, we shall prove the fundamental lemma for τ and for a subdivision τ'' of the triangulation τ' satisfying the conditions of the geometrical lemma. *Instead of τ'' we shall again write τ'.*

So along with the triangulation τ and the triangulation τ' following it we are given a sequence of triangulations

$$\tau_0 = \tau, \ \tau_1, \ \tau_2, \ \cdots, \ \tau_m, \ \cdots,$$

of which each succeeding one is a subdivision of its predecessor. In each triangulation τ_m there is singled our a finite closed subcomplex K_m such that

$$K_0 \subset K_1 \subset K_2 \subset \cdots \subset K_m \subset \cdots,$$

and the triangulation τ' is their sum: $\tau' = \bigcup_m K_m$. Hence it follows that each

---

* The open kernel of a triangulation K (which in the case at hand is finite) is the largest subcomplex of that triangulation whose body is open in S^n.

** By the closure of a (non-closed) subcomplex of a given triangulation we understand the combinatorial closure.

simplex $t \in \tau'$, which becomes for some m a simplex of the triangulation τ_m (and indeed of its finite closed subcomplex K_m), already does not undergo any further subdivision on the transition from τ_m to τ_{m-1} and so forth, and enters in an invariant form into all of these triangulations.

We shall denote by σ^m $(m = 1, 2, 3, \cdots)$ any canonical shift of the triangulation τ_m into the triangulation τ_{m-1} (thus, we shall write σ^m instead of $\sigma^{\tau_m}_{\tau_{m-1}}$). Let t be any simplex of the triangulation τ'. There exists a smallest m such that $t \in K_m$ and accordingly t does not undergo further subdivision. The vertices of the simplex t belonging to the complex τ_m remain in place under all the shifts $\sigma^{m+1}, \sigma^{m+2}, \cdots$; also the simplex t itself remains in place. As to the shifts $\sigma^m, \sigma^{m-1}, \cdots, \sigma^1$, under the shift σ^m the simplex t goes over into some simplex $\sigma^m t \in \tau_{m-1}$, which under the shift σ^{m-1} goes into some simplex $\sigma^{m-1}\sigma^m t \in \tau_{m-2}$, and so forth. Ultimately, as a result of the successive shifts, the simplex t goes into some simplex $\sigma^1 \sigma^2 \cdots \sigma^{m-1} \sigma^m t \in \tau$. Thus the sequence of shifts σ_m uniquely determines some canonical shift $\sigma = \sigma^{\tau'}_{\tau}$ of the triangulation τ' into τ. This shift σ determines also an adjoint homomorphism $s = s^{\tau'}_{\tau}$ of the the group $L\tau$ into the group $L\tau'$, and accordingly a certain definite ∇-extension sz of the cycle z of the triangulation τ.

On the other hand, the canonical shifts $\sigma^1, \sigma^2, \cdots, \sigma^m, \cdots$ successively determine ∇-cycles belonging respectively to the triangulations $\tau_1, \tau_2, \cdots, \tau_m, \cdots$, namely

$$s_1 z = s^{\tau}_{\tau_1} z, \ s_2 z = s^{\tau_1}_{\tau_2} s_1 z, \cdots, \ s_m z = s^{\tau_{m-1}}_{\tau_m} s_{m-1} z, \cdots .$$

On each simplex $t \in \tau'$ all the ∇-cycles $s_m z$, beginning at some point, take on one and the same value, and that value, as is easily seen, coincides with the value on that simplex of the ∇-cycle sz. Thus, considering ∇-cycles as functions defined on the set of oriented simplexes of a given complex, we may write

$$sz = \lim_{m \to \infty} s_m z. \tag{6}$$

For each $s_m z$ we choose in τ_m a star cycle $D^* s_m z$, and in τ' we choose the star cycle $D^* sz$. For each simplex $t^p \in \tau'$ there exists an m such that all the star simplexes t^p in τ' belong in an invariant way to the triangulations $\tau_m, \tau_{m+1}, \cdots$. Accordingly, the barycentric stars adjoint to the simplex t^p in the triangulation τ' are also barycentric stars adjoint to that simplex in all the triangulations $\tau_m, \tau_{m+1}, \cdots$, and on them the cycles $D^* s_m z, D^* s_{m+1} z, \cdots$ take on one and the same value, equal to the value of the cycle $D^* sz$. In other words, in complete analogy with (6) we may write

$$D^* sz = \lim_{m \to \infty} D^* s_m z. \tag{7*}$$

Our problem is the proof of the fundamental lemma

$$Dsz \sim sDz \text{ in } \tau',$$

where on the right s is a Δ-extension (uniquely defined) from τ to τ', and on the left is a ∇-extension from τ to τ', defined only up to homology in τ'. The point of the plan which we have adopted for the proof of the fundamental lemma, i.e., the use of the triangulation τ' satisfying the conditions of the geometrical lemma, is that, as ∇-extensions z we may choose not just any ∇-extension of the cycle z, but indeed the special ∇-extension which we have constructed.

These preliminary considerations complete the following important proposition on the cycles $s_m z$. In view of the combinatorial lemma, the cycle $D^* s_m z$ goes into a cycle $D^* s_{m-1} z$ under the shift $\sigma_{(1)}^m$, which is a canonical shift of a barycentric subdivision of the triangulation τ_m into a barycentric subdivision of the triangulation τ_{m-1}. Then, on the complex K_{m-1} (which, as we recall, consists of simplexes not undergoing further subdivision and entirely entering into τ') the shift $\sigma_{(1)}^m$ is an identity mapping, and the cycles $D^* s_m z$, $D^* s_{m-1} z$ coincide. Therefore their difference lies on $\tau_m - K_{m-1}$, and the homology, realized by the shift $\sigma_{(1)}^m$, is a homology in $\tau_m - K_{m-1}$. Thus $D^* s_m z \sim D^* s_{m-1} z$ in $\tau_m - K_{m-1}$. This is the auxiliary formula which we had in mind.

5. Second step in the proof of the fundamental lemma: final construction. Thus we need to construct a chain x of the complex τ' for which

$$\Delta x = sDz - Dsz. \tag{8}$$

This chain will be constructed as a sum of an infinite sequence of chains $x = \sum\limits_{m} x_m$, it being clear that this sum, in order to have a meaning, must be obtained with a choice of its terms such that on each simplex $t \in \tau'$ only a finite number of terms x_m can be different from zero.

We now proceed to the construction of the chain x.

A canonical shift of the barycentric subdivision $\tau_{(1)}$ of the triangulation τ into the same triangulation carries the cycle $D^* z$ into a definite cycle Dz. In the same way a canonical shift of a barycentric subdivision of the complex τ' into that complex carries the cycle $D^* s_1 z$ into the cycle $Ds_1 z$. Here, because of the combinatorial lemma, the following homology holds: $s_{\tau_1}^\tau Dz \sim Ds_1 z$ in τ_1. Subjecting both of its sides to an extension from τ_1 to τ', we obtain the homology $sDz \sim sDs_1 z$ in τ'. Accordingly, we have a chain x_1 in τ', for which

$$\Delta x_1 = sDz - sDs_1 z. \tag{9_1}$$

This is the first point of our construction. We turn to the second.

Now we carry out a canonical shift of the barycentric subdivision of the triangulation τ_2 into the triangulation τ_2, such that the shift coincides, on the barycentric subdivision of the complex K_1, with the shift just defined (the first point of the construction). Here, the cycle $D^* s_2 z$ goes into a definite cycle $Ds_2 z$ of the triangulation τ_2, coinciding on K_1 with the cycle $Ds_1 z$. In view of the combinatorial lemma there exists a chain y_2, lying (according to the auxiliary formula at the end of subsection 4) outside the open kernel of the complex K_1, for which

$$\Delta y_2 = s_{\tau_2}^{\tau_1} Ds_1 z - Ds_2 z.$$

Operating on both sides of this equality with the extension operator from τ_2 to τ', we obtain a chain x_2 of the triangulation τ', and the equality

$$\Delta x_2 = sDs_1 z - sDs_2 z, \tag{9_2}$$

where x_2 lies outside the open kernel of the polyhedron $\widetilde{K}_1$. Continuing this process, we obtain for any m a chain x_m of the triangulation τ' which lies outside the open kernel of the polyhedron $\widetilde{K}_{m-1}$ and satisfies the equality

$$\Delta x_m = sDs_{m-1} z - sDs_m z. \tag{9_m}$$

This sequence of appropriately chosen shifts of the barycentric subdivisions of the triangulations τ_m into the τ_m's themselves defines a certain shift of the barycentric subdivision $\tau'_{(1)}$ of the triangulation τ' into the triangulation τ', and thus a definite cycle

$$Dsz = \lim_{m \to \infty} Ds_m z. \tag{7}$$

Since each simplex of the triangulation τ' is eventually in the open kernel of some K_m, and the corresponding chain x_{m+1} lies outside that open kernel, on each simplex of the triangulation τ' only a finite number of chains x_m can be different from zero; therefore the sum of all these chains is defined:

$$x = \sum_{m=1}^{\infty} x_m.$$

We shall prove that the chain x realizes the desired homology

$$\Delta x = sDz - Dsz. \tag{10}$$

We take an arbitrary $(n-p)$-dimensional simplex $t \in \tau'$. We find an m sufficiently large that t lies in the open kernel of K_m. Adding the equations (9_1), (9_2), $\cdots$, (9_m), we obtain

$$\Delta(x_1 + \cdots + x_m) = sDz - sDs_m z. \tag{11}$$

But on t the value of the chains Δx and $\Delta(x_1 + \cdots + x_m)$ are the same: as a matter of fact, if $i \geq m+1$ all the chains x_i lie outside the open kernel of K_m, which means that all the boundaries of these chains lie outside the open kernel of K_m.

Therefore, comparing (10) and (11), we see that we need only prove the equation

$$(sDs_m z \cdot t) = (Dsz \cdot t).$$

But $(sDs_m z \cdot t) = (Ds_m z \cdot t)$, since the simplex t on the transition from τ_m to τ' is not subdivided, and by formula (7) we have

$$(Ds_m z \cdot t) = (Dsz \cdot t).$$

Formula (10) is thus proved.

We turn to the proof of the geometrical lemma, with which we shall have completed the proof of the fundamental lemma, and thus the isomorphism $\nabla^p_{\text{ext}} A = \Delta^{n-p}_{\text{ext}} A$.

Suppose we are given a triangulation τ' following the triangulation τ.

We represent τ' in the form of a sum of a sequence of finite closed subcomplexes Q_m, where each complex Q_m is contained in the open kernel of the complex Q_{m+1}; moreover, we suppose that each complex Q^m consists of n-dimensional simplexes and their faces (this means that Q_m is the closure of its open kernel, and makes it possible, in particular, to speak of the boundary of the complex Q_m). We choose any Q_m; through all the $(n-1)$-dimensional elements of the complex Q_m we pass planes carrying them. These planes decompose the simplexes of the complexes τ and τ' into convex polyhedra; we thus obtain a star-finite complex of convex polyhedra which are subdivisions of the triangulations τ and τ'; on subdividing these complexes of convex polyhedra simplicially, we obtain triangulations θ_m and θ'_m, which are subdivisions of the triangulations τ and τ'. Here the complex Q_m, subdivided by the triangulation θ'_m, turns out to be a subcomplex not only of that triangulation, but also of the triangulation θ_m (the complex Q_m is subdivided in the same way by the triangulations θ_m and θ'_m). We take the complex Q_1; it is important for us that its boundary is subdivided in the same way in θ_2 and in θ'_2. Take any subdivision of the complex Q_1 that coincides on its boundary* with the subdivision of this boundary in θ_2 (in θ'_2). The complex Q_1 subdivided in this way we will denote by K_1; the triangulation τ'' we shall take on Q_1 to coincide with K_1, and we shall take the subdivision τ_1 of the triangulation τ to be equal to K_1 on Q_1 and equal to θ_2 outside of Q_1.

We turn to the "ring" $Q_2 - Q_1$. Its outer boundary (i.e., the boundary of the complex Q_2) is subdivided in the same way by the triangulations θ_3 and θ'_3. Now we construct a subdivision of the ring $Q_2 - Q_1$, already subdivided by the triangulation τ_1, which coincides on the outer boundary of the ring $Q_2 - Q_1$ with

---

*As this subdivision we could of course have taken the subdivision in θ'_2. However, for analogy with what happens later we choose simply a central subdivision of the simplexes adjoining the boundary.

the subdivision of this boundary in the triangulation θ_3', and which does not sub-divide the interior boundary of the ring at all (i.e., the boundary of the complex Q_1 subdivided in τ_1). The resulting subdivision of the ring $Q_2 - Q_1$ we take on that ring to coincide with the triangulation τ'', which thus turns out to be deter-mined on the whole complex Q_2; that complex, subdivided in the triangulation τ'', we will denote by K_2. The subdivision τ_2 of the triangulation τ we take to be equal to K_2 on Q_2 and θ_3 outside Q_2.

We pass in this way from ring to ring, and thus define stepwise a triangulation τ'' as the sum of the appropriate subdivisions of the rings $Q_{m+1} - Q_m$, and also the corresponding triangulation τ_m.

The geometrical lemma is proved.

§4. The isomorphism Γ of the group $\Delta_{\mathrm{ext}}^{q+1} A$ onto the group $\Delta^q B$

1. **Construction of the isomorphism Γ.** This consists in putting into corre-spondence to each exterior Δ-cycle u^{q+1} of the set A, lying on the triangulation τ on the neighborhood λ of the set A, some Δ-cycle* $z^q = \Gamma u^{q+1}$ of the set B; here the cycle Γu^{q+1} depends not only on the cycle u^{q+1}, but also on the par-ticular representation of the triangulation τ on which it lies, in the form of a sum of an increasing sequence of its closed finite subcomplexes, and also on certain shifts applied in the construction of the cycle Γu^{q+1}. However it turns out that the homology class of the cycle Γu^{q+1}, i.e., the element of the group $\Delta^q B$ deter-mined by it, does not depend on the choice of these elements; moreover, to exte-rior cycles u_1^{q+1} and u_2^{q+1}, homologous to each other around A, there corresponds one and the same element of the group $\Delta^q B$, so that the operator Γ turns out to be a homomorphism of the group $\Delta_{\mathrm{ext}}^{q+1} A$ into the group $\Delta^q B$, concerning which it will further be proved that it is an isomorphism and moreover an isomorphism onto the whole group $\Delta^q B$. With this we shall bring to completion both the construc-tion of the Sitnikov isomorphism M between the groups $\nabla^p A$ and $\Delta^q B$, and the proof of the duality law realizing it.

We proceed to the execution of this plan.

Thus, suppose we are given a triangulation τ of the neighborhood λ of the set A and on that triangulation a (generally speaking, infinite) Δ-cycle u^{q+1}. We represent the triangulation τ in the form of a sum of increasing finite closed subcomplexes Q_k, with Q_k contained in the open kernel of the complex Q_{k+1}.

Denote by u_k^{q+1} the piece of the cycle u^{q+1} lying in Q_k, and put

$$z_k^q = \Delta u_k^{q+1} ,$$

---

*By Δ-cycle and Δ-homology we always understand cycle and homology in the sense of Sitnikov.

$$x_k^{q+1} = u_{k+1}^{q+1} - u_k^{q+1}.$$

Then $\Delta x_k^{q+1} = z_{k+1}^q - z_k^q$, so that $z^q = \{z_k^q, x_k^{q+1}\}$, or in detail,

$$z^q = \begin{bmatrix} z_1^q, & z_2^q, & \cdots, & z_k^q, & \cdots \\ x_1^{q+1}, & x_2^{q+1}, & \cdots, & x_k^{q+1}, & \cdots \end{bmatrix},$$

is a Δ-cycle (of the whole space S^n). The cycles z_k^q and the chains x_k^{q+1} lie in λ, but the distances of their vertices from the compactum $\psi = S^n - \lambda$ converge to zero as k increases. Therefore, putting into correspondence with each vertex of the cycle u^{q+1} the closest point to it of the compactum $\psi \subseteq B$, we obtain an infinitely small shift of the cycle $z^q = \{z_k^q, x_k^{q+1}\}$ into a Δ-cycle of the compactum ψ, which we again denote by $z^q = \{z_k^q, x_k^{q+1}\}$. This Δ-cycle is by definition the cycle Γu^{q+1}:

$$\Gamma u^{q+1} = \{z_k^q, x_k^{q+1}\}.$$

The homology class defined by the cycle Γu^{q+1} — an element of the group $\Delta^q B$ — does not depend on the choice of the complexes Q_k.

The proof of this assertion will begin with an auxiliary definition. We shall call a *cofinal section* of the Δ-cycle $z = \{z_k, x_k\}$ any Δ-cycle of the form

$$\begin{bmatrix} z_{k1}, & z_{k2}, & \cdots, & z_{kh}, & \cdots \\ \xi_1, & \xi_2, & \cdots, & \xi_h, & \cdots \end{bmatrix}$$

where the z_{kh} form a subsequence of the sequence $\{z_k\}$ and

$$\xi_h = x_{kh} + x_{kh+1} + \cdots + x_{kh+1-1}.$$

From the definition of Δ-cycle and Δ-homology it results without difficulty that each Δ-cycle, lying on a compactum $\psi \subseteq B$, is homologous on that same compactum to each of its cofinal sections. Therefore for the proof of our assertion it is sufficient to convince ourselves of the fact that for every two sequences of complexes

$$Q_1', Q_2', \cdots, Q_k', \cdots$$

and

$$Q_1'', Q_2'', \cdots, Q_k'', \cdots,$$

satisfying the conditions set at the beginning of this section, we may find a third sequence

$$Q_1, Q_2, \cdots, Q_k, \cdots,$$

also satisfying these conditions and containing as subsequences some subsequence of the sequence Q_k' and some sebsequence of the sequence Q_k''. To construct such a subsequence Q_k is easy: we put $Q_1 = Q_1'$. Further, we denote by Q_2 the first among the complexes Q_k'' of the second sequence whose open kernel contains the complex Q_1 (the existence of such a Q_k'' easily follows from the fact that the body of the finite triangulation Q_1 is a compactum, contained in the sum of the increasing open kernels of all the Q_k''). Then we denote by Q_3 the first among the complexes Q_k' whose open kernel contains Q_2, and so forth, always choosing the complex Q_k in turn from the complexes Q_k' and Q_k''.

Thus, the homology class of the cycle Γu^{q+1} is uniquely determined by that cycle u^{q+1} itself.

2. Proof of the fact that the operator Γ generates a completely defined homomorphism of the group $\Delta^{q+1}_{\text{ext}}$ into the group $\Delta^q B$. This is contained in the proof of the following two assertions.

a) If the triangulation τ' follows τ, then always

$$\Gamma u^{q+1} \sim \Gamma s_{\tau'}^{\tau} u^{q+1} \text{ in } B.$$

b) If the exterior cycle u^{q+1} of the set A is homologous to zero around A, then $\Gamma u^{q+1} \sim 0$ in B.

Proof of assertion a). We take a sequence of increasing finite triangulations $Q_k \subset \tau$ and $Q_k' \subset \tau'$, $k = 1, 2, 3, \cdots$, such that the body $\widetilde{Q}_k'$ of the complex Q_k' is contained in the open kernel $(\widetilde{Q}_k)$ of the body $\widetilde{Q}_k$ of the complex Q_k. These bodies are finite polyhedra, and therefore the difference $\widetilde{Q}_k - (\widetilde{Q}_k')$ is also a finite polyhedron. Taking such a subdivision of the complex Q_k that we have on Q_k' a subdivision of the complex Q_k', it is easy to see that the piece v_k^{q+1} of the (infinite) cycle* u^{q+1}, lying on the thus subdivided "ring" $\widetilde{Q}_k - (\widetilde{Q}_k')$, is a finite chain, having as its boundary the dif-ference of the subdivisions of the finite cycles $z_k^q - z_k'^q$; the schema of this is shown in Figure 1. Taking the prism of the shifts π_k^{q+1} and π'^{q+1}_k of the subdivided cycles z_k^q and $z_k'^q$ into the non-subdivided cycles, we obtain

$$\Delta(v_k^{q+1} - \pi_k^{q+1} + \pi'^{q+1}_k) = z_k^q - z_k'^q.$$

Thus, the finite chains

$$y_k^{q+1} = v_k^{q+1} - \pi_k^{q+1} + \pi'^{q+1}_k,$$

$$\Delta y_k^{q+1} = z_k^q - z_k'^q$$

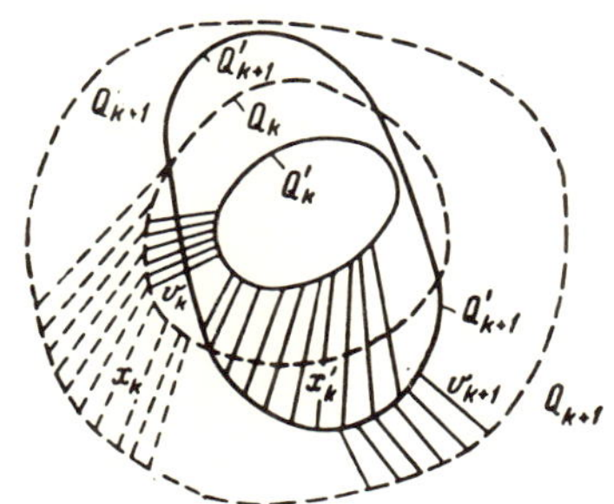

Figure 1

____

* We denote by $\{z_k^q, x_k^{q+1}\}$ the cycle Γu^{q+1}, constructed by means of the sequence $\{Q_k\}$, and by $z'^q = \{z_k'^q, x_k'^{q+1}\}$ the cycle $\Gamma s_{\tau'}^{\tau} u^{q+1}$, constructed by means of the se-quence $\{Q_k'\}$.

(the "films" between the cycles z_k^q and z'^q_k) are constructed; for sufficiently large k they lie in an arbitrarily close neighborhood of the boundary of the open set $\lambda' = \tilde{\tau}'$, and therefore their union, under an infinitely small shift, goes into a true chain of the compactum

$$\psi = S^n - \lambda' \subseteq B.$$

For the completion of the proof of proposition a) we still need to construct finite chains x_k^{q+2} bounded by the (finite) cycles

$$y_{k+1}^{q+1} - y_k^{q+1} - x_k^{q+1} + x'^{q+1}_k$$

and lying in neighborhoods of the boundary of λ which contract as k increases. We take a triangulation K of the finite polyhedron $\tilde{Q}_{k+1} - (\tilde{Q}'_k)$, and a subdivision of the cycle $y_{k+1} - y_k - x_k + x'_k$ so small that it may be canonically shifted into the complex K. Since for sufficiently large k the complex K lies in an arbitrarily close neighborhood of the boundary of λ, our object will be achieved if we prove that under this shift the subdivided cycle $y_{k+1} - y_k - x_k + x'_k$ goes to zero.

Under our shift the subdivided chain $v_{k+1}^{q+1} + x'^{q+1}_k$ goes into a piece of the cycle u^{q+1} lying on K and subdivided by that complex (see Figure 2). Since each of the subdivided prisms π_k^{q+1} and π'^{q+1}_k (as any $(q+1)$-dimensional chain lying on a q-dimensional polyhedron — in the case at hand lying on the body of the cycle z_k^q, respectively z'^q_k) goes under this shift into zero, then also the chain

$$y_{k+1}^{q+1} + x'^{q+1}_k = v_{k+1}^{q+1} + x_k^{q+1} - \pi_k^{q+1} + \pi'^{q+1}_k$$

goes under our shift into a piece of the cycle u^{q+1} lying on K. But also the chain

$$y_k^{q+1} + x_k^{q+1} = v_k^{q+1} + x_k^{q+1} -$$
$$- \pi_k^{q+1} + \pi'^{q+1}_k$$

goes into this same piece under our shift, from which it follows that the chain $y_{k+1}^{q+1} - y_k^{q+1} - x_k^{q+1} + x'^{q+1}_k$ goes into zero, with which all is proved.

Proof of assertion b). Suppose that the infinite cycle u^{q+1} (of the triangulation τ) bounds in τ the infinite chain v^{q+2}:

$$\Delta v^{q+2} = u^{q+1}.$$

We shall show that then the Sitnikov cycle Γu^{q+1} is also homologous to zero in the compactum $\psi = S^n - \tilde{\tau} \subseteq B$. The piece of the chain v^{q+2}, lying on the finite triangulation Q_k, will

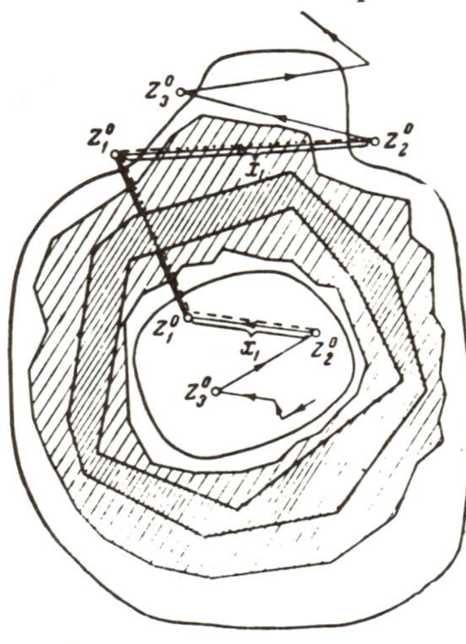

Figure 2

be denoted by v_k^{q+2}. Putting

$$w_k^{q+1} = u_k^{q+1} - \Delta v_k^{q+2},$$
$$x_k^{q+2} = v_k^{q+2} - v_{k+1}^{q+2},$$

we have

$$\Delta w_k^{q+1} = \Delta u_k^{q+1} = z_k^q,$$

$$\Delta x_k^{q+2} = \Delta v_k^{q+2} - \Delta v_{k+1}^{q+2} = (u_k^{q+1} - w_k^{q+1}) - (u_{k+1}^{q+1} - w_{k+1}^{q+1}) =$$

$$= (w_{k+1}^{q+1} - w_k^{q+1}) - (u_{k-1}^{q-1} - u_k^{q+1}),$$

i.e.,

$$\Delta x_k^{q+2} = w_{k+1}^{q+1} - w_k^{q+1} - x_k^{q+1}.$$

The chains w_k^{q+1} and x_k^{q+2}, for sufficiently large k, lie on an arbitrarily close neighborhood of the boundary of the set $\widetilde{r}$, and, accordingly, under an infinitely small shift give us true cycles of the compactum $\psi = S^n - \lambda \subseteq B$, which we shall again denote by $\{w_k^{q+1}\}$, $\{x_k^{q+2}\}$. These chains realize a Sitnikov homology

$$\Gamma u^{q+1} \sim 0 \text{ in } \psi \subseteq B,$$

with which assertion b) is proved.

Thus, indeed we have *a homomorphism of the group* $\Delta_{\text{ext}}^{q+1} A$ *into the group* $\Delta^q B$.

3. The homomorphism Γ is an isomorphism. Suppose that the Sitnikov cycle Γu^{q+1} is homologous to zero in B. We shall show that the infinite cycle u^{q+1}, lying on the triangulation r of some neighborhood λ of the set A, is homologous to zero around A, i.e., that its extension $s_{r'}^r u^{q+1}$ in some triangulation r', following r, is homologous to zero in r'.

In view of the construction itself of the cycle $\Gamma u^{q+1} = \{z_k^q, x_k^{q+1}\}$, we have an increasing sequence of finite triangulations $Q_k \subset r$, while the finite chains x_k^{q+1} and the cycles z_k^q have been obtained as the results of a shift onto the compactum $\psi = S^n - \lambda$ of the pieces x'^{q+1}_k of the cycle u^{q+1} lying on the finite complexes ("rings") $Q_{k+1} - Q_k$, and respectively of the boundaries z'^q_k of pieces lying on Q_k. We are given that $\Gamma u \sim 0$ in B, i.e., the Δ-cycle Γu^{q+1} of the compactum ψ bounds in the Sitnikov sense on some compactum ψ', $\psi \subseteq \psi' \subseteq B$. In other words, there exist ϵ_k-chains y_k^{q+1} on ψ' bounded by the cycles z_k^q and ϵ_k-chains x_k^{q+2} on ψ', $\epsilon_k \to 0$, bounded by the cycles $y_{k+1}^{q+1} - y_k^{q+1} - x_k^{q+1}$. We may suppose that for sufficiently large k all of the chains x_k^{q+1}, x_k^{q+2}, y_k^{q+1}, z_k^q, as well as the prisms of the shifts mentioned above, consist of nondegenerate spherical simplexes.*

Disregarding in the sequence $Q_1, Q_2, \cdots$ a finite number of initial terms, we may assume that the conditions laid down on smallness have been satisfied, beginning with $k = 1$. As always, we shall denote by u_k^{q+1} the piece of the cycle u^{q+1} lying on Q_k.

By π_k^{q+1}, for any k, we denote the prism of the shift carrying $z_k'^q$ into z_k^q. Then the cycle

$$u_1^{q+1} - y_1^{q+1} - \pi_1^{q+1}$$

bounds in S^n some chain x_0^{q+2}. We denote by π_k^{q+2} the prism of the shift of the chain $x_k'^{q+1}$ into x_k^{q+1} and consider the infinite chain

$$x^{q+2} = x_0^{q+2} - \sum_{k=1}^{\infty} (x_k^{q+2} - \pi_k^{q+2}). \tag{1}$$

The simplexes of this chain may also be taken to be nondegenerate simplexes in S^n. Among them only a finite number may have diameters exceeding any pre-assigned $\epsilon > 0$, and only finitely many of them can lie at a distance larger than ϵ from the compactum $\psi' = S^n - \lambda'$. Here *the chain* x^{q+2}, *which is generally speaking not a chain of any triangulation, has the property,* fundamental for all that follows, *of local finiteness* with respect to the open set $\lambda' = S^n - \psi' \supseteq A$. This property, approximating the chain x^{q+2} to chains which consist of simplexes of some triangulation, consists in the fact that each point $a \in \lambda'$ has a neighborhood intersecting not more than a finite number of simplexes of the chain x^{q+2}. From this property of local finiteness it follows that each simplex of any triangulation τ' of the set λ' can intersect only a finite number of simplexes of the chain x^{q+2}, and therefore, in its turn, it follows that each triangulation τ' of the set λ' has a subdivision θ' having the property that each simplex $t' \in \theta'$, intersecting with any simplex of the chain x^{q+2}, entirely lies on it. In order to obtain such a subdivision θ' it suffices to take all possible intersections of simplexes of the complex τ' with the simplexes of the chain x^{q+2} and their faces, which gives a complex of convex polyhedra, of which any simplicial subdivision may be taken as the desired triangulation.

We shall now prove that

$$\Delta x^{q+2} = u^{q+1}.$$

Indeed, we have:

$$\Delta x_0^{q+2} = u_1^{q+1} - y_1^{q+1} - \pi_1^{q+1},$$

---

* To this end, taking the radius of S^n as unity, it is sufficient to require that the skeletons of all of these simplexes have diameters less than $\pi/2$; but for sufficiently large k these skeletons have diameters less than any $\epsilon > 0$ given in advance.

$$\Delta x_k^{q+2} = y_{k+1}^{q+1} - y_k^{q+1} - x_k^{q+1},$$

$$\Delta \pi_k^{q+2} = - x_k^{q+1} + x'^{q+1}_k + \pi_k^{q+1} - \pi_{k+1}^{q+1}.$$

Introducing these expressions into formula (1), we obtain, after some evident sim-
plifications:

$$\Delta x^{q+2} = u_1^{q+1} + \sum_{k=1}^{\infty} x'^{q+1}_k = u^{q+1}.$$

We take any triangulation τ' of the neighborhood $\lambda' = S^n - \psi'$ of the set A, following τ. As we have seen, there exists a subdivision θ' of the triangulation τ'
such that each simplex $t' \in \theta'$, intersecting with any simplex t^{q+2} of the chain
x^{q+2}, entirely lies on that simplex t^{q+2}. Evidently θ' follows τ.

Now we define the operation of *reduction of the chain x^r of the space S^n by
a triangulation θ'* which has the property that each simplex t'^r of the triangulation θ', intersecting any simplex t^r of the chain x^r, entirely lies on t^r. The
reduction of a chain, which we shall denote by Jx^r, we shall define by giving it
on any simplex $t'^r \in \theta'$ lying on the simplexes $t_1^r, \cdots, t_s^r$ of the chain x^r,
oriented the same as t'^r, the value equal to the sum of the values of the chain x^r
on the simplexes $t_1^r, \cdots, t_s^r$ (of which there is a finite number) .*

Now we shall show that the reduction operator and the boundary operator Δ
commute with each other. In view of the linearity of both of these operators, it is
sufficient to prove their commutability for chains x^r consisting of a single simplex t^r. First of all, the boundary of the reduced chain Jt^r, consisting, generally
speaking, already of an infinite number of simplexes, lies on the boundary of the
simplex t^r. This follows from the fact that every $(r-1)$-dimensional simplex of
the chain Jt^r, lying inside a simplex t^r, has a solid spherical neighborhood in
the open set $\widetilde{\theta}' = \lambda'$ and thus is a face of the two likewise oriented r-dimensional
simplexes of Jt^r adjoining it.

Each $(r-1)$-dimensional simplex entering in ΔJt^r, and, by what has been
proved, lying on the boundary of the simplex t^r, is a face of only one r-dimensional
simplex of Jt^r adjoining it, and therefore enters in $J\Delta t^r$ with the same coefficient
(equal to $+1$ or -1) as ΔJt^r.

Thus, reducing the chain x^{q+2} by the triangulation θ', we obtain a chain
Jx^{q+2} of that triangulation, while

---

    *If $r < n$ and the chain x^r is taken in general position (so that the intersection of two
of its simplexes has a dimension less than r), then each simplex $t'^r \in \theta'$ can lie only on
one simplex t^r of the chain x^r, and then $(Jx^r \cdot t'^r) = (x^r \cdot t^r)$. This case may be regarded
as general.

    Remark. The reduction operator for a chain x^r is a generalization of the operation of
Δ-extension and goes into the latter if x^r is a chain of some triangulation τ and the
"reducing" triangulation θ follows τ.

$$\Delta J x^{q+2} = J \Delta x^{q+2} = J u^{q+1}.$$

But θ' follows τ and the operation of reduction by the triangulation θ' of the cycle u^{q+1} lying on τ is identical to the operation $s^{\tau}_{\theta'}$ of extension of that cycle. Therefore the cycle $s^{\tau}_{\theta'} u^{q+1} = J u^{q+1}$, bounding in θ' the chain $J x^{q+2}$, is homologous to zero in θ', and therefore the cycle u^{q+1} is homologous to zero around A, and the assertion that the operator Γ is an isomorphism of the group $\Delta^{q+1}_{\mathrm{ext}} A$ into the group $\Delta^q B$ is proved.

4. **Remark.** From considerations analogous to those just given, it results that every finite Δ-cycle u^{q+1}, lying on some triangulation τ (of an open set λ), is homologous to zero[*] in τ (from which it follows that the finite cycles u^{q+1} around A define the null element of the group $\Delta^{q+1}_{\mathrm{ext}} A$). Indeed, since S^n is acyclic in dimension $q+1$, then there exists a finite chain x^{q+2}, bounded by the cycle u^{q+1}. There exists a triangulation τ', which is a subdivision of the triangulation τ and has the property that each simplex t'^{q+2} of that triangulation, intersecting any simplex of the chain x^{q+2}, lies entirely on that simplex. Carrying out the operation of reduction of the chain x^{q+2} by the triangulation τ', we obtain the (generally speaking, infinite) chain $x'^{q+2} = J x^{q+2}$ of the triangulation τ', bounded by the cycle $s^{\tau}_{\tau'} u^{q+1}$, so that $s^{\tau}_{\tau'} u^{q+1} \sim 0$ in τ'. But then $u^{q+1} \sim 0$ in τ.

Analogous considerations for infinite cycles do not go through, because the infinite chain x^{q+2}, bounded by the cycle u^{q+1} in the space S^n, though it indeed exists (for example, a cone constructed over u^{q+1}), does not satisfy the condition of local finiteness relative to the set λ (see page 166).

5. **The operator Γ maps the group $\Delta^{q+1}_{\mathrm{ext}} A$ onto the entire group $\Delta^q B$.** Suppose that $z^q = \{z^q_i, x^{q+1}_i\}$ is any Δ-cycle lying on some compactum $\psi \subseteq B$. We shall construct an infinite Δ-cycle u^{q+1}, lying on some triangulation τ of the neighborhood $\lambda = S^n - \psi$ of the set A, such that $\Gamma u^{q+1} \sim z_q$ in B.

In the absence of any statement to the contrary we shall consider the cycles z^q_i and the chains x^{q+1}_i as simplicial cycles and chains in S^n whose simplexes are nondegenerate spherical simplexes. In view of the acyclicity of the sphere S^n in the dimension q, there exists a finite chain x^{q+1}_0 bounded by the cycle z^q_1.

We consider the infinite cycle $x^{q+1} = \sum\limits_{i=0}^{\infty} x^{q+1}_i$. For any $\epsilon > 0$, all of the chains x^{q+1}_i, beginning at some point, are contained in an ϵ-neighborhood of the compactum ψ, whence (since all of the x^{q+1}_i are finite chains) it immediately follows that the chain x^{q+1} satisfies the condition of local finiteness relative to

---

[*] I.e., bounds some, generally speaking, *infinite* chain of the triangulation τ.

λ, and each simplex of any triangulation τ of the set λ intersects only a finite number of simplexes of the cycle x^{q+1}. Therefore we may again construct a subdivision τ' of the triangulation τ such that each simplex of the triangulation τ', intersecting any simplex of the cycle x^{q+1}, entirely lies on it. Accordingly, we may apply to the cycle x^{q+1} the operation of reduction by the triangulation τ', which gives us the cycle $u^{q+1} = Jx^{q+1}$ of the triangulation τ'. We shall show that $\Gamma u^{q+1} \sim z^q$ in ψ. This will achieve our goal and the proof of the duality law will be complete.

We shall take for the basis of the construction of the cycle Γu^{q+1} a sequence of complexes Q_k which, aside from the ususl conditions, satisfies also the following: all the chains x_i^{q+1} for $i \geq k$ lie outside Q_k. Since x^{q+1} is a cycle, then for any k we have

$$z_k^q = \Delta \sum_{i=0}^{k-1} x_i^{q+1} = -\Delta \sum_{i=k}^{\infty} x_i^{q+1},$$

and since x_i^{q+1}, $i \geq k$ lie outside Q_k, then the z_k^q also lie outside the complex Q_k. We shall take a triangulation of the polyhedron which is the sum of the bodies $\widetilde{x}_i^{q+1}$ of the chains x_i^{q+1}, $i \leq k-1$, which on Q_k will be a subdivision of the complex Q_k. The chain $\sum_{i=0}^{k-1} x_i^{q+1}$, subdivided in this triangulation, may be represented in the form of the sum of two terms, of which the first is a piece lying on Q_k and coinciding, as one sees without difficulty, with the subdivided piece u_k^{q+1} of the chain u^{q+1}, and the second is a piece lying outside Q_k, which we shall denote by y_k^{q+1}. Since the boundary of the whole chain $\sum_{i=0}^{k-1} x_i^{q+1}$ is z_k^q, therefore

$$\text{subd } z_k^q = \text{subd } \Delta u_k^{q+1} + \Delta y_k^{q+1},$$

i.e.,

$$\Delta y_k^{q+1} = \text{subd } z_k^q - \text{subd } \Delta u_k^{q+1}.$$

But the elements z'^q_k, x'^{q+1}_k of the cycle $\Gamma u = \{z'^q_k, x'^{q+1}_k\}$ are exactly $z'^q_k = \Delta u_k^{q+1}$, $x'^{q+1}_k = u_{k+1}^{q+1} - u_k^{q+1}$, so that the chains y_k^{q+1} are the films between the subdivisions z_k^q and z'^q_k. Denoting by π_k^{q+1} and π'^{q+1}_k the prisms of the shifts of the subdivided cycles z_k^q, z'^q_k into the original cycles, we have

$$\Delta(y_k^{q+1} - \pi_k^{q+1} + \pi'^{q+1}_k) = z_k^q - z'^q_k.$$

It remains only to construct the chains x_k^{q+1} bounded by the cycles

$$w_k^{q+1} = (y_{k+1}^{q+1} - \pi_{k+1}^{q+1} + \pi'^{q+1}_{k+1}) - (y_k^{q+1} - \pi_k^{q+1} + \pi'^{q+1}_k) - x_k^{q+1} + x'^{q+1}_k$$

and lying in neighborhoods of the set ψ as close as desired for sufficiently large

k (the chains y_k^{q+1}, and also the w_k^{q+1} evidently satisfy the last condition).

The carrier of the cycle w_k^{q+1} is the polyhedron $\bigcup_{i=0}^{k} \tilde{x}_i^{q+1} - (Q_k)$. We shall denote by K any triangulation of this polyhedron, and choose a subdivision of the cycle w_k^{q+1} so fine that it may be canonically shifted into the complex K. Our object will be achieved if we show that under this shift the subdivided cycle w_k^{q+1} goes into zero. We shall do this, repeating the considerations of subsection 2 of this section (page 164) in the following way:

Under our shift the subdivided chain $y_{k+1}^{q+1} + x'_k{}^{q+1}$ goes into a piece of the chain $\sum_{i=0}^{k} x_i^{q+1}$ lying in and subdivided by K; but into this same piece there also goes the chain $y_k^{q+1} + x_k^{q+1}$, entering into the cycle w_k^{q+1} with a minus sign. Since the prisms entering into this cycle (being $(q + 1)$-dimensional chains lying on q-dimensional simplexes), under the canonical shift go into zero, our assertion is proved. With this, the proof of the duality law is complete.

§5. Spectral duality law

1. The groups $\nabla^r X$ and $\Delta^r X$ are limit groups of the direct spectra:

$$\nabla^r X = \varinjlim \{\nabla^r \alpha, \pi_\beta^\alpha\},$$

$$\Delta^r X = \lim \{\Delta^r \phi, E_\phi^\phi\} \tag{1}$$

which we shall in this section call respectively the r-dimensional ∇-spectrum and the r-dimensional Δ-spectrum of X (over a coefficient domain $\mathfrak{A}$ that is given once and for all). The Sitnikov duality law, whose proof constituted the content of the preceding section of this chapter, may be strenghthened in the following way:

Spectral duality law. *Let A and B be two complementary sets in S^n and $q = n - p - 1$. Then the p-dimensional ∇-spectrum of the set A and the q-dimensional Δ-spectrum of the set B, after an appropriate multiplication, contain isomorphic cofinal sections.*

This theorem (first proved by Sitnikov in $[17a]^*$ is indeed a strengthening of his duality law, since from the possibility of mapping one spectrum into the other by multiplication and passage to cofinal sections there evidently follows an isomorphism of the limit groups of both spectra. More generally, it is natural to call two spectra *equivalent*, if one of them may be obtained from the other by a finite number of passages, each of which is a passage to a spectrum which is a cofinal spectrum of the given one, or conversely a passage to a spectrum containing the given one as a cofinal section ("cofinal extension of one spectrum into another").

---

*See, besides, Chapter VI, which finishes up in a sense the really basic ideas involved in the first duality law of Chapter II.

Since multiplication constitutes a special case of cofinal extension, the most essential part of the spectral duality law may be formulated as follows:

the p-dimensional ∇-spectrum of A and the q-dimensional Δ-spectrum of B are equivalent to each other.

The proof of the spectral duality law is based on the concepts of canonical covering and canonical triangulation, and also on the concept of canonical neighborhood of a set. These are the same concepts on which the proof of the isomorphism $\delta^p A = D^p A$ in §1 of Chapter II was based. We shall formulate in the form of a special lemma the fundamental properties of canonical coverings of triangulations and of neighborhoods; as proved in Chapter II, §§1.4 and 1.5:

Lemma 1. *For each covering α and each neighborhood λ of the set A we can find a canonical triangulation τ such that the canonical covering α' excised by it follows α, and some canonical neighborhood λ' corresponding to it is contained in λ.*

We turn to the proof itself. The group $\nabla^p_{\text{ext}} A$ is the limit group of the direct spectrum

$$\{\nabla^p \tau, \, s^\tau_{\tau'} \}, \qquad\qquad (1_{\text{ext}})$$

where the τ are all possible triangulations of all possible neighborhoods of the set A.

All the groups $\nabla^p \tau$, where τ are all possible triangulations of one and the same neighborhood λ, are isomorphic to each other. Therefore we may, by identifying all of these groups, obtain a single group $\nabla^p(\lambda)$. More simply, we shall do this as follows. We shall call a cycle lying on any of the triangulations of the set λ a *cycle of the set λ*. Two ∇-cycles z^p_1 and z^p_2, lying on the triangulations τ_1 and τ_2 of the set λ, will be said to be homologous to one another in λ if there exists a triangulation τ of the set λ, following both τ_1 and τ_2, such that

$$s^{\tau_1}_\tau z^p_1 \sim s^{\tau_2}_\tau z^p_2 \text{ in } \tau.$$

The classes thus defined, of mutually homologous ∇-cycles, are the elements of the groups $\nabla^p(\lambda)$.

If λ and λ' are two neighborhoods of the set A, then the homomorphism $s^\lambda_{\lambda'}$ is defined as follows. We take any triangulation τ of the neighborhood λ and a triangulation τ' of the neighborhood λ' which follows τ. In each homology class $\zeta^p \in \nabla^p(\lambda)$ we take any ∇-cycle z^p lying* on τ. Then $s^\lambda_{\lambda'} \zeta^p$ is defined as the

---

* One could prove that in each of the homology classes ζ^p just defined there lies one and only one homology class of an arbitrary triangulation τ of the open set λ: with this one establishes a "natural isomorphism" between the groups $\nabla^p(\lambda)$ and $\nabla^p \tau$, and also a natural isomorphism between the groups of two different triangulations of the set λ; analogously one constructs also an isomorphism between $\nabla^p \tau$ and $\nabla^p \lambda$, so that $\nabla^p \tau = \nabla^p \lambda = \nabla^p(\lambda)$.

class $\zeta'^p \in \nabla^p(\lambda')$, containing the ∇-cycle $s^\tau_{\tau'} z^p$. The proof of the correctness of these definitions may be left to the reader. These definitions allow one to speak of the spectrum

$$\{\nabla^p\lambda, \, s^\lambda_{\lambda'}, \}. \tag{2}$$

Since the spectra (1_{ext}) and (2) are equivalent, then the limit group of the latter is the group $\nabla^p_{ext} A$. Now we leave in the spectrum (2) only the canonical neighborhoods. Since one and the same canonical neighborhood may be retracted onto several canonical polyhedra, we carry out a multiplication of the partially ordered set of all canonical neighborhoods, taking each of them a number of times equal to the number of canonical triangulations onto whose bodies the given neighborhood retracts. Then we regard each canonical neighborhood as related to one definite canonical triangulation onto whose body it retracts. The operation thus carried out on the spectrum (2) (i.e., leaving in it only the canonical neighborhoods and the just-described multiplication) turns the spectrum (2) into a new spectrum, which again we shall call the spectrum (2). In the spectrum so obtained we now carry out a weakening of the order, which consists of regarding a canonical neighborhood λ' as following the canonical neighborhood λ if $\lambda' \subset \lambda$, and if, moreover, the canonical triangulation α' corresponding to the neighborhood λ' follows the canonical triangulation* λ, corresponding to the neighborhood λ: recall that we have agreed to regard each canonical neighborhood as corresponding to one definite canonical triangulation. The spectrum obtained as a result we shall denote by

$$\{\nabla^p\lambda, \, s^\lambda_{\lambda'}, \}; \tag{2_{can}}$$

we shall always keep in mind that it has been obtained from the original spectrum (2) by multiplication and passage to cofinal section (the further weakening of the order which we have carried out may be absorbed in the passage to the cofinal section).

Now we pass in the spectrum (1) to a cofinal section, by leaving in this spectrum only the canonical coverings. Then we carry out a multiplication by taking each canonical covering a number of times equal to the number of canonical neighborhoods corresponding to the canonical triangulations excising the given canonical covering. This gives us a spectrum, whose elements already lie in a one-to-one relation with the elements of the spectrum (2_{can}). Carrying over into the spectrum just obtained the order from the spectrum (2_{can}), which means a weakening of the original order,** we obtain a spectrum which we shall denote by

$$\{\nabla^p\alpha, \, \pi^\alpha_\beta\}. \tag{1_{can}}$$

____

*It is easily verified that this weakening of the order is legitimate, i.e., it does not destroy the directedness of the given partially ordered set, and therefore reduces to a cofinal section of it.

**See the preceding footnote.

Thus, the spectrum (1_{can}) is obtained from the original spectrum(1) by multiplication and passage to a cofinal section.

2. The fundamental step in the proof of the spectral duality law is

Lemma 2. *The spectra* (1_{can}) *and* (2_{can}) *are isomorphic to each other.* [*]

Since between the elements of both spectra we have already established a one-to-one relation, preserving the order, Lemma 2 follows from the two following propositions, of which the first asserts an isomorphism of the corresponding groups of both spectra, and the second that that isomorphism preserves the projections.

Lemma 3. *A retracting mapping of the canonical neighborhood* λ *(corresponding to the canonical triangulation* α) *onto the canonical polyhedron* $\tilde{\alpha}$ *generates an isomorphism* $\bar{g}$ *of the group* $\nabla^P \alpha$ *onto the group* $\nabla^P \lambda$.

Lemma 4. *Let* $\lambda' > \lambda$ *be two canonical neighborhoods corresponding to the canonical triangulations* $\alpha' > \alpha$. *Let* $\bar{g}$ *and* $\bar{g}'$ *be the isomorphism of the group* $\nabla^P \alpha$ *onto the group* $\nabla^P \lambda$ *and the isomorphism of the group* $\nabla^P \alpha'$ *onto* $\nabla^P \lambda'$, *established in Lemma 3. Then for any element* ζ *of the group* $\nabla^P \alpha$ *we have*

$$s^\lambda_{\lambda'}\, \bar{g}\, \zeta = \bar{g}'\, s^\alpha_{\alpha'}\, \zeta.$$

Proof of Lemma 3. We take any triangulation τ of the neighborhood λ. Replacing, if necessary, this triangulation by an appropriate subdivision of it, we may from the very beginning assume that each simplex $t \in \tau$ intersecting any simplex of the canonical triangulation α entirely lies on it. Then the triangulation τ defines some subdivision σ of the canonical triangulation α. We denote by g_θ, $0 \leq \theta \leq 1$, the retracting deformation of the polyhedron $\tilde{\tau}$ onto the polyhedron $\tilde{\sigma}$. We approximate the end result of this deformation, i.e., the mapping g_1, by a simplicial mapping g defined on some subdivision τ' of the complex τ. The subdivision τ' and the simplicial approximation g of the mapping g_1 corresponding to it are constructed in the usual way: we consider the principal stars oe_i of the triangulation σ. We denote by O_i the complete preimage of the set oe_i under the mapping g_1; we obtain a covering $\omega = \{O_i\}$ of the polyhedron $\tilde{\tau}$. The subdivision τ' we take to be so fine that each star oe' of that subdivision is contained in at least one of the sets O_i. If we put into correspondence with the vertex e' of the triangulation τ' the vertex e_i of the triangulation τ, we obtain a simplicial mapping g of the complex τ' into σ which is the desired approximation of the mapping g_1. The triangulation τ' generates a subdivision σ' of the triangulation σ, and the mapping g onto σ' is the canonical shift of σ' into σ.

---

[*] Since the below-following proof of Lemma 2 does not depend on Chapter I, it gives a new proof of the isomorphism $\nabla^P A = \nabla^P_{ext} A$.

The simplicial mapping g generates a homomorphism $\bar{g}$ of the group $\nabla^P\sigma$ (and, accordingly, of the group $\nabla^P\alpha$) into the group $\nabla^P\tau'$ (and, accordingly, into $\nabla^P\lambda$). We shall prove that $\bar{g}$ is in fact the desired isomorphic mapping of the group $\nabla^P\sigma$ onto the group $\nabla^P\lambda$.

(a) The mapping $\bar{g}$ is an isomorphism. Let z be a ∇-cycle of the complex σ and suppose that $\bar{g}z \sim 0$ in τ'. We shall prove that $z \sim 0$ in σ. Since the excision operator J (under the passage from the given complex to a closed subcomplex) commutes with the operator ∇, then the cycle $J\bar{g}z$, i.e., the piece of the cycle $\bar{g}z$ lying on σ', is homologous to zero on σ'. But the mapping g onto σ' is a canonical shift of σ' into σ, so that the cycle Jgz is a ∇-extension of the ∇-cycle z into the complex σ', and, if this extension is bounding in σ', then also $z \sim 0$ in σ.

(b) The isomorphism $\bar{g}$ is a mapping onto the whole group $\Delta^P\lambda$. Choose any element $\zeta \in \nabla^P\lambda$ and a ∇-cycle $w^P \in \zeta$ lying on the triangulation τ. It suffices to prove that $\bar{g}Jw \sim s_{\tau'}^{\tau}w$ in τ'. The cycle $s_{\tau'}^{\tau}w$ is the image of the cycle w under the mapping adjoint to the canonical shift f (of the triangulation τ' into τ), and $\bar{g}Jw$ is the image of the same ∇-cycle w under the mapping $\bar{g}$, adjoint to the mapping g. But f and g are homotopic to the identity mapping, and therefore homotopic to each other. Therefore, for the proof of the homology $\bar{g}Jw \sim s_{\tau'}^{\pi}$ in τ', i.e., for the completion of the proof of Lemma 3, it is sufficient to prove the following known proposition:

Lemma 5. *Suppose given two mutually homotopic simplicial mappings f_0 and f_1 of the complex K' into the complex K''. Then for any ∇-cycle z^P of the complex K'', we have*

$$\bar{f}_0 z \sim \bar{f}_1 z \ \text{in} \ K'.$$

Proof of Lemma 5. We take a cylinder of height 1, constructed on K', i.e., the product of K' with the segment $0 \leq \theta \leq 1$. The given deformation of the mapping f_0 into f_1 defines a mapping F' of the cylinder Q into K''. For convenience we may suppose that in the strata $0 \leq \theta \leq 1/10$ the mapping F' coincides with f_0, and that in the layers $9/10 \leq \theta \leq 1$ with f_1. This makes it possible to approximate the mapping F' by a simplicial mapping which corresponds on the lower base of the cylinder Q with f_0 and on the upper base with f_1. We take a covering Ω of the cylinder Q by means of preimages under the mapping F' of principal stars of the complex K''. The strata Q_0 and Q_1, of thickness $1/10$, adjoining the bases of the cylinder, we triangulate as prisms. Because of the condition which the mapping F' satisfies in these strata, the stars of the above triangulations of them turn out to be subordinate to the covering Ω. Outside of these strata we choose a triangulation of the cylinder so fine that its stars are subordinate to Ω; moreover, we require that this triangulation induces a subdivision of the strata Q_0 and Q_1

adjoining the bases $\theta = 1/10$ and $\theta = 9/10$. We subdivide the simplexes of this triangulation of the strata Q_0 and Q_1 centrally with respect to the already given subdivision of their boundaries. We do not subdivide the simplexes of the bases any further. As a result we obtain a new triangulation K of the cylinder with stars subordinate to Ω and such that one may construct a simplicial mapping F of the triangulation K into K'' which is a simplicial approximation of the mapping F'. Here the mapping will coincide on the bases of the cylinder Q respectively with f_0 and f_1. Moreover, we may suppose that the triangulation is a subdivision of the natural cell decomposition Q' of the cylinder into cells which are prisms of height 1 over the simplexes of the complex K'.

We take a ∇-cycle $\overline{F}z$ of the complex K, i.e., the image of a ∇-cycle of the complex K'' under the mapping $\overline{F}$ adjoint to the simplicial mapping F. This ∇-cycle $\overline{F}z$ excises on the bases of the cylinder Q the cycles $\overline{f_0}z$ and $\overline{f_1}z$. The cycle $\overline{F}z$ is homologous in K to the ∇-subdivision (extension) of some cycle Z of the prismatic complex Q'. Accordingly, also the ∇-cycles $\overline{f_0}z$ and $\overline{f_1}z$ excised by the bases are homologous on these bases to the corresponding excisions of the J_0Z and J_1Z of the same cycle Z. But these excisions, considered as cycles of the complex K', are homologous to each other in this complex,[*] which means that also $\overline{f_0}z \sim \overline{f_1}z$ in K', with which Lemma 5, and accordingly also Lemma 3, are proved.

Proof of Lemma 4. We approximate the mappings g_1 and g_1' (of the neighborhoods λ and λ' on $\tilde{\alpha}$ and $\tilde{\alpha}'$) by simplicial mappings of some triangulations τ and τ' of the neighborhoods λ and λ'. Here we assume that τ' follows τ. We take any ∇-cycle $z \in \zeta$ lying on α. We have to prove that $s_{\tau'}^{\tau}\overline{g}z \sim \overline{g}'s_{\alpha'}^{\alpha}z$ in τ'. To this end we choose a canonical shift $f_{\alpha}^{\alpha'}$ of the complex α' into α and a canonical shift $f_{\tau}^{\tau'}$ of the complex τ' into τ. The mappings $gf_{\tau}^{\tau'}$ and $f_{\alpha}^{\alpha'}g'$ of the polyhedron $\tilde{\tau}'$ into $\tilde{\alpha}$ are homotopic in $\tilde{\tau}$ to the identity mapping and accordingly homotopic to each other in $\tilde{\alpha}$. Therefore, in view of Lemma 5, the ∇-cycles $s_{\tau'}^{\tau}\overline{g}z$ and $\overline{g}'s_{\alpha'}^{\alpha}z$ as images of the ∇-cycle z under mappings adjoint to the mutually homotopic mappings $gf_{\tau}^{\tau'}$ and $f_{\alpha}^{\alpha'}g'$, will be homologous to each other in τ', with which Lemma 4 and thus Lemma 2 are proved.

3. The spectral duality law now may be proved in a few words. Indeed, the isomorphism D (Chapter III, §3), in conjunction with the commutability of the operators $s_{\tau'}^{\tau}$ and D, realizes an isomorphism between the spectra $\{\nabla^p\lambda, s_{\lambda'}^{\lambda}\}$ and $\{\Delta^{q+1}\lambda, s_{\lambda'}^{\lambda}\}$, and the isomorphism Γ (Chapter III, §4), if we keep in mind

---

[*] The chain realizing this homology takes, on each $t^{p-1} \ni K'$, a value equal to the value of the cycle Z on the prism $t^p \in Q'$, constructed over t^{p-1} (compare with Part I, Chapter III, §3, pages 43–44).

$$\Gamma u^{q+1} \sim \Gamma s^{\tau}_{\tau'} u^{q+1} \quad \text{in} \quad \psi' = S^n - \overset{\sim}{\tau}',$$

realizes an isomorphism between the spectra $\{\Delta^{q+1}\lambda, \ s^{\lambda}_{\lambda'}\}$ and $\{\Delta^q\psi, \ E^{\psi}_{\psi'}\}$. In all of these spectra one makes multiplications and passages to cofinal sections in accordance with the analogous operations carried out in the desired spectrum $\{\nabla^p\lambda, \ s^{\lambda}_{\lambda'}\}$.

CHAPTER IV

Further relations between groups defined for the same or for mutually complementary point sets.

Complete theorem on isomorphism of duality

§1. Fundamental property of the Sitnikov isomorphism M

1. The homomorphism f of the group $\nabla^r Y$ into the group $\nabla^r X$, adjoint to the continuous mapping f of the space X into the space Y, is applied constantly in contemporary topological investigations. We shall have to deal with it in the case of a simplicial mapping f of polyhedra, and also in the special case when $X \subseteq Y$ and f is the identity mapping, the adjoint homomorphism $\overline{f}$ then being the so-called excision homomorphism J, first introduced and studied in detail in my paper [22]. Now we shall need the homomorphism $\overline{f}$ for an arbitrary continuous mapping f of the space X into the space Y. We shall give a definition of this homomorphism.

Let $fX = Y_0 \subseteq Y$; suppose $z \in \mathfrak{z} \in \nabla^r Y$. The cycle z is a ∇-cycle lying on some covering β of the space Y. This covering excises a covering β_0 of the space Y_0, whose elements are the intersections with Y_0 of the elements of the covering β. The excision operator $J^{\beta}_{\beta_0}$ carries a ∇-cycle z_0 into a ∇-cycle z_0 of the nerve β_0, which takes on each simplex $t_0 \in \beta_0$ the value which the cycle z takes on the same simplex t_0.

We shall denote by $\alpha_0 = f^{-1}\beta_0$ the covering of the space X whose elements are the preimages under the mapping of the elements of the covering β_0. Since f is a mapping onto all of Y_0, then the nerves α_0 and β_0 stand in a natural isomorphism relation, carrying the cycle z_0 from β_0 onto α_0.

The resulting ∇-cycle of the covering α_0 will be denoted by $\overline{f}z$. Its homology class in the group $\nabla^r X$ is by definition $\overline{f}\mathfrak{z}$. It is easy to verify that the definition of the mapping $\overline{f}$ of the group $\nabla^r Y$ into the group $\nabla^r X$ just given is correct (i.e., does not depend on the elements of arbitrariness entering into it), and that $\overline{f}$ is a homomorphism of the group $\nabla^r Y$ into $\nabla^r X$.

2. The scalar product of any infinite chain u^r_{α} and the finite chain x^r_{α}, lying on the nerve α of any covering of the space X, is defined, as usual, under the assumption that one of the two chains u^r_{α}, x^r_{α} (it makes no difference which)

is taken over a discrete group $\mathfrak{A}$, and that the other is taken over the bicompact group $\mathfrak{B}$, where $\mathfrak{B} \mid \mathfrak{A}$.

This definition of a scalar product for chains goes over, as usual, to the elements of the groups $\nabla^r(a, \mathfrak{A})$ and $\bar{\delta}^r(a, \mathfrak{B})$ (see Chapter I, §3.5) and consequently, by continuity, to the "ideal" elements of the latter group; furthermore, it automatically reduces to the definition of the scalar product $(\mathfrak{z} \cdot \zeta)$ for the elements $\mathfrak{z}$, ζ respectively of the groups $\nabla^r(X, \mathfrak{A})$ and $\bar{\delta}^r(X, \mathfrak{B})$.

In exactly the same way (and as a consequence of the absence of ideal elements even somewhat more simply) one defines also the scalar product of elements of the groups $\nabla^r(X, \mathfrak{B})$ and $\delta^r(X, \mathfrak{A})$.

Now we may formulate and prove

3. The fundamental property of the isomorphism M. Let $\mathfrak{z}$ be any element of the group $\nabla^p(A, \mathfrak{A})$, respectively of the group $\nabla^p(A, \mathfrak{B})$, and ζ any element of the group $\bar{\delta}^p(A, \mathfrak{B})$, respectively of the group $\delta^p(A, \mathfrak{A})$. Then

$$(\mathfrak{z} \cdot \zeta) = \mathfrak{b}(M\mathfrak{z}, \zeta), \tag{1}$$

where, as always, we denote by $\mathfrak{b}$ the linking coefficient of Δ-cycles and their homology classes.

Before turning to the proof, we shall explain this formula. Taking any ∇-cycle $u^p \in \mathfrak{z}$ and projective (sliding) cycle $z^p \in \zeta$, we denote by $\mathfrak{b}(M\mathfrak{z}, \zeta)$ the linking coefficient $\mathfrak{b}(Mu^p, z^p)$ of the cycles $z^q = Mu^p$ and z^p. If now ζ is an ideal element of the group $\bar{\delta}^p(A, \mathfrak{B})$, then the linking coefficient on the right (the same as the scalar product on the left) is defined by continuity.

Thus, formula (1) may be replaced by

$$(u^p \cdot z^p) = \mathfrak{b}(Mu^p, z^p), \tag{1'}$$

where both the left and right sides do not change when the corresponding cycles run through their homology classes.

Proof of formula (1'). From the cycles u^p and z^p we turn to the exterior cycle $J^{-1}u^p$ and the sliding cycle $J^{-1}z^p = \{z_\tau^p\}$ (see Chapter III, §1). Here the ∇-cycle $J^{-1}u^p$ lies on the triangulation τ of some neighborhood λ of the set A. The scalar product $(J^{-1}u^p \cdot J^{-1}z^p)$ is defined on the triangulation τ, while

$$(J^{-1}u^p \cdot J^{-1}z^p) = (u^p \cdot z^p). \tag{2}$$

Indeed, we may always take as the a serving in the definition of the scalar product $(u^p \cdot z^p)$ some canonical covering, * and as τ the triangulation of a canonical neighborhood, retracting, by means of a continuous mapping f, onto the nerve

---

* The definition of a canonical covering and of a canonical neighborhood may be found in Chapter II, §1. See also Chapter III, §5.

$\widetilde{\alpha}$. Then one and the same cycle z_τ^p may be regarded as an element z_α^p of the projective cycle z^p and as an element z_λ^p of the sliding cycle $J^{-1}z^p$. As to the ∇-cycle $J^{-1}u^p$, we may write $J^{-1}u^p = \overline{f}u^p$, where by f we understand the operator adjoint to the operator of the continuous mapping f. But then

$$(u^p \cdot z_\tau^p) = (u^p \cdot fz_\tau^p) = (\overline{f}u^p \cdot z_\tau^p),$$

which means, as well, the equality (2).

To the ∇-cycle $J^{-1}u^p$ there corresponds the star Δ-cycle $D^*J^{-1}u^p$, lying on the complex of barycentric stars of the triangulation τ and having with the cycle z_τ^p an intersection equal to the scalar product $(J^{-1}u^p \cdot z_\tau^p)$.

The cycle $\Gamma DJ^{-1}u^p = \{z_k^q, x_k^{q+1}\}$ is obtained from the cycle $D^*J^{-1}u^p$ in the following way: we take an increasing sequence of finite subcomplexes τ_k of the triangulation τ, and define the z_k^q as the boundaries of the pieces w_k of the cycle $D^*J^{-1}u^p$ lying on the complexes τ_k. Then for sufficiently large k we have

$$D^*J^{-1}u^p \times z_\tau^p = w_k \times z_\tau^p = \mathfrak{b}(z_k^q, z_\tau^p)$$

and consequently

$$(u^p \cdot z^p) = \mathfrak{b}(z_k^q, z_k^p),$$

as we were required to prove.

$$\S 2.\text{ The fundamental homomorphisms } H \text{ and } \overline{h}.$$

The nonlinkability groups

1. **The fundamental homorphism** h of the group $\Delta_c^p(A, \mathfrak{A})$ (respectively of the group $\Delta^p(A, \mathfrak{A})$) into the group $\delta^p(A, \mathfrak{A})$ was defined in Chapter I, $\S 2.5$. We shall determine the kernel $N_{\Delta c}^p(A, \mathfrak{A})$ of the homomorphism h of the group $\Delta^p(A, \mathfrak{A})$ into $\delta^p(A, \mathfrak{A})$. The group $\Delta^p(A, \mathfrak{A})$ maps onto $\Delta_c^p(A, \mathfrak{A})$ under a natural homomorphism, consisting in that each Sitnikov cycle is replaced by its first line. Therefore the question comes down to the determination of the kernel of the homomorphism h of the group $\Delta_c^p(A, \mathfrak{A})$ into $\delta^p(A, \mathfrak{A})$. On the other hand, the group $\delta^p(A, \mathfrak{A})$ is naturally isomorphic to the group $D^p(A, \mathfrak{A})$, so that the homomorphism h consists simply in that we consider each true cycle of the set A as a sliding cycle. Accordingly, the kernel of the homomorphism h of the group $\Delta_c^p(A, \mathfrak{A})$ (respectively of the group $\Delta^p(A, \mathfrak{A})$) into the group $\delta^p(A, \mathfrak{A}) = D^p(A, \mathfrak{A})$ consists of those homology classes (in the Vietoris, or, respectively, in the Sitnikov sense) whose elements are cycles homologous to zero in some neighborhood λ of the set A (in the case of a Sitnikov cycle the last condition must naturally be satisfied by the first line of the cycle). But by the Alexander-Pontrjagin duality law, a cycle is homologous to zero in the given open set λ if and only if it has a null linking coefficient with every Vietoris cycle lying on the complementary, closed set $\psi = S^n - \lambda$. Therefore the desired *kernel of the homomorphism h is the*

subgroup of the group $\Delta^p_c(A, \mathfrak{A})$, *respectively of* $\Delta^p(A, \mathfrak{A})$, *consisting of all elements of that group having a null linking coefficient with every element of the group* $\Delta^q_c(B, \mathfrak{B})$. Therefore the group $N^p_{\Delta_c}(A, \mathfrak{A})$ is also called *the second* (or *augmented*) *nonlinkability group of the set* A (over the coefficient domain $\mathfrak{A}$). The designation "augmented" nonlinkability group will be clarified in the following section.

2. **The homomorphism** $\bar{h}$ **of the group** $\Delta^p_c(A, \mathfrak{B}) = \Delta^p(A, \mathfrak{B})$ **into the group** $\overline{\delta}^p(A, \mathfrak{B})$. This is constructed quite analogously to the homomorphism h.

If the group $\Delta^p_c(A, \mathfrak{B})$ already is identified with the group $\delta^p_c(A, \mathfrak{B})$ isomorphic to it (see Chapter I, §2.5), then every element of the group $\Delta^p_c(A, \mathfrak{B}) = \delta^p_c(A, \mathfrak{B})$ is simply an element of the group $\overline{\delta}^p(A, \mathfrak{B})$, being "proper" and not "ideal"; the inclusion homomorphism defined by this identification is the desired homomorphism $\bar{h}$. The kernel of the homomorphism h of the group $\Delta^p_c(A, \mathfrak{B}) = \Delta^p(A, \mathfrak{B})$ into the group $\overline{\delta}^p(A, \mathfrak{B})$ is donated by $N^p_{\Delta_c}(A, \mathfrak{B})$.

We shall now prove the following proposition.

1. *The group* $N^p_{\Delta_c}(A, \mathfrak{B})$ *may be defined as the subgroup of the group* $\Delta^p_c(A, \mathfrak{B})$, *all of whose elements have null linking coefficients with all the elements of the group* $\Delta^q_c(B, \mathfrak{A})$.

This new definition of the group $N^p_{\Delta_c}(A, \mathfrak{B})$ makes it possible to call this group the *second* (or *augmented*) nonlinkability group of the set A (over the coefficient domain $\mathfrak{B}$).

The proof of Proposition 1 is based on the following lemma:

Lemma. *The group* $\overline{\delta}^p(A, \mathfrak{B})$ *stands in a natural isomorphism relationship with the limit group* $\overline{D}^p(A, \mathfrak{B})$ *of the inverse spectrum*

$$\{\overline{\Delta}^p(\lambda, \mathfrak{B}), E^{\lambda'}_\lambda\},$$

where the λ *are the neighborhoods of the set* A, *ordered by inclusion, and the* $E^{\lambda'}_\lambda$ *(for* $\lambda' \subset \lambda$*) are, as always, the usual inclusion homomorphisms.*

The proof of the isomorphism $\overline{\delta}^p(A, \mathfrak{B}) = \overline{D}^p(A, \mathfrak{B})$ runs parallel to the proof of the isomorphism $\delta^p(A, \mathfrak{A}) = D^p(A, \mathfrak{A})$, constituting the content of the invariance theorem of Chapter II. It is based on the consideration of canonical coverings and canonical neighborhoods. The only point in the proof requiring any special attention consists in establishing the fact that, for a canonical α and corresponding λ, the factorization by nonlinking cycles (in the passage from $\Delta^p_c(\lambda, \mathfrak{B})$ to $\overline{\Delta}^p(\lambda, \mathfrak{B})$) and the factorization by cycles of the nerve α having null scalar products with every ∇-cycle of that nerve (the passage from $\Delta^p(\alpha, \mathfrak{B})$ to $\overline{\delta}^p(\alpha, \mathfrak{B})$) correspond to each other; but this follows from the fundamental property of the isomorphism M, proved in the preceding section.

From the lemma just proved it follows that the homomorphism $\bar{h}$ may be condered as an inclusion homomorphism of the group $\Delta^p_c(A, \mathfrak{B})$ into the group

$\bar{D}^p(A, \mathfrak{B}) = \bar{\delta}^p(A, \mathfrak{B})$ in the sense that every true cycle $z^p \in \zeta^p \in \Delta^p_c(A, \mathfrak{B})$ is a cycle lying in any neighborhood λ of the set A, so that every element $\zeta^p \in \Delta^p_c(A, \mathfrak{B})$ defines a true element of the group $\zeta^p_\lambda \in \bar{\Delta}^p(\lambda, \mathfrak{B})$, while these ζ^p_λ evidently form a thread, i.e., an element of the group $\bar{D}^p(A, \mathfrak{B})$.

In this interpretation the kernel of the isomorphism $\bar{h}$ of the group $\Delta^p(A, \mathfrak{B}) = \Delta^p_c(A, \mathfrak{B})$ into the group $\bar{\delta}^p(A, \mathfrak{B})$ consists of these elements which are classes of true cycles, homologous to zero in every neighborhood of the set A, i.e., having a null linking coefficient with every true cycle of the set B (with every element of the group $\Delta^q_c(B, \mathfrak{A})$.

Proposition 1 is proved.

It clears up the terminology "augmented nonlinkability group", as applied to the group $N^p_{\Delta c}(A, \mathfrak{B})$. Indeed, the group $N^p_\Delta(A, \mathfrak{B})$ (the "first nonlinkability group") was defined (in Chapter I, §3.2) as the subgroup of the group $\Delta^p(A, \mathfrak{B}) = \Delta^p_c(A, \mathfrak{B})$ consisting of all the elements having null linking coefficients with every element of the group $\delta^q(B, \mathfrak{A}) = D^q(B, \mathfrak{A})$, i.e., with every sliding cycle (thus *a fortiori* with every true cycle) of the set B, while in the definition of the group $N^p_{\Delta c}(A, \mathfrak{B})$ we require only that the linking coefficient with every true cycle of the set B should be zero. Therefore

$$N^p_{\Delta c}(A, \mathfrak{B}) \supseteq N^p_\Delta(A, \mathfrak{B}).$$

It still remains to consider the group $N^p_\Delta(A, \mathfrak{A})$ that consists of all the elements of the group $\Delta^p(A, \mathfrak{A})$ which have null linking coefficients with every (proper, and by continuity also to every more general) element of the group $\bar{\delta}^q(B, \mathfrak{B})$. The group $N^p_\Delta(A, \mathfrak{A})$ is naturally called: the *(first) nonlinkability group over a discrete coefficient domain.*

Interpreting the homomorphism $\bar{h}$ of the group $\Delta^q_c(B, \mathfrak{B})$ into the group $\bar{\delta}^q(B, \mathfrak{B})$ as an inclusion homomorphism, we immediately see that the requirement that the linking coefficient of a given element of the group $\Delta^p(A, \mathfrak{A})$ with every element of the group $\bar{\delta}^q(B, \mathfrak{B})$ should be zero is more rigid than the requirement that the linking coefficient with every element of the group $\Delta^q_c(B, \mathfrak{B})$ should be zero. Therefore

$$N^p_{\Delta c}(A, \mathfrak{A}) \supseteq N^p_\Delta(A, \mathfrak{A}),$$

whence the designation "augmented nonlinkability group" which we gave to the group $N^p_{\Delta c}(A, \mathfrak{A})$.

3. **Resumé of the definitions of nonlinkability groups.** We have two "simple" nonlinkability groups, $N^p_\Delta(A, \mathfrak{A})$ and $N^p_\Delta(A, \mathfrak{B})$, and two augmented ones, $N^p_{\Delta c}(A, \mathfrak{A})$ and $N^p_{\Delta c}(A, \mathfrak{B})$. The groups $N^p_{\Delta c}(A, \mathfrak{A})$ and $N^p_{\Delta c}(A, \mathfrak{B})$ consist of those elements of the group $\Delta^p(A, \mathfrak{A})$, respectively $\Delta^p(A, \mathfrak{B})$, which have null

linking coefficients with all elements of the group $\Delta_c^q(B, \mathfrak{B})$, respectively $\Delta_c^q(B, \mathfrak{A})$. The groups $N_{\Delta_c}^p(A, \mathfrak{A})$ and $N_{\Delta_c}^p(A, \mathfrak{B})$ are topological invariants of the set A, since they are respectively kernels of the homomorphisms:

$$h \text{ of the group } \Delta^p(A, \mathfrak{A}) \text{ into the group } \delta^p(A, \mathfrak{A}),$$

$$\overline{h} \text{ of the group } \Delta^p(A, \mathfrak{B}) \text{ into the group } \overline{\delta}^p(A, \mathfrak{B}).$$

The group $N_\Delta^p(A, \mathfrak{B})$ is the subgroup of the group $\Delta_c^p(A, \mathfrak{B}) = \Delta^p(A, \mathfrak{B})$, consisting of all the elements of that group which have null linking coefficients with every element of the group $\delta^q(B, \mathfrak{A}) = D^q(B, \mathfrak{A})$.

The group $N_\Delta^p(A, \mathfrak{B})$ is also a topological invariant of the set A, since it consists of all the elements of the group $\Delta_c^p(A, \mathfrak{B})$ having a null scalar product with every element of the group $\nabla_c^p(A, \mathfrak{A})$.

Finally, the group $N_\Delta^p(A, \mathfrak{A})$ consists of all elements of the group $\Delta^p(A, \mathfrak{A})$ having null linking coefficients with all elements of the group $\overline{\delta}^q(B, \mathfrak{B})$.

This group is a topological invariant of the set A. We shall show (in §3 of this chapter) that *it coincides with the group* $H_c^p(A, \mathfrak{A})$ (see Chapter I, §2.3), i.e., *with the kernel of the natural homomorphism of the group* $\Delta^p(A, \mathfrak{A})$ *onto the group* $\Delta_c^p(A, \mathfrak{A})$.

4. ∇-**nonlinkability groups.** We have already met one of these in Chapter I, §3.5. This was the group $N_\nabla^p(A, \mathfrak{A})$, which was defined as a limit group

$$N_\nabla^p(A, \mathfrak{A}) = \varinjlim \{N_\nabla^p(\alpha, \mathfrak{A}), \pi_\beta^\alpha\},$$

where $N_\nabla^p(\alpha, \mathfrak{A})$ is the subgroup of the group $\nabla^p(\alpha, \mathfrak{A})$ consisting of all those elements which have a null scalar product with every element of the group $\Delta^p(\alpha, \mathfrak{B})$. It is easy to see that *the group* $N_\nabla^p(A, \mathfrak{A})$ *may be directly defined as the subgroup of the group* $\nabla^p(A, \mathfrak{A})$ *consisting of all elements having a null scalar product with every element of the group* $\overline{\delta}^p(A, \mathfrak{B})$.

The group $N_\nabla^p(A, \mathfrak{B})$ is defined as the subgroup of the group $\nabla^p(A, \mathfrak{B})$, consisting of all elements having a null scalar product with every element of the group $\delta^p(A, \mathfrak{A})$.

The groups $N_\nabla^p(A, \mathfrak{A})$ and $N_\nabla^p(A, \mathfrak{B})$ are naturally called (simple) ∇-*non-linkability groups*. Along with them there are also defined the *"augmented"* ∇-nonlinkability groups, namely:

The group $N_{\nabla_c}^p(A, \mathfrak{A})$ consists of all elements of the group $\nabla^p(A, \mathfrak{A})$ having null scalar products with every element of the group $\Delta^p(A, \mathfrak{B}) = \Delta_c^p(A, \mathfrak{B})$; the latter group is conceived of as included by means of the operator $\overline{h}$ in the group $\overline{\delta}^p(A, \mathfrak{B})$, so that

$$N_{\nabla_c}^p(A, \mathfrak{A}) \supseteq N_\nabla^p(A, \mathfrak{A}).$$

The group $N^p_{\nabla_c}(A, \mathfrak{B})$ consists of those elements of the group $\nabla^p(A, \mathfrak{B})$ having null scalar products with all the elements of the group $\Delta^p(A, \mathfrak{A})$, or, what is the same thing, with all the elements of the group $\Delta^p_c(A, \mathfrak{A})$. These elements are conceived of as included, under the operator h, in the group $\delta^p(A, \mathfrak{A})$, which makes it possible to show the validity of the inclusion

$$N^p_{\nabla_c}(A, \mathfrak{B}) \supseteq N^p_{\nabla}(A, \mathfrak{B}).$$

All the ∇-nonlinkability groups are defined in invariant terms, and therefore are topological invariants of the set A.

We have seen that *both the nonlinkability groups and the ∇-nonlinkability groups of the set A are dualizable invariants of that set.*

§3. Complete theorem on isomorphism of duality;

the Sitnikov diagram

Theorem 1. *Under the isomorphism M between the groups $\nabla^p(A, \mathfrak{A})$ and $\Delta^q(B, \mathfrak{A})$, the group $N^p_{\nabla}(A, \mathfrak{A})$ maps into the group $H^q_c(B, \mathfrak{A})$.*

Before turning to the proof of this proposition, we shall deduce a series of consequences from it. They make clear the fundamental character of the theorem.

First of all, it follows from Theorem 1 that under the isomorphism M the factor-group $\nabla^p(A, \mathfrak{A}) - N^p_{\nabla}(A, \mathfrak{A})$ maps onto the factor-group $\Delta^q(B, \mathfrak{A}) - H^q_c(B, \mathfrak{A})$, i.e., onto the group $\Delta^q_c(B, \mathfrak{A})$, so that using the notation $\nabla'^p(A, \mathfrak{A}) = \nabla^p(A, \mathfrak{A}) - N^p_{\nabla}(A, \mathfrak{A})$ introduced in Chapter I, §3.5, we have the isomorphism

$$\nabla'^p(A, \mathfrak{A}) = \Delta^q_c(B, \mathfrak{A}), \tag{1}$$

which contains *the dualizability of the group Δ^r_c*. But also in Chapter I, §3.5, there was established the duality

$$\overline{\delta}^p(A, \mathfrak{B}) \mid \nabla'^p(A, \mathfrak{A}),$$

which makes it possible to rewrite the isomorphism (1) in the form of a duality

$$\overline{\delta}^p(A, \mathfrak{B}) \mid \Delta^q_c(B, \mathfrak{A}). \tag{2}$$

Theorem 1 and formulas (1) and (2) are sometimes amalgamated under the name of *the second Sitnikov duality law* (by the first Sitnikov duality law we understand the isomorphism $\nabla^p A = \Delta^q B$).

Finally, as an easy consequence of Theorem 1 and the fundamental property of the isomorphism M (see §1 of this chapter), there is the isomorphism

$$N^p_{\Delta}(A, \mathfrak{A}) = H^p_c(A, \mathfrak{A}). \tag{3}$$

Indeed, according to Theorem 1, the group $N^p_{\nabla}(A, \mathfrak{A})$ goes under the isomorphism M into the group $H^q_c(B, \mathfrak{A})$. But the group $N^p_{\nabla}(A, \mathfrak{A})$ consists of those elements

u of the group $\nabla^p(A, \mathfrak{A})$ for which $(u \cdot \zeta) = 0$ for any $\zeta \in \overline{\delta}{}^p(A, \mathfrak{B})$. Then, by the fundamental property of the isomorphism M,

$$(u \cdot \zeta) = \mathfrak{b}(Mu, \zeta).$$

Accordingly, the group $H_c^q(B, \mathfrak{A})$, being the image of the group $N_\nabla^p(A, \mathfrak{A})$ under this isomorphism, consists of all of those elements $\mathfrak{z}$ of the group $\Delta^q(B, \mathfrak{A})$ for which $\mathfrak{b}(\mathfrak{z}, \zeta) = 0$ for any $\zeta \in \overline{\delta}{}^p(A, \mathfrak{B})$. This means that $H_c^q(B, \mathfrak{A}) = N_\Delta^q(B, \mathfrak{A})$ and the isomorphism (3) is thus proved.

We turn to the proof of Theorem 1. We begin with the special case when the set A is open and accordingly $B = S^n - A$ is a compactum. Let τ be any triangulation of the open set A. We apply to this special case the constructions of §§1 and 2 of Chapter III, concerning the groups $\nabla^p A$ and $\nabla^p_{\mathrm{ext}} A$. Since a triangulation τ', following τ, in this case turns simply into a subdivision of the triangulation τ, and the operator $s_{\tau'}^\tau$ is an isomorphism of the group $\nabla^p \tau$ onto the group $\nabla^p \tau'$, the group $\nabla^p_{\mathrm{ext}} A$ in our case turns out to be isomorphic to the group $\nabla^p \tau$, and the isomorphism J turns into an isomorphism of the group $\nabla^p \tau$ onto the group $\nabla^p A$, so that M becomes an isomorphism of the group $\nabla^p \tau$ onto the group $\Delta^q B$. From the fundamental property of the isomorphism M there results — now for any $\xi \in \nabla^p(\tau, \mathfrak{A})$ and $\eta \in \Delta^p(\tau, \mathfrak{B})$ — the equality *

$$(\xi \cdot \eta) = \mathfrak{b}(M\xi, \eta),$$

from which it follows that under this isomorphism the subgroup $N_\nabla^p(\tau, \mathfrak{A})$ of the group $\nabla^p(\tau, \mathfrak{A})$ goes into the subgroup $N_{\Delta c}^q(B, \mathfrak{A})$ of the group $\Delta^q(B, \mathfrak{A})$ consisting of all the elements ζ having a null linking coefficient with any $\eta \in \Delta^p(\tau, \mathfrak{B}) = \Delta^p(A, \mathfrak{B})$. We shall show that in our case this subgroup is also the group $H_c^q(B, \mathfrak{A})$. Indeed, if the cycle ** $z \in \zeta \in \Delta^q(B, \mathfrak{A})$ has a null linking coefficient with all elements of the group $\Delta^p(\tau, \mathfrak{B})$, then for any finite triangulation $\tau_k \subset \tau$ this cycle, having a null linking coefficient with all $y \in \eta \in \Delta^p(\tau_k, \mathfrak{B})$ will by the classical Alexander-Pontrjagin theorem be homologous to zero in $\mu = S^n - \widetilde{\tau}_k$ and consequently will be homologous to zero in any neighborhood of the compactum B. But then the cycle z will be homologous to zero in B itself in the weak sense, i.e., of Vietoris homology, i.e., we will have $z \in \zeta \in H_c^q(B, \mathfrak{A})$. Conversely, if $z \in \zeta \in H_c^q(B, \mathfrak{A})$, then z bounds in any neighborhood of the compactum B, and therefore will have a null linking coefficient with any (finite) cycle of the triangulation τ, i.e., $z \in \zeta \in N_{\Delta c}^q(B, \mathfrak{A})$.

---

*We identify the element η of the group $\Delta^p(\tau, \mathfrak{B}) = \Delta_c^p(A, \mathfrak{B})$ with the element of the group $\overline{\delta}{}^p(A, \mathfrak{B})$ corresponding to it under the homomorphism h.

**We identify the Sitnikov cycle z with its first line, i.e., we consider the Vietoris cycle corresponding to it.

Thus, in the case of an open A, Theorem 1 is proved.

We turn to the case of a general $A \subset S^n$. Suppose that α is a canonical covering of the set A and that $\lambda = \overset{\sim}{\tau}$ is a canonical neighborhood retracting onto the nerve $\overset{\sim}{\alpha}$. Put $\psi = S^n - \lambda$. As we have already said in §1 of this chapter, under the natural isomorphisms

$$\nabla^P (\alpha, \mathfrak{A}) = \nabla^P (\tau, \mathfrak{A}),$$

$$\Delta^P (\alpha, \mathfrak{B}) = \Delta^P (\tau, \mathfrak{B}),$$

scalar products are preserved. Therefore it follows that under the natural isomorphism between the groups $\nabla^P (\lambda, \mathfrak{A})$ and $\nabla^P (\alpha, \mathfrak{A})$, the groups $N^P_\nabla (\lambda, \mathfrak{A})$ and $N^P_\nabla (\alpha, \mathfrak{A})$ correspond to one another. But the group $N^P_\nabla (\lambda, \mathfrak{A})$, from the special case of Theorem 1 proved above, goes into the group $H^q_c (\psi, \mathfrak{A})$. Therefore we may say that under the natural isomorphism M between the groups $\nabla^P (\alpha, \mathfrak{A})$ and $\Delta^q (\psi, \mathfrak{A})$, there is realized an isomorphism between the groups $N^P_\nabla (\alpha, \mathfrak{A})$ and $H^q_c (\psi, \mathfrak{A})$.

We now turn to the spectral duality law (Chapter III, §5). In it we have an isomorphism M of the spectrum $\{\nabla^P (\alpha, \mathfrak{A}), \pi^\alpha_\beta\}$ under canonical coverings α of the set A and the spectrum $\{\Delta^q (\psi, \mathfrak{A}), E^\psi_{\psi'}\}$ under compacta ψ complementary to canonical neighborhoods of the set A. In these spectra one may single out the groups $N^P_\nabla (\alpha, \mathfrak{A})$ and $H^q_c (\psi, \mathfrak{A})$, which stand, from what has been proved, in a natural isomorphic relation with each other. Under the same projections π^α_β and $E^\psi_{\psi'}$ these groups are easily seen to form two isomorphic spectra whose limit groups are exactly the groups $N^P_\nabla (A, \mathfrak{A})$ and $H^q_c (B, \mathfrak{A})$, these latter thus turning out to be isomorphic because of the fundamental isomorphism M.

Theorem 1 is proved.

2. **Dualizability of the remaing nonlinkability groups. Fundamental theorem on the isomorphism of duality ("the Sitnikov diagram").** From the fundamental property of the isomorphism M it follows that under the isomorphism M between the groups $\nabla^P (A, \mathfrak{B})$ and $\Delta^q (B, \mathfrak{B})$, the group $N^P_\nabla (A, \mathfrak{B})$ goes into the group $N^q_\Delta (B, \mathfrak{B})$. On the other hand, recalling that every Vietoris cycle (and therefore every Sitnikov cycle) under the isomorphism h, respectively $\bar{h}$, may be considered as a projective cycle $z \in \zeta \in \delta^P (A, \mathfrak{A})$, respectively $z \in \zeta \in \bar{\delta}^P (A, \mathfrak{B})$, we deduce from the same property of the isomorphism M that it carries the group $N^P_{\nabla_c} (A, \mathfrak{B})$ into the group $N^q_{\Delta_c} (B, \mathfrak{B})$, and the group $N^P_{\nabla_c} (A, \mathfrak{A})$ into the group $N^q_{\Delta_c} (B, \mathfrak{A})$.

Theorem 2. *All the nonlinkability groups and ∇-nonlinkability groups, simple and augmented, are dualizable topological invariants.*

As the sum total of all of these investigations there is *the complete theorem on the isomorphism of duality*, which we shall call the *Sitnikov diagram of duality*.

In this, along with the usual symbol $P \mid Q$ expressing the duality of the groups P and Q, we will require the symbol $\overset{P}{\underset{Q}{\parallel}}$, expressing isomorphism of the groups P and Q, and the symbol $\overset{Q}{\underset{R}{\vdash}} \!\!—P$, which means that the group P maps homomorphically into R with kernel Q.

The Sitnikov diagram

$$\overline{\Delta^P}(A, \mathfrak{B}) \qquad \Big| \qquad \delta^q(B, \mathfrak{A})$$

$$h \uparrow$$

$$N^q_{\nabla}(B, \mathfrak{B}) \subseteq N^q_{\nabla_c}(B, \mathfrak{B}) \subseteq \nabla^q(B, \mathfrak{B}) \qquad \Delta^q(B, \mathfrak{A}) \supseteq N^q_{\Delta_c}(B, \mathfrak{A}) \supseteq N^q_{\Delta}(B, \mathfrak{A}) = H^q_c(B, \mathfrak{A})$$

$$\parallel \qquad \parallel \qquad \parallel \qquad \parallel \qquad \parallel \qquad \parallel$$

$$N^P_{\Delta}(A, \mathfrak{B}) \subseteq N^P_{\Delta_c}(A, \mathfrak{B}) \subseteq \Delta^P(A, \mathfrak{B}) \quad \nabla^P(A, \mathfrak{A}) \supseteq N^P_{\nabla_c}(A, \mathfrak{A}) \supseteq N^P_{\nabla}(A, \mathfrak{A})$$

$$\downarrow \overline{h}$$

$$\overline{\delta^P}(A, \mathfrak{B}) \qquad \Big| \qquad \Delta^q_c(B, \mathfrak{A}).$$

§4. Extension of the homomorphism $\overline{h}$ to the group $\overline{\Delta}^q$ and its adjointness to the homomorphism h; the example of E. F. Miščenko

In §2 we defined the homomorphism $\overline{h}$ as the natural "inclusion" homomorphism of the group $\Delta^q(B, \mathfrak{B}) = \Delta^q_c(B, \mathfrak{B})$ into the group $\overline{\delta}^q(B, \mathfrak{B})$, while the kernel of this homomorphism turned out to be the augmented nonlinkability group $N^q_{\Delta_c}(B, \mathfrak{B})$. Since $N^q_{\Delta}(B, \mathfrak{B}) \subseteq N^q_{\Delta_c}(B, \mathfrak{B})$, it follows from the theorem of E. Noether that the homomorphism $\overline{h}$ defines in a natural way a continuous homomorphism, of the same name, of the topologized factor-group $\Delta'^q(B, \mathfrak{B}) = \Delta^q_c(B, \mathfrak{B}) - N^q_{\Delta}(B, \mathfrak{B})$, which may be carried over by continuity also into the bicompact supplement $\overline{\Delta}^q(B, \mathfrak{B})$ of the group $\Delta'^q(B, \mathfrak{B})$. So we have *a homomorphism $\overline{h}$ of the group $\overline{\Delta}^q(B, \mathfrak{B})$ into the group $\overline{\delta}^q(B, \mathfrak{B})$* and a homomorphism h of the group $\Delta^P_c(A, \mathfrak{A})$ into the group $\delta^P(A, \mathfrak{A})$. Here $\delta^P(A, \mathfrak{A}) \mid \overline{\Delta}^q(B, \mathfrak{B})$ and $\Delta^P_c(A, \mathfrak{A}) \mid \overline{\delta}^q(B, \mathfrak{B})$. Since these dualities are linking dualities and the homomorphism h and $\overline{h}$ are inclusion homomorphisms, these homomorphisms turn out to be adjoint: for any $\xi \in \Delta^P_c(A, \mathfrak{A})$, $\overline{\eta} \in \overline{\Delta}^q(B, \mathfrak{B})$ we have $(\xi \cdot \overline{h}\,\overline{\eta}) = (h\xi \cdot \overline{\eta}) = \mathrm{o}(\xi, \overline{\eta})$. Denoting by $\overline{X}^q_1(B, \mathfrak{B})$ the image of the group $\overline{\Delta}^q(B, \mathfrak{B})$ under the homomorphism $\overline{h}$, by $Y^P_1(A, \mathfrak{A})$ the image of the group $\Delta^P_c(A, \mathfrak{A})$ under the homomorphism h, and noting that the kernels of these homomorphisms are respectively the groups $[N^q_{\Delta_c}(B, \mathfrak{B})]$ (closures in $\overline{\Delta}^q(B, \mathfrak{B})$) and $N^p_{\Delta_{cc}}(A, \mathfrak{A})$,* we may draw the diagram

$$h \left| \begin{array}{c} \Delta^p_c(A,\mathfrak{A}) \supseteq N^p_{\Delta_{cc}}(A,\mathfrak{A}) \perp \overline{X}^q_1(B,\mathfrak{B}) \subseteq \overline{\delta}^q(B,\mathfrak{B}) \\ \hline \delta^p(A,\mathfrak{A}) \supseteq Y^p_1(A,\mathfrak{A}) \perp [N^q_{\Delta_c}(B,\mathfrak{B})] \subseteq \overline{\Delta}^q(B,\mathfrak{B}) \end{array} \right| \overline{h},$$

which is an application to the present special case of the diagram expressing the fundamental lemma on adjoint homomorphisms, ** first proved by me in the paper [16].

From the adjointness of the homomorphisms h and $\overline{h}$ it follows that *every property of the set A which may be expressed by properties of the homomorphism*

---

     *By $N^p_{\Delta_{cc}}(A, \mathfrak{A}) \subseteq \Delta^p_c(A, \mathfrak{A})$ we denote the image of the group $N^p_{\Delta_c}(A, \mathfrak{A}) \subseteq \Delta^p(A, \mathfrak{A})$ under the natural homomorphism of the group $\Delta^p(A, \mathfrak{A})$ onto the group $\Delta^p_c(A, \mathfrak{A})$, i.e., $N^p_{\Delta_{cc}}(A, \mathfrak{A}) = N^p_{\Delta_c}(A, \mathfrak{A}) - H^p_c(A, \mathfrak{A})$.

     ** The lemma in question is as follows: Let σ be a homomorphism of the group X into the group Y, and $\overline{\sigma}$ a homomorphism of the group $\overline{Y}\,|\,Y$ into the group $\overline{X}\,|\,X$, adjoint to the isomorphism σ (i.e., $(\sigma x \cdot \overline{y}) = (x \cdot \overline{\sigma}\,\overline{y})$ for any $x \in \overline{X}; \overline{y} \in \overline{Y}$). Suppose that X_0 is the kernel, and Y_1 the image, of the group X under the homomorphism σ. Let $\overline{Y}_0$ be the kernel and $\overline{X}_1$ be the image of the group $\overline{Y}$ under the homomorphism $\overline{\sigma}$.

     We describe this state of affairs in the form of a diagram:

$$\sigma \left| \begin{array}{cc} X \supseteq X_0 & \overline{X}_1 \subseteq \overline{X} \\ \hline Y \supseteq Y_1 & \overline{Y}_0 \subseteq \overline{Y} \end{array} \right| \overline{\sigma} \,. \tag{1}$$

The lemma asserts that in this case there exist annihilations $X_0 \perp \overline{X}_1$ and $Y_1 \perp \overline{Y}_0$ and a duality $\overline{X}_1\,|\,Y_1$. Moreover, from the theorem on annihilators, taken from the elements of the theory of characters, we also have the dualities $(X_0\,|\,\overline{X} - \overline{X}_1)$ and $(Y - Y_1)\,|\,\overline{Y}_0$. In other words, the diagram (1) may always be supplemented to the diagram

$$\sigma \left| \begin{array}{cc} X \supseteq X_0 \vdash \overline{X}_1 \subseteq \overline{X} \\ \hline Y \supseteq Y_1 \vdash \overline{Y}_0 \subseteq \overline{Y} \end{array} \right| \overline{\sigma}$$

$$X_0\,|\,(\overline{X} - \overline{X}_1); \quad (Y - Y_1)\,|\,\overline{Y}_0,$$

where $\perp$ denote annihilation, and the diagonal stroke duality. For the proof it is sufficient to give a proof of any one of the annihilations mentioned, for example of the fact that $X_0 \perp \overline{X}_1$. To this end we choose arbitrarily $x \in X_0, \overline{y} \in \overline{Y}$. Then $(x \cdot \overline{\sigma}\overline{y}) = (\sigma x \cdot \overline{y}) = 0$ (since $x \in X_0$). On the other hand, if $\overline{x} \in X_0$, i.e., if $\sigma x \neq 0$, then there exists a $\overline{y} \in \overline{Y}$ such that $(\sigma x \cdot \overline{y}) \neq 0$, and accordingly $(x \cdot \overline{\sigma}\overline{y}) \neq 0$. The annihilation $X_0 \perp \overline{X}_1$ is thus proved. The whole lemma is proved.

     Remark. If we are not mistaken, the above diagram, taken from our paper [22], is the first case of the formulation of a theorem as a diagram of this sort, a method which has undergone an extraordinary development in contemporary algebraic topology.

h, is dualizable. Clearly every property of the homomorphism $\bar{h}$ is dualizable.
Hence, in particular, there follows not only the dualizability of the kernels
$N^p_{\Delta c \underline{c}}(A, \mathfrak{A}) = N^p_{\Delta c}(A, \mathfrak{A}) - H^p_c(A, \mathfrak{A})$ and $[N^q_{\Delta c}(B, \mathfrak{B})]$ of the homomorphisms h
and $\bar{h}$, but also the dualizability of the factor-group

$$\mathfrak{M}^p(A, \mathfrak{A}) = \delta^p(A, \mathfrak{A}) - Y^p_1(A, \mathfrak{A}),$$

called the "sliding growth" (in dimension p over the coefficient domain $\mathfrak{A}$) of
the set A. It follows from the annihilator theorem (see the preceding footnote)
that

$$\mathfrak{M}^p(A, \mathfrak{A}) \mid [N^q_{\Delta c}(B, \mathfrak{B})].$$

This sliding growth may be nontrivial (i.e., may be a non-null group). In
other words, the homomorphism h may not be a homomorphism onto the
whole group $\delta^p(A, \mathfrak{A})$. The example of a set A realizing this possibility
was first constructed by E. F. Miščenko [13b]. We shall present this example.

In view of the isomorphism between the groups $\delta^p(A, \mathfrak{A})$ and $D^p(A, \mathfrak{A})$, the
question comes down to the construction of a set A for which, in the group
$D^p(A, \mathfrak{A})$, there exists an element which is a class of sliding cycles, none of
which is a cycle with a compact carrier. We construct in three-dimentional Euclidean space R^3 (completed by a unique point at infinity) a set A for which the
group $\Delta^1_c A$ is a null-group, while the group $D^1 A$ is different from zero (i.e., in
A there exists a one-dimensional sliding cycle which is not homologous to zero).
The coefficient domain is the group of integers.

We suppose that we are given in R^3 a system of coordinates x, y, z. We
shall denote by K_0 a broken line, consisting of segments lying in the plane
$z = 0$ and given there by the conditions:

$$x = 0, \frac{1}{2} \le y \le 1,$$

$$y = 1, \ 0 \le x \le 2,$$

$$x = 2, -1 \le y \le 1,$$

$$y = -1, 0 \le x \le 2,$$

$$x = 0, -1 \le y \le -\frac{1}{2}.$$

Further, for all $i = 1, 2, 3, \cdots$ we denote by II_i the segment lying in the plane
$z = 0$ and given by the conditions

$$x = \frac{1}{i}, \ -\frac{1}{2} \le y \le \frac{1}{2}.$$

Put

$$K = K_0 \cup \bigcup_{i=1}^{\infty} \text{II}_i.$$

The set K is represented in Figure 3. We denote now by Q_j the closed square (i.e., the interior of the square along with its boundary) lying in the plane $z = i/j$ and bounded there by the lines

$$x = 0, \; y = \frac{1}{2}, \; x = 1, \; y = -\frac{1}{2}.$$

Finally, we define the set A as the sum

$$A = K \cup \bigcup_{j=1}^{\infty} Q_j.$$

It is easy to see that $\Delta_c^1 A = 0$. We shall prove that in A there is a one-dimensional sliding cycle not homologous to zero. Let $\{\lambda\}$ be the system of all neighborhoods of the set A. We take arbitrarily one of them, namely, λ_α. If ϵ is the minimum of the distances of the points $\left[0, \frac{1}{2}, 0\right]$ and $\left[0, -\frac{1}{2}, 0\right]$ from the boundary of λ_α, then the ends of the segment $\Pi_{i(\alpha)}$, for $i(\alpha) > \frac{1}{\epsilon}$, may be respectively joined in

λ_α with the points $\left[0, \frac{1}{2}, 0\right]$, $\left[0, -\frac{1}{2}, 0\right]$ by the segments c_α and c'_α. The polyhedral one-dimensional cycle whose body is the set-theoretical sum of the sets K_0, $\Pi_{i(\alpha)}$, c_α and c'_α, will be denoted by z'_α. It lies in λ_α (see Figure 3). It is not difficult to see that the system $\{z_\lambda^1\}$ is not a sliding cycle in K; however it is one in A. Indeed, let $\lambda_\beta \subset \lambda_\alpha$ and z_β^1, z_α^1 be cycles of the system $\{z_\alpha^1\}$, lying respectively in λ_β and in λ_α. We denote by δ the mini

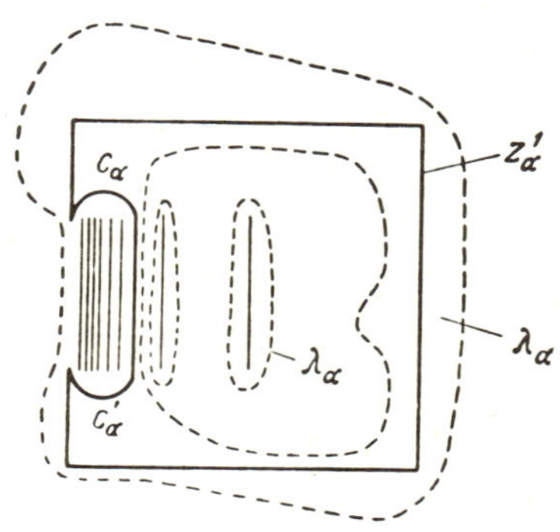

Figure 3

mum of the distances from z_β^1 and z_α^1 to the boundary of λ_α. Then the homology $z_\beta^1 \sim z_\alpha^1$ in λ_α is realized by means of any square Q_j, given only that $j > \frac{1}{\delta}$.

We observe that, using the construction just presented, one may construct a sliding (projective) cycle of arbitrary dimension which is not homologous to a compact cycle.

Supplement to Chapter IV

In this supplement we give a new proof of the theorem on the dualizability of the groups $\Delta_c^q(X, \mathfrak{A})$, found by N. A. Berikašvili [23] (independently of K. Sitnikov).

1. Let K be any simplicial complex and $\mathfrak{B}$ a bicompact group. Then the bicompact groups $\Delta^P(K_\alpha, \mathfrak{B})$, where K_α are all possible finite subcomplexes of the complex K, form a direct spectrum: one may take $K_\alpha < K_\beta$ when K_α is a subcomplex of K_β, and, as the projections $\Delta_\alpha^P \to \Delta_\beta^P$, one may take the natural inclu

sion homomorphisms. The limit group (see page 120) of such a direct spectrum of bicompact groups we denote by $\overline{E}{}^p(K, \mathfrak{B})$.

The second duality law for closed sets

$$\overline{\Delta}{}^p(G, \mathfrak{B}) \mid \Delta_c^q(F, \mathfrak{A}), \quad G = S^n - F, \tag{1}$$

may be written now in the form

$$E^p(K, \mathfrak{B}) \mid \Delta_c^q(F, \mathfrak{A}), \tag{2}$$

where K is any triangulation of the open set G. That assertion evidently follows from the following facts, which we now know: a) the limit group of a spectrum of groups is isomorphic to the limit group of each cofinal subspectrum; b) each compact subset of the body of any triangulation is contained in the body of some one of its finite subcomplexes, and the body of each finite closed subcomplex is compact; c) if K is a finite triangulation, then $\Delta^p(K) = \Delta_c^p(\widetilde{K})$ (see Part I, page 59).

The scalar product of the duality (2) may be described in the following way: let z^p be a finite cycle of the triangulation K over the coefficient domain $\mathfrak{B}$, and $z^q = \{z_n^q\}$ a true cycle of the compactum F over the coefficient domain $\mathfrak{A}$. Then the linking coefficient $\mathfrak{d}(z^p, z^q) = \mathfrak{d}(z^p, z_n^q)$ is defined (n sufficiently large). We shall take $(\xi^p \cdot \zeta^q) = \mathfrak{d}(z^p, z^q)$ for $z^p \in \xi^p$, $z^q \in \zeta^q$. We extend this multiplication over all elements by continuity.

We shall also need the following fact. Let K and L be two triangulations lying in S^n. Suppose further that each principal star of the triangulation L lies in some principal star of the triangulation K. Under this subordination, as usual, there is defined a simplicial mapping of L into K (i.e., to a vertex $a \in L$ one puts into correspondence one of the vertices $b \in K$ for which $O_L a \subset O_K b$). If z^p is any finite chain of the triangulation L, and z_1^p denotes its image in K under the simplicial mapping just set up, then

$$z^p - z_1^p \sim 0 \text{ in } \widetilde{K},$$

i.e., there exists a $(p + 1)$-dimensional chain x^{p+1} of the space S^n (possibly with degenerate simplexes), with a body in $\widetilde{K}$, such that $\Delta x^{p+1} = z^p - z_1^p$.

This assertion is easily proved by constructing prisms and taking into account the following fact: if σ is a closed simplex of the complex L, then the minimal convex closed set, containing both sets $\widetilde{\sigma}$ and $\widetilde{\pi\sigma}$, lies in $\widetilde{K}$. This last assertion is proved as follows. The set $\widetilde{\sigma}$ lies in the union of the closed simplexes of K having $\pi\sigma$ as their face. If $x' \in \widetilde{\pi\sigma}$ and $x \in \widetilde{\sigma}$ are any points, then they lie in some closed simplex of K. Accordingly, the spherical segment joining x and x' lies entirely in some closed simplex of K. The union of all possible such segments is the minimal closed convex set containing $\widetilde{\sigma}$ and $\widetilde{\pi\sigma}$; it is contained in $\widetilde{K}$.

2. The group $\overline{E}^p(K, \mathfrak{B})$ defined above is a group of Δ-type in the sense of Chapter I, $\S 1$, which is proved by the following discussion.

If ϕ is a simplicial mapping of the abstract simplicial complex K into the abstract simplicial complex L, then the continuous homomorphism ϕ^* of the group $\overline{E}^p(K, \mathfrak{B})$ into the group $\overline{E}^p(L, \mathfrak{B})$ is defined in the following way. Let $\{\Delta^p(K_\alpha), i_\beta^\alpha\}$ and $\{\Delta^p(L_\rho), i_\tau^s\}$ be direct spectra, by means of which respectively the groups $\overline{E}^p(K, \mathfrak{B})$ and $\overline{E}^p(L, \mathfrak{B})$ are defined. We unite these two spectra into a single spectrum, taking $K_\alpha < L_\rho$ if $\phi(K_\alpha) \subset L_\rho$, and, as the corresponding projection, taking the continuous homomorphism $\Delta^p(K_\alpha) \to \Delta^p(L_\rho)$ generated by ϕ. Thus we obtain a direct spectrum of bicompact groups in which the first spectrum is a subspectrum and the second is a cofinal spectrum. Accordingly, the limit group of this direct spectrum stands on the one hand in a natural isomorphism with the group $\overline{E}^p(L, \mathfrak{B})$ and on the other is mapped into, in a natural way, by the group $\overline{E}^p(K, \mathfrak{B})$. This defines ϕ^*. Let $z^p \in K$ be a finite cycle. Then it is easy to see that if $z^p \in \xi \in \overline{E}^p(K, \mathfrak{B})$ we have $\phi z^p \in \phi^* \xi$, so that the homomorphism ϕ^* may also be defined as follows: if the element $\xi \in \overline{E}^p(K, \mathfrak{B})$ contains the cycle z^p, then $\phi^* \xi$ is the element of $\overline{E}^p(L, \mathfrak{B})$ containing ϕz^p. For the remaining $\xi \in \overline{E}^p(K, \mathfrak{B})$ $\phi^* \xi$ the definition is by continuity.

From the second definition of the homomorphism ϕ^*, it is immediately clear that: 1) if ϕ is the identity mapping of the complex into itself, then ϕ^* is also the identity homomorphism; 2) if ϕ is a simplicial mapping of K into L, and ϕ_1 is a simplicial mapping of L into P, then $(\phi_1 \phi)^* = \phi_1^* \phi^*$; if we assume that ϕ and ϕ_1 are combinatorially close simplicial mappings of the complex K into the complex L, we have the equation $\phi^* = \phi_1^*$ (so that for any finite cycle z^p of K the cycle $\phi z^p - \phi_1 z^p$ of the complex L bounds a finite chain, which means that ϕz^p and $\phi_1 z^p$ belong to one and the same element of the group $\overline{E}^p(L, \mathfrak{B})$).

Thus the group $\overline{E}^p(K, \mathfrak{B})$ indeed is a group of Δ-type.

3. We select an arbitrary space A. For the nerve α of an open covering of A we consider the group $\overline{E}^p(\alpha, \mathfrak{B})$; for $\alpha < \beta$ we take a homomorphism of the group $\overline{E}^p(\beta, \mathfrak{B})$ into the group $\overline{E}^p(\alpha, \mathfrak{B})$, corresponding to the projection ϖ_α^β. We obtain an inverse spectrum of bicompact groups, whose limit group is a topological invariant of the space A. We denote it by $\overline{E}^p(A, \mathfrak{B})$.

Our object is the proof of the following duality theorem:

Theorem 1. *If A and B are complementary sets of the space S^n, then the following duality holds:*

$$\overline{E}^p(A, \mathfrak{B}) \mid \Delta_c^q(B, \mathfrak{A}). \tag{3}$$

First we prove the duality

$$\overline{E}_{\text{ext}}^p(A, \mathfrak{B}) \mid \Delta_c^q(B, \mathfrak{A}), \tag{4}$$

where $\overline{E}^p_{\text{ext}}(A, \mathfrak{B})$ is some "exterior" group of the set A, to the definition of which we turn now.

We consider all possible triangulations K lying in S^n, with bodies which are open in S^n and contain the set A. We shall take $K < K'$ if the covering of the set $\widetilde{K}'$ by principal stars K' is subordinate to the analogous covering of the set $\widetilde{K}$. As usual, in general this subordination does not uniquely define a projection (simplicial mapping) of the nerve of the system of principal stars of the polyhedron K' into the nerve of the principal stars of the polyhedron K (i.e., of the complex K' into the complex K), but all of them are combinatorially close.

The groups $\overline{E}^p(K, \mathfrak{B})$ and the continuous homomorphisms for $K > K'$, corresponding to the projections referred to above, form an inverse spectrum of bicompact groups, given only that the set of triangulations is directed. But this is in fact the case. Indeed, for any K and K' the set $\widetilde{K} \cap \widetilde{K}'$ is open in S^n and contains A. Let L be a triangulation of it. Sets of the form $O_{Ke} \cap O_{K'e'}$ define an open covering of the space $\widetilde{L}$; there exists a subdivision L' of L (page 133) such that the covering formed by its principal stars is subordinate to that covering. Accordingly, L' is subordinate both to K and to K'.

The limit group of the spectrum described above is the group $\overline{E}^p_{\text{ext}}(A, \mathfrak{B})$.

We turn to the proof of the duality $\overline{E}^p_{\text{ext}}(A, \mathfrak{B}) \,|\, \Delta^q_c(B, \mathfrak{A})$. If K is a triangulation of an open set, containing A, then $F_K = S^n - \widetilde{K}$ is a compact subset of B. We carry out a multiplication in the spectrum, defining $\Delta^q_c(B, \mathfrak{A})$, as follows: we shall consider a group $\Delta^q_c(F, \mathfrak{A})$ of this spectrum to be repeated as many times as there are distinct triangulations of the set $S^n - F$. The limit group of the spectrum thus obtained remains the group $\Delta^q_c(B, \mathfrak{A})$, and the spectrum itself turns out to be adjoint to the spectrum defining the group $\overline{E}^p_{\text{ext}}(A, \mathfrak{B})$.

In order to prove this adjointness and along with it the duality (4) we must show that: 1) $\overline{E}^p(K, \mathfrak{B}) \,|\, \Delta^q_c(F, \mathfrak{A})$ for $\widetilde{K} = S^n - F$; 2) the homomorphisms of the spectra are adjoint. The first is the duality law for closed sets, and the second, taking into account the remark at the end of subsection 1, is derived as follows: let $K < L$, and let ϕ be the corresponding projection. Suppose further that $F_K = S^n - \widetilde{K}$ and $F_L = S^n - \widetilde{L}$. If z^p is any finite cycle of L, and $z^q = \{z^q_n\}$ is a cycle of F_K, then, since z^p and ϕz^p are mutually homologous in $\widetilde{K}$, we have the equality

$$\mathfrak{b}(z^p, z^q) = \mathfrak{b}(\phi z^p, z^q).$$

This means that

$$(\phi^*\xi \cdot \zeta) = (\xi \cdot \psi^*\zeta),$$

where $z^p \in \xi \in \overline{E}^p(L, \mathfrak{B})$, $z^q \in \zeta \in \Delta^q_c(F_K, \mathfrak{A})$, and ϕ^* and ψ^* are homomorphisms of the spectra under consideration corresponding to the pair $K < L$; in view of the continuity this equality holds for all $\xi \in \overline{E}^p(L, \mathfrak{B})$ and $\zeta \in \Delta_c^q(F^K, \mathfrak{A})$. Thus the duality (4) is proved. For the proof of the duality (3) it remains to prove the isomorphism

$$\overline{E}^p(A, \mathfrak{B}) = \overline{E}^p_{\text{ext}}(A, \mathfrak{B}).$$

This will be done below.

4. In this subsection, by a spectrum of complexes we shall understand a directed set of complexes α along with simplicial mappings (projections) $\mathfrak{w}_\alpha^\beta$, defined for $\alpha < \beta$ and mapping β into α. The projections are subjected to the following conditions: 1) projections corresponding to one and the same pair $\alpha < \beta$ are combinatorially close; 2) the identity mapping of the complex α into itself is a projection; 3) if $\mathfrak{w}_\alpha^\beta$ and $\mathfrak{w}_\beta^\gamma$ are projections, then also $\mathfrak{w}_\alpha^\gamma = \mathfrak{w}_\alpha^\beta \mathfrak{w}_\beta^\gamma$ is a projection.

We shall say that the spectrum S is a cofinal subspectrum in the spectrum T if the following conditions are satisfied: 1) each complex of S is a subcomplex in T; 2) if $\alpha < \beta$ in S, then $\alpha < \beta$ in T, and the corresponding projection in S is a projection in T; to each complex of T there is a following complex of S.

Further, two spectra S and T will be said to be equivalent if there exists a finite sequence of spectra $S = S_1, S_2, \cdots, S_k = T$ such that for each i, either S_i is a cofinal subspectrum in S_{i+1}, or conversely S_{i+1} is a cofinal subspectrum in S_i.

For the topological space X the nerves of its open coverings and of its projections, defined as usual by subordination, evidently form a spectrum of complexes. A spectrum of complexes is also formed by triangulations of open sets in S^n containing the set A; here the ordering $K < K'$ and the projections are defined as in the preceding subsection (in the definition of the group $\overline{E}^p_{\text{ext}}(A, \mathfrak{B})$). This spectrum will be called a polyhedral spectrum of the set A.

We have the following (see Chapter II, §1, and Chapter III, §§2, 5):

Theorem 2. *The spectrum of the nerves and the polyhedral spectrum of a set lying in S^n are equivalent.*

Proof. Let $A \subset S^n$. We denote by T_1 the nerve spectrum, and by T_5 the polyhedral spectrum. The spectra T_2, T_3, T_4, connecting these spectra will be described as follows.

T_4 is the spectrum of exterior nerves of the set A, i.e., one considers a system $\gamma = \{u\}$ of sets open in S^n such that $\bigcup u \supset A$. The nerve (with respect to S^n) of this system is a complex of the spectrum T_4. Further, we take $\gamma < \gamma_1$ if γ_1 is subordinate to γ, i.e., for $u_1 \in \gamma_1$ there exists a $u \in \gamma$ satisfying $u \subset u_1$. The simplicial mappings defined by subordination are taken as the projections. This completes the description of the spectrum T_4. It is easy to see that T_5 is a cofinal subspectrum in T_4. Indeed, each triangulation is the nerve of a covering of its body by principal stars and we have defined the projections and ordering in T_5 by subordination; further, for $\gamma = \{u\}$ there exists a triangulation K of the set $\bigcup u$ (see page 132) such that its covering by principal stars is subordinate to γ.

We define the spectrum T_3 as a subspectrum of the spectrum T_4, corresponding to those exterior coverings $\gamma = \{u\}$ which excise in A coverings similar to themselves, i.e., when the nerve γ in S^n and the nerve of the covering $\{u \cap A\}$ may be identified. The spectrum T_3 is cofinal in T_4: for an exterior covering $\gamma = \{u_i\}$ we consider the interior covering $\{u'_i\}$, $u'_i = A \cap u_i$ excised by it; taking account of the fact that γ is a star-finite covering, we consider the exterior covering $\{O_i\}$, similar to the covering $\{u'_i\}$ (it exists, as we have proved on page 151); the exterior covering $\{O_i \cap u_i\}$ follows $\{u_i\}$ and belongs to T_3.

We shall take the complexes of T_2 to be the nerves of the coverings $\gamma = \{u\}$ of the space A. The coverings may be interior or exterior but they must excise in A coverings similar to themselves. Then $\gamma < \gamma'$ if for any $u \in \gamma'$ there exists $v \in \gamma$ satisfying the condition $u \cap X \subset v$. The correspondence

$$u \to v, \quad u \in \gamma', \quad v \in \gamma, \quad u \cap A \subset v$$

defines a projection of the spectrum T_2. The spectrum T_2 defined in this way evidently contains the spectra T_1 and T_3 as cofinal subspectra. The proof of the theorem is thus complete.

Let us remark that, as is evident from the above proof, the following theorem is also true.

Theorem 3. *Let Y be a metric space and X a subspace of Y. Then the spectrum of the nerves of the coverings of the space X and the spectrum of the nerves of its outer (with respect to Y) coverings are equivalent.*

From Theorem 2 follows the desired isomorphism

$$\overline{E}^p(A, \mathfrak{B}) = \overline{E}^p_{\text{ext}}(A, \mathfrak{B}).$$

For let $\{\alpha, \omega_\alpha^\beta\}$ be the spectrum of the complexes; then the system $\{\overline{E}^p(\alpha, \mathfrak{B}), \omega_\alpha^\beta\}$ forms the inverse spectrum of the bicompact groups and equivalent spectra of groups correspond to equivalent spectra of complexes.

CHAPTER V

Special classes of sets.

Duality regions

§1. The conditions (r^p) and (a^p).

1. The condition (r^p). We say that the set A satisfies the condition (r^p) or is an (r^p)-*set* (over a given arbitrary coefficient domain) [*] if each p-dimensional true cycle of that set of the given coefficient domain which is homologous to zero in every neighborhood of the set A is (compactly) homologous to zero in the set A itself.

This condition, evidently, is equivalent to the condition that *the fundamental homomorphism h of the group $\Delta^p_c A$ into the group $\delta^p A = D^p A$ should be an isomorphism*. Therefore the class of (r^p)-sets is, first, topologically invariant, and secondly, dualizable (because of the general assertion of Chapter IV, §4 on the adjointness of the homomorphisms h and $\bar{h}$).

2. The condition (a^p). The set A satisfies the condition (a^p) or is an (a^p)-*set* if for any neighborhood λ of the set A one may find a neighborhood λ' of that same set such that each p-dimensional cycle lying in λ' is homologous in λ to some true cycle of the set A. We now shall prove the *topological invariance of this condition*. To this end we shall prove that the set (a^p) is equivalent to the following, invariantly formulated

Condition (A^p). For any covering ω of the set A there exists a covering ω' of the set A such that for any cycle $z^p_{\omega'}$ of the nerve ω' one may find a compact projective cycle [**] $z^p_\phi = \{z^p_{\phi\,a}\}$, satisfying the homology

$$\mathfrak{D}^{\omega'}_\omega \cdot z^p_{\omega'} \sim z^p_{\phi\omega} \text{ in } \omega. \tag{1}$$

Remark 1. If (A^p) is satisfied, with the covering ω given and the covering ω' selected according to this condition, then for any covering ω'', following ω', and any cycle $z^p_{\omega''}$, we may select $z^p_\phi = \{z^p_{\phi\,a}\}$ so that

$$\mathfrak{D}^{\omega''}_\omega z^p_{\omega''} \sim z^p_{\phi\omega} \text{ in } \omega.$$

Indeed, it is sufficient to put $z^p_{\omega'} = \mathfrak{D}^{\omega''}_{\omega'} z^p_{\omega'}$, and to select for that $z^p_{\omega''}$ a compact projective cycle z^p_ϕ, according to (1). Then we have

---

[*] The topology in the coefficient group, if there is any, is not taken into account, so that the coefficient domain may be considered to be a discrete group.

[**] Denoting this compact cycle by z^p_ϕ, we understand by ϕ any of its carriers (where, of course, to different ω' there correspond, generally speaking, different z^p_ϕ, lying on different ϕ's).

$$\eth_{\omega}^{\omega''} z^P_{\omega''} = \eth_{\omega}^{\omega'} \eth_{\omega'}^{\omega''} z^P_{\omega''} = \eth_{\omega}^{\omega'} z^P_{\omega'} \sim z^P_{\phi\omega} \text{ in } \omega.$$

Now suppose that the condition (A^p) is satisfied. We shall show that then the condition (a^p) is also satisfied. Suppose that λ is given. We choose any triangulation τ of the set λ. The canonical complex τ' becomes, as we know (Chapter II, §1.2), the nerve of the adjoint canonical covering ω. To this covering ω we assign ω' according to the condition (A^p). We choose a triangulation θ strictly following τ and such that the canonical covering ω_θ adjoint to θ' follows ω'.

We assert that the canonical neighborhood* $\lambda'_{\theta'} = \lambda'$ satisfies the condition (a^p) with respect to the already given λ. Indeed, let $z^P_{\lambda'}$ be a cycle lying in λ'. It is homologous in λ', and therefore a fortiori in λ, to some cycle of the canonical complex θ'.

So it is sufficient to prove that each cycle $z^P_{\theta'}$ of the complex θ' is homologous in λ to some true cycle of the set A. However, since the covering $\omega_{\theta'}$ follows ω', then, in view of the remark to condition (A^p), there exists a compact projective cycle $z^P_{\theta'}$ such that for the given $z^P_{\theta'}$ we have

$$\eth_{\omega}^{\theta'} z^P_{\theta'} \sim z^P_{\phi\omega} \text{ in } \omega. \tag{2}$$

Since $\tilde\theta' \subseteq \tilde\tau'$, and the projection $\eth_{\omega}^{\theta'}$ of the cycle $z^P_{\theta'}$ is realized by means of a shift in the polyhedron $\tilde\tau'$, it follows from (2) that

$$z^P_{\theta'} \sim z^P_{\phi\omega} \text{ in } \tilde\tau' \subseteq \tilde\tau = \lambda.$$

We shall consider only canonical complexes γ following the nerve $\omega = \tau'$. Then the coverings γ pick out in the thread $\{z^P_{\phi\alpha}\} = z^P_\phi$ cycles $z^P_{\phi\gamma}$ which define a pure cycle (of the compactum $\phi \subseteq A$)

$$(z^P_1, z^P_2, \cdots, z^P_k, \cdots),$$

all the components z^P_k of which are homologous in $\tilde\tau'$ to the cycle $z^P_{\phi\omega} \sim z^P_{\theta'}$, and therefore a fortiori $z^P_k \sim z^P_{\theta'}$ in λ, whereby our assertion is proved.

Now suppose that condition (a^p) is satisfied. We shall prove that the condition (A^p) is also satisfied. We choose an arbitrary covering ω. We need to select a covering ω' so that the pair of coverings ω, ω' satisfy the condition (A^p), in the sense that for each cycle $z^P_{\omega'}$ there exists a compact projective cycle $z^P_\phi = \{z^P_{\phi\alpha}\}$ such that $\eth_{\omega}^{\omega'} z^P_{\omega'} \sim z^P_{\phi\omega}$ on ω. It suffices to choose such an ω' for each *canonical* ω. Indeed, under the assumption that for canonical ω we have

---

    * If θ' is a canonical complex, then we shall always denote by $\lambda'_{\theta'}$ the corresponding canonical neighborhood of the set A (see Chapter II, §1.5).

been able to solve this problem, we consider any covering ω_1 and select a canonical ω following ω_1. Selecting for ω a covering ω' such that the pair of coverings ω, ω' satisfy condition (A^p), we have for any cycle $z^p_{\omega'}$ a compact cycle z^p_ϕ such that

$$\mathfrak{D}^{\omega'}_{\omega} z^p_{\omega'} \sim z^p_{\phi\omega} \text{ in } \omega.$$

But then

$$\mathfrak{D}^{\omega'}_{\omega_1} z^p_{\omega'} \sim \mathfrak{D}^{\omega}_{\omega_1}\mathfrak{D}^{\omega'}_{\omega} z^p_{\omega'} \sim \mathfrak{D}^{\omega}_{\omega_1} z^p_{\phi\omega} \text{ in } \omega_1,$$

i.e., the pair ω_1, ω' also satisfies the condition (A^p). So, let the given ω be a canonical covering, whose nerve τ'_ω is the derived triangulation of the triangulation τ_ω of some neighborhood λ_ω of the set A. For the corresponding canonical neighborhood $\lambda'_{\tau'_\omega} = \lambda$ we select λ' in accordance with condition (a^p). We select a triangulation $\tau_{\lambda'}$ of the set λ' following τ_ω, and its derived triangulation $\tau'_{\lambda'}$. We assert that the corresponding canonical covering $\omega' = \omega_{\tau'\lambda''}$ is the one desired. Indeed, each cycle $z^p_{\omega'}$, being a cycle of the open set λ', satisfies a homology of the form

$$z^p_{\omega'} \sim z^p_\phi \text{ in } \lambda, \tag{3}$$

where z^p_ϕ is a true cycle (on some compactum $\phi \subseteq A$). Recall that λ retracts onto the polyhedron $\widetilde{\tau}'_\omega$. Since the cycles z^p_ϕ, $z^p_{\omega'}$ lie respectively on $\phi \subseteq A \subseteq \widetilde{\tau}'_\omega$ and on $\widetilde{\tau}'_{\lambda'} \subseteq \widetilde{\tau}_\omega$, which means that they remain fixed under the retraction just mentioned, this retraction carries the homology (3) into the homology

$$z^p_{\omega'} \sim z^p_\phi \text{ in } \widetilde{\tau}'_\omega. \tag{4}$$

Since $z^p_{\omega'}$ goes into $\mathfrak{D}^{\omega'}_{\omega} z^p_{\omega'}$, by means of a shift in $\widetilde{\tau}'_\omega$, it follows from (4) that

$$\mathfrak{D}^{\omega'}_{\omega} z^p_{\omega'} \sim z^p_\phi \text{ in } \tau'_\omega. \tag{5}$$

We take a projective cycle $\{z^p_{\phi\gamma}\}$, homologous to the true cycle z^p_ϕ in the sense that, for any covering γ, the homology class $\mathfrak{z}^p_{\phi\gamma}$ of the cycle $z^p_{\phi\gamma}$ is the homology class of the true cycle z^p_ϕ in the nerve γ (see Part I, page 56). For every sufficiently fine canonical nerve γ following the nerve ω, the cycles $z^p_{\phi\gamma}$ are arbitrarily fine cycles of the polyhedron $\widetilde{\tau}'_\omega$, while the projections $\mathfrak{D}^{\gamma}_{\omega}$ are realized by shifts within the limits of the same polyhedron, so that always also $z^p_{\phi\gamma} \sim z^p_{\phi\omega}$ on $\widetilde{\tau}'_\omega$. Since in view of (5) the cycle $\mathfrak{D}^{\omega'}_{\omega} z^p_{\omega'}$ is homologous on $\widetilde{\tau}'_\omega$ to the cycle $z^p_{\phi\gamma}$ for sufficiently small γ, therefore $\mathfrak{D}^{\omega'}_{\omega} z^p_{\omega'} \sim z^p_{\phi\omega}$ in $\widetilde{\tau}'_\omega$, and thus also in τ'_ω, as we were required to prove.

Remark 2. A natural strengthening of the condition (a^p) is the following condition:

Condition (ua^p). In the set A there is contained a compactum ϕ such that for each neighborhood $\lambda \supseteq A$ one may choose a neighborhood $\lambda' \supseteq A$ so that every cycle lying in λ' is homologous in λ to some true cycle of the compactum ϕ.

The topological invariance of the condition (ua^p) is proved by establishing the equivalence of that condition to the following condition:

(UA^p). In the set A there is contained a compactum ϕ such that for each covering ω of the set A one may choose a covering ω' so that each cycle $z^p_{\omega'}$, satisfies a homology

$$\partial^{\omega'}_{\omega} z^p_{\omega'} \sim z^p_{\phi\omega},$$

where $z^p_\phi = \{z^p_{\phi\alpha}\}$ is some projective cycle lying on ϕ.

The proof of the equivalence of conditions (ua^p) and (UA^p) goes through in the same way as that of the equivalence of the conditions (a^p) and (A^p).

3. $(r^p a^p)$-sets. Sets satisfying at the same time conditions (r^p), (a^p) are called $(r^p a^p)$-sets (over the coefficient domain $\mathfrak{A}$).

Theorem 1. *Let A be an $(r^p a^p)$-set over the coefficient domain $\mathfrak{A}$. Select in A cycles and homologies over the group $\mathfrak{A}$, and in B over the group $\mathfrak{B}$. Then each p-dimensional true cycle of the set A, not bounding in A, links with some q-dimensional true cycle in B, and each q-dimensional true cycle of the set B, not bounding in B, links with some p-dimensional true cycle in A, so that the groups $N^p_{\Delta c}(A, \mathfrak{A})$ and $N^q_{\Delta c}(B, \mathfrak{B})$ are null-groups, and the group $\overline{\Delta}^q(B, \mathfrak{B})$ coincides with the bicompact extension $\overline{\Delta}^q_c(B, \mathfrak{B})$ of the group $\Delta^q_c(B, \mathfrak{B})$.*

Indeed, it follows from condition (r^p) that each true cycle z^p_A not bounding in A links with some true cycle in B. Suppose that the true cycle z^q_B of the set B does not bound in B. Suppose that ψ is the carrier of the cycle z^q_B. Then $\lambda = S^n - \psi \supseteq A$. We select $\lambda' \subseteq \lambda$ so that every cycle z^p_λ, lying in λ' is homologous in λ to some true cycle of the set A. Then $\psi \subseteq \psi' = S^n - \lambda'$ and z^q_B does not bound in ψ'. Therefore in λ' there exists a cycle $z^p_{\lambda'}$ linking with z^q_B. This cycle, in view of condition (a^p), is homologous in λ to some true cycle of the set A which links with z^q_B, as we were required to prove.

Theorem 2.[*] *For any $(r^p a^p)$-set the natural homomorphism h of the group $\Delta^p_c A$ into the group $\delta^p A = D^p A$ is an isomorphism onto the whole group $\delta^p A$.*

Remark 1. From this theorem and the one preceding, it follows that for $(r^p a^p)$-sets A the first duality law takes the form

$$\Delta^p_c(A, \mathfrak{A}) \mid \overline{\Delta}^q_c(B, \mathfrak{B}).$$

---

[*] This theorem and the proof of it given here are due to E. F. Miščenko [13a]. The analogous theorem in the special case of homology retracts (see the following section) was earlier proved by me in [10].

Remark 2. From the proof of Theorem 2 below it will be clear that for an (a^P)-set the homomorphism h is a homomorphism onto the whole group $\delta^P A$.

Proof of Theorem 2. We consider the inverse spectrum

$$\{\Delta_c^P(A, \lambda), E_\lambda^{\lambda'}\}, \tag{1}$$

where $\Delta^P(A, \lambda)$ is the factor group of the group of all true cycles of the set A modulo the subgroup of cycles bounding in λ, and $E_\lambda^{\lambda'}$ is the natural inclusion homomorphism of the group $\Delta_c^P(A, \lambda')$ into the group $\Delta_c^P(A, \lambda)$ for $\lambda' \subset \lambda$. It is easily seen that if A is an (r^P)-set, then the limit group of the spectrum (1) is the group $\Delta_c^P A$:

$$\Delta_c^P A = \varprojlim \{\Delta_c^P(A, \lambda), E_\lambda^{\lambda'}\}. \tag{2}$$

We shall call by the name *sliding true cycle* of the set A a system ("thread") $z^P = \{z_{c\lambda}^P\}$ whose elements are true cycles $z_{c\lambda}^P$ of the set A, set into correspondence with the neighborhoods λ of that set, while if $\lambda' \subset \lambda$ we have $z_{c,\lambda'}^P \sim z_{c,\lambda}^P$ in λ. We shall say that a sliding true cycle is bounding if $z_{c,\lambda}^P \sim 0$ in λ for any λ. The group (2) may be regarded as the factor group of the group of all sliding true cycles modulo the subgroup of bounding cycles.

Further, by the definition of the group $D^P A$ we have

$$\delta^P A = D^P A = \varprojlim \{\Delta_c^P \lambda, E_\lambda^{\lambda'}\}. \tag{3}$$

We shall show that the natural homomorphism of the group (2) into the group (3) is an isomorphism onto all of that group. This will constitute a proof of Theorem 2.

Suppose that $\{z_\lambda^P\}$ is any sliding cycle of the set A. By definition, we have for $\lambda' \subset \lambda$

$$z_{\lambda'}^P \sim z_\lambda^P \text{ in } \lambda. \tag{4}$$

We choose a λ. Since A satisfies the condition (a^P), we may select a λ' for that λ so that any cycle lying in λ' is homologous in λ to some true cycle of the set A. We choose such a λ'. Then, in particular, the cycle $z_{\lambda'}^P$ is homologous in λ to some true cycle of the set A, which we shall denote by $z_{c,\lambda}^P$. We have the homology

$$z_{\lambda'}^P \sim z_{c,\lambda}^P \text{ in } \lambda; \tag{5}$$

from it and from the homology (4) follows the homology

$$z_{c,\lambda}^P \sim z_\lambda^P \text{ in } \lambda. \tag{6}$$

Carrying out this construction for any neighborhood λ of the set A, we obtain a system $\{z_{c,\lambda}^P\}$. We shall prove that that system is a sliding true cycle. Indeed, let $\lambda' \subset \lambda$. Then

$$z^P_{c,\lambda'} \sim z^P_{\lambda'},$$

which along with (6) and (4) yields

$$z^P_{c,\lambda'} \sim z^P_{c,\lambda} \text{ in } \lambda.$$

By the same construction the cycle $\{z^P_{c,\lambda}\}$ is homologous to the cycle $\{z^P_\lambda\}$, which is what was required to be proved.

§2. Homology retracts

1. The condition (ur^P).

Definition. *We shall say that the set A satisfies the condition (ur^P), or is a (ur^P)-set, if there exists a neighborhood λ of the set A such that every p-dimensional true cycle of the set A bounding in λ also bounds in A* (the condition (ur^P) this becomes the condition that the condition r^P be satisfied uniformly). The coefficient domain is any group $\mathfrak{A}$.

Remark 1. Suppose that the coefficient domain is a finite cyclic group Π_m. Then every compactum A is an (ar)-set (i.e., at the same time an (a^P)-set and an (r^P)-set for any p): here the condition (ur^P) is satisfied by all finitely-connected modulo m compacta [*] and only those. Since every retract is a (ur)- and even a $(ur)\,(a)$-set, and every open set of S^n is a retract, it follows from what has been said that the complement to a (ur)- and even to a $(ur)\,(a)$-set may not be a (ur)-set.

We shall prove the *topological invariance of the condition* (ur^P). To this end we shall prove that it is equivalent to the following

Condition (UR^P). *There exists a covering ω of the set A such that every compact projective cycle $z^P_A = \{z^P_\alpha\}$ satisfying the homology $z^P_\omega \sim 0$ in ω is compactly homologous to zero in A.*

We shall prove first of all that from (UR^P) follows (ur^P).

We select a covering ω as is indicated in (UR^P), and any canonical covering ω' following it which is adjoint to the derived triangulation τ' of the triangulation τ of the neighborhood λ_0 of the set A.

We assert that the canonical neighborhood $\lambda = \lambda'_{\tau'}$ may be taken as the neighborhood λ mentioned in the formulation of the condition (ur^P). Suppose that z^P_ϕ is any true cycle of the set A (with the carrier ϕ), homologous to zero in λ. We need to prove that $z^P_\phi \sim 0$ in A. We select a compact projective cycle $\{z^P_{\phi\,\alpha}\}$, homologous in A to the true cycle z^P_ϕ.

Since

$$z^P_\phi \sim 0 \text{ in } \lambda = \lambda'_{\tau'},$$

---

[*] I.e., having a finite group $\Delta^P(A, \Pi_m)$.

therefore

$$z^P_\phi \sim 0 \text{ in } \tau'.$$

We choose a sequence of canonical complexes

$$\tau'_1, \tau'_2, \cdots, \tau'_k, \cdots$$

such that τ'_1 follows τ', and (for any $k \geq 1$) τ'_{k+1} follows τ'_k, and such that the diameters of the simplexes of the complex τ'_k tend to zero as $k \to \infty$. Then for the canonical coverings α_k adjoint to the triangulations τ'_k, the projections $\overset{\sim}{\omega}{}^{\alpha_k}_{\omega'}$ are realized by shifts in the polyhedron $\overset{\sim}{\tau}{}^{i}$ and therefore

$$z^P_{\phi \alpha_k} \sim z^P_{\phi \omega'} \text{ in } \overset{\sim}{\tau}{}'.$$

On the other hand, the true cycle $\{z^P_{\phi \alpha_k}\}$ is homologous to the true cycle $\{z^P_\phi\}$ in ϕ, so that always $z^P_{\phi \alpha_k} \sim 0$ in $\overset{\sim}{\tau}{}'$, which means that also $z^P_{\phi \omega'} \sim 0$ in $\overset{\sim}{\tau}{}'$ and $z^P_{\phi \omega'} \sim 0$ in τ'. We assume the condition (UR^P) satisfied, from which it follows that the projective cycle $\{z^P_{\phi a}\}$ is compactly homologous to zero in A, and therefore the true cycle z^P_ϕ, homologous to it in ϕ, is also homologous to zero in A, as was required to be proved.

Now we suppose that condition (ur^P) is satisfied, and prove that the condition (UR^P) is also satisfied. We choose a λ satisfying the requirements in condition (ur^P). We choose a canonical neighborhood $\lambda' = \lambda'_\tau$ contained in λ and a canonical covering $\omega = \omega_\tau$. We assert that the covering ω satisfies the condition (UR^P). Indeed, suppose that $\{z^P_{\phi a}\}$ is a compact projective cycle satisfying the homology

$$z^P_{\phi \omega} \sim 0 \text{ in } \omega. \tag{1}$$

We choose a sequence of canonical complexes

$$\tau_1, \tau_2, \cdots, \tau_k, \cdots,$$

the diameters of whose simplexes converge to zero as $k \to \infty$, while τ_1 follows τ and each τ_{k+1} follows τ_k.

We obtain a true cycle

$$z^P_\phi = (z^P_1, z^P_2, \cdots, z^P_k, \cdots),$$

where $z^P_k = z^P_{\phi \alpha_k}$. This true cycle is homologous to the projective cycle $\{z^P_{\phi a}\}$. Moreover, all the z^P_k are homologous to the cycles $z^P_{\phi \omega}$ in the polyhedron $\overset{\sim}{\tau}$, and, in view of (1), they are homologous to zero in this polyhedron and therefore, a fortiori, in λ. So, $z^P_\phi \sim 0$ in λ, and therefore, by the condition (ur^P),

$$z^P_\phi \sim 0 \text{ in } A,$$

which means that also $\{z^P_{\phi a}\} \sim 0$ in A, as was required.

Definition. (ur^P) (a^P)-*sets, i.e., sets satisfying at the same time conditions*

(ur^p) *and* (a^p), *are called p-retracts (with respect to the given coefficient domain* $\mathfrak{A}$). *They form a topologically invariant class of sets.*

Sets which are p-retracts for any p are called simply homology retracts (over the given coefficient domain $\mathfrak{A}$) *or* (ur) (a)-*sets.*

Remark 2. Sets which are neighborhood retracts in the usual sense of the word are evidently also homology retracts with respect to any coefficient domain. However, we shall see that there exist sets which are not neighborhood retracts but are nevertheless homology retracts with respect to any coefficient domain; for example, all skin polyhedra * and all of their topological images. On the other hand, it is known (see [3]) that every (even infinite) polyhedron is a neighborhood retract. But it is easy to construct infinite polyhedra which do not satisfy the condition (ua^p); such an infinite polyhedron (of dimension zero) is, for example, any set consisting of an infinite number of isolated points.

2. **The second form of definition of p-retracts; the groups** $\delta^p(\Gamma:\Gamma')$ **and** $\Delta_c^q(\Phi':\Phi)$. The definition of p-retracts may be given in another form, of which we shall immediately make use. To this end we consider two open sets Γ and Γ' of the set S^n, with $\Gamma \subseteq \Gamma'$. We shall then say that the group $\Delta_c^p(\Gamma:\Gamma')$ is the factor-group of all the cycles of dimension p lying in Γ by the subgroup of all cycles bounding in Γ'. The group $\Delta_c^p(\Gamma:\Gamma')$ may also be defined

a) as the subgroup $\Delta_c^p(\Gamma':\Gamma)$ of the group $\Delta_c^p\Gamma'$, whose elements are homology classes containing cycles lying in Γ;

b) as the factor-group $\Delta_c^p(\Gamma:\Gamma')$ of the group $\Delta_c^p\Gamma$ modulo the subgroup $H^p(\Gamma:\Gamma')$ of all elements $\mathfrak{z}^p \in \Delta_c^p\Gamma$, which are homology classes of cycles homologous to zero in Γ'.

Remark. Thus our notations ** satisfy the symmetry condition $\Delta_c^p(\Gamma:\Gamma') = \Delta_c^p(\Gamma':\Gamma)$.

Now suppose that Γ', Γ are neighborhoods of the set A. Since every true cycle of the set A lies in Γ and two true cycles homologous to one another in A are a fortiori homologous to one another in Γ', we have a natural homomorphism of the group $\Delta_c^p A$ into the group $\Delta_c^p(\Gamma:\Gamma')$, generated by the identity mapping of the set A in $\Gamma \subseteq \Gamma'$.

Theorem 1. *In order that the set* A *should be a p-retract with respect to the given coefficient domain* $\mathfrak{A}$, *it is necessary and sufficient that there exist a*

---

* A skin polyhedron (lying in a given S^n) is the sum of a finite number of pairwise disjoint simplexes (of whatever dimension) lying in that S^n. It is easy to see that skin polyhedra (in the given S^n) form the smallest class of sets closed with respect to the operations of (finite) addition and subtraction and containing all convex polyhedra lying in the given S^n.

** Here Δ_c^p may be replaced throughout by δ^p, so that for an open set $\Delta_c^p\Gamma = \delta^p\Gamma$.

$\lambda_0 \supseteq A$ *such that for each* $\lambda \subseteq \lambda_0$ *one may select* $\lambda' \subseteq \lambda$ *so that the natural homomorphism of the group* $\Delta_c^p A$ *into the group* $\Delta_c^p(\lambda':\lambda)$ *is an isomorphism of the group* $\Delta_c^p A$ *onto the group* $\Delta_c^p(\lambda':\lambda)$.

Proof. Indeed, suppose that conditions (ur^p) and (a^p) are satisfied. We choose a λ_0 such that every true cycle z_A^p, homologous to zero in λ_0 is homologous to zero in A. We then select $\lambda \subseteq \lambda_0$ arbitrarily. Thereupon we take a neighborhood λ' of the set A such that every p-dimensional cycle lying in λ' is homologous in λ to a true cycle of the set A. We assert that the natural homomorphism of the group $\Delta_c^p A$ into the group $\Delta_c^p(\lambda':\lambda)$ is under these conditions an isomorphism onto the group $\Delta_c^p(\lambda':\lambda)$. Indeed, in view of the very choice of the neighborhood λ_0, this homomorphism is an isomorphism, while in view of the choice of λ' each homology class $\mathfrak{z}_{\lambda',\lambda}^p \in \Delta_c^p(\lambda':\lambda)$ contains true cycles of the set A.

Conversely, if there exists a λ_0 as described in Theorem 1, then, writing $\lambda = \lambda_0$ and choosing λ' so that the identity mapping of the set A into λ' generates an isomorphism of the group $\Delta_c^p A$ onto the group $\Delta_c^p(\lambda':\lambda)$, we see that every true cycle of the set A not homologous to zero in A is not homologous to zero in λ_0, so that the condition (ur^p) is satisfied. Now taking any $\lambda \subseteq \lambda_0$ and choosing λ' so that we have a natural isomorphism of $\Delta_c^p A$ onto $\Delta_c^p(\lambda':\lambda)$, we see that every homology class $\mathfrak{z}_{\lambda',\lambda}^p \in \Delta_c^p(\lambda':\lambda)$ contains a true cycle of the set A, i.e., that the condition (a^p) is satisfied for these λ and λ'.

In exactly the same way as for the groups $\Delta_c^p(\Gamma:\Gamma')$, one defines the groups $\Delta_c^q(\Phi':\Phi)$ for closed sets $\Phi' \subseteq \Phi \subseteq S^n$: the group $\Delta_c^q(\Phi':\Phi)$ is the factor-group of the group of all q-dimensional true cycles of the compactum Φ' over the subgroup of all cycles homologous to zero in Φ. As in the case of open sets, the group $\Delta_c^q(\Phi':\Phi)$ may also be defined as follows:

a) as the subgroup $\Delta_c^q(\Phi:\Phi')$ of the group $\Delta_c^q\Phi$, consisting of all homology classes containing as elements true cycles of the set Φ';

b) as the factor-group $\Delta_c^q(\Phi':\Phi)$ of the group $\Delta_c^q\Phi'$ modulo the subgroup $H^q(\Phi':\Phi)$ consisting of the homology classes $\mathfrak{z}_{\Phi'}^q \in \Delta^q\Phi'$ whose elements are cycles homologous to zero in Φ. Again the notations are symmetric.

Finally, if $\Phi \subset \Gamma$, where Φ is closed and Γ is open in S^n, we shall denote by $\Delta_c^p(\Phi:\Gamma)$ the factor-group of the group of all p-dimensional true cycles of the complex Φ over the subgroup of cycles bounding in Γ. This group may be defined as the factor-group $\Delta_c^p(\Phi:\Gamma)$ of the group $\Delta_c^p\Phi$ modulo the subgroup defined by cycles bounding in Γ, and as the subgroup $\Delta_c^p(\Gamma:\Phi)$ of the group $\Delta_c^p\Gamma$ whose elements are homology classes of the set Γ, containing true cycles of the compactum Φ.

One may prove without difficulty the following theorem:

Theorem 2. *Suppose given* $\phi \subset \lambda' \subset \lambda$; $\mu = S^n - \phi$; $\psi = S^n - \lambda$, $\psi' = S^n - \lambda'$, *where, as always, ϕ is closed and λ and λ' are open in S^n. Then*

$$\Delta_c^p(\lambda':\lambda) \mid \Delta_c^q(\psi:\psi'),$$

$$\Delta_c^p(\lambda:\phi) \mid \Delta_c^q(\psi:\mu),$$

the coefficient domain on the left being $\mathfrak{A}$ and on the right $\mathfrak{B}$.

For the proof, for example, of the first of these dualities, it is sufficient to observe that under the duality $\Delta_c^q\psi \mid \Delta_c^p\lambda$ the annihilator of the subgroup $H^q(\psi:\psi')$ in the group $\Delta_c^q\lambda$ is exactly the group $\Delta_c^p(\lambda:\lambda')$. Accordingly, we have the duality

$$\Delta_c^p(\lambda:\lambda') \mid (\Delta_c^q\psi - H^q(\psi:\psi')),$$

i.e.,

$$\Delta_c^p(\lambda:\lambda') \mid \Delta_c^q(\psi:\psi').$$

The condition (J^q). We shall say that the set B satisfies the condition (J^q) (or is a (J^q)-set) over the given coefficient domain if there exists a compactum $\psi_0 \subseteq B$ with the property that for any $\psi \supseteq \psi_0$ we may find a $\psi' \supseteq \psi$ such that the natural homomorphism of the group $\Delta_c^q(\psi:\psi')$ into the group $\Delta_c^q B$ is an isomorphism[*] onto the whole group $\Delta_c^q B$.

This condition is equivalent to the following pair of conditions:

a) there exists a ψ_0 such that every q-dimensional true cycle of the set B is homologous in B to some cycle of the set ψ_0;

b) for each $\psi \supseteq \psi_0$ one may choose a $\psi' \supseteq \psi$ such that if any true cycle z^q of the set ψ bounds in B, then it bounds also in ψ'.

In order to convince ourselves of this equivalence, we denote by $\Delta_c^q(\psi:B)$ the factor-group of the group of all q-dimensional true cycles of the set ψ modulo the subgroup of cycles bounding in B. From the fact that there exists a $\psi' \supseteq \psi$ such that the group $\Delta^q(\psi:\psi')$ is naturally isomorphic to the group $\Delta_c^q B$, it follows that for any choice of $\psi'' \supseteq \psi'$ we have a natural isomorphism of the group $\Delta_c^q(\psi:\psi'')$ onto the group $\Delta_c^q B$, and further a natural isomorphism of the group $\Delta_c^q(\psi:B)$ onto the group $\Delta_c^q B$, from which it follows that each q-dimensional true cycle of the set B is homologous in B to some cycle of the compactum ψ.

Further, for any $\psi \supseteq \psi_0$, choosing for ψ a ψ' such that the natural homomorphism of the group $\Delta^q(\psi:\psi')$ into $\Delta_c^q B$ is an isomorphism, we see that every cycle of the set ψ, homologous to zero in B, bounds also in ψ'.

---

[*] We are dealing with an algebraic isomorphism, since there is as yet in $\Delta_c^q B$ no topology.

Conversely, if for given ψ, ψ', $\psi' \supseteq \psi \supseteq \psi_0$ the conditions (a) and (b) are satisfied, then in view of condition (b) the natural homomorphism of the group $\Delta_c^q(\psi:\psi')$ into $\Delta_c^q B$ is an isomorphism, while in view of (a) this isomorphism is a mapping onto the whole group $\Delta_c^q B$.

Remark. There is a striking analogy between the definition of (J^q)-sets on the one hand and p-retracts on the other (particularly if one chooses the second definition of p-retracta). Sets satisfying the condition (J^q) may be called "*interior q-retracts.*" The analogy in question is accentuated by the following proposition (corresponding to the fact that every retract in the usual sense of the word is a p-retract for all p):

Theorem 3. *If the compactum $\psi_0 \subseteq B$ may be chosen so that there exists a retracting deformation[*] of the set B into the compactum ψ_0, then, for any q and any coefficient domain, B is a (J^q)-set.*

Indeed, the natural homomorphism of the group $\Delta_c^q \psi_0$ into the group $\Delta_c^q B$ is in our case an isomorphism onto the group $\Delta_c^q B$, since each cycle of the compactum ψ_0, bounding (some "film") in B, bounds (the image of that "film") in ψ_0, and at the same time each cycle of the set B is homologous (and even homotopic) in B to some cycle of the compactum ψ_0. Suppose, further, that we are given a compactum $\psi \supseteq \psi_0$. Then we denote by ψ' a compactum which is the tract of the compactum ψ under the retracting deformation of the set B in ψ_0. Any cycle z_ψ of the compactum ψ is homologous in ψ' to its image z_0 under the retraction of B into ψ_0; if z_0 bounds in B, then z_0 bounds in ψ_0 and so z_ψ bounds in ψ'.

3. **Duality theorem for homology retracts.** We begin with the following proposition:

Theorem 4. *If A is a p-retract over the coefficient domain $\mathfrak{A}$, then B satisfies the condition (J^q) over the coefficient domain $\mathfrak{B}$.*

Indeed, let us choose λ_0 so that for any $\lambda \subseteq \lambda_0$ we can choose a $\lambda' \subseteq \lambda$ satisfying the condition

$$\Delta_c^p(\lambda':\lambda) = \Delta_c^p A \quad \text{(natural isomorphism!)}. \tag{2}$$

We have to prove that then the natural homomorphism of the group $\Delta_c^q(\psi:\psi')$ into the group $\Delta_c^q B$ is an isomorphism onto that group. In other words, it is required to prove two assertions:

a) each true cycle z_ψ^q of the set ψ bounding in B bounds also in ψ';

b) each true cycle z_B^q of the set B is homologous in B to some cycle of the

---

[*] A retracting deformation of the set B into the compactum ψ_0 is, as is known, a deformation of the set B onto itself, which carries the set B into the compactum ψ_0, leaving all the points of that compactum invariant.

compactum ψ.

For the proof of the first assertion we observe that every true cycle z^q_ψ of the set ψ not bounding in ψ' links with some cycle $z^p_{\lambda'}$, lying in λ', and from the definition of λ' is homologous in λ to some true cycle z^p_A of the set A.

Thus z^q_ψ turns out to link with z^p_A and therefore cannot bound in B.

For our further developments the following remark is essential. Denoting by $Z^q\psi$ the group of all q-dimensional true cycles of the compactum ψ, and by $H^q(\psi:B)$ the subgroup of the group $Z^q\psi$ consisting of those cycles which bound in B, we conclude from what has just been proved that the factor-group

$$\Delta^q_c(\psi:B) = Z^q\psi - H^q(\psi:B)$$

is identical to the group $\Delta^q_c(\psi:\psi')$ (i.e., both of the groups consist of the same elements, with the same addition operation), which makes it possible for us to regard the topology on $\Delta^q_c(\psi:\psi')$ as carried over to $\Delta^q_c(\psi:B)$, and on the basis of (2) to write the duality

$$\Delta^q_c(\psi:B) \mid \Delta^p_c A \tag{3}$$

for any $\psi \supseteq \psi_0$.

Turning to the proof of the assertion (b), we replace it by the following stronger assertion.

b_0). Every true cycle z^q_B is homologous in B to some true cycle z^q_0 of the compactum ψ_0.

We shall prove assertion (b_0). Suppose that we are given a true cycle z^q_B lying on the compactum ψ, concerning which we always may assume that it contains ψ_0, so that (3) holds. The groups $\Delta^q_c(\psi_0:B)$ and $\Delta^q_c(\psi:B)$, as bicompact groups, dual to one and the same discrete group $\Delta^p_c A$, are topologically isomorphic to each other. On the other hand, the natural homomorphism of the group $\Delta^q_c(\psi_0:B)$ into the group $\Delta^q_c(\psi:B)$ is an isomorphism into that group, and since this isomorphism preserves scalar products (which are linking coefficients), it is an isomorphism onto the whole group $\Delta^q_c(\psi:B)$. Hence it follows that the homology class $\mathfrak{z}^q_{\psi B} \in \Delta^q_c(\psi:B)$ containing the given cycle z^q_B contains some homology class $\mathfrak{z}^q_{\phi,B} \in \Delta^q_c(\psi_0:B)$, so that the cycle z^q_B is homologous in B to some cycle of the compactum ψ_0.

Thus Theorem 4 is proved.

We turn to the fundamental result of this section.

Theorem 5 (duality theorem). *If A is a p-retract over the coefficient domain $\mathfrak{A}$, then the group $\Delta^q_c(B, \mathfrak{B})$ coincides with the group $\Delta^q(B, \mathfrak{B})$ and, receiving from it its topology, turns out to be bicompact and dual to the group $\Delta^p_c(A, \mathfrak{A})$, so that*

$$\Delta_c^p (A,\, \mathfrak{A}) \mid \Delta_c^q (B,\, \mathfrak{B}).$$

Proof. From Theorem 2 of the preceding section we have $\Delta_c^p(A,\, \mathfrak{A}) = \delta^p(A, \mathfrak{A})$. Moreover, according to Theorem 1 of the preceding section we have $\overline{\Delta}{}^q (B,\, \mathfrak{B}) = \overline{\Delta}_c^q (B,\, \mathfrak{B})$, so that we have the duality

$$\Delta_c^p (A,\, \mathfrak{A}) \mid \overline{\Delta}_c^q (B,\, \mathfrak{B}). \tag{4}$$

Since, by what has just now been proved, B satisfies the condition (J^q), the compacta $\psi' \subseteq \psi \subseteq B$ may be chosen so that the group $\Delta^q (\psi' : \psi)$ is naturally (algebraically) isomorphic to the group $\Delta_c^q (B,\, \mathfrak{B})$. We shall show that putting $\lambda = S^n - \psi$, $\lambda' = S^n - \psi'$, we have a natural isomorphism $\Delta_c^p (A,\, \mathfrak{A}) = \Delta_c^p (\lambda : \lambda')$. Next to the duality (4) we write the duality

$$\Delta_c^p (\lambda : \lambda') \mid \Delta_c^q (\psi' : \psi). \tag{5}$$

We have natural isomorphisms of the group $\Delta_c^p (A,\, \mathfrak{A}) = D^p (A,\, \mathfrak{A})$ into the group $\Delta_c^p (\lambda : \lambda')$ and of the group $\Delta^q (\psi : \psi')$ onto the group $\Delta_c^q (B,\, \mathfrak{B}) \subseteq \overline{\Delta}_c^q (B,\, \mathfrak{B})$. Since these isomorphisms preserve the scalar product, all the assertions of the theorem follow from the following elementary proposition.

Lemma. *Suppose we are given two pairs*

$$\mathfrak{A}' \mid \mathfrak{B}' \quad and \quad \mathfrak{A}'' \mid \mathfrak{B}''$$

and algebraic isomorphisms π of the group $\mathfrak{A}'$ into the group $\mathfrak{A}''$ and ϖ of the group $\mathfrak{B}''$ into the group $\mathfrak{B}'$. If these isomorphisms preserve the scalar product, then both of them are isomorphisms "onto", and the second is a topological isomorphism.

Indeed, we shall prove first that π is an isomorphism onto the whole group $\mathfrak{A}''$. To this end we consider any element b'' of the annihilator (in $\mathfrak{B}''$) of the group $\pi(\mathfrak{A}')$. The element $b' = \varpi b''$ annihilates the group $\mathfrak{A}'$. Therefore $b' = 0$. Accordingly, $b'' = 0$, from which $\pi(\mathfrak{A}') = \mathfrak{A}''$. It remains to be proved that the isomorphism ϖ is a topological isomorphism of the group $\mathfrak{B}''$ onto the whole group $\mathfrak{B}'$. But each element b'' of the group $\mathfrak{B}''$ is a character of the group $\mathfrak{A}''$, namely the character

$$b''(x'') = (b'' \cdot x''),$$

where x'' runs through the whole group $\mathfrak{A}''$. If (recalling that π is an isomorphism of the group $\mathfrak{A}'$ onto the whole group $\mathfrak{A}''$) we put

$$b'(x') = b''(x'') = (b'' \cdot x'')$$

for $\pi x' = x''$ and x'' running through the whole group $\mathfrak{A}''$, we obtain a topological isomorphism $b'' \leftrightarrow b'$ of the group $\mathfrak{B}''$ onto $\mathfrak{B}'$. But according to our hypotheses this isomorphism coincides with the isomorphism ϖ, so that our assertion follows.

The lemma, and with it Theorem 5, is proved.

§3. Two-sided homology retracts.
Miščenko elementary duality region

Definition 1. *A set $A \subset S^n$ satisfying over all coefficient domains and for all dimensions p the set of conditions (J^p), (ur^p), (a^p) is said to be a (J) (ur) (a)-set, or a two-sided homology retract.*

It follows from the preceding that the class of two-sided homology retracts is a topologically invariant class of sets.

Theorem 1. *If the set $A \subseteq S^n$ is a two-sided homology retract, then the set $B = S^n - A$ is also a two-sided homology retract.*

Before proving this theorem, we introduce the following

Definition 2. *The set $\mathfrak{G}$ of sets lying in various spaces S^n is said to be a duality region if it satisfies the following two conditions:*

1. If the set $A \subseteq S^n$ is contained in $\mathfrak{G}$, then the set $B = S^n - A$ is also contained in $\mathfrak{G}$.

2. If the set $A \subseteq S^n$ is contained in $\mathfrak{G}$, then every set $A' \subseteq S^{n'}$ homeomorphic to the set A is also contained in $\mathfrak{G}$.

In other words, *a duality region is any collection of sets containing along with every given set $A \subseteq S^n$ its complement $B = S^n - A$ and any topological image of the set A lying in any space $S^{n'}$.*

The duality region $\mathfrak{G}$ is said to be *elementary* if, given any $\mathfrak{A} \mid \mathfrak{B}$ and any dimension p, for every $A \subseteq S^n$, $A \in \mathfrak{G}$, the following *elementary duality law* is valid:

$$\Delta_c^p (A, \mathfrak{A}) \mid \Delta_c^q (B, \mathfrak{B})$$

under the assumption that the group $\Delta_c^q (B, \mathfrak{B})$ is taken in an appropriate topology.

From the topological invariance of (J) (ur) (a)-sets, and from Theorem 1, it follows that the class of all two-sided homology retracts is a duality region. In view of Theorem 5 of the preceding section it turns out that this duality region is elementary. Finally, it is sufficiently extensive, since the following theorem holds:

Theorem 2. *All skin polyhedra are two-sided homology retracts.*

We begin with Theorem 2. It follows from the following elementary property of skin polyhedra:

Theorem 3. *For every skin polyhedron $A \subseteq S^n$, one can find a set $\lambda_0 \supseteq A$ open in S^n and a finite polyhedron $\phi_0 \subseteq A$ such that there exists a retracting deformation of the set λ_0 into the polyhedron ϕ_0, which, if it is considered in A, turns out to be a deformation of the set A into itself.*

We shall deduce Theorem 2 from Theorem 3.

First of all, in view of Theorem 3 of the previous section, it follows from the condition of the present Theorem 3 that A is a (JP)-set (for any p and any coefficient domain). We shall prove (again for any p and any coefficient domain) that the set A satisfies the conditions (ur^p) and (a^p). To this end we observe first of all that, given any $\lambda \subseteq \lambda_0$, for every point $x \in A$ one may find a neighborhood Ox so small that in the course of the whole retracting deformation it remains inside λ (the existence of such a neighborhood follows from the fact that under our deformation the point x describes a compactum (continuous curve) $\phi_x \subseteq A \subseteq \lambda$). Taking the union of the neighborhoods for all $x \in A$, we obtain a neighborhood λ' of the set A which in the course of the whole deformation remains inside the arbitrarily given λ. Hence it follows that every cycle lying in λ' is homologous (and even homotopic) in λ to some true cycle of the polyhedron ϕ_0, i.e., that the condition (a^p) (and in addition the condition (ua^p)) is satisfied. It remains to be proved that every true cycle of the set A bounding in λ_0 bounds also in A. This last assertion need only be proved for true cycles of the polyhedron ϕ_0. But for these last, it is obviously true that an even stronger assertion holds: every cycle of the polyhedron ϕ_0 bounding in λ_0 bounds in ϕ_0 itself.

Let A be a skin polyhedron and K a triangulation of the space S^n such that A is the body of a subcomplex P of that triangulation. The complex $K - P$ we shall denote by Q. Its body is the skin polyhedron B. We denote by $K_{(1)}$ the barycentric subdivision of the complex K. Each vertex of the complex $K_{(1)}$ is either the center of mass of some simplex of the complex P, or else is the center of mass of some simplex of the complex Q. In the first case, we shall call this vertex red, and in the second case blue. Among the simplexes of the complex $K_{(1)}$ we distinguish:

    a) red, for which all the vertices are red;

    b) blue, for which all the vertices are blue;

    c) plaid, for which at least one vertex is red and at least one blue.

All the red simplexes are contained in A, and all the blue simplexes in B. As to the plaid simplexes, each of them belongs to A or to B, according to the color of its leading vertex.

Each face of a monochromatic (i.e., red or blue) simplex is a simplex of the same color, and therefore the set of red simplexes form a triangulation P_1 of a certain polyhedron $\phi \subseteq A$, and the blue simplexes a triangulation Q_1 of some polyhedron $\psi \subseteq B$. In particular, all the red vertices are contained in ϕ, and all the blue in ψ.

Each plaid simplex has a certain red and a certain blue face, which are mutually opposite faces of that simplex. Thus each plaid simplex, along one of its own faces (the "red "one), adjoins the polyhedron ϕ, and along the opposite blue face adjoins the polyhedron ψ. If we add to ϕ all the plaid simplexes, we obtain a set $\lambda = S^n - \psi$ which (as a complement to the closed set ψ) is a neighborhood of the set A (in the same way, if we adjoin to ψ all of the plaid simplexes, then we obtain a neighborhood $\mu = S^n - \phi$ of the set B). We now decompose each plaid simplex into open rectilinear segments, each joining a certain point of the blue face of that simplex with some point of its red face. We shall call these segments plaid. Plaid (open) segments do not have common points. Now we subject the set λ to the following deformation into itself: each point $x \in \lambda$, belonging to some plaid simplex, we cause to slide, in the course of one unit of time, along the plaid segment containing x into the red endpoint of that segment. But if the point x belongs to the set ϕ, we leave it invariant. This deformation is the desired retracting deformation of the set λ. It carries λ into ϕ, leaving all of the points of ϕ invariant, and moves the points of the set A only within the limits of that set.

We turn to the proof of Theorem I.

The group of coefficients may be assumed discrete, since nowhere in the conditions (J^p), (ur^p), (a^p) is the topology involved, even if the group of coefficients has been supplied with one.

So we suppose that the set A satisfies the conditions (J^p), (ur^p), (a^p) over a discrete coefficient group $\mathfrak{A}$ for any $p \leq n$. It is required to prove that the set B satisfies the same conditions, over the same coefficient group $\mathfrak{A}$ as well as for any p.

We choose a character group $\mathfrak{B} \mid \mathfrak{A}$. The group $\mathfrak{B}$ is bicompact, but if it is considered without topology, then in view of our assumptions the set A satisfies all of the conditions (J^p), (ur^p), (a^p) over that group. This allows us in the given case and in analogous cases simply to say that the set A satisfies the conditions (J^p), (ur^p), (a^p) *over the bicompact coefficient group* $\mathfrak{B}$. Therefore it follows that Theorem 1 will be proved as soon as we have proved the proposition

$1_{\mathfrak{B}}$. If the set A satisfies the conditions (J^p), (ur^p), (a^p) over the bicompact coefficient group $\mathfrak{B}$, then the set B satisfies the conditions (J^q), (ur^q), (a^q) over the discrete coefficient group $\mathfrak{A} \mid \mathfrak{B}$.

The proof of this proposition requires a number of lemmas.

Lemma 1. *If A satisfies the conditions (a^p) and (J^p) over the group $\mathfrak{B}$, then B satisfies the condition (ur^q) over the group $\mathfrak{A}$.*

Lemma 2. *If A satisfies the conditions (ur^p) and (J^p) over the group $\mathfrak{B}$, then B satisfies the condition (a^q) over the group $\mathfrak{A}$.*

Proof of Lemma 1. In view of the condition (J^p) satisfied by the set A, we can find a compactum $\phi_0 \subseteq A$ such that every p-dimensional true cycle z_A^p of the set A is homologous in A to some cycle of the compactum ϕ_0. Further, in view of the condition (a^p) we may choose a neighborhood λ_0 of the set A such that for any closer neighborhood $\lambda \subseteq \lambda_0$ there exists a $\lambda' \subseteq \lambda$ such that every cycle lying in λ' is homologous in λ to some true cycle of the set A. Thus *every cycle lying in λ' is homologous in λ to some true cycle of the compactum ϕ_0* (coefficient domain $\mathfrak{B}$).

We shall show that each true cycle z_B^q of the set B bounding in $\mu_0 = S^n - \phi_0$ bounds also in B (coefficient domain $\mathfrak{A}$). Indeed, we denote by $\psi \subseteq B$ a compactum which is a carrier of the cycle z_B^q. We may always suppose (if necessary, after replacing the compactum ψ by a larger compactum $\psi' \subseteq B$) that $\lambda = S^n - \psi \subseteq \lambda_0$.

Therefore one may, for this λ, choose $\lambda' \subseteq \lambda$ as indicated above. The cycle z_B^q, being homologous to zero in μ_0, has a null linking coefficient with every cycle lying on ϕ_0, and accordingly, in view of the choice of λ', also with every cycle of the set λ'.

According to the second duality law for closed sets (Chapter II, §2), we have, putting, as always, $\psi' = S^n - \lambda'$,

$$\Delta_c^p(\psi', \mathfrak{A}) \mid \overline{\Delta}^q(\lambda', \mathfrak{B}).$$

The homology class $\zeta^p \in \Delta_c^p(\psi', \mathfrak{A})$ containing our cycle z_B^p annihilates the the everywhere dense subgroup $\Delta'^q(\lambda', \mathfrak{B})$ of the group $\overline{\Delta}^q(\lambda', \mathfrak{B})$. Therefore ζ^p annihilates the entire group $\overline{\Delta}^q(\lambda', \mathfrak{B})$, i.e., $\zeta^p = 0$ and $z_B^p \sim 0$ in ψ', i.e., a fortiori $z_B^p \sim 0$ in B.

Lemma 1 is proved.

Proof of Lemma 2.

We shall find a neighborhood λ of the set A such that each p-dimensional true cycle of the set A bounding in λ bounds also in A; such a neighborhood λ exists in view of condition (ur^p), which the set A satisfies by hypothesis. Now suppose we are given an arbitrary neighborhood μ_0 of the set B, sufficiently close that $\phi = S^n - \mu_0$ contains the compactum ϕ_0 of the condition (J^p) satisfied for A. Then there exists in A a compactum $\phi' \supseteq \phi$ such that every p-dimensional cycle of the compactum ϕ homologous to zero in A is homologous to zero also in ϕ' (coefficient domain $\mathfrak{B}$).

Under this choice of ϕ, ϕ', λ, the following natural isomorphism (in fact an identity) holds:

$$\Delta_c^p(\phi : \phi') = \Delta_c^p(\phi : \lambda).$$

Putting $\psi = S^n - \lambda$, $\mu' = S^n - \phi'$, $\mu = S^n - \phi$, we have a natural isomorphism of the group $\Delta_c^q(\mu : \psi)$ onto a subgroup of the group $\Delta_c^q(\mu : \mu')$, and the duality

$$\Delta_c^q(\mu' : \mu) \mid \Delta_c^p(\phi : \phi'),$$

$$\Delta_c^q(\mu : \psi) \mid \Delta_c^p(\phi : \lambda).$$

So we have the conditions of the lemma of §2.3, from which we deduce the existence of a natural isomorphism between the groups $\Delta_c^q(\mu' : \mu)$ and $\Delta_c^q(\mu : \psi)$. But this isomorphism means that any q-dimensional cycle of the set μ' is homologous in μ to some cycle of the compactum ψ, i.e., that the set B satisfies the condition (a^q) over the coefficient domain $\mathfrak{A}$.

Lemma 2 is proved.

Applying Lemmas 1 and 2 under the conditions of Theorem 1, we immediately establish the fact that under these conditions the set B satisfies the conditions (ur^q), (a^q) over the coefficient domain $\mathfrak{A}$. It remains to be proved that the set B satisfies also the condition (J^q).

We turn to this proof.

We choose a neighborhood λ_0 of the set A according to the condition (ur^p); by λ we denote any neighborhood of the set A contained in λ_0, and we choose $\lambda' \subseteq \lambda$ so that every cycle lying in λ' will be homologous in λ to some true cycle of the set A. Our object is achieved if we show that under such a choice of neighborhoods λ_0, λ, λ', and with the usual notations $\psi_0 = S^n - \lambda_0$, $\psi = S^n - \lambda$, $\psi' = S^n - \lambda'$, the natural homomorphism of the group $\Delta_c^q(\psi : \psi')$ into the group $\Delta_c^q B$ is an isomorphism onto that group. To this end we must prove two assertions:

$1°$. Every true cycle z_ψ^q of the set ψ, bounding in B, bounds also in ψ'.

$2°$. Every true cycle of the set B is homologous in B to some cycle of the compactum ψ (coefficient domain $\mathfrak{A}$).

Proof of assertion $1°$. Suppose that z_ψ^q is a true cycle of the set ψ not bounding in ψ'. We shall prove that it does not bound in B either. By the second duality law for closed sets there is in the group $\overline{\Delta}^p(\lambda', \mathfrak{B})$ an element $\xi_{\lambda'}^p$ such that $\mathfrak{b}(z_\psi^q, \xi_{\lambda'}^p) \neq 0$. Accordingly, there is also a cycle $z_{\lambda'}^p$ lying in λ' such that

$$\mathfrak{b}(z_\psi^q, z_{\lambda'}^p) \neq 0.$$

But, by the choice of λ', the cycle $z_{\lambda'}^p$ is homologous in λ (i.e., outside the carrier ψ of the cycle z_ψ^q) to some true cycle z_A^p of the set A, and therefore $\mathfrak{b}(z_\phi^q, z_A^p) \neq 0$ and the cycle z_ψ^q does not bound in B. Assertion $1°$ is proved.

Proof of assertion $2°$. This requires new auxiliary propositions. We denote the closure of the subgroup $\Delta_c^p(\lambda : \lambda') \subset \Delta_c^p \lambda$ in the group $\overline{\Delta}^p(\lambda, \mathfrak{B}) = \overline{\Delta}_c^p(\lambda, \mathfrak{B})$ by $\overline{\Delta}_c^p(\lambda : \lambda')$.

Lemma 3. *We have the duality*

$$\Delta^q_c(\psi:\psi') \mid \Delta^p_c(\lambda:\lambda')$$

(the coefficient domain on the left is $\mathfrak{A}$, on the right, $\mathfrak{B}$).

We shall prove this. We represent the group $\Delta^p_c(\psi:\psi')$ as a factor-group $\Delta^q_c\psi - H^q(\psi:\psi')$, where $H^q(\psi:\psi')$ is a subgroup consisting of classes of cycles bounding in ψ'.

We shall prove that the group $H^q(\psi:\psi')$ is the annihilator in the group $\Delta^q_c\psi$ of the subgroup $\Delta^p(\lambda:\lambda')$ of the group $\Delta^p_c(\lambda, \mathfrak{B}) \mid \Delta^q_c(\psi, \mathfrak{A})$.

a) Suppose that $z^q_\psi \in H^q(\psi:\psi')$. Then for any cycle $z^p_{\lambda'}$ of the set λ' we have $\mathfrak{b}(z^q_\psi, z^p_{\lambda'}) = 0$. If now z'^p_λ is a cycle entering into any class which is an element of the group $\Delta^p_c(\lambda:\lambda')$, then z'^p_λ is homologous in λ, i.e., outside ψ, to some cycle of the set λ', and therefore $\mathfrak{b}(z^q_\psi, z'^p_\lambda) = 0$. But this means that z^q_ψ annihilates the everywhere dense subgroup $\Delta^p_c(\lambda:\lambda')$ of the group $\overline{\Delta}^p(\lambda:\lambda')$. Accordingly, z^q_ψ annihilates the whole group $\overline{\Delta}^p(\lambda:\lambda')$.

b) We now suppose that the cycle $z^q_\psi \in \zeta^q_\psi \in \Delta^q_c(\psi, \mathfrak{A})$ does not bound in ψ'. By the second duality law for closed sets, there is in the group $\Delta^p(\lambda', \mathfrak{B})$ an element $\zeta^p_{\lambda'}$ such that $(\zeta^q_\psi \cdot \zeta^p_{\lambda'}) \neq 0$. Since the group $\Delta'^p_c(\lambda', \mathfrak{B})$ is everywhere dense in the group $\overline{\Delta}^p(\lambda', \mathfrak{B})$, there is a cycle $z^p_{\lambda'}$ in λ' such that $\mathfrak{b}(z^q_\psi, z^p_{\lambda'}) \neq 0$. In view of the fact that $z^p_{\lambda'}$ enters into one of the classes which are elements of the group $\Delta^p_c(\lambda:\lambda') \subseteq \overline{\Delta}^p(\lambda:\lambda')$, the cycle z^q_ψ does not annihilate the group $\overline{\Delta}^p_c(\lambda:\lambda')$. Thus we have proved that the group $H^q(\psi:\psi')$ is indeed the annihilator in the group $\Delta^q_c\psi$ of the group $\overline{\Delta}^p_c(\lambda:\lambda')$, i.e., that

$$\Delta^q_c(\psi, \mathfrak{A}) \supseteq H^q(\psi:\psi') \perp \overline{\Delta}^p_c(\lambda:\lambda') \subseteq \overline{\Delta}^p(\lambda, \mathfrak{B}).$$

But in view of the elementary theorem of the theory of characters ("the annihilation theorem"), it therefore follows that the duality needed by us follows:

$$\Delta^q_c(\psi:\psi') = (\Delta^q_c(\psi, \mathfrak{A}) - H^q(\psi:\psi')) \mid \overline{\Delta}^p_c(\lambda:\lambda').$$

Lemma 3 is proved.

We return to the proof of Assertion $2°$, completing also the proof of Theorem 1.

We consider the group

$$\Delta^q_c(\psi:B) = \Delta^q_c(\psi, \mathfrak{A}) - H^q(\psi:B),$$

where $H^q(\psi:B)$ denotes, as always, the subgroup of the group $\Delta^q_c(\psi, \mathfrak{A})$ consisting of those homology classes whose elements are cycles bounding in B. From the Assertion $1°$ proved above, it follows that the group $\Delta^q_c(\psi:B)$ coincides with the group $\Delta^q_c(\psi:\psi')$, so that in view of Lemma 3 we may write

$$\Delta^q_c(\psi:B) \mid \overline{\Delta}^p_c(\lambda':\lambda). \tag{1}$$

On the other hand, since A is an (ur^p) (a^p)-set over the coefficient domain $\mathfrak{B}$, we have the natural algebraic isomorphism

$$\Delta^p_c(\lambda':\lambda) = \Delta^p_c(A,\mathfrak{B}). \tag{2}$$

But the group $\Delta^p_c(\lambda':\lambda)$, as a subgroup of the topological group $\overline{\Delta}^p(\lambda':\lambda)$, is provided with a topology. Using the isomorphism (2) to carry this topology from $\Delta^p_c(\lambda':\lambda)$ to $\Delta^p_c(A,\mathfrak{B})$, and completing the group $\Delta^p_c(A,\mathfrak{B})$ in this topology, we obtain a natural topological isomorphism

$$\Delta^p_c(\lambda':\lambda) = \overline{\Delta}^p_c(A,\mathfrak{B}).$$

From this and from the duality (1) it follows that

$$\Delta^q_c(\psi:B)\,|\,\Delta^p_c(A,\mathfrak{B})$$

for any $\psi \supseteq \psi_0$.

We now take any cycle z^q_B of the set B over the coefficient domain $\mathfrak{A}$. Let ψ be a carrier for it. We may always assume that $\psi \supseteq \psi_0$. Then from the dualities

$$\Delta^q_c(\psi:B)\,|\,\overline{\Delta}^p_c(A,\mathfrak{B}),$$

$$\Delta^q_c(\psi_0:B)\,|\,\overline{\Delta}^p_c(A,\mathfrak{B})$$

and from the fact that the natural isomorphism of the group $\Delta^q(\psi_0:B)$ into the group $\Delta^q(\psi:B)$ preserves the scalar product, it follows that this isomorphism is an isomorphism of the group $\Delta^q_c(\psi_0:B)$ onto the whole group $\Delta^q(\psi:B)$. This means that an element of the group $\Delta^q_c(\psi:B)$, which is a class containing the cycle z^q_B, contains some true cycle z^q_ψ of the compactum ψ_0, i.e., that $z^q_B \sim z^q_\psi$ in B, whereby Assertion 2° is proved, and along with it Theorem 1.

Thus *the collection of all two-sided homology retracts forms an elementary duality region, containing all skin polyhedra. This duality region is appropriately called a Miščenko duality region.*

§4. Maximal Sitnikov elementary duality region

1. The concept of duality region is so simple and natural that the problem of studying the entire collection of duality regions and among them, in particular, the elementary regions, represents one of the most interesting questions of the topology of general (i.e., generally speaking, non-closed) point sets. What has been done in this direction is very little. Nothing is known even of the cardinality of the set of all duality regions. Each given collection of point sets naturally determines a minimal duality region containing it. Thus, for example, one may speak of the minimal duality region containing all finite sets, all closed sets, or equivalently all open sets, all polyhedra, all skin polyhedra, and so forth. It is not known which of the duality regions thus obtained coincide, and which are

distinct from one another. It is even unknown how many different elementary duality regions there are. It is obvious that under inclusion all possible duality regions form a partially ordered set. However, about the structure of this partially ordered set absolutely nothing is known.

With such a state of affairs, a particular interest attaches to the characterization given by Sitnikov of a natural maximal elementary duality region (i.e., an elementary duality region $\mathfrak{S}_0$, such that every elementary duality region is a subset of the duality region $\mathfrak{S}_0$).[*] *It turns out that the duality region $\mathfrak{S}_0$ consists of sets, which in the terminology of Sitnikov are at the same time quasi-open and quasi-closed.* We turn to the definition of these sets.

2. **Quasi-open sets.** Sitnikov calls the set $A \subseteq S^n$ *quasi-open if for all dimensions p and for all coefficient groups $\mathfrak{A} \mid \mathfrak{B}$ one may introduce into the group* $\nabla^p (A, \mathfrak{B})$ *a topology such that we have the duality*

$$\Delta_c^p (A, \mathfrak{A}) \mid \nabla^p (A, \mathfrak{B})$$

realized by the scalar product (defined naturally for elements $\xi \in \Delta_c^p (A, \mathfrak{A})$ and $\eta \in \nabla^p (A, \mathfrak{B})$ of these groups).

Theorem 1. *In order that the set $A \subseteq S^n$ should be quasi-open, it is necessary and sufficient that for every dimension p and for every $\mathfrak{A} \mid \mathfrak{B}$ the group $\Delta_c^p (A, \mathfrak{A})$ should be dual in the sense of linking to an appropriately topologized group $\Delta_c^q (B, \mathfrak{B})$.*

The proof of this proposition immediately follows from the Sitnikov isomorphism M between the group $\nabla^p (A, \mathfrak{B})$ and the group $\Delta^q (B, \mathfrak{B})$, coinciding in the case of the coefficient group $\mathfrak{B}$ with the group $\Delta_c^q (B, \mathfrak{B})$. Here, from the property of the isomorphism M, scalar multiplication in the duality

$$\nabla^p (A, \mathfrak{B}) \mid \Delta_c^p (A, \mathfrak{A})$$

goes into the linking coefficient in the duality

$$\Delta_c^q (B, \mathfrak{B}) \mid \Delta_c^p (A, \mathfrak{A})$$

and conversely.

3. **Quasi-closed sets.** *The set $A \subseteq S^n$ is said to be quasi-closed if for any bicompact group $\mathfrak{B}$ the natural inclusion homomorphism of the group $\Delta_c^p (A, \mathfrak{B})$ into the group $\overline{\delta}^p (A, \mathfrak{B})$ is an isomorphism onto all of that group.*

Theorem 2. *In order that the set $A \subseteq S^n$ should be quasi-closed, it is necessary and sufficient that the group $\Delta_c^p (A, \mathfrak{B})$, over any bicompact coefficient group*

---

[*] After the existence of at least one elementary duality region has been proved, the existence of a (natural) maximal elementary duality region $\mathfrak{S}_0$ is evident: $\mathfrak{S}_0$ is simply the set-theoretical sum of all elementary duality regions.

$\mathfrak{B}$ *and for any dimension* p, *taken in the appropriate topology, should be dual in the sense of linking with the group* $\Delta_c^p(B, \mathfrak{A})$.

Proof. In view of the results of Chapter IV, §3, we have the linking duality

$$\overline{\delta}^p(A, \mathfrak{B}) \mid \Delta_c^q(B, \mathfrak{A}).$$

The requirement of linking duality

$$\Delta_c^p(A, \mathfrak{B}) \mid \Delta_c^q(B, \mathfrak{A})$$

is equivalent to the requirement that the groups $\Delta_c^p(A, \mathfrak{B})$ and $\overline{\delta}^p(A, \mathfrak{B})$ should not only be isomorphic, but that this isomorphism should in fact be realized by the natural inclusion of the group $\Delta_c^p(A, \mathfrak{B})$ into the group $\overline{\delta}^p(A, \mathfrak{B})$.

4. **Proof of Sitnikov's theorem.** It is easy to see that the complement of a quasi-open set is quasi-closed, and conversely. Consequently the sets which are at the same time quasi-open and quasi-closed form a duality region, evidently elementary. On the other hand, if for any set $A \subseteq S^n$, for all dimensions p and for all possible coefficient groups $\mathfrak{A} \mid \mathfrak{B}$ the linking dualities

$$\Delta_c^p(A, \mathfrak{A}) \mid \Delta_c^q(B, \mathfrak{B}),$$

$$\Delta_c^p(A, \mathfrak{B}) \mid \Delta_c^q(B, \mathfrak{A})$$

are realized, then from Theorems 1 and 2 the set A is at the same time quasi-open and quasi-closed. Therefore it follows that the sets which are both quasi-open and quasi-closed form a maximal elementary duality region, as was to be proved.

CHAPTER VI

Simplicial spectra and the theorem
on homeomorphisms of point sets

§1. Remark on invariance theorems.
Geometrical spectra

1. **On invariance theorems.**[*] If we return to the proof of the first general duality theorem (Chapter II, §1), we see that the proof there consisted of two essentially different parts: of the proof of the isomorphism $D^p(A, \mathfrak{A}) = \nabla_c^q(B, \mathfrak{A})$ (or the duality $D^p(A, \mathfrak{A}) \mid \Delta^q(B, \mathfrak{B})$ equivalent to it), and of the proof of the so-called invariance theorem $\delta^p A = D^p A$. The first part is essentially nothing more than the Pontrjagin spectral duality theorem, connecting the Betti groups of mutually complementary sets, of which one is open and the other closed.

The second part, the invariance theorem, asserts the equivalence of the

---

[*] This subsection is presented here only to form a connection with what has gone before. The further exposition is not based on it, and the reader may begin his study of the theorem on homeomorphism with the next subsection.

"exterior groups," the simplest of which is the group $D^p A$, defined as the limit of the corresponding homology groups of neighborhoods of the set A, and of projections, relating already to the set A itself and defined as the limit of the groups of nerves of its coverings. The very same analysis, only in a more complicated setting, also holds in the central duality law of Sitnikov, where the invariance theorem is the assertion of the isomorphism $\nabla^p_{\text{ext}} A = \nabla^p A$. A particularly striking analogous analysis runs through the proof of the so-called spectral duality law (Chapter III, §5).

A simple analysis of the proof of the isomorphism $\delta^p A = D^p A$ given in §1 of the second chapter shows that at the basis of this proof (as well as of the essential part of the proof of the spectral duality law) stands the following simple but significant geometrical fact.

In the directed set of triangulations,[*] lying in S^n and covering the given set $A \subseteq S^n$, there exists a cofinal section consisting of all canonical triangulations, which is moreover a cofinal section of some multiplication of the directed set of nerves of all coverings of the set A. Here, if the canonical triangulation τ_β follows the canonical triangulation τ_α, then the corresponding covering β is subordinate to the canonical covering α and the canonical shift[**] of τ_β into τ_α generates the usual projection of the nerve β into the nerve α.

The body of each canonical triangulation is a retract of some neighborhood of the set A, where these neighborhoods may be so chosen that their totality coincides with the set of all neighborhoods of the set A, ordered by inclusion.

2. **Generalized simplicial mappings; geometrical spectra.** The fundamental role played, not only in the last proposition but throughout absolutely all of the constructions dealt with in this book, by directed sets of triangulations forces us to study them in detail. Directed sets of triangulations along with appropriately defined mappings, i.e., projections of the triangulation $\tau_\beta > \tau_\alpha$ into the triangulation τ_α, constitute what is called a geometrical spectrum. We shall now give a rigorous definition of this concept, and to this end we must define "generalized simplicial mappings."

Suppose that we are given two (generally speaking abstract) complete star-finite complexes β and α, and suppose that to each simplex $T_\beta \in \beta$ we have put in correspondence a simplex $\sigma^\beta_\alpha T_\beta = T_\alpha$ such that the resulting mapping of

---

[*] When we speak of a directed set of triangulations we always mean the natural order between triangulations in which the triangulation τ' is considered to follow the triangulation τ, $\tau' > \tau$, if every simplex of the triangulation τ' is contained in one (and of course only one) simplex of the triangulation τ.

[**] I.e., a simplicial mapping putting into correspondence with each vertex of the triangulation τ_β some vertex of its carrier in the triangulation τ_α.

the complex β into the complex α satisfies the following condition, which it is natural to call the *condition of simpliciality* (or of continuity):

If the simplexes $T_{\beta_1}, \cdots, T_{\beta_r}$ of the complex β have in β a "combinatorial sum" (i.e., are the faces of some simplex $T_\beta \in \beta$), then the simplexes $T_{\alpha_1} = \sigma_\alpha^\beta T_{\beta_1}, \cdots, T_{\alpha_r} = \sigma_\alpha^\beta T_{\beta_r}$ have the same property in the complex α. Here, denoting by $T_{\beta_1} + \cdots + T_{\beta_s}$ the combinatorial sum in β of the simplexes $T_{\beta_1}, \cdots, T_{\beta_s}$ (i.e., the simplex of smallest dimension in β having all the simplexes $T_{\beta_1}, \cdots, T_{\beta_s}$ among its faces), we require that the following condition be safisfied:

$$\sigma_\alpha^\beta(T_{\beta_1} + \cdots + T_{\beta_r}) = \sigma_\alpha^\beta T_{\beta_1} + \cdots + \sigma_\alpha^\beta T_{\beta_r}.$$

The mapping σ_α^β of the complex β into the complex α, satisfying this simpliciality condition, is called a generalized simplicial mapping.

Evidently a generalized simplicial mapping is completely defined if one knows the images of the vertices of the complex β under this mapping, and if in addition the simpliciality condition is satisfied for the vertices, i.e., if $e_{\beta_0}, \cdots$, $\cdots, e_{\beta_r}$ form in β the skeleton of some simplex t_β, then the simplexes $\sigma_\alpha^\beta e_{\beta_0}, \cdots, \sigma_\alpha^\beta e_{\beta_r}$ have in α a combinatorial sum; then, setting

$$\sigma_\alpha^\beta(e_{\beta_0}, \cdots, e_{\beta_r}) = \sigma_\alpha^\beta e_{\beta_0} + \cdots + \sigma_\alpha^\beta e_{\beta_r}$$

for each simplex $T_\beta = (e_{\beta_0}, \cdots, e_{\beta_r})$, we extend the mapping σ_α^β (as a mapping of the set of all vertices of the complex β) to a mapping of the set of all elements of that complex. Evidently the analogy with the usual simplicial mappings is complete: a generalized simplicial mapping differs from the ordinary one only in that the image of a vertex of the complex β is not, generally speaking, a vertex (i.e., a zero-dimensional simplex) but rather a simplex of arbitrary dimension of the complex α. The characterizing condition of simpliciality is one and the same in both cases.

One of the most important examples of generalized simplicial mappings is the natural mapping ϖ_α^β into the triangulation α of the triangulation β following α. To each simplex $T \in \beta$ this mapping assigns its carrier in the triangulation α.

Now suppose we are given any directed (by the natural ordering, as always) set of triangulations τ lying in S^n. Considering this set along with the natural mappings ("projections") $\varpi_\tau^{\tau'}$, for any $\tau', \tau, \tau' > \tau$, we obtain the arbitrary geometrical spectrum

$$\Sigma = \{\tau, \varpi_\tau^{\tau'}\}$$

in S^n. The following examples show how natural this concept is.

3. **Examples of geometrical spectra.**

a) The combinatorial spectrum of the triangulation τ_0 (or of the "polyhedron with a definite triangulation τ_0") is defined as the directed set of all triangulations which are subdivisions of the given triangulation τ_0, as always with the natural projections. Here it is evident that of two triangulations τ and τ' which are subdivisions of one and the same triangulation τ_0 the triangulation τ' follows τ only when τ' is a subdivision of the triangulation τ.

If the triangulation τ_0 is finite (i.e., if the polyhedron τ_0 is a compactum), then the whole combinatorial spectrum of that triangulation consists of finite complexes.

b) **The complete geometrical spectrum of the set** A is defined as the directed set of all triangulations in S^n covering the set A, along with the natural projections. If we retain in this spectrum only canonical triangulations, we obtain the canonical spectrum of the set A in S^n.

c) **The finite geometrical spectrum of the compactum** $A \subseteq S^n$ is defined as the directed set of all finite triangulations in S^n covering the set A, along with the natural projections.

§2. Abstract spectra.

The homeomorphism problem

1. **Definition of an abstract (or general) spectrum.** By an abstract spectrum Σ we understand any directed set of star-finite complete simplicial complexes in which for any complexes β, α, with say β following α, there is defined a generalized simplicial mapping ("projection") π^{β}_{α} of the complex β into α such that the following condition of weakened transitivity is satisfied: if

$$\gamma > \beta > \alpha$$

in the directed set Σ, then for any vertex $e_\gamma \in \gamma$ the simplex $\pi^{\beta}_{\alpha}\pi^{\gamma}_{\beta}e_\gamma$ becomes a (proper or improper) face of the simplex $\pi^{\gamma}_{\alpha}e_\gamma$, which is written as follows:

$$\pi^{\beta}_{\alpha}\pi^{\gamma}_{\beta}e_\gamma \leq \pi^{\gamma}_{\alpha}e_\gamma. \tag{1}$$

Remark. Identifying simplexes with their skeletons, which in abstract complexes is acceptable and very natural, we may write condition (1) in the form of a simple inclusion

$$\pi^{\beta}_{\alpha}\pi^{\gamma}_{\beta}e_\gamma \subseteq \pi^{\gamma}_{\alpha}e_\gamma. \tag{1'}$$

Then for any simplex $t_\gamma \in \gamma$ we evidently have

$$\pi^{\beta}_{\alpha}\pi^{\gamma}_{\beta}t_\gamma \subseteq \pi^{\gamma}_{\alpha}t_\gamma.$$

2. **Transitive spectra. Open spectrum of a space (point set).** If for any $\gamma > \beta > \alpha$ of Σ we have satisfied not only (1), but also the condition of complete transitivity

$$\pi^{\beta}_{\alpha}\pi^{\gamma}_{\beta}e_\gamma = \pi^{\gamma}_{\alpha}e_\gamma, \tag{2}$$

then the spectrum is said to be transitive. It is easy to see that every geometrical spectrum is transitive. The most important example of an abstract spectrum that does not satisfy the condition of complete transitivity is the open spectrum of a point set (space) A. It consists of the nerves of all possible star-finite coverings of the set A, while the nerve β, by definition, follows the nerve α if the covering β is subordinate to the covering α. The projections are defined as follows: if $\beta > \alpha$, and if e_β is any element of the covering (vertex of the nerve) β, then we define $\omega_\alpha^\beta e_\beta$ as a simplex $t_\alpha \in \alpha$ whose vertices correspond to all of those elements of the covering α which contain the given element e_β of the covering β. It is easy to verify that the condition of weakened transitivity is satisfied here, while simple examples show that in general strong transitivity does not hold.

3. Cofinality. The spectrum $\Sigma' = \{\alpha', \omega_\alpha^{\beta'}\}$ is said to be a section of the spectrum $\Sigma = \{\alpha, \omega_\alpha^\beta\}$ if every complex which is an element of the spectrum Σ', is also an element of the spectrum Σ, while if $\beta' > \alpha'$ in Σ', then $\beta' > \alpha'$ in Σ (but in general not conversely!) and if $\beta' > \alpha'$ in Σ', the projection $\omega_{\alpha'}^{\beta'}$ in Σ' coincides with the projection $\omega_{\alpha'}^{\beta'}$ in Σ.

Remark 1. If corresponding projections of the two spectra Σ and Σ' coincide, i.e., if the equation $'\omega_{\alpha'}^{\beta'} e_{\beta'} = \omega_{\alpha'}^{\beta'} e_{\beta'}$ holds, then we say that Σ' is a *section in the strong sense* of Σ.

We shall say that the spectrum $\Sigma' = \{\alpha', \omega_{\alpha'}^{\beta'}\}$ is a cofinal section of the spectrum $\Sigma = \{\alpha, \omega_\alpha^\beta\}$ or that it may be cofinally augmented to the spectrum Σ if Σ' is a section of the spectrum Σ and if the following condition is also satisfied:

Cofinality condition: to each $\alpha \in \Sigma$ there corresponds in Σ' an α' which follows α in Σ.

Examples. a) The canonical spectrum of any set A is a geometrical (and therefore transitive) spectrum forming a cofinal section both of the geometrical spectrum of the set A and of the open spectrum of that set.

b) We shall call a *finite open spectrum* of a given space A a section of an open spectrum of that set, formed by the finite open coverings of that space (with preservation of order and projections, defined for the entire open spectrum of the space A). If the space A is compact, then its finite open spectrum constitutes a cofinal section of the open spectrum of the compactum A.

c) If $A \subseteq S^n$ is compact, then in the complete geometrical spectrum of the set A a finite geometrical spectrum forms a cofinal section.

4. The problem of homeomorphism for polyhedra. More than half a centruy ago, M. Dehn and Heegaard stated the following hypothesis, which came to be

called the fundamental hypothesis of combinatorial topology:

In order that two finite polyhedra $\widetilde{\tau}_0$ and $\widetilde{\tau}_0'$, given in their triangulations τ_0 and τ_0', should be homeomorphic, it is necessary (it is obviously sufficient) that the triangulations τ_0 and τ_0' have isomorphic subdivisions τ_1 and τ_1'. This hypothesis was proved for finite polyhedra of dimension less than or equal to three, while in the general case of n-dimensional polyhedra it has been recently disproved. In view of this state of affairs the following homeomorphism theorems may have a certain interest.

5. **The two main results of this chapter: "Homeomorphism theorem for polyhedra" and "general homeomorphism theorem."**

In what follows we shall prove the following

Theorem on homeomorphism of polyhrdra. *In order that the triangulations τ_0 and τ_0' should be triangulations of two homeomorphic polyhedra $\widetilde{\tau}_0$ and $\widetilde{\tau}_0'$, it is necessary and sufficient that in the combinatorial spectra of these triangulations there exist cofinal sections which may be cofinally augmented to one and the same abstract spectrum.*

Remark 2. This theorem remains true if one considers only cofinal sections in the strong sense.

Finally, we may also formulate a general theorem on homeomorphism of arbitrary point sets, whose proof will be the basic content of this chapter (the proof of the theorem on the homeomorphism of polyhedra is developed along lines quite parallel to the proof of the general theorem, and it is obtained from the latter automatically).

General theorem on the homeomorphism of arbitrary point sets. *Suppose there are given two point sets $A_1 \subseteq S^n1$ and $A_2 \subseteq S^n2$. In order that these sets should be mutually homeomorphic, it is necessary and sufficient that their complete geometric spectra Σ_1 in S^n1 and Σ_2 in S^n2 should contain cofinal sections, σ_1 and σ_2 respectively, which may be cofinally augmented to one and the same abstract spectrum Σ.*

Remark 3. It is not known whether in the formulation of the general homeomorphism theorem one can replace cofinal sections by cofinal sections in the generalized sense.

Further, it is not known whether in the general homeomorphism theorem (or in the homeomorphism theorem for polyhedra) one can require that the abstract spectrum Σ is transitive. Thus it is not yet known whether the following hypothesis is valid:

Hypothesis on homeomorphism of polyhrdra. If τ_0 and τ_0' are triangulations of two homeomorphic polyhedra, then the combinatorial spectra of these triangulations contain cofinal sections which may be cofinally augmented to one and the same *transitive* spectrum.

One may also state the analogous hypothesis for arbitrary point sets.

In this connection a historical survey is appropriate. The general theorem on homeomorphism was stated by me in the same formulation as that just now given to it, in the papers [18] and [19] and in the reports given at the Moscow Mathematical Society on 11 May 1954 and to the International Congress of Mathematicians in Amsterdam in September 1954. In the paper [19] there was also given a complete proof of the theorem. But although the formulation of the theorem in the papers and reports mentioned is the same as now, the content of the theorem was different. The point is that by a cofinal section of a spectrum I understood not the self-evident cofinality which has been defined above but a certain different "complicated" cofinality, expounded in detail in the papers mentioned and now seen to be superfluous. In the conclusion of §7 of the paper [19] I wrote: "The clarity of the theorem on homeomorphism of point sets, which constitutes the fundamental content of this paper, is much diminished by the tedious concept of cofinality on which it is based. It would be extremely desirable to replace the concept of cofinality used in the homeomorphism theorem by a simpler concept that leads to analogous results". With the appearance of the now customary concept of cofinality, as defined above, the problem arose of proving the homeomorphism theorem with this simpler concept, i.e., of proving the same theorem as we prove below. My own efforts on this problem were unsuccessful and it remained for my student, I. Švedov, to construct the proof.

Now we find ourselves in an analogous difficulty: the homeomorphism theorem has been proved under the assumption that the concept of abstract spectrum has been understood in a more complicated form, in the sense that the transitivity condition is taken in the weakened sense. Again it is natural to ask whether one cannot get rid of this defect by proving the theorem for transitive spectra, which is what all geometrical spectra turn out to be. But the problem thus arising — the proof of the "homeomorphism hypothesis" is substantially more difficult: how are we to choose the spectrum Σ, to which cofinal sections of the spectra Σ_1 and Σ_2 are to be augmented, if not in the form of some sort of modification of an open spectrum, the same for both of the homeomorphic sets? But open spectra strongly resist the requirement of "complete transitivity." Therefore the problem of proving the homeomorphism hypothesis, even for polyhedra, seems to me very difficult.

§3. **Plan of proof of the homeomorphism theorem**
and the first part of the proof (necessity)

1. First of all we shall prove (what in essence is already known to us) that
the canonical spectrum is a cofinal section of some multiplication of the open
spectrum of the set $A \subseteq S^n$. Therefore the necessary condition immediately follows in the homeomorphism theorem. Indeed, if the sets A_1 and A_2 are homeomorphic to each other, then they have the same open spectrum. The canonical
spectra σ_1 and σ_2 of the sets A_1 and A_2 are cofinal sections in the strong
sense of the complete geometrical spectra (in S^{n_1} and S^{n_2} respectively) of the
sets A_1 and A_2, and they may be cofinally augmented to some (as we shall see,
one and the same) multiplication Σ of the open spectra of these sets.

The proof of sufficiency is based on the fact that for each spectrum its
space is completely and uniquely defined. Thus one can prove the following propositions.

Theorem 1. *If the geometrical spectra Σ_1 and Σ_2 of the sets A_1 and A_2 in
the spheres S^{n_1} and S^{n_2} have cofinal sections σ_1 and σ_2, both of which are
cofinal sections of one and the same abstract spectrum, then the spaces of the
spectra Σ_1 and Σ_2 are homeomorphic.*

Theorem 2. *The space defined by the complete geometrical spectrum of the
set A is homeomorphic to the set A.*

From these two theorems it follows immediately that the conditions given in
the general homeomorphism theorem for the homeomorphism of two sets is sufficient.
Thus one obtains the second part of the proof of this theorem.

We proceed to the execution of this plan. First of all we prove that the canonical spectrum of the spectrum A, which is evidently (in the strong sense) a cofinal
section of the complete geometrical spectrum of that set, forms at the same time
a cofinal section of some multiplication of the open spectrum of the set A.*
Indeed, canonical complexes are nerves of the corresponding canonical coverings.
Counting each covering as many times as there are canonical complexes excising
it, and leaving the remaining complexes of the open spectrum without change, we
obtain the desired multiplication of the open spectrum. Since in the transition
from canonical triangulations to canonical coverings the projections can be at
most strengthened, our assertion is therefore proved. We remark further that if we
have two multiplications Σ' and Σ'' of one and the same spectrum Σ, and if for
each $\alpha \in \Sigma$ we unite into one set all the elements of the spectrum Σ' corresponding

---

*If A is the polyhedron of a triangulation τ_0, then the combinatorial spectrum of τ_0
is in the strong sense a cofinal section of the open spectrum of the polyhedron $A = \tilde{\tau}_0$.

to the element $\alpha \in \Sigma$ and all the elements of the spectrum Σ'' corresponding to the same α, we obtain a new multiplication containing both Σ' and Σ'', evidently as cofinal sections. This multiplication will be called the union of the two given multiplications.

Now suppose $A_1 \subseteq S^{n_1}$ and $A_2 \subseteq S^{n_2}$ are homeomorphic to each other. Then the canonical spectrum σ_1 of the set A_1, being a cofinal section of the complete geometrical spectrum Σ_1 of the set A_1 in S^{n_1}, is a cofinal section of some multiplication Σ' of the common open spectrum Σ_0 of both sets A_1 and A_2. In the same way, the canonical spectrum σ_2 of the set A_2 in S^{n_2} is a cofinal section of the multiplication Σ'' of the spectrum Σ_0.

We shall denote by Σ the union of the two multiplications Σ' and Σ''. Then both σ_1 and σ_2 are cofinal sections of the spectrum Σ, so that the first part of the theorem on homeomorphism (necessity) is proved.

In the case of polyhedra A_1 and A_2, given respectively in triangulations τ_1 and τ_2, these considerations become simpler: it is easy to see that the combinatorial spectrum of the triangulation τ_1 (as also the combinatorial spectrum of the triangulation τ_2), being a cofinal section of the corresponding complete geometrical spectrum, also forms a cofinal section of the open spectrum Σ_2. Thus we may further say that the combinatorial spectra Σ_1 and Σ_2 of homeomorphic triangulated polyhedra $\tilde{\tau}_1$ and $\tilde{\tau}_2$ may be themselves cofinally augmented to one and the same spectrum, and indeed to the common open spectrum of both polyhedra.

§4. The space of an abstract spectrum.[*]

Completion of the proof of the homeomorphism theorem

1. **Introductory remarks and notations.** In this section we shall identify sets of vertices of a complex with the largest subcomplex spanned by these vertices, i.e., with the largest closed subcomplex which contains only vertices from the given set.

We shall always denote by e_α an arbitrary vertex of the complex α.

We assume, as we evidently may, that any two different complexes of a spectrum Σ have no common vertices (and therefore no common elements).

Let $\Sigma = \{\alpha, \, \mathfrak{d}_\alpha^\beta\}$ be any spectrum, α a complex of Σ and Σ' a subset of the set of all complexes belonging to Σ. We denote by ξ a set of vertices which all belong to various complexes of Σ. The set $\alpha\xi$ of all of the vertices of ξ lying in α is then, from the conventions above, also a subcomplex of α, and we may regard ξ as the system of all of these subcomplexes.

---

*Editor's note: At the request of the author the translation of this section was made from a German revision of his article, prepared by him and to be published as: P. Alexandroff, *Die topologishchen Dualitätssatze*. II. *Nichtabgeschlossene Mengen*. VEB Deutscher Verlag der Wissenschaften, Berlin.

We denote by $\Sigma'\xi$ the subset of ξ consisting of all of the vertices of ξ which lie in complexes of Σ'. Let E_β be a set of vertices of β. Then by $\omega_\alpha^\beta E_\beta$ we mean the set of all of the vertices of α which for some e_β of E_β are vertices of the simplex $\omega_\alpha^\beta e_\beta$. We denote by $\omega\xi$ the union of the sets $\omega_\alpha^\beta \beta\xi$, where $\beta \geq \alpha$ runs through all complexes of Σ (we shall take the identity mapping ω_α^α to be a projection of the spectrum). By the open star $U_\alpha(\beta)$ of a complex β of α we understand as usual the set of simplexes of α which contain a vertex belonging to β. The closed star of β, the combinatorial hull $[U_\alpha(\beta)]$ of $U_\alpha(\beta)$, will be denoted by $S_\alpha^1(\beta)$. This last definition will be generalized by defining, for $k = 1, 2, \cdots, n, \cdots$, the complex $S_\alpha^k(\beta)$ recursively through the equation $S_\alpha^k(\beta) = S_\alpha^1[S_\alpha^{k-1}(\beta)]$. Here we put $S_\alpha^0(\beta) = \beta$. For the set of vertices of the complex $S_\alpha^k(\alpha\xi)$ we introduce the notation $S_\alpha^k(\xi)$. $S_\alpha^k(\xi)$ may then be regarded, with our conventions, as a subcomplex enclosing $S_\alpha^k(\alpha\xi)$. Finally, we write $S^k(\xi)$ for the union of all the sets $S_\alpha^k(\xi)$, where α runs through all of Σ.

The smallest non-negative integer i with the property that for each complex α of Σ and each vertex e_α of $\alpha\xi$ the entire complex $\alpha\xi$ is contained in $S_\alpha^i(e_\alpha)$ will be called the *width* of ξ and will be denoted by $F\xi$. The set ξ is said to be bounded if $F\xi$ is finite.

2. **Canals and threads of a spectrum.** Let $\Sigma = \{\alpha, \omega_\alpha^\beta\}$ again be an arbitrary spectrum and let ξ be a set of vertices of the sort given above. We say that ξ constitutes a *canal* if the two following conditions are satisfied.

A. The set $\alpha\xi$ is not empty, no matter how α is chosen from Σ. We shall say in this case that ξ goes through all of the complexes of Σ.

B. The system ξ is closed under projection, i.e., $\omega\xi \subseteq \xi$.

Remark. If we regard ξ only as a spectrum, then we shall always assume that the projections belonging to ξ are obtained by restricting the corresponding projections belonging to Σ.

The following three properties of a canal are equivalent.

C. The canal ξ is minimal, i.e., ξ contains no proper subset which is also a canal.

D. To any vertex e_α of ξ there corresponds in the spectrum Σ a complex β following α such that for any vertex e_β of $\beta\xi$ the point e_α is a vertex of the simplex $\omega_\alpha^\beta e_\beta$.

D'. If $e_{\alpha_1}, \cdots, e_{\alpha_2}$ are finitely many vertices of ξ, then there is in Σ a complex $\alpha = \alpha(e_{\alpha_1}, \cdots, e_{\alpha_2})$ with the property that for each complex β following α and for each vertex e_β of $\beta\xi$ the simplex $\omega_i^\beta e_\beta$ contains the vertex e_{α_i} $(i = 1, \cdots, s)$.

Proof. First we will show that D follows from C.

We suppose that D is not satisfied, i.e., that ξ contains a vertex e_α which has the property that for any complex $\beta > \alpha$ there is a point e_β in $\beta\xi$ whose image $\omega^\beta_\alpha e_\beta$ does not contain the point e_α as a vertex. The points e_β with this property will be called "bad" vertices. Let ξ' be the set of all bad vertices. Then it follows from the weakened transitivity condition that the set $\omega\xi'$ goes through all the complexes of Σ and is closed under projection, and thus forms a canal. This new canal is contained in ξ, while the point e_α does not belong to it. This conclusion is obviously in contradiction with property C.

Now we shall prove that D$'$ follows from D.

Let ξ be a canal with property D and $e_{\alpha_1}, \cdots, e_{\alpha_s}$ finitely many vertices of ξ. To each of these vertices e_{α_i} we make correspond a complex $\beta_i > \alpha_i$ so that for each vertex e_{β_i} of $\beta_i\xi$ the point e_{α_i} belongs to $\omega^{\beta_i}_{\alpha_i} e_{\beta_i}$. Now let β be an arbitrary complex of Σ which follows all of these complexes β_i, and let e_β be a vertex of $\beta\xi$. Then we have $\omega^\beta_{\alpha_i} e_\beta \supseteq \omega^{\beta_i}_{\alpha_i} \omega^\beta_{\beta_i} e_\beta$. Since $\omega^\beta_{\beta_i} e_\beta$ is contained in $\beta_i\xi$, e_{α_i} belongs to $\omega^{\beta_i}_{\alpha_i}\omega^\beta_{\beta_i} e_\beta$ and therefore *a fortiori* to $\omega^\beta_{\alpha_i} e_\beta$.

Thus it is proved that D$'$ follows D.

Finally, we show that each canal ξ' which is contained in a canal ξ with the property D$'$ coincides with the latter. Let e_α be an arbitrary vertex belonging to ξ. Since the canal ξ satisfies the condition D$'$ and therefore *a fortiori* satisfies condition D, there is a complex β following α such that every vertex of $\beta\xi$, and therefore also every vertex of the (nonempty) set $\beta\xi'$ is mapped by ω^β_α onto a simplex of α which contains the vertex e_α. Since ξ' is closed as a canal under projection, e_α must belong to ξ', so that the equation $\xi' = \xi$ is proved.

In what follows we shall call minimal canals *threads*.

In addition to the properties just proved, threads also have the following property:

E. Every α-component of a thread consists of the vertices of a simplex which belongs to α, or, as our conventions allow us to say, is itself a simplex of α. It follows from this in particular that all threads are bounded and that their width is equal to 0 or 1.

In order to prove this, consider the finite set $E_\alpha = \{e^1_\alpha, \cdots, e^s_\alpha\}$ of endpoints of $\alpha\xi$. Since ξ satisfies the condition D$'$, there is in the spectrum Σ a complex β following α such that the image $\omega^\beta_\alpha e_\beta$ of any vertex e_β belonging to $\beta\xi$ (taken as a skeleton) contains the set E_α. As a side of $\omega^\beta_\alpha e_\beta$, E_α is itself the skeleton of a simplex. In order to complete the proof, we need only to refer to the fact that $\alpha\xi$, because of the star-finiteness of α, contains only finitely many vertices.

226   P. S. ALEKSANDROV

3· **The space $\tilde{\Sigma}$ of a spectrum Σ.** The points of the space $\tilde{\Sigma}$ will be by definition all of the threads belonging to the spectrum Σ.

Let ξ be a thread and e a vertex of ξ. We define a neighborhood $O_e\xi$ of ξ as the set of all the threads which contain the vertex e. By distinguishing these neighborhoods the set $\tilde{\Sigma}$ is made into a topological T_1-space. In order to prove this, it suffices to prove the following two statements:

1°. In the intersection of two neighborhoods $O_{e_\alpha}\xi$ and $O_{e_\beta}\xi$ of a point there is contained a further neighborhood $O_{e_\gamma}\xi$ of ξ.

2°. The intersection of all neighborhoods of a point ξ of $\tilde{\Sigma}$ contains no point other than ξ.

In order to prove the first assertion it suffices to choose the vertex e_γ from ξ in such a way that γ follows α and β and that e_α is contained in $\omega_\alpha^\gamma e_\gamma$ and e_β is contained in $\omega_\beta^\gamma e_\gamma$. Then if ξ' is any thread of $O_{e_\gamma}\xi$, it follows that e_γ and therefore also $e_\alpha \in \omega_\alpha^\gamma e_\gamma$ and $e_\beta \in \omega_\beta^\gamma e_\gamma$ lie in ξ'. The thread ξ' therefore lies in both $O_{e_\alpha}\xi$ and $O_{e_\beta}\xi$, and the asserted inclusion $O_{e_\gamma}\xi \subseteq O_{e_\alpha}\xi \cap O_{e_\beta}\xi$ is proved.

In order to prove Assertion 2°, we suppose that ξ' is an element of all the neighborhoods $O_e\xi$ of the thread ξ. It follows directly from this that every vertex e_α belonging to ξ also belongs to ξ'. Because of the minimality of ξ', ξ cannot be properly contained in ξ', so that ξ' and ξ must coincide.

4· **Proof of Theorem 1 of §3.** First we shall refer to a property of the geometrical spectrum which is very important for us. Let τ be any triangulation and τ' be a double barycentric subdivision of it. Then for every vertex e' of τ' one can find a vertex e of τ such that the body of the closed star $S^1_{\tau'}(e')$ is contained in the body of the open star $U_\tau(e)$. For each simplex t' of τ' which has e' as a vertex, e is also a vertex of the image $\omega_\tau^{\tau'} t'$.

More generally we say that a complex β following α in the spectrum $\Sigma = \{\alpha,\ \omega_\alpha^\beta\}$ in Σ is transformed into α by a star-projection if to each vertex e_β of β there corresponds a vertex e_α of α such that e_α is a vertex of all the simplexes $\omega_\alpha^\beta e'_\beta$ for which e'_β belongs to $S^1_\beta(e_\beta)$. A spectrum Σ is said to be an *S-spectrum* if to each complex α of Σ there corresponds a β of Σ such that β is mapped into α by a star-projection in Σ.

Remark 1. Let $\delta > \gamma > \beta > \alpha$ be four complexes in Σ and suppose that γ is carried by a star-projection into β. Then the projection ω_α^δ is also a star-projection.

Lemma 1. *Let $\alpha_0 < \alpha_1 < \cdots < \alpha_k$ be a sequence of complexes of a spectrum $\Sigma = \{\alpha,\ \omega_\alpha^\beta\}$ and suppose that all the projections $\omega_{\alpha_{i-1}}^{\alpha_i}$ $(i = 1, 2, \cdots, k)$ are star-projections. Then for each vertex e_{α_k} of α_k one can find a vertex e_{α_0} such that e_{α_0} is a vertex of the image $\omega_{\alpha_0}^{\alpha_k} e'_{\alpha_k}$ of each vertex e'_{α_k} of $S^k_{\alpha_k}(e_{\alpha_k})$.*

The proof of this lemma is easily given by induction on the length k of the chain of complexes under consideration.

Lemma 2. *Every bounded canal of an S-spectrum $\Sigma = \{\alpha,\, \vartheta_\alpha^\beta\}$ contains exactly one thread.*

Proof. Let ξ be a bounded canal of width $F\xi \leq n$. We refer to a vertex e_α of ξ as distinguished if there exists a complex β following α such that e_α is a vertex of the image $\vartheta_\alpha^\beta e_\beta$ of each vertex e_β of $\beta\xi$. The set of distinguished vertices will be denoted by ξ'. We shall show that $\vartheta\xi'$ is a thread. Let α_0 be any complex of Σ. We can select in Σ a sequence $\alpha_0 < \alpha_1 < \cdots < \alpha_n$ such that $\vartheta_{\alpha_{i-1}}^{\alpha_i}$ is always a star-projection. Since $F\xi \leq n$, there is in $\alpha_n\xi$ a vertex e_{α_n} whose star $S_{\alpha_n}^n(e_{\alpha_n})$ completely contains $\alpha_n\xi$. Applying Lemma 4.1, we can find in α_0 a vertex e_{α_0} which is a vertex of the image $\vartheta_{\alpha_0}^{\alpha_n} e'_{\alpha_n}$ of every vertex e'_{α_n} of $\alpha_n\xi$. This means that e_{α_0} is a distinguished vertex. The set $\vartheta\xi'$ therefore goes through all the complexes of the spectrum Σ and is thus a canal.

From the definition of distinguished vertex it follows that every canal contained in ξ must contain all of these distinguished vertices. The canal $\vartheta\xi'$ is therefore minimal and consequently is the only minimal canal contained in ξ. Thus Lemma 2 is proved.

Let $\Sigma' = \{\alpha',\, '\vartheta_{\alpha'}^{\beta'}\}$ be a cofinal section of the spectrum $\Sigma = \{\alpha,\, \vartheta_\alpha^\beta\}$. We assign to each canal ξ of Σ' the canal $\vartheta\xi$ of Σ and to each canal η of Σ the canal $\Sigma'\eta$ of Σ'. We have thus defined two operators ϑ and Σ' for canals, some of whose properties we now present.

Lemma 3. *For each bounded canal η of the spectrum Σ we have*

$$F(\Sigma'\eta) \leq F(\vartheta(\Sigma'\eta)) \leq F\eta.$$

The proof of this inequality is trivial.

Lemma 4. *For each bounded canal ξ of Σ' we have*

$$F(\vartheta\xi) \leq 2F\xi + 1.$$

Proof. We assume that $F\xi \leq n$. Let α be any complex of Σ and e_α^1 and e_α^2 two vertices of $\alpha\vartheta\xi$. Then there are vertices $e_{\beta_1'}$ and $e_{\beta_2'}$ of ξ for which e_α^i is a vertex of $\vartheta_\alpha^{\beta_i'} e_{\beta_i'}$ $(i = 1, 2)$. Choose in Σ' a complex γ' following β_1' and β_2' in Σ', and in $\gamma'\xi$ a vertex $e_{\gamma'}$. Then

$$e_{\beta_i'} \in S_{\beta_i'}^n(\,'\vartheta_{\beta_i'}^{\gamma'} e_{\gamma'}) \qquad (i = 1, 2). \tag{1}$$

We observe that for two arbitrary complexes $\beta > \alpha$ of Σ and any subcomplex $\hat\beta$ of β the following inclusion holds:

$$\vartheta_\alpha^\beta(S_\beta^k(\hat\beta)) \subseteq S_\alpha^k(\vartheta_\alpha^\beta\hat\beta) \qquad (k \geq 0). \tag{2}$$

From relations (1) and (2) and from the weakened transitivity condition for projections it follows that

$$e_\alpha^i \in \eth_\alpha^{\beta'_i} e_{\beta'_i} \subseteq \omega_\alpha^{\beta'_i} (S_{\beta'_i}^n ('\eth_{\beta'_i}^{\gamma'} e_{\gamma'})) \subseteq$$

$$\subseteq S_\alpha^n (\eth_\alpha^{\beta'_i} ('\eth_{\beta'_i}^{\gamma'} e_{\gamma'})) \subseteq S_\alpha^n (\eth_\alpha^{\beta'_i} (\eth_{\beta'_i}^{\gamma'} e_{\gamma'})) \subseteq S_\alpha^n (\eth_\alpha^{\gamma'} e_{\gamma'})$$

$$(i = 1, 2).$$

The image $\eth_\alpha^{\gamma'} e_{\gamma'}$ is a simplex belonging to α, so that e_α^1 belongs to $S_\alpha^{2n+1}(e_\alpha^2)$.

Proposition 1. *Let* $\Sigma_1 = \{\alpha, \eth_{\alpha 1}^{\alpha 2}\}$ *and* $\Sigma_2 = \{\beta, \pi_{\beta 1}^{\beta 2}\}$ *be two S-spectra, each of which contains a cofinal section* $\Sigma_1' = \{\alpha', '\eth_{\alpha'1}^{\alpha'2}\}$, *respectively* $\Sigma_2' = \{\beta', '\pi_{\beta'1}^{\beta'2}\}$. *Suppose that these cofinal sections are at the same time cofinal sections of a spectrum* $\Sigma = \{\gamma, \eth_{\gamma 1}^{\gamma 2}\}$. *Then the spaces of the spectra* Σ_1 *and* Σ_2 *are homeomorphic.*

Remark 2. The spaces of the spectra Σ_1, Σ_1', Σ_2 and Σ_2' are not always homeomorphic to each other.

Remark 3. Since geometrical spectra are certainly *S*-spectra, Theorem 1 follows directly from Proposition 1.

Proof of Proposition 1. First we shall set up a one-to-one relation between the points of the two spaces $\tilde{\Sigma}_1$ and $\tilde{\Sigma}_2$.

Suppose then that ξ is an arbitrary thread of the spectrum Σ_1. From Lemmas 3 and 4 it follows, in connection with the property E of threads, that the width $\overline{F\xi}$ of the canal $\bar{\xi} = \pi(\Sigma_2'(\eth(\Sigma_1'\xi)))$ in the spectrum Σ_2 is at most equal to 7. From Lemma 2 $\bar{\xi}$ thus contains exactly one thread belonging to Σ_2, which we will denote by $f(\xi)$. In just the same way we assign to each thread η of Σ_2 the thread $g(\eta)$ of the canal $\bar{\eta} = \eth(\Sigma_1'(\omega(\Sigma_2'\eta)))$.

Now we shall show that the superposed mapping gf is the identity. To this end we choose an arbitrary thread ξ of Σ_1 and put $\eta = f(\xi)$. From

$$\eta \subseteq \bar{\xi} = \pi(\Sigma_2'(\omega(\Sigma_1'\xi)))$$

follows

$$\bar{\eta} = \eth(\Sigma_1'(\omega(\Sigma_2'\eta))) \subseteq \bar{\bar{\xi}} = \eth(\Sigma_1'(\omega(\Sigma_2'\bar{\xi})))$$

and thereupon

$$g(\eta) \subseteq \bar{\bar{\xi}}. \tag{1}$$

The inclusion

$$\Sigma_2'(\omega(\Sigma_1'\xi)) \subseteq \Sigma_2'\bar{\xi}$$

has the consequence that for each complex γ of Σ the γ-component

$$\gamma\{\omega(\Sigma_1'\xi) \cap \omega(\Sigma_2'\bar{\xi})\}$$

is not empty. In particular the α'-components $(\alpha' \in \Sigma_1')$

$$\alpha'\{\Sigma'_1(\omega(\Sigma'_1\xi)) \cap \Sigma'_1(\omega(\Sigma''_2\xi))\}$$

are not empty, so that there results

$$\alpha(\xi' \cap \overline{\overline{\xi}}) \neq \emptyset \text{ for all } \alpha \in \Sigma_1, \tag{2}$$

where we have put

$$\xi' = \mathfrak{H}(\Sigma'_1(\omega(\Sigma'_1\xi))).$$

From

$$\Sigma'_1\xi \subseteq \Sigma'_1(\omega(\Sigma'_1\xi))$$

and

$$\xi \subseteq \mathfrak{H}(\Sigma_1\xi)$$

we obtain the inclusion

$$\xi \subseteq \xi'. \tag{3}$$

Once again employing Lemmas 3 and 4, we see that the canal ξ' has at most the width 7, while $\overline{\overline{\xi}}$ is not wider than 31. Taking into account that from (2) the α-components of ξ' and $\overline{\overline{\xi}}$ always have a common vertex, it results that the canal $\xi' \cup \overline{\overline{\xi}}$ has at most the width 38 and is therefore bounded. From (1) and (3) both the threads $g(\eta)$ and ξ are contained in this last canal. In view of Lemma 2 it therefore follows that $\xi = g(\eta)$.

That the mapping fg is also the identity may be proved in a quite analogous way.

It remains to be proved that g is continuous.

To this end we consider an arbitrary thread η_0 of the spectrum Σ_2 and the neighborhood $O_{e_\alpha}\xi$ of $\xi = g(\eta_0)$ represented by the vertex $e_\alpha (\alpha \in \Sigma_1)$. In the spectrum Σ'_1 we choose a sequence $\alpha'_0 < \alpha'_1 < \cdots < \alpha'_7$ in which the projections $\omega^{\alpha'_i}_{\alpha'_{i-1}}$ are star-projections $(i = 1, 2, \cdots, 6)$ and whose first complex α_0 follows α in such a way that e_α is a vertex of each simplex $\mathfrak{H}^{\alpha'_0}_\alpha e_{\alpha'_0}$ given only that $e_{\alpha'_0}$ belongs to $\alpha'_0\xi$. Let β' be a complex contained in Σ'_2 which follows α'_7 in Σ, and $O_{e_{\beta'}}\eta_0$ the neighborhood of η_0 represented by a vertex $e_{\beta'}$ of $\beta'\eta_0$. For each thread η of $O_{e_\beta'}\eta_0$ we have

$$e'_\beta \in \beta'\eta \subseteq \beta'(\Sigma'_2\eta) \subseteq \beta'(\omega(\Sigma'_2\eta)).$$

Hence it follows that

$$\omega^{\beta'}_{\alpha'_7} e_{\beta'} \subseteq \alpha'_7(\omega(\Sigma'_2\eta)) \subseteq \alpha'_7(\Sigma'_1(\omega(\Sigma'_2\eta))) \subseteq \alpha'_7\overline{\eta}, \tag{4}$$

where as usual we have put

$$\overline{\eta} = \mathfrak{H}(\Sigma'_1(\omega(\Sigma'_2\eta))).$$

As we already know, the canal η has to most the width 7. From the relation

(4) it therefore follows for every vertex $e_{\alpha'_7}$ of $\mathfrak{W}^{\beta'}_{\alpha'_7} e_\beta$ that

$$S^7_{\alpha'_7}(e_{\alpha'_7}) \supseteq \alpha'_7 \bar{\eta} \supseteq \alpha'_7(g(\eta)). \tag{5}$$

Recalling Lemma 1, there is in the complex α'_0 a vertex $e_{\alpha'_0}$ which belongs to all the images $\mathfrak{W}^{\alpha'_7}_{\alpha'_0} e'_{\alpha'_7}$ of the vertices $e'_{\alpha'_7}$ of $S^7_{\alpha'_7}(e_{\alpha'_7})$. From this and the inclusion (5) it follows that $e_{\alpha'_0}$ belongs to $\alpha'_0(g(\eta))$. Since this holds for all the threads η of O_{e_β}, η_0, we obtain in particular

$$e_{\alpha'_0} \in \alpha'_0(g(\eta_0)) = \alpha'_0\xi.$$

The complex α'_0 of Σ'_1 was so chosen that e_α belongs to $\mathfrak{W}^{\alpha'_0}_\alpha e_{\alpha'_0}$ and thus to $g(\eta)$. We obtain therefore the inclusion $g(O_{e_\beta}, \eta_0) \subseteq O_{e_\alpha}\xi$. which we are trying to prove.

In the same way one proves the continuity of f. Thus Proposition 1 and also Theorem 1 are proved.

5. **Proof of Theorem 2 of §3.** The space $\widetilde{\Sigma}$ of the complete geometrical spectrum Σ of a subset A of S^n is homeomorphic to A.*

By a simplex here we always understand a geometrical simplex contained in S^n. We shall denote this simplex by $\widetilde{t}$ and its skeleton by t.

Now we shall establish a one-to-one relation between the threads of the spectrum Σ and the points of A.

Let κ_0 be any point of the set A. We choose in each triangulation α of Σ (thus in each of the triangulations covering the set A) the carrier $\widetilde{t}_\alpha$ of κ_0 and form the union

$$\xi_0 = \bigcup_{\alpha \in \Sigma} t_\alpha \tag{1}$$

of the corresponding skeletons. The requirement $e_\alpha \in t_\alpha$ is evidently equivalent to the fact that $\widetilde{t}_\alpha$ is contained in the open star $o_\alpha e_\alpha$ of e_α, and therefore also equivalent to the inclusion $\kappa_0 \in o_\alpha e_\alpha$. In t_α therefore lie exactly those vertices e_α of α whose open stars $o_\alpha e_\alpha$ contain the point κ_0. Thus we may also define the system ξ_0 as the set of all of the vertices e_α of complexes α belonging to Σ whose open stars $o_\alpha e_\alpha$ contain the point κ_0.

We shall show that ξ_0 is a thread of the spectrum Σ. The requirements A and B are certainly fulfilled, so that we need only show that ξ_0 also has the property D.

Indeed, let e_α be a vertex of ξ_0. Then e_0 is a vertex of the carrier $\widetilde{t}_\alpha$ of

---

* We give here essentially the proof of I. Švedov, which somewhat simplifies my original proof.

   In the same way one can prove that a polyhedron A is homeomorphic to the space of the combinatorial spectrum of any triangulation of A.

κ_0 in α. We denote by β a subdivision of α in which the carrier $\tilde{t}_\beta$ of κ_0 is contained with all of its vertices in t_α (since we regard the boundary points of a simplex as not belonging to the simplex, κ_0 must be an interior point of $\tilde{t}_\alpha$). All the vertices of t_β therefore have in α the carrier $\tilde{t}_\alpha$ and thereupon are mapped by $\mathfrak{w}_\alpha^\beta$ onto t_α. Hence $e_\alpha \in \mathfrak{w}_\alpha^\beta e_\beta$ for all e_β of $t_\beta = \beta \xi_0$.

We define a mapping ϕ of A into $\tilde{\Sigma}$ by setting

$$\xi_0 = \phi(\kappa_0).$$

To prove Theorem 2 it will suffice to prove that this mapping ϕ is a homeomorphism of the set $\tilde{A}$ onto the space $\tilde{\Sigma}$.

Let ξ_0 be an arbitrary thread of the spectrum Σ. Then the closed hulls of all the open stars whose centers belong to ξ_0 form, as is easily seen, a centered system of sets in S^n. Since among these stars there appear stars with arbitrarily small diameter, the intersection of the system consists of exactly one point, which we denote by

$$\kappa_0 = f(\xi_0).$$

This point lies in the body of each of the triangulations covering the set A and therefore must lie in A. Now we wish to show that κ_0 lies not only in the closed hull $[o_\alpha e_\alpha]$ of all stars with centers e_α belonging to ξ_0 but indeed in the stars $o_\alpha e_\alpha$ themselves. To this end we choose a triangulation $\alpha_0 > \alpha$, existing because of requirement C, for which each vertex e_{α_0} of ξ_0 contains the point e_α in its projection $\mathfrak{w}_\alpha^{\alpha_0} e_{\alpha_0}$. Here we may suppose that κ_0 is a vertex of the triangulation α_0. If this is not already the case, then we pass from α_0 to a subdivision in which the carrier $\tilde{t}_{\alpha_0}$ of κ_0 is centrally subdivided relative to κ_0. This subdivision, which certainly contains κ_0 as a vertex and also has the property stated above for α_0, may be then substituted for α_0.

Now we shall show that

$$\kappa_0 \in \xi_0. \tag{2}$$

Indeed, let γ be a triangulation following α_0 with the property that an open star of γ which contains κ_0 in its closed hull lies with its boundary points in the open star $o_{\alpha_0}\kappa_0$. Thus if e_γ is any vertex of γ belonging to ξ_0, we obtain

$$\kappa_0 \in [o_\gamma e_\gamma] \subseteq o_{\alpha_0}\kappa_0.$$

The carrier of the point e_γ in α_0 is thus a simplex among whose vertices the point κ_0 appears, so that we obtain the relation

$$\kappa_0 \in \mathfrak{w}_{\alpha_0}^\gamma e_\gamma \subseteq \xi_0.$$

The statement (2) is thus demonstrated. From this and the special choice of α_0 it results that e_α belongs to $\mathfrak{w}_\alpha^{\alpha_0} \kappa_0$, i.e., that κ_0 lies in $o_\alpha e_\alpha$.

We have thus proved that the point $\kappa_0 = f(\xi_0)$ is the single point which lies in all the open stars $o_\alpha e_\alpha$ whose centers e_α are vertices belonging to ξ_0.

The thread $\phi(\kappa_0)$, which is known to consist of all the vertices of complexes α belonging to Σ and containing κ_0 in their stars $o_\alpha e_\alpha$, will be denoted by ξ_0'.

From what has just been proved it follows that

$$\xi_0 \subseteq \xi_0',$$

from which, taking account of the minimality of ξ_0', there results immediately the equation

$$\xi_0 = \xi_0'.$$

The two mappings, ϕ of A into $\widetilde{\Sigma}$ and f of $\widetilde{\Sigma}$ into A, are thus constructed in such a way that for each ξ_0 of $\widetilde{\Sigma}$ the equation

$$\phi f(\xi_0) = \xi_0$$

holds. Immediately from the definition follows the further equation

$$f\phi(\kappa_0) = \kappa_0,$$

which is satisfied for each point κ_0 of A. The mappings ϕ and f are thus inverse to each other and one-to-one, and they map A onto all of $\widetilde{\Sigma}$ and $\widetilde{\Sigma}$ onto all of A. Further, the two mappings are continuous, as can be seen at once from the fact that the sets $o_\alpha e_\alpha$ and $O_{e\alpha}$ form bases in the spaces A and $\widetilde{\Sigma}$ respectively, and are carried into one another by the mappings ϕ and f.

Thus Theorem 2 and consequently the theorem on homeomorphism are proved.

BIBLIOGRAPHY

[1] C. Kuratowski, *Sur un théorème fondamental concernant le nerf d'un système d'ensembles*, Fund. Math. 20 (1933), 191–196.

[2] E. Čech, *Théorie générale de l'homologie dans un éspace quelconque*, Fund. Math. 19 (1932), 149–183.

[3] G. S. Čogošvili, *The duality law for retracts*, Dokl. Akad. Nauk SSSR 51 (1946), 90–97. (Russian)

[4] ———, *On homology approximations and laws of duality for arbitrary sets*, Mat. Sb. (N.S.) 28 (70) (1951), 89–118. (Russian)

[5] P. S. Aleksandrov, *On the dimension of normal spaces*, Proc. Roy. Soc. London Ser. A 189 (1947), 11–39.

[6] ———, *On the dimension of bicompact spaces*, Dokl. Akad. Nauk SSSR 26 (1940), 627–630. (Russian)

[7] C. H. Dowker, *Mapping theorems for non-compact spaces*, Amer. J. Math. 69 (1947), 200–242.

[8] ———, *An extension of Alexandrov's mapping theorem*, Bull. Amer. Math. Soc. 54 (1948), 386–391.

[9] E. Hemmingsen, *Some theorems in dimension theory for normal Hausdorff spaces*, Duke Math. J. 13 (1946), 495–504.

[10] P. S. Aleksandrov, *General duality theorems for nonclosed sets*, Mat. Sb. (N.S.) 21 (63) (1947), 161–232. (Russian)

[11] S. Kaplan, *Homology properties of arbitrary subsets of Euclidean spaces*, Trans. Amer. Math. Soc. 62 (1947), 248–271.

[12] P. S. Aleksandrov, *Untersuchungen über Gestalt und Lage abgeschlossener Mengen beliebiger Dimension*, Ann. of Math. (2) 30 (1928), 101–187.

[13] E. F. Miščenko, a) *On the homology theory of non-closed sets*, Mat. Sb. (N.S.) 29 (71) (1951), 587–592; b) *On some questions of combinatorial topology of non-closed sets*, Mat. Sb. (N.S.) 32 (74) (1953), 219–224. (Russian)

[14] K. A. Sitnikov, a) *The duality laws for non-closed sets*, Dokl. Akad. Nauk SSSR 81 (1951), 359–362; b) *New duality relations for non-closed sets*, Dokl. Akad. Nauk SSSR 82 (1952), 925–928. (Russian)

[15] ———, a) *On continuous deformations of non-closed sets*, Dokl. Akad. Nauk SSSR 82 (1952), 845–848; b) *On the dimension of non-closed sets*, Dokl. Akad. Nauk SSSR 83 (1952), 31–34. (Russian)

[16] P. S. Aleksandrov, *On some consequences of Sitnikov's second duality law*, Dokl. Akad. Nauk SSSR 96 (1954), 885–887. (Russian)

[17] K. A. Sitnikov, a) *Combinatorial topology of non-closed sets. I*, Mat. Sb. (N.S.) 34 (76) (1954), 3–54 (Russian); English transl., Amer. Math. Soc. Transl. (2) 15 (1960), 245–295; b) *Combinatorial topology of non-closed sets. II*, Mat. Sb. (N.S.) 37 (79) (1955), 385–434 (Russian); English transl., Amer. Math. Soc. Transl. (2) 15 (1960), 297–349.

[18] P. S. Aleksandrov, *On the homeomorphism of point sets*, Dokl. Akad. Nauk SSSR 97 (1954), 757–760. (Russian)

[19] ———, *On homeomorphism of point sets*, Trudy Moskov. Mat. Obšč. 4 (1955), 405–420. (Russian)

[20] A. I. Švedov, *Proof of a theorem on the homeomorphism of polyhedra and point sets*, Dokl. Akad. Nauk SSSR 122 (1958), 566–569. (Russian)

[21] F. Hausdorff, *Grundzüge der Mengenlehre*, 1914.

[22] P. S. Aleksandrov, *On homology properties of complexes and closed sets*, Izv. Akad. Nauk SSSR Ser. Mat. 6 (1942), 227–282 (Russian); Eng. transl., Trans. Amer. Math. Soc. 54 (1943), 286–339.

[23] N. A. Berikašvili, *On duality theorems for arbitrary sets*, Soobšč. Akad. Nauk Gruzin. SSR 15 (1954), 407–414. (Russian)

Translated by:

J. M. Danskin

ON CONTINUOUS DECOMPOSITIONS OF BICOMPACTA

V. I. PONOMAREV

1. The following basic fact is proven in section 4 of the present article. If a continuous decomposition of a bicompactum X is given, then the set of all components of all the elements of the decomposition also is a continuous decomposition of the bicompactum X. This theorem, known for the case where X is compact (see, for example, [1], pp. 141−142), was proven in this special case under an essential supposition that a definite metric was introduced in the compact space X. Hence proofs known heretofore for the case of arbitrary bicompacta are not applicable. In the general proof given in the present work we employ the famous theorem of M. R. Shura-Bura on components of bicompacta (see, for example, [2], p. 97).

It immediately follows from our theorem that each continuous map f of a bicompactum X onto a bicompactum Y decomposes into a superposition $f(x) = \phi(\psi(x))$ of two maps, the first (i.e., ψ) of which is monotone, and the second, zero-dimensional.

In section 3 we introduce a new concept of a completely continuous decomposition of a bicompactum and prove that a continuous map of a bicompactum (onto a bicompactum) is open if and only if the decomposition corresponding to this map is completely continuous.

In section 4 we prove that a decomposition $X = \bigcup A$ of a bicompactum into pair-wise disjoint closed sets is completely continuous if and only if the set of all elements of A of the given decomposition is closed in the space of all closed sets of the bicompactum X.

The general concept of a continuous decomposition was, as well known, introduced and investigated in 1925 by P. S. Alexandrov in the works [3] and [4];* an analogous concept under more special suppositions was simultaneously introduced by the American mathematician R. L. Moore [6].

2. **Theorem 1.** *Let $\{A\}$ be a continuous decomposition of a bicompactum R. Each element $A \in \{A\}$ of the decomposition decomposes into its components. The thus obtained new decomposition $\{k\}$ of all of R is continuous.*

Proof. We denote by k_α the components of the element A_α of the decomposition. Let $A_\alpha \in \{A\}$, $k_\alpha \subseteq A_\alpha$ and the neighborhood Uk_α with respect to R be arbitrarily chosen. (According to the theorem of M. R. Shura-Bura) the set k_α is contained in

---

*A detailed exposition of the basic facts of the theory of continuous decompositions can be found in [5], chapter 1, paragraph 5, and chapter 2, paragraph 2.

an A_α-open-closed[*] set $H_\alpha \subseteq Uk_\alpha \cap A_\alpha$. We construct an R-open set $U_1 k_\alpha \subseteq Uk_\alpha$, such that $U_1 k_\alpha \cap A_\alpha = H_\alpha$. Since $U_1 k_\alpha \cap A_\alpha$ is closed in A_α, $U_1 k_\alpha \cap A_\alpha = [U_1 k_\alpha] \cap A_\alpha$. The R-open set $U_1 k_\alpha \cup (R \setminus [U_1 k_\alpha]) = VA_\alpha$ is some R-neighborhood of the element A_α of the decomposition. Since the decomposition $\{A\}$ is continuous, there exists a distinguished[**](with respect to decomposition $\{A\}$) neighborhood $V_1 A_\alpha \subseteq VA_\alpha$.

The theorem will be proven as soon as we are certain that the R-open set $V_1 A_\alpha \cap U_1 k_\alpha$ is a distinguished (with respect to decomposition $\{k\}$) neighborhood Ok_α of the set $k_\alpha \in \{k\}$, which is contained in Uk_α. We prove the latter assertion. If some $k \in \{k\}$, $k \subseteq A$, intersects Ok_α, then the set A intersects $V_1 A_\alpha$ and, consequently, is contained in $V_1 A_\alpha$, and so much the more for this reason $k \subseteq V_1 A_\alpha \subseteq VA_\alpha = U_1 k_\alpha \cup (R \setminus [U_1 k_\alpha])$. But since k is connected, $k \cap U_1 k_\alpha \neq \Lambda$, $k \subset U_1 k_\alpha \subseteq Uk_\alpha$, which was to be proven. From Theorem 1 we have

Theorem 2. *Every continuous map f of a bicompactum X onto a bicompactum Y can be represented in the form of a composition $f(x) = \phi(\psi(x))$ where ψ is a monotone map of the bicompactum X onto some bicompactum K, and ϕ is a zero-dimensional map of the bicompactum K onto Y.*

Proof. Indeed, let $\{A\}$ be a continuous decomposition of X, corresponding to a continuous map f. Breaking down each element of the decomposition $\{A\}$ into components we get, as was just proven, a continuous decomposition $\{k\}$ of the entire bicompactum X. The map ψ is a natural map of X onto the bicompactum K which is the decomposition space for the decomposition $\{A\}$, i.e., the map which associates to point $x \in X$ the element of decomposition $\{k\}$ containing it. The map ϕ associates to each point of the space K, i.e., to each element of the decomposition $\{k\}$, the element of decomposition $\{A\}$ containing it. By this means we obtain a bicompactum Y which is the space $Z\{A\}$ for the decomposition $\{A\}$, where for each point $x \in X$ we have $f(x) = \phi(\psi(x))$. The map ψ is monotone because the inverse image of each point $(k) = \psi(x) \in K$ is a continuum $k \subseteq X$. The map ϕ of K onto Y is zero-dimensional since the inverse image in K of the point $y \equiv (A)$ is zero-dimensional, being the set of all components of A.

3. Throughout the remaining part of the work by a covering of any set in space R we shall mean an open covering (i.e., a covering the elements of which are open sets of space R). The union of the sets which are elements of a covering α is called the *body* of that covering and is denoted by $\tilde{\alpha}$.

---

*Translator's note. That is, H_α is open-closed in A_α. Thus if B and C are subsets of a space W, we say that B is C-open if and only if B is a subset of C and is an open set in the space C, the topology of which is induced by that of W. This terminology is used throughout this article to avoid an awkward construction if the terms were translated literally.

**A set $E \subseteq R$ is called "*distinguished*" (for a given decomposition) if it is the union of some elements of this decomposition. We obtain a distinguished neighborhood $V_1 A_\alpha$ by taking the union of all elements of the decomposition lying in VA_α.

Definition. A decomposition $\{A\}$ of a bicompactum R is called *completely continuous* if for each element $A_0 \in \{A\}$ and its finite open covering $\alpha_0 = (O_1, O_2, \cdots$ $\cdots, O_s)$ there exists a neighborhood UA_0 of this element A_0 such that for $A \cap UA_0 \neq \Lambda$ we have

a) $A \subset \tilde{\alpha}_0$,
b) $A \cap O_i \neq \Lambda$ $(i = 1, 2, \cdots, s)$.

In what follows we shall need

Lemma. *In order that a decomposition $\{A\}$ be completely continuous, it is necessary and sufficient that for any element $A_0 \in \{A\}$ of the decomposition and its finite covering α_0, the union of all $A \in \{A\}$ that are contained in $\tilde{\alpha}_0$ and that intersect each element of the covering α_0 be open in space R.*

The sufficiency of the condition contained in the lemma is obvious. We prove its necessity. Let an element $A_0 \in \{A\}$ of the decomposition and its finite open covering $\alpha_0 = (O_1, \cdots, O_s)$ be arbitrarily chosen. We denote by Γ the union of all elements of the decomposition which intersect each element of covering α_0 and lie in $\tilde{\alpha}_0$. For any point $x \in \Gamma$ we pick the element A_x of the decomposition containing it and a neighborhood UA_x such that for $A \cap UA_x$ one has a) $A \subset \tilde{\alpha}_0$ and b) $A \cap O_i \neq \Lambda$ $(i = 1, \cdots, s)$. Then $UA_x \subseteq \Gamma$, i.e., the set Γ entirely consists of internal points.

A basic result of this point is

Theorem 3. *In order that a decomposition $\{A\}$ of a bicompactum R be completely continuous it is necessary and sufficient that it be generated by an open map on this bicompactum.*

Proof. Necessity. Let $\{A\}$ be a completely continuous decomposition of a bicompactum R. The map which generates this decomposition $\{A\}$ is defined as single valued: the natural continous map f of R onto the bicompactum Z, where Z is the space of the decomposition $\{A\}$. Elements of decomposition $\{A\}$ are points of space Z; open sets in Z are images of distinguished (for $\{A\}$) open sets of R under f. Now we prove that f is open. For this it is necessary to prove that any point $x_0 \in R$ and its neighborhood Ux_0, its image $f(Ux_0)$ is an open subset in Z. But $f(Ux_0)$ is the set of all elements of decomposition $\{A\}$ that intersect Ux_0; this set is open in Z if the union Γ of elements A composing it is open in R. But the set Γ indeed is open in R. In fact, let $A_0 \subset \Gamma$ and α_0 be a covering of set A_0 such that $Ux_0 = O_1 \in \alpha_0$ (one can, for example, pick as α_0 a covering consisting of a set $O_1 = Ux_0$ and a set $O_2 = R \setminus [U_1 x_0]$, where $U_1 x_0$ is a neighborhood of x_0 satisfying the condition $[U_1 x_0] \subset Ux_0$). Then the union of the elements of decomposition $\{A\}$ which are contained in $\tilde{\alpha}_0$ and intersect each element of covering α_0, is (as follows from the lemma) a distinguished neighborhood of element A_0 of decomposition $\{A\}$ that is contained in Γ; thus Γ is open in R.

Sufficiency. Suppose the decomposition $\{A\}$ of a bicompactum R is generated by an open map f onto a bicompactum Y. Then $Y = Z$ is bicompact and the decomposition $\{A\}$ is continuous. We prove that $\{A\}$ is completely continuous. Pick an arbitrary $A_0 \in \{A\}$ and its finite covering $\alpha_0 = (U_1, U_2, \cdots, U_s)$. Since decomposition $\{A\}$ is continuous, there exists a distinguished neighborhood OA_0 of set A_0 contained in $\mathfrak{A}_0$. We examine a covering $\beta_0 = (O_1, O_2, \cdots, O_s)$, where $O_1 = OA_0 \cap U_1$, $O_2 = OA_0 \cap U_2$, $\cdots$, $O_s = OA_0 \cap U_s$, and sets

$$\Gamma_1, \Gamma_2, \cdots, \Gamma_s,$$

where $\Gamma_i (i = 1, \cdots, s)$ is the union of all elements of the decomposition that intersect O_i. For any $i = 1, \cdots, s$ the set $G_i \subset Z$, consisting of all $A \subseteq \Gamma_i$ (as an image of the open set $O_i \subseteq X$ for the open map f), is open in Z, and this means that Γ_i itself is open in R. The open set $\bigcap_{i=1}^{s} \Gamma_i$ contains the set A_0 and is a (distinguished) neighborhood O^*A_0 of this set such that every element $A \subset O^*A_0$ intersects each element of the covering α_0 and is contained in $\mathfrak{A}_0$: therefore the decomposition A is completely continuous.

4. We shall denote the set of all non-empty closed sets of a bicompactum R by $F(R)$. The following topology is induced in this set: a neighborhood of a point $(\Phi_0) \in F(R)$ is the set of all those closed sets $\Phi \subseteq R$ which belong to the body and which intersect each element of some fixed covering α of set Φ_0. The set $F(R)$ with this defined topology turns out to be a bicompactum.[*]

---

[*]This theorem has been proved by many authors (cf., for example [7], page 161). We quote a proof due to V. M. Ivanova. It is based on the concept of a pseudobasis of a topological space and the so-called Alexander lemma. An open pseudobasis of a space is a system of open sets such that upon supplementing it with all finite intersections we obtain a basis for the space. Analogously, a closed pseudobasis is a system of closed sets, the supplementing of which by finite unions gives us a closed basis of the space (i.e., a system of closed sets from which by intersections alone one can obtain all closed sets of the space). The Alexander lemma reads: if from every covering of a space R by means of the elements of some open pseudobasis one can single out a finite covering of R, then R is bicompact (the proof of this proposition is given, for example, in [8], page 139). It is evident that the Alexander lemma can be formulated in the following way: if every centered system of sets which are elements of a given closed pseudobasis of space R has a non-empty intersection, then space R is bicompact. Basing ourselves on this proposition, we proceed to the proof of $F(R)$ being compact for any bicompactum R. Let Γ be an arbitrary non-empty open set in R. By $D_1(\Gamma)$ we denote the set of all $(F) \in F(R)$ such that $F \subseteq \Gamma$, and by $D_2(\Gamma)$ the set of all $(F) \in F(R)$ such that $F \cap \Gamma \neq \Lambda$. It is easy to see that sets $D_1(\Gamma)$ and $D_2(\Gamma)$ are open in $F(R)$ and that these sets, which are constructed for all possible open $\Gamma \subseteq R$, form a pseudobasis for the space $F(R)$. Consequently, all possible $F(R) \backslash D_1(\Gamma) = D_2(R \backslash \Gamma)$ and $F(R) \backslash D_2(\Gamma) = D_1(R \backslash \Gamma)$, i. e., all possible $D_2(F)$ and $D_1(F)$, where F runs through the collection of all closed sets of space R, form a closed pseudobasis S for space $F(R)$. Let us suppose now that $F(R)$ is not bicompact, and we pick the smallest cardinal number m such that there exists a centered syatem Σ of sets from S having an empty intersection. This system can be ordered according to type ζ equal to the smallest ordinal number $\omega(m)$ of potency m: $\Sigma = \{B_\xi\}$, $\xi < \zeta$. Then $\bigcap_{\xi < \zeta} B_\xi = \Lambda$, while $\bigcap_{\xi < \eta} B_\xi \neq \Lambda$ for any $\eta < \zeta$. We produce a contradiction to this statement. Each

A basic result of this section is

Theorem 4. *In order that a decomposition $\{A\}$ of a bicompactum R be completely continuous, it is necessary and sufficient that the set of elements of decomposition $\{A\} = \mathfrak{B}$, considered as points of the bicompactum $F(R)$, be closed in $F(R)$.*

Proof. Sufficiency. Let set $\mathfrak{B} = \{A\}$ of elements of the decomposition be closed in $F(R)$. Pick arbitrary $A_0 \in \{A\}$ and α_0 its finite covering in space R. The set of closed sets contained in $\tilde{\alpha}_0$ and intersecting with each element of α_0 form a neighborhood of point A_0 in $F(R)$, which we shall denote by OA_0. Now for each element of decomposition $A \in \{A\}$ which is a point of the set $\mathfrak{B} \backslash OA_0$ that is closed in $F(R)$, we pick such a covering α_A (in space R) that the body of this covering does not intersect some neighborhood $V_A A_0$ (in space R) of the element A_0 of the decomposition. We denote by Γ_A the neighborhood (in space $F(R)$) of the point $A \in \mathfrak{P} \backslash OA_0$ determined by the covering α_A. These Γ_A form a covering $\gamma = \{\Gamma_A\}$ of the closed set $\mathfrak{B}_0 \backslash OA_0 \subset F(R)$; we single out from it a finite covering $\Gamma_{A_1}, \Gamma_{A_2}, \cdots, \Gamma_{A_s}$ of the set $\mathfrak{B}_0 \backslash OA_0$, Let us consider the corresponding coverings $\alpha_{A_1}, \alpha_{A_2}, \cdots, \alpha_{A_s}$ and the neighborhoods $V_{A_1} A_0, V_{A_2} A_0, \cdots, V_{A_s} A_0$ of the set A_0 in R. The neighborhood $VA_0 = V_{A_1} A_0 \cap V_{A_2} A_0 \cap \cdots \cap V_{A_s} A_0$ is the desired neighborhood of the element of decomposition A_0 in R. In fact, no element of the decomposition that intersects VA_0 can lie in any of the sets $\tilde{\alpha}_{A_1}, \tilde{\alpha}_{A_2}, \cdots$ $\cdots, \tilde{\alpha}_{A_s}$ since $\tilde{\alpha}_{A_1} \cap VA_0 = \Lambda, \cdots, \tilde{\alpha}_{A_s} \cap VA_0 = \Lambda$; consequently A cannot be a point of any of the sets $\Gamma_{A_1}, \Gamma_{A_2}, \cdots, \Gamma_{A_s}$. Since $\Gamma_{A_1}, \Gamma_{A_2}, \cdots, \Gamma_{A_s}$ form a covering of the set $\mathfrak{B}_A \backslash OA_0$, our element of the decomposition A (as a point of space $F(R)$) is contained in OA_0; in other words, from the fact that $A \cap VA_0 \neq \Lambda$ in R, it follows that A intersects each element of covering α_0 and is contained by $\tilde{\alpha}_0$, which was to be proved.

Necessity. Suppose the decomposition $R = \bigcup A$ is completely continuous. We must show that the set $\mathfrak{A} = \{A\}$ of all elements of this decomposition is closed in $F(R)$, i.e., that it is bicompact in the topology induced by space $F(R)$. to this end we must prove that from any covering $\Omega = \{\omega^*\}$ of set $\mathfrak{A}$ in $F(R)$ it is possible

---

B_ξ is either some $D_1(F_\xi)$ or some $D_2(F_\xi)$. Supposing in the former case $F_\xi' = F_\xi$, and in the second $F_\xi = R$, for any $\xi < \zeta$ we may write $B_\xi = D_1(F_\xi) \cap D_2(F_\xi')$. Since system $\{B_\xi\}$ is centered in $F(R)$, $\{F_\xi\}$ is centered in R and $\bigcap\limits_{\xi < \zeta} F_\xi = \Phi \neq \Lambda$. Let us pick an arbitrary $\theta < \zeta$ and examine the system $\{F_\xi\}_{\xi < \zeta} \cup (F_\theta')$ (consisting of all F_ξ and more from F_θ'). We shall prove that it is also centered. First of all, for any $\xi_1, \cdots, \xi_s < \zeta$ we have $B_{\xi_1} \cap \cdots$ $\cdots \cap B_{\xi_s} \cap B_\theta \neq \Lambda$. And so much the more, $D_1(F_{\xi_1}) \cap \cdots \cap D_1(F_{\xi_s}) \cap D_2(F_\theta') \neq \Lambda$. But then there exists an $F \subseteq R$ which intersects F_θ' and is contained in all $F_{\xi_1}, \cdots, F_{\xi_s}$, meaning $\Lambda \neq F \cap F_\theta' \subseteq F_{\xi_1} \cap \cdots \cap F_{\xi_s} \cap F_\theta'$. Thus system $\{F_\xi\}_{\xi < \zeta} \cup (F_\theta')$ is centered in R for any $\theta < \zeta$. But then $\Phi \cap F_\theta' \neq \Lambda$ for any $\theta < \zeta$, consequently $(\Phi) \in \bigcap\limits_{\xi < \zeta} D_1(F_\xi) \cap \bigcap\limits_{\xi < \zeta} D_2(F_\xi') = \bigcap\limits_{\xi < \zeta} B_\xi \neq \Lambda$. The thus obtained contradiction proves the theorem.

to pick a finite covering of this set. Without loss of generality we restrict ourselves to coverings, the elements of which belong to the basis $F(R)$ of the space that is applied by virtue of its definition. This means that each ω^* is a neighborhood of some element A of decomposition $\{A\}$, defined by some finite covering ω of set $A \subset R$ and consists of all those $F \in F(R)$ that are closed sets intersecting every element of covering ω and lying in the body $\widetilde{\omega}$ of that covering. Since the decomposition $\{A\}$ is completely continuous, the union of all the elements of the decomposition that lie in $\widetilde{\omega}$ and intersect every element ω is an open set σ in R. Sorting out all $\omega^* \in \Omega$ we obtain a covering $\Sigma = \{\sigma\}$ of R. From this finite covering Σ we pick a finite covering $\sigma_1, \sigma_2, \cdots, \sigma_s$. We now consider the corresponding $\omega_1, \omega_2, \cdots, \omega_s$ and $\omega_1^*, \omega_2^*, \cdots, \omega_s^*$; we prove that $\omega_1^*, \omega_2^*, \cdots, \omega_s^*$ cover all of $\mathfrak{A}$.

Indeed, let A be an arbitrary element of the decomposition $\{A\}$. We pick any point $x \in A$: it is contained in some σ_i $(1 \leq i \leq s)$; but then (since σ_i is a distinguished set) all of the set A is contained in this σ_i, consequently, according to the definition of the set σ_i set A intersects each element of covering ω_i and lies in $\widetilde{\omega}_i$, but this means that $(A) \in \omega_i^* \subseteq \neq F(R)$. Therefore from covering Ω of the set $\mathfrak{A}$ we picked a finite covering with which fact that set $\mathfrak{A}$ is bicompact and closed in $F(R)$ is proven. The entire Theorem 4 is proven.

BIBLIOGRAPHY

[1] G. T. Whyburn, *Analytic topology,* Amer. Math. Soc., Providence, R. I., 1942.

[2] L. C. Pontrjagin, *Continuous groups,* 2nd ed., GITTL, Moscow, 1954. (Russian)

[3] P. Alexandroff, *Über stetige Abbildungen kompakter Räume,* Proc. Acad. Amsterdam 28(1925), 97.

[4] ———, *Über stetige Abbildungen kompakter Räume,* Math. Ann. 96 (1926), 555–571.

[5] P. Alexandroff and H. Hopf, *Topologie. I,* Springer, Berlin, 1935.

[6] R. L. Moore, *Concerning upper semi-continuous collections of continua,* Trans. Amer. Math. Soc. 27(1925), 416–428.

[7] E. Michael, *Topologies on spaces of subsets,* Trans. Amer. Math. Soc. 71 (1951), 152–182.

[8] J. L. Kelley, *General topology,* Van Nostrand, New York, 1955.

Translated by:

R. B. Paine

ALGEBRAIC METHODS IN SET-THEORETIC TOPOLOGY

G. S. ČOGOŠVILI

More and more the application of algebraic methods in the topology of polyhedra has led to the penetration of these methods into the seemingly inappropriate field of set-theoretic topology. The first significant class of spaces to which the algebraic methods were applied was the class of *compact* spaces. The methods created for the purpose by P. S. Aleksandrov [2a], E. Vietoris [4], S. Lefschetz [11a] and others have not only *algebraized* set-theoretic topology but have also *geometrized* it, reducing the study of general spaces to the study of polyhedra. Thus for example, the P. S. Aleksandrov's *approximation method*, which has proved so fruitful, was developed further in the works of C. Kuratowski [10a], E. Čech [17], N. Steenrod [16a] and others. It reduces the study of a space to the consideration of a directed set of *finite coverings* of the space, directed by refinement, and the *spectrum of complexes*, the so-called *nerves*, which are associated in a definite way with these coverings. The homology groups and cohomology groups of these complexes generate inverse and direct spectra whose limit groups, called *Čech groups*, are taken as the homology and cohomology groups of the space.

This method led to a new concept of combinatorial topology as the study of geometric figures of the *most varied degree of generality*, by way of considering their coverings, i. e., by representing the figures as sums of a *finite* number of sufficiently 'small' parts and reducing the problem to the investigation of a definite *finite* scheme characterizing the structure of the covering (i. e., characterizing the neighborhood relationship and intersections of the elements of the covering [2b]).

But subsequent investigations, which were devoted to a broader class of spaces than the compacta, showed that if one has in view really general, i. e., *noncompact* spaces, then both the method and the concepts based on it require modification and enrichment.

In the first place the condition of finiteness mentioned above is too restrictive. Homology groups of a space based on finite coverings can fail to express appropriately the connectivity properties of the space. (C. H. Dowker [7a] established that such a one-dimensional homology group of the straight line has the power of continuum.) As a result, these homology groups behave badly from all points of view, from the points of view of homotopy, duality and dimension theory. The reason is that the family of finite coverings of a noncompact space is not cofinal in the family of all coverings. Therefore in the general case we must make use of *infinite* coverings and polyhedra.

Moreover, in the recently constructed *algebraic topology of non-closed sets,* with which we shall chiefly be concerned, since it is more general and topical and also less well known, we need new specific approximations in addition to those mentioned above. These are the *approximations to a given set from within by closed subsets (compacta) or from without by open sets (infinite polyhedra)* . Approximations by open and closed sets were used earlier but either they were not interrelated, each being considered separately (N. Steenrod, K. Borsuk, S. Eilenberg), or else they were used to approximate a closed set or even, in the study of local properties, a point (P. Aleksandrov, S. Lefschetz), or again it was the other way round, approximation being from within by open sets and from without by closed sets (H. Cartan) or else finally, only approximations were considered (E. Syčeva) that led to spectra that were double but with the same name.

A typical situation which we have in the case of *nonclosed sets* (for example in the duality laws with whose investigation this field of topology arose and to a large extent took its form) consists just in requiring the *simultaneous approximation of a given set from within by closed subsets and of its complement from without by open sets.* This leads to *double mixed spectra,* for example to a direct spectrum of inverse spectra, in which lies the algebraic novelty of the problem. This leads to considering *direct spectra of compact groups.* On the other hand, as we shall see, this approximation is to a large extent *sufficient.* In particular, all the duality laws so far obtained for nonclosed sets (and almost all kinds have already been obtained) have been obtained explicitly or implicitly by this method of approximation by closed and open sets starting from a definite duality law for closed sets. And the explicit application of this method simplifies the problem and makes it clearer.

This method found its first and most complete application in the Alexander-Pontrjagin *duality law* ([1], [13a, b, c]). A most important relation between the homology groups (denoted below by the symbol Δ) and the cohomology groups (∇) of a closed set F_a and its complement G_a in the n-dimensional spherical space S^n is expressed by the following diagram (where D is a discrete coefficient group, C is the compact group dual to it, and duality is denoted by a line: $C \mid D$):

$$\frac{H^r_\Delta(F_a,\, C)}{H^r_\nabla(F_a,\, D)} \ \ \middle|\ \ \backsim\ \ \frac{H^{n-r-1}_{\Delta f}(G_a,\, D)}{H^{n-r-1}_{\nabla \infty}(G_a,\, C)} \ . \tag{1}$$

Here $\sim$ denotes the isomorphism and a line denotes the duality of the groups which are at opposite sides of the symbol. All four groups can be taken as Čech *groups*, based on finite coverings in the case of F_a and on infinite coverings in the case of G_a. For G_a they coincide with the *singular groups of the set G_a* and with the ordinary *polyhedral groups* of any triangulation of G_a, the chains being taken finite (symbol f) and the cochains infinite (∞). These groups also coincide with the *Vietoris groups* defined by means of spectra [1b], [6], [8], [3d]. To each covering we associate a complex which we call the *vietorisian* of the coverings. Its vertices are the points of the given space, and a simplex is a finite set of vertices belonging to some common element of the covering. If one covering refines another, then the vietorisian of the first is a subcomplex of the second, and there is an induced homomorphism of their groups. Thus there arise spectra whose limit groups are the Vietoris groups. They coincide with the Čech groups for a broad class of spaces, in particular for the *paracompact* and hence for the *separable metric spaces*. For compacta they coincide with the usual metrically defined Vietoris groups [4], [11b].

The isomorphisms and the horizontal dualities in (1) are different forms of the Alexander-Pontrjagin duality law. They are all equivalent, for each of them together with the vertical dualities implies the others. The duality of the upper line is called the *basic* or *first form*.

We wish to extend to an arbitrary pair (X, Y) of mutually complementary sets in S^n the relations which are given by (1) for (F_a, G_a). (Cf. [18a, b] where the discussion which follows was carried out for the basic form.) We consider the directed set of all subsets F_a of X which are closed in S^n, in increasing order $(F_a < F_b \longleftrightarrow F_a \subset F_b)$, and the isomorphic directed set of their complements $G_a = S^n \setminus F_a$ in decreasing order $(G_a < G_b \longleftrightarrow G_a \supset G_b)$, the latter being the set of all open sets containing Y. We shall consider the diagram (1) for every element a of the family $\{a\}$. For $a < b$ the inclusions $F_a \subset F_b$ and $G_b \subset G_a$ induce transitive homomorphisms (continuous in the case of the co-efficient group C) so that the respective groups in diagram (1) form in this way respective spectra. Diagram (1) for indices a and b with $a < b$ together with all the induced homomorphisms forms a three dimensional diagram, a cube on whose faces and diagonal planes all homomorphisms commute. For example, on the upper face $(\pi_b^a h_{\Delta a}, h_{\Delta b}) = (h_{\Delta a}, \sigma_a^b h_{\Delta b})$ for $h_{\Delta a} \in H_\Delta^r(F_a, C)$, $h_{\Delta b} \in H_{\Delta f}^{n-r-1}(G_b, D)$. From this there follows isomorphism $\sim$ and conjugacy in the resulting spectra, as expressed in the diagram

$$\frac{\{H^r_{\Delta}(F_a, C), \pi^a_b\}}{\{H^r_{\nabla}(F_a, D), \rho^b_a\}} \quad \begin{matrix}|\\ \Downarrow \\ |\end{matrix} \quad \frac{\{H^{n-r-1}_{\Delta f}(G_a, D), \sigma^b_a\}}{\{H^{n-r-1}_{\nabla \infty}(G_a, C)\omega^a_b\}} \; . \tag{2}$$

There remains only to construct limit groups of the spectra of diagram (2) so that these groups shall satisfy relations analogous to (1) and shall be topological invariants of the respective sets.

The essence of the problem of constructing the limit groups consists in the fact that the first and the fourth spectra in (2) are *direct spectra of compact groups*. This new kind of spectrum arises here because the first spectrum, for example, is essentially a double spectrum, namely a direct spectrum of inverse spectra represented by the groups $H^r_{\Delta}(F_a, C)$.

Firstly we consider the usual *non-topological* limit groups of the spectra (2). The limit groups $H^r_{\Delta c}(X, C)$ and $H^{n-r-1}_{\nabla \infty e}(Y, C)$ of the first and fourth spectra will be isomorphic to each other as also will be the limit groups $H^{n-r-1}_{\Delta fe}(Y, D)$ and $H^r_{\nabla c}(X, D)$ of the second and third spectra. From the way they originate, the groups with subscript c *are called groups with compact supports or inner groups*, and the groups with subscript e are called *outer groups*.

Now we multiply the groups $H^r_{\Delta c}(X, C)$ and $H^{n-r-1}_{\nabla \infty e}(Y, C)$ by $H^{n-r-1}_{\Delta fe}(Y, D)$ taking as product of two elements the product of their corresponding coordinates. Then in the first two groups we distinguish the annihilators $\overset{\circ}{H}{}^r_{\Delta c}(X, C)$ and $\overset{\circ}{H}{}^{n-r-1}_{\nabla \infty e}(Y, C)$ of $H^{n-r-1}_{\Delta fe}(Y, D)$, which will also be isomorphic. Hence we have the isomorphism of pairs of groups,

$$(H^r_{\Delta c}(X, C), \overset{\circ}{H}{}^r_{\Delta c}(X, C)) \sim (H^{n-r-1}_{\nabla \infty e}(Y, C), \overset{\circ}{H}{}^{n-r-1}_{\nabla \infty e}(Y, C)). \tag{3}$$

In each of these pairs we form the factor group $\widetilde{H}{}^r_{\Delta c}(X, C)$, respectively $\widetilde{H}{}^{n-r-1}_{\nabla \infty e}(Y, C)$ of the first group with respect to the second, and we introduce in these factor groups the Pontrjagin (i. e., compact-open) topology from the groups $H^{n-r-1}_{\Delta fe}(Y, D)$ topologized discretely. Then it turns out that $H^r_{\nabla c}(X, D)$ (respectively $H^{n-r-1}_{\Delta fe}(Y, D)$) is the character group of $\widetilde{H}{}^r_{\Delta c}(X, C)$ (respectively of $\widetilde{H}{}^{n-r-1}_{\nabla \infty e}(Y, C)$), and the latter is an everywhere dense subgroup of the character group of $H^r_{\nabla c}(X, D)$ (respectively of $H^{n-r-1}_{\Delta fe}(Y, D)$). We call such a relation between these groups *generalized duality*. It follows that the groups $\widetilde{H}{}^r_{\Delta c}(X, C)$ and $\widetilde{H}{}^{n-r-1}_{\nabla \infty e}(Y, C)$ have compact completions $H^r_{\Delta c}(X, C)$ and $H^{n-r-1}_{\nabla \infty e}(Y, C)$ isomorphic to the character groups of $H^r_{\nabla c}(X, D)$ and $H^{n-r-1}_{\Delta fe}(Y, D)$. Thus we arrive at the diagram

$$\frac{H^r_{\Delta c}\,(X,\ C)}{H^r_{\nabla_c}\,(X,\ D)}\ \biggm|\ \oint\ \frac{H^{n\,-r-1}_{\Delta fe}\,(Y,\ D)}{H^{n\,-r-1}_{\nabla_{\infty e}}(Y,\ C)}\ . \tag{4}$$

Hence the limit groups of the spectra (2) are constructed and it is shown that for arbitrary X, Y of S^n they satisfy relations analogous to (1). Now we must verify the invariance of these groups.

The invariance of the inner groups can give rise to doubts only in connection with the fact that in their definition (e. g., in the definition of $H^r_{\Delta c}\,(X,\ C)$) we used conjugate spectra connected with the outer groups (e. g., $H^{n-r-1}_{\Delta fe}(Y,\ D)$). But the given spectrum determines its conjugate spectrum uniquely, and therefore one can construct an *abstract* conjugate spectrum and not necessarily use the homology conjugate spectrum, i. e., the second spectrum of (2). In general *all this holds for arbitrary spectra and not only for the above concrete homology spectra, which indeed gives the theory of arbitrary spectra of this kind.*

Moreover, recently N. A. Berikašvili [3d] gave a construction of the same limit group of a discrete spectrum of compact groups without using the conjugate spectrum. For this one defines the *compact direct sum* ΣC_α of the compact groups C_α of a direct spectrum $\{C_\alpha,\ \pi^\alpha_\beta\}$, connected with the limit group $C = \varinjlim \{C_\alpha,\ \pi^\alpha_\beta\}$ of this spectrum in the sense indicated above. It will be the limit group, in the sense referred to above, of the direct spectrum of groups which are the finite sums of groups C_α with inclusion homomorphisms. But now, on the other hand, we can first define ΣC_α and then C. That is, one takes the ordinary direct sum (group of finite forms) $\widetilde{\Sigma} C_\alpha$ of the groups C_α with the following topology. In this topology 1) $\widetilde{\Sigma} C_\alpha$ has a compact completion, 2) the identity map $C_\alpha \to \widetilde{\Sigma} C_\alpha$ is continuous, and 3) each homomorphism of $\widetilde{\Sigma} C_\alpha$ into any compact group G continuous on each C_α is continuous on $\widetilde{\Sigma} C_\alpha$. Such a topology exists and is unique. It is the compact completion of $\widetilde{\Sigma} C_\alpha$ which is ΣC_α. And the limit group C is the factor group $\overline{\Sigma C_\alpha}/C_0$ where C_0 is the subgroup generated by elements of the form $c_\alpha - \pi^\alpha_\beta c_\alpha,\ c_\alpha \in C_\alpha$.

The invariance of the outer groups was first proved for *infinite polyhedra* [18c] and then for *neighborhood retracts* [18d] which are more general than infinite polyhedra. In this case the problem is simplified, for the *outer groups turn out to be groups with compact supports.*

The invariance of the outer group for arbitrary Y was established by P. S.

Aleksandrov in his well known paper [2c]. The outer groups turned out to be isomorphic with the Čech groups (and hence with the Vietoris groups) $H^{n-r-1}_{\Delta fg}(Y, D)$ and $H^{n-r-1}_{\nabla\infty g}(Y, C)$ based on infinite coverings,

$$H^r_{\Delta fe}(Y, D) \sim H^r_{\Delta fg}(Y, D), \quad H^r_{\nabla\infty e}(Y, C) \sim H^r_{\nabla\infty g}(Y, C). \tag{5}$$

In these groups, introduced in [18a], the chains are finite and the cochains infinite, and it is not assumed that chains and cochains lie on compact supports. Such chains and cochains and the groups based on them will be called *general*. The isomorphism is established by bringing in a third group based on so-called canonical coverings of Y, with homomorphisms induced by normal displacements. The isomorphisms (5) were also established by S. Kaplan [8] by bringing in groups based on the coverings of all neighborhoods of Y. The proof is based on a generalized Čech lemma asserting, for a given covering of Y, the existence of an outer covering of Y (i. e., a covering of a neighborhood of Y) intersecting in the given covering and isomorphic with it. Here the space containing Y can be any completely normal space. Under various other hypotheses, the generalization of the Čech lemma was given by me in [18a] and by Ju. M. Smirnov (cf. [14b], p. 15). Smirnov also announced [15] an immediate derivation of relations (5) from this lemma. The outer and inner coverings, isomorphic according to the lemma, obviously form a cofinal subfamily in the families of all outer, respectively inner, coverings of Y, whence the isomorphism of their limit groups follows using single properties of spectra. Moreover Ju. M. Smirnov considers a direct set of sets analogous to the so-called nested systems of axiomatic homology theory [19] and proves for their Čech groups a theorem from which follows both (5) and also a proposition containing the Alexander–Pontrjagin law in the closed case. N. A. Berikašvili [3a] also proved (5), constructing a spectrum based simultaneously on the inner and outer coverings of Y.

Thus the basic form of the Alexander–Pontrjagin law is fully generalized to an *arbitrary* set in S^n:

$$H^r_{\Delta c}(X, C) \mid H^{n-r-1}_{\Delta fg}(Y, D). \tag{6}$$

The other forms equivalent to it were considered by A. N. Kolmogorov and P. S. Aleksandrov (the so-called second form : the isomorphism of the second diagonal of (4)), by K. A. Sitnikov (the third form : the isomorphism of the main diagonal), etc. All these forms *together* occur in [18b], formula 3. 3:3. One transfers these results from S^n to manifolds just as in the closed case. Already a simple comparison of the duality (6) with the smaller number of uncoordinate and considerably less general results obtained in this direction by purely set-theoretic methods

convincingly shows the power and importance of algebraic methods in set-theoretic topology of even the most general spaces.

P. S. Aleksandrov pointed out an important consequence of (6). Taking the
set Y compact we obtain a duality law for a *closed* set with respect to a *discrete*
coefficient group D. Now on the previous pattern we can take this proposition as
starting point and approximate from within by closed sets and from without by
open sets, as was done by N. A. Berikašvili [3a] and K. A. Sitnikov [14c]. We
obtain a duality law for arbitrary sets

$$H^r_{\Delta c}(X, D) \mid H^{n-r-1}_{\Delta fg}(Y, C),\tag{7}$$

where, *in contrast to* (6), *the group with compact supports is over a discrete coefficient group and the general group is over a compact one.* One obtains this general group if one takes on the nerves or vietorisians of the inner or outer coverings (just as above one shows that the inner and outer coverings lead to the same
results for the vietorisians as well as for the nerves) the groups which are the
limits of direct spectra of the compact homology groups of the finite subcomplexes,
and then goes to the limit in the inverse spectrum they form with respect to refinement [3a].

In (6) and (7) in the groups on the left, the cycles and the chains they bound
lie on compacta, and in the group on the right, cycles and homologies are general.
A *mixed* construction can exist, with *cycles compact but homologies general.*
These mixed groups $H^r_{\Delta m}(X, D)$ and $H^r_{\Delta m}(X, C)$ are clearly images by natural
inclusion homomorphisms of the groups $H^r_{\Delta c}(X, D)$ and $H^r_{\Delta c}(X, C)$ in the
groups $H^r_{\Delta fg}(X, D)$ and $H^r_{\Delta fg}(X, C)$, respectively. The question of whether
in fact these groups are a new invariant of the set X was answered positively by
E. F. Miščenko [12]. P. S. Aleksandrov [2a] showed the dualizability of these
groups; they turned out to be dual to each other (cf. [3d]):

$$H^r_{\Delta m}(X, D) \mid H^r_{\Delta m}(Y, C).\tag{8}$$

Relations of duality for compact cycles with general homologies were also obtained by S. Kaplan [8], but they were expressed in terms of numbers (the equality of the numbers of compact cycles independent with respect to general homology
for mutually complementary subsets of S^n) and are weaker than the duality (8).

One should remark that all the dualities (6), (7), (8) are *dualities of linking.*
The factoring in the limit groups of direct spectra e. g. in (6) with respect to the
annihilator (the so-called *non-linking subgroup*) $H^r_{\Delta c}(X, C)$, plays an essential
role here.

The Čech group based on infinite coverings is needed in homotopy questions also. By means of it, Dowker [7a] succeeded in giving a generalization of the *Hopf and Brušlinsky classification theorems* and also of related theorems. That is, he proved that if R is a paracompact normal space of dimension $\leq n\,(n \geq 1)$, then the elements of the n-dimensional integral Čech cohomology group of R are in one-to-one correspondence with the classes of homotopic mappings of R in S^n. In the case S^1 (the case of Brušlinsky) the dimension requirement on R can be dropped and the correspondence becomes an isomorphism.

In view of their importance, it was necessary to define a general form of these groups, the *relative Čech groups of a pair* (R, A) when A is an *arbitrary* set in R. This definition also is due to Dowker [7b]. It is interesting to us also because it is based on the same *process of approximations to a set A by open sets containing it.* Indeed, as distinct from the closed case, where one selects for A the subfamily of all the elements of the covering of R which intersect A, Dowker selects for A *any* subfamily of the covering of R whose union contains A. But this union is a neighborhood of A and the group obtained is the group of R relative to neighborhoods of A. Naturally such an approach had to be taken in using a relative Čech group with a non-closed set A, in particular in trying to generalize the *Alexander–Kolmogorov duality law* [1a], [9] from the closed case to the general one. There have been two such attempts. One of them [18a] consisted of studying some special properties of boundary operators occuring in the above law. The class of sets A contained, for example, all locally compact subsets of locally compact spaces R. For them relative groups of a definition kind were indeed taken relative to a certain family of neighborhoods of A cofinal in the family of all its neighborhoods. As a result of another approach [18b, e] in the case of an arbitrary set A of a locally compact space R, those *relations between images and kernels of the extension and intersection homomorphisms* of spaces R, A and $R \setminus A$, which were established for closed sets A by P. S. Aleksandrov [2d] and which, generalizing the Alexander-Kolmogorov relation, express the situation properties of the sets A and $R \setminus A$ in R, have been extended to the case of an arbitrary set in locally compact space R.

On becoming acquainted with the results of [18a], L. S. Pontrjagin advised sticking to generalized duality, for the process of going over to ordinary duality by means of completion does not have a unique inverse, and this might weaken the relation. Whether this is in fact so is still not known. K. A. Sitnikov [14a] and N. Ja. Vilenkin [5] remarked that also the application of the factoring used above waters down the results: the isomorphism of the pair (3) asserts more than the duality (6). Hence if one does not wish to give the relations the classical form they have for closed sets, and in particular to express them in terms of linking, then it is better to stop with the isomorphism (3). This conclusion was reached as a result of investigations connected with generalizing the *duality law of N. Steenrod* [16b]. If the Alexander–Pontrjagin law (1) solves the problem of the dualizability of the homology groups $H^r_{\Delta f}(G_a, D)$ with *finite* cycles of a (triangulable) open set G_a, the Steenrod law solves the problem of the dualizability of the outer basic group of G_a, the homology group $H^r_{\Delta\infty}(G_a, D)$ with *infinite* cycles. It asserts the isomorphism of this group for any D with the group $H^r_R(F_a, D)$ of *regular cycles* of F_a:

$$H^r_R(F_a, D) \sim H^r_{\Delta\infty}(G_a, D). \tag{9}$$

In order to generalize (9) to arbitrary sets X, $Y = S^n \setminus X$, we again consider (9) for all elements of the isomorphic directed sets $\{F_a\} \sim \{G_a\}$, $F_a \subset X$ and, in the resulting isomorphic direct spectra (with inclusion homomorphisms for $H^r_R(F_a, D)$ and the intersection homomorphisms for $H^r_{\Delta\infty}(G_a, D)$ commuting with the isomorphism (9)), we go to the limit:

$$H^r_{Rc}(X, D) \sim H^r_{\Delta\infty e}(Y, D). \tag{10}$$

To prove the invariance of the outer groups $H^r_{\Delta\infty e}(Y, D)$, Sitnikov proceeds from the Poincaré law (for infinite chains) $H^r_{\Delta\infty}(G_a, D) \sim H^{n-r}_{\nabla\infty}(G_a, D)$ and by verifying the corresponding commutativity and going to the limit over the directed set $\{G_a\}$, $G_a \supset Y$, he generalizes it:

$$H^r_{\Delta\infty e}(Y, D) \sim H^{n-r}_{\nabla\infty e}(Y, D). \tag{11}$$

After this the invariance of the group $H^{n-r}_{\nabla\infty e}(Y, D)$ is shown in the usual way: it is isomorphic with the Čech group $H^{n-r}_{\Delta\infty}(Y, C)$ with finite chains on the *compact* coefficient group C, $C \mid D$, dual to the group $H^{n-r}_{\nabla\infty}(Y, D)$, N. A. Berikašvili [3b] takes, as a group of $(n - r)$-chains on the nerve of an (infinite) covering of Y, the compact direct sum of the group C taken as many times as there are $(n - r)$-simplexes in the nerve, and from this he constructs the Čech group $H^{n-r}_{\Delta f}(Y, C)$.

Thus the generalized Steenrod law takes the form

$$H^r_{Rc}(X, D) \sim H^{n-r}_{\nabla\infty}(Y, D) \mid H^{n-r}_{\Delta f}(Y, C). \tag{12}$$

To connect it with the generalized Alexander–Pontrjagin law, we take the *weak homology group* [16b] with compact carriers $\widetilde{H}^r_{Rc}(X, D)$. Then the factor group of $H^r_{Rc}(X, D)$ with respect to $\widetilde{H}^r_{Rc}(X, D)$ is isomorphic with the group

$H^{r-1}_{\Delta c}(X, D)$, and if D allows of a compact topology then $\tilde{H}^r_{Rc}(X, D) = 0$ and $H^r_{Rc}(X, D) \sim H^{r-1}_{\Delta c}(X, D)$. From this and from (12) we obtain the *third form of the Alexander–Pontrjagin law*, and hence, as above, the isomorphism of pairs (3), etc. For closed sets, from the investigations of S. Eilenberg and S. MacLane [20] we even have a converse: *From the Alexander–Pontrjagin law there follows the Steenrod law for an arbitrary coefficient group.* It is an interesting problem to decide whether this is so in the general case. The investigation of N. A. Berikašvili [3c] in this direction led to a new generalization of the Steenrod law, but did not answer the question. Here was needed a detailed investigation of the behavior of the functors Hom and Ext for spectra, which is partly new.

K. A. Sitnikov proves the isomorphism (11) for groups which are a variant of the group of regular cycles but without making use of the Steenrod law for closed sets. By means of his groups K. A. Sitnikov succeeded in generalizing the Aleksandrov *obstruction theorem* from the closed case to the general one. He even gave a *homology characterization of the dimension* of arbitrary sets by means of relative cocycles based on *infinite* coverings. This characterization connects better with other propositions of the theory than the characterization based on finite coverings. These and other results of K. A. Sitnikov [14d], together with the results of P. S. Aleksandrov [2e], C. H. Dowker [7a] and others, permit us to consider that the algebraic methods, used with such success for compacta in P. S. Aleksandrov's homological dimension theory, fully keep their power also for the investigation of the dimension of general spaces.

The coefficient groups above have been assumed to be either discrete or compact. The problem of constructing the preceding theory for more *general topological coefficient groups* lies in the theory of discrete products and sums and in the theory of spectra of such groups. Considerable progress has been achieved recently in this direction in works of J. Braconnier, N. Vilenkin, S. Kaplan, H. Leptin, and others. Applications to homology theory and especially to the duality laws, were given by N. Vilenkin [5]. He systematically applies his Betti pairs (analogous to (3)), non-closed subgroups and non-open homomorphisms, and by the method of approximation by closed and open sets establishes duality theorems for a broad class of coefficient groups. But Vilenkin's limit of the spectrum is not the same as that used above, and that is why the relations he obtains are different even in the case of compact and discrete coefficient groups.

The exceptional role of algebra in topology became especially clear after it turned out that one of the fundamental topological tools, the homology group, appears to be indeed an algebraic construction. Now it is convenient to become familiar with it first in algebra and only later in topology. This algebraization of the concept of homology and other successful applications of algebra in topology made possible the Eilenberg-Steenrod *axiomatic homology theory* [19], which is one of the outstanding recent achievements in topology.

Our work in the foundations of algebraic topology has aimed at verifying the Eilenberg-Steenrod axioms for various homology groups with *compact coefficient groups*. The point is that on the one hand it has turned out to be possible to *use the theory of spectra* to define, not only Čech groups, but also other groups, the Vietoris groups [4], singular groups [6], [18e], etc. On the other hand the definition of a direct limit group for a spectrum of compact groups, which did not exist earlier (cf. [19], p. 188, 232, etc; [6], p. 393), has given rise to the possibility of considering certain homology groups with compact coefficient groups and investigating them from the point of view of the above axioms. For *singular homology groups* this was done in [18e]. For the same groups over any *locally compact coefficient group* defined by the Vilenkin-Kaplan limit procedure, this was done by G. T. Abdušelišvili. For Čech groups with finite chains based on infinite coverings, figuring in (12), it was done by N. A. Berikašvili [3b]. The creation of the axiomatic theory of spectra is also due to Berikašvili [3d]. He found axioms which limit groups of spectra should satisfy in order that the Čech groups constructed with them should satisfy the Eilenberg-Steenrod axioms. For direct spectra of discrete (compact) groups these axioms are as follows. Let such a spectrum a correspond to the discrete (compact) limit group a^∞, and let this correspond to a spectrum b and subspectrum a, $b \supset a$, a homomorphism $p(a, b): a^\infty \to b^\infty$, such that 1) if $a \subset b \subset c$, then $p(a, c) = p(a, b) \cdot p(b, c)$, 2) if the subspectrum a is cofinal in b then $p(a, b)$ is an isomorphism onto, 3) if a consists of a single group a then $a^\infty = a$. For a^∞ and $p(a, b)$ satisfying these conditions, the properties are sufficient for their use in homology theory. The author gives two more axioms which make the groups a^∞ coincide with the limit groups used above, but there exist other limit groups satisfying axioms 1)–3). On the basis of this he proves the following *abstract duality theorem,* expressing in the most general form the meaning in these questions of the method of approximation by closed and open sets. If in a finite polyhedron P we have a duality law for certain homology groups of closed sets F_a and their complements G_a expressed as an isomorphism, and if this isomorphism commutes with the homomorphisms induced for the groups in question by inclusions $F_a \subset F_b$ and $G_b \subset G_a$, then for any definition of limit groups satisfying axioms 1)–3), the group with compact supports of any X in P and the Čech group of $Y = P \backslash X$, based on these definitions, are isomorphic.

In spite of such generality, further algebraization, namely in the direction of lattice theory, seems to us to be possible. It is the next task in this part of topology.

BIBLIOGRAPHY

[1]   (a) J. W. Alexander, *A proof and extension of the Jordan-Brouwer separation theorem,* Trans. Amer. Math. Soc. 23 (1922), 333–349.

    (b) ———, *On the connectivity ring of an abstract space,* Ann. of Math. (2) 37 (1936), 698–708.

[2]   (a) P. S. Aleksandrov, *Untersuchungen über Gestalt und Lage abgeschlossener Mengen beliebiger Dimension,* Ann. of Math. 30 (1928), 101–187.

    (b) ———, *Duality theorems in combinatorial topology,* Symposium in honor of 30th anniversary of the October Revolution, Vol. 1, pp. 134–180. Moscow. 1947. (Russian)

    (c) ———, *General duality theorems for non-closed sets in n-dimensional space,* Mat. Sb. (N. S.) 21 (63)(1947), 161–232 (Russian); German transl. Sowjetwissenschaft 1948, no. 1, 176–243.

    (d) ———, *On homological situation properties of complexes and closed sets,* Trans. Amer. Math. Soc. 54 (1943), 286–339.

    (e) ———, *On some consequences of Sitnikov's second duality law,* Dokl. Akad. Nauk SSSR 96 (1954), 885–887. (Russian)

    (f) ———, *The present status of dimension theory,* Uspehi Mat. Nauk 6 (1951), no. 5 (45), 43–68; Amer. Math. Soc. Transl. (2) 1 (1955), 1–26.

[3]   (a) N. A. Berikašvili, *On direct and inverse spectra of topological groups,* Soobšč. Akad. Nauk Gruzin. SSR 15 (1954), 257–264. (Russian)

    (b) ———, *On homology groups of a space with compact coefficient group,* ibid. 16 (1955), 753–760. (Russian)

    (c) ———, *On the generalized duality theorem of Steenrod,* ibid. 17 (1956), 385–392. (Russian)

    (d) ———, *On the axiomatic spectral theory and the duality rules for arbitrary sets,* Akad. Nauk Gruzin. SSR Trudy Tbiliss. Mat. Inst. Razmadze 24 (1957), 409–484. (Russian)

[4]   E. Vietoris, *Über den höheren Zusammenhang kompakter Raume und eine Klasse von zusammenhangstreuen Abbildungen,* Math. Ann. 97 (1927), 454–472.

[5]   N. Ja. Vilenkin, *Generalized normal divisors of topological groups and their application to combinatorial topology,* Trudy Moskov. Mat. Obšč. 3 (1954), 15–88. (Russian)

[6] W. Hurewicz, J. Dugundji and C. H. Dowker, *Continuous connectivity groups in terms of limit groups*, Ann. of Math. (2) 49 (1948), 391–406.

[7] (a) C. H. Dowker, *Mapping theorems for non-compact spaces*, Amer. J. Math. 69 (1947), 200–242.

(b) ———, *Čech cohomology theory and the axioms*, Ann. of Math. (2) 51 (1950), 278–292.

[8] S. Kaplan, *Homology properties of arbitrary subsets of Euclidean spaces*, Trans. Amer. Math. Soc. 62 (1947), 248–271.

[9] A. N. Kolmogorov, *Cycles relatifs. Théorème de dualité de M. Alexander*, C. R. Acad. Sci. Paris 202 (1936), 1641–1643.

[10] (a) C. Kuratowski, *Sur un théorème fondamental concernant le nerf d'un système d'ensembles*, Fund. Math. 20 (1933), 191–196.

(b) ———, *Topologie*, Vol. I, 3ème éd., Monogr. Mat. Tom XX, Warsaw, 1952.

[11] (a) S. Lefschetz, *Topology*, New York, 1930.

(b) ———, *Algebraic topology*, Amer. Math. Soc., New York, 1942.

[12] E. F. Miščenko, *On some problems of the combinatorial topology of non-closed sets*, Mat. Sb. (N. S.) 32 (74) (1953), 219–224. (Russian)

[13] (a) L. S. Pontrjagin, *Über den algebraischen Inhalt topologischer Dualitä-tssatze*, Math. Ann. 105 (1931), 165–205.

(b) ———, *Topological duality theorems*, Uspehi Mat. Nauk 2 (1947), no. 2 (18), 21–44. (Russian)

(c) ———, *The general topological theorem of duality for closed sets*, Ann. of Math. 35 (1934), 904–914.

(d) ———, *Topological groups*, Moscow, 1938; English transl., Princeton Univ. Press, Princeton, N. J., 1939.

[14] (a) K. A. Sitnikov, *The duality law for non-closed sets*, Dokl. Akad. Nauk SSSR 81 (1951), 359–362. (Russian)

(b) ———, *Combinatorial topology of non-closed sets. I. The first duality law; spectral duality*, Mat. Sb. (N. S.) 34 (76) (1954), 3–54 (Russian); Amer. Math. Soc. Transl. (2) 15 (1960), 245–295.

(c) ———, *New duality relations for nonclosed sets*, Dokl. Akad. Nauk SSSR 96 (1954), 925–928. (Russian)

(d) ———, *Combinatorial topology of non-closed sets. II. Dimension*, Mat. Sb. (N. S) 37 (79) (1955), 385–434 (Russian); Amer. Math. Soc. Transl. (2) 15 (1960), 297–349.

[15] Ju. M. Smirnov, *The Betti groups of the intersection of an infinite number of sets*, Dokl. Akad. Nauk SSSR 76 (1951), 29–32. (Russian)

[16] (a) N. E. Steenrod, *Universal homology groups*, Amer. J. Math. 58 (1936), 661–701.

(b) ———, *Regular cycles of compact metric spaces*, Ann. of Math. (2) 41 (1940), 833–851.

[17] E. Čech, *Théorie générale de l'homolgie dans un espace quelconque*, Fund. Math. 19 (1932), 149–183.

[18] (a) G. S. Čogošvili, *Duality relations in topological spaces*, Dissertation, Moscow, 1945. (Russian)

(b) ———, *On homological approximations and laws of duality for arbitrary sets*, Mat. Sb. (N. S.) 28 (70) (1951), 89–118. (Russian)

(c) ———, *Théorème de dualité pour le polyèdre infini*, C. R. Acad. Sci. Paris 221 (1945), 15–17.

(d) ———, *The duality law for retracts*, Dokl. Akad. Nauk SSSR 51 (1946), 87–90. (Russian)

(e) ———, *On the application of direct spectra of bicompact groups in homology theory*, Soobšč. Akad. Nauk Gruzin. SSR 15 (1954), 655–662. (Russian)

[19] S. Eilenberg and N. Steenrod, *Foundations of algebraic topology*, Princeton Univ. Press, Princeton, N. J., 1952.

[20] S. Eilenberg and S. MacLane, *Group extensions and homology*, Ann. of Math. (2) 43 (1942), 757–831.

Translated by:
C. H. Dowker

INTRINSIC HOMOLOGY THEORY*

V. A. ROHLIN

The problems from which intrinsic homology theory arose can be formulated quite simply. Let M^n be a closed n-dimensional manifold. Is it possible for it to be the boundary of an $(n + 1)$-dimensional manifold? And if M^n is oriented is it possible for it to be the oriented boundary of an oriented $(n + 1)$-manifold?

Ten years ago these problems seemed to be unapproachable. Today to a large extent they are solved. This paper is devoted to a survey of results in this area obtained in this short time.

1. **Terminology.** By *manifold* we mean a smooth (class C^∞) compact manifold either closed or with boundary. We do not assume that it is connected; a manifold can consist of any finite number of components. To orient a manifold means to orient each of its components. The boundary of an $(n + 1)$-manifold is a *closed n-manifold*. An orientation of a manifold is naturally transmitted to its boundary, so that the boundary of an oriented manifold is an oriented manifold.

We say that a manifold M^n *bounds* mod 2 if it admits a smooth (class C^∞) homeomorphism onto the boundary of some manifold M^{n+1}. We say that an *oriented* manifold M^n *bounds* if it admits an orientation preserving smooth homeomorphism onto the boundary of some oriented manifold M^{n+1}. It is clear that only a closed manifold can bound mod 2. The problems posed above can now be formulated as follows:

Let M^n be a closed manifold; does it bound mod 2?

Let M^n be an oriented closed manifold; does it bound?

In the interest of brevity we will call these "problem $\mathfrak{N}$" and "problem $\mathfrak{O}$". It should be emphasized that it is not known a priori whether these geometrical problems are topological in the strict sense, i.e., is the property of a manifold of bounding mod 2 for example a topological property or does it depend on the smooth structure. As Milnor showed [1], there are manifolds which are homeomorphic but not by a smooth homeomorphism.

2. **Classical results.** Long ago it was noticed that *if a manifold M^n bounds* mod 2 *then its Euler characteristic* $\chi(M^n)$ *is even.* The proof is so simple that it can be given here. If n is odd then $\chi(M^n) = 0$. If n is even, then let M_1^{n+1} be a manifold with boundary M^n. We triangulate M_1^{n+1} and double it, i. e., paste it together with another copy of itself along the boundary M^n. The closed manifold

---

*Expository lecture given at Moscow Mathematical Society on December 2, 1958.

M^{n+1}, so obtained, clearly satisfies the relations

$$\alpha_i (M^{n+1}) = 2\alpha_i (M_1^{n+1}) - \alpha_i (M^n) \qquad (i = 0, 1, \ldots),$$

where α_i is the number of i-dimensional simplices. It follows from these relations that

$$\chi (M^{n+1}) = 2\chi (M_1^{n+1}) - \chi (M^n).$$

But $\chi(M^{n+1}) = 0$. Hence $\chi(M^n) = 2\chi(M_1^{n+1})$.

This theorem shows that for every even n there are closed n-manifolds which do not bound mod 2, for example, real projective space $PR(n)$ whose characteristic is 1.

Every connected closed 1-manifold is a circle. Hence, *every closed 1-manifold bounds* mod 2 *and every oriented closed 1-manifold bounds.*

Every connected orientable closed 2-manifold is a sphere with handles. Every connected nonorientable closed 2-manifold is either a projective plane with handles or a Klein bottle with handles. It follows from this and the theorem about the Euler characteristic that *every oriented closed 2-manifold bounds* and that *a closed 2-manifold bounds* mod 2 *if and only if its Euler characteristic is even.*

Naturally these proofs being "classical" do not take into account completely the definitions of the preceding paragraph and require some supplementation. First of all the theorem about the characteristic was not proved for smooth but for triangulable manifolds. As Cairnes showed [2], smooth manifolds are triangulable. Further, the classification of 1-manifolds and 2-manifolds on which we leaned is known as a topological classification and not as a classification with respect to smooth homeomorphisms. In order to establish the correctness of its use it would be necessary to show that for closed 1 and 2-manifolds, homeomorphic implies smoothly homeomorphic. This theorem is almost obvious for 1-manifolds and is "known" for 2-manifolds but its proof has not been published to my knowledge. In any case the desired solution of problems $\mathfrak{D}$ and $\mathfrak{N}$ for 2-manifolds can be justified without the classification theorem.

3. **Stiefel-Whitney and Pontrjagin characteristic classes.** These classes are themselves the subjects of a vast theory. Here we will say a few things about them which are necessary to understand what follows.

We begin with two auxiliary propositions. Let m and n be integers such that $0 \leq m \leq n$, and let $E(m, n)$ be the space of real rectangular matrices of m rows and n columns. Let $A(m, n)$ be the subspace of $E(m, n)$ of matrices of rank m, and let $B(m, n)$ be the subspace of $E(m, n)$ of matrices of ranks m and $m - 1$. We will need the theorems:

(A) *If $r < n - m$, then the homotopy group $\pi_r(A(m, n))$ is trivial. If $r = n - m$, then $\pi_r(A(m, n))$ is a cyclic group, either free or of order 2.*

(B) *If $r < 2(n - m) + 3$, then the group $\pi_r(B(m, n))$ is trivial. If $r = 2(n - m) + 3$ and $n - m$ is even, then $\pi_r(B(m, n))$ is a free cyclic group.*

We turn to the definition of the characteristic classes. Let M^n be a closed manifold and let k be an integer such that $1 \leq k \leq n$. We triangulate M^n and denote by K^r the r-skeleton of this triangulation (i. e., subcomplex consisting of all simplices of dimension $\leq r$) and we set $m = n - (k - 1)$. The problem considered by Stiefel and Whitney when defining their characteristic classes is to construct on as large a part of M^n as possible a system of m linearly independent continuous tangent vector fields. For brevity we will call such systems *m-systems*.

It is not hard to show that an m-system can always be constructed on K^{k-1}. In fact, let τ^{r+1} be an arbitrary simplex of the triangulation. We assume that an m-system has already been constructed on its boundary σ^r, and we investigate whether it can be extended to the interior of the simplex τ^{r+1}. For this we construct some n-system F on τ^{r+1} (which, as is easily seen, is always possible) and then at each point $x \in \sigma^r$ we express the vectors of our m-system in terms of the vectors of F. The result will be a matrix with m rows and n columns of rank m defined at each point $x \in \sigma^r$, which varies continuously with x. In other words we get a continuous map ϕ of the sphere σ^r into $A(m, n)$, and it is clear that our m-system can be extended to all of τ^{r+1}, if and only if the map ϕ is homotopically trivial. But by Theorem (A) for $r < k - 1 = n - m$ every map of σ^r into $A(m, n)$ is homotopically trivial. Hence, for $r < k - 1$ every m-system defined on σ^r can be extended to τ^{r+1}. Whence it is clear that an m-system can always be constructed on K^{k-1}: it can be given arbitrarily on K^0 and then extended consecutively over $K^1, K^2, \cdots, K^{k-1}$.

However if $r = k - 1$ an m-system defined on σ^r cannot always be extended to τ^{r+1}. If the sphere σ^r, or what comes to the same thing the simplex τ^{r+1}, is oriented, ϕ determines an element $\overline{\phi}$ of the cyclic group $\pi_r(A(m, n))$ which is either free or of order 2. It is easy to see that $\overline{\phi}$ is not changed if we choose instead of F another n-system connected to F (at each point $x \in \tau^k$) by a matrix of positive determinant. But in the case of negative determinant $\overline{\phi}$ can be changed by a sign, and in order to remove this dependence of $\overline{\phi}$ on the choice of the n-system F, which is undesirable for us, we reduce it mod 2, i. e., we replace it by 0 if it is divisible by 2 in the group $\pi_r(A(m, n))$, and by 1 otherwise. Naturally this reduction does not change anything if $\pi_r(A(m, n))$ has order 2.

Now let F_m be any m-system defined on K^{k-1}. Since it is defined on the boundary of each k-simplex, by virtue of what has just been said it assigns to each simplex τ^k an integer mod 2. It turns out that the function of k-simplices so obtained is a cocycle mod 2 and that its cohomology class is independent of the choice of m-system F_m and of the triangulation of the manifold and is determined by the manifold M^n and the integer k. It is called the k-dimensional Stiefel-Whitney (characteristic) class of M^n and is denoted by W_k.

The simplest Stiefel-Whitney class is W_1. It is not hard to show that it is 0 if and only if M^n is orientable.

The Pontrjagin classes can be defined analogously. Let M^n and K^r be the same as above, but let k be such that $4 \leq 4k \leq n$ and let $m = n - 2(k - 1)$. By m-system we will now mean a system of m continuous tangent vector fields whose rank is never less than $m - 1$. Such a system can always be constructed on K^{4k-1}. In fact, let the m-system be constructed on the boundary σ^r of the simplex τ^{r+1} and let F be any system of n linearly independent continuous tangent vector fields defined on τ^{r+1}. Expressing the first system on σ^r by means of the second we obtain a continuous map ϕ of the sphere σ^r into $B(m, n)$. The homotopy triviality of this map is necessary and sufficient for the extension of our m-system from σ^r to τ^{r+1}. If $r < 4k - 1 = 2(n - m) + 3$, then by Theorem (B) every map of σ^r into $B(m, n)$ is homotopically trivial. Hence, for $r < 4k - 1$ every m-system defined on σ^r can be extended to τ^{r+1}. Hence an m-system can always be constructed on K^{4k-1}.

We turn to the case $r = 4k - 1$. If the simplex τ^{r+1} and hence also the sphere σ^r is oriented then the map ϕ determines an element $\bar{\phi}$ of the free cyclic group $\pi_r(B(m, n))$. It is possible to show that this element does not depend on the choice of the system F. Choosing a generator in the group $\pi_r(B(m, n))$ we can identify this group with the group of integers. Then $\bar{\phi}$ becomes an integer.

Let F_m be any m-system defined on K^{4k-1}. By what has been said it assigns to each oriented simplex τ^{4k} an integer. It turns out that the function of $4k$-simplices so obtained is an integral cocyle whose cohomology class is independent of F_m and of the triangulation of M^n and is determined by M and k. It is called the $4k$-dimensional Pontrjagin (characteristic) class and is denoted by P_{4k}.

It should be said that these definitions are different from the original definitions of Stiefel, Whitney, and Pontrjagin. In the definitions of Stiefel [3] and Whitney [4] the group $\pi_{k-1}(A(m, n))$ was not reduced mod 2 but was used as is. Pontrjagin constructed [5, 6] many other classes along with the classes P_{4k}. The classes P_{4k} were singled out by Wu [7]; it turned out that all the classes constructed by Pontrjagin were generated by means of simple homology operations from the classes P_{4k} and W_k. The definition given here of the classes P_{4k} is taken from my note [8].

The most obvious difference between the classes W_k and P_{4k} is that the former are defined for all dimensions and are classes mod 2, and the latter are defined only for dimensions divisible by 4 and are integral classes. But there are deeper differences. As early as 1950 Thom [9] had shown that the classes W_k are topological invariants of the manifold, i.e., do not depend on its smooth structure. Right afterwards Wu [10] obtained formulas computing W_k by means of homological invariants of the manifold (Steenrod squares). Thus the classes W_k are even homotopy invariants of the manifold. On the other hand, as is clear from examples constructed by Dold [11], the classes P_{4k} are not homotopy invariants. There have been serious attempts to prove their topological invariance but still only partial results have been obtained. For example, the topological (and homotopy) invariance of the classes P_{4k} reduced mod 12 has been proved [12]; the topological (and homotopy) invariance of the class P_4 for 4-manifolds has been proved (cf. §5 below); the topolgical invariance of the class P_4 considered as a weak cohomology class has been proved for 5-manifolds [13]; the combinatorial invariance of all the classes P_{4k} considered as weak cohomology classes has been proved [14].

4. **Pontrjagin's theorems.** Let M^n be a closed manifold and $k_1, \cdots, k_s$ a system of non-negative integers such that $k_1 + \cdots + k_s = n$. We denote by ρ_n the product of the cohomology classes $W_{k_1}, \cdots, W_{k_s}$ and by μ^n the basic homology class of M^n. Since ρ_n is an n-dimensional cohomology class mod 2, and μ^n is an n-dimensional homology class mod 2, their scalar product (ρ_n, μ^n) is defined. This is a residue mod 2 defined by M^n and $k_1, \cdots, k_s$. It is called a *characteristic residue* of M^n and is denoted by $\rho_n[M^n]$.

Now let M^n be an *oriented* closed manifold whose dimension is divisible by 4, and let $k_1, \cdots, k_s$ be a system of non-negative numbers such that $4k_1 + \cdots$ $\cdots + 4k_s = n$. We denote by ρ_n the product of the cohomology classes $P_{4k_1}, \cdots$ $\cdots, P_{4k_s}$ and by μ^n the basic homology class of the oriented manifold M^n. Now ρ_n is an integral n-dimensional cohomology class, μ^n is an integral n-dimensional homology class and their scalar product (ρ_n, μ^n) is an integer. It is defined by the oriented manifold M^n and the system $k_1, \cdots, k_s$ is called a *characteristic number* of M^n and is denoted by $\rho_n[M^n]$.

Thus every closed manifold has characteristic residues, and every oriented closed manifold whose dimension is divisible by 4 has characteristic numbers.

As examples of characteristic residues we have $W_n[M^n]$ (for which $s = 1$, $k_1 = n$) and $W_1^n[M^n]$ (for which $s = n$, $k_1 = \cdots = k_s = 1$). It can be shown that $W_n[M^n]$ is equal to the Euler characteristic of M^n reduced mod 2. For 2-manifolds these residues satisfy the relation $W_1^2[M^2] = W_2[M^2]$. We will consider examples of characteristic numbers later.

The following theorems (in equivalent form) were published by Pontrjagin [5] in 1947:

If the manifold M^n bounds mod 2, *then all its characteristic residues are* 0.

If the oriented manifold M^n bounds, then all its characteristic numbers are 0.

Naturally the second theorem has no provision for manifolds whose dimension is not divisible by 4. Since the residue $W_n[M^n]$ is the Euler characteristic of M^n reduced mod 2, the first theorem contains the theorem on the characteristic of §2.

5. **3-manifolds and 4-manifolds. The groups $\mathfrak{N}^n$ and $\mathfrak{O}^n$.** Since the characteristic of a closed 3-manifold is 0, we have $W_3[M^3] = 0$. If M^3 is orientable, then $W_1 = 0$, and so $W_1^3[M^3] = W_1 W_2[M^3] = 0$. Thus all the characteristic residues of an orientable 3-manifold are 0. Since there are no characteristic numbers for a 3-manifold, the theorems of Pontrjagin do not contradict the conjecture that *every oriented closed 3-manifold bounds.* This conjecture was made by Pontrjagin [5] in 1947.

In 1951 I succeeded in proving this conjecture [15]. The proof turned out to be rather complicated. Roughly, it amounts to imbedding the manifold M^3 in Euclidean space R^5, spanning a cone over it and then gradually smoothing it out until it becomes a 4-manifold with boundary M^3. There is the added difficulty that it is not known whether every M^3 can be imbedded in R^5. This is avoided by replacing M^3 by another manifold M_1^3 which is known to be contained in R^5 and which is related to M^3 in such a way that if M_1^3 bounds, then so does M^3.

In the same note [15] there is a second theorem according to which *every closed 3-manifold bounds* mod 2. It was derived comparatively easily from theorems on oriented 3-manifolds. It follows from this that all the characteristic residues of a closed 3-manifold are 0 even in the non-orientable case. This last result however could be obtained more easily by means of Wu's formula according to which for any closed 3-manifold $W_2 = W_1^2$ and $W_1 W_2 = 0$.

Turning to 4-manifolds we meet with completely new phenomena. Consider, for example, real projective space $PR(4)$ and the complex projective plane $PC(2)$. As is easy to show, $W_4[PR(4)] = 1$, $W_1^4[PR(4)] = 1$, $W_4[PC(2)] = 1$, $W_1^4[PC(2)] = 0$. We now construct the manifold $PR(4) + PC(2)$, of which $PR(4)$ and $PC(2)$ are components. Clearly,

$$W_4[PR(4) + PC(2)] = 0, \quad W_1^4[PR(4) + PC(2)] = 1.$$

Thus, the manifolds $PR(4)$ and $PC(2)$ not only fail to bound separately, but even together, and in this sense differ not only from manifolds which bound mod 2 (for example from spheres) but even from one another. This property leads us, strictly speaking, out of the realm of problem $\mathfrak{N}$ which only singles out manifolds which do not bound mod 2, not making further distinctions among them. Thus we proceed to the following formulation.

We say that closed manifolds M^n and N^n are *intrinsically homologous* mod 2 or are *cobordant* mod 2 if their sum $M^n + N^n$, i.e., the manifold consisting of disjoint copies of M^n and N^n, bounds mod 2. In particular, manifolds which bound mod 2 are said to be *intrinsically homologous to* 0 mod 2 (origin is a 0-manifold not having any components). This relation divides the closed n-manifolds into disjoint classes called *intrinsic homology classes* mod 2. Addition of manifolds leads to addition of classes, as a result of which the collection of classes becomes a commutative group known as the *n-dimensional intrinsic homology group,* which we denote by $\mathfrak{N}^n$. The zero of $\mathfrak{N}^n$ is the class of manifolds which bound mod 2. All of its elements have order 2. "Problem $\mathfrak{N}$" can now be formulated in a more general form as the problem of computing the group $\mathfrak{N}^n$.

Clearly, under addition of manifolds, the characteristic residues add. Since they are 0 for manifolds intrinsically homologous to 0, each characteristic residue determines a homomorphism of $\mathfrak{N}^n$ into the group of residues mod 2.

Analogous definitions can be given for *oriented* manifolds. To each oriented closed manifold M^n corresponds the oppositely oriented manifold $-M^n$ and to two oriented closed manifolds M^n and N^n corresponds their sum $M^n + N^n$, consisting of disjoint copies of M^n and N^n. Oriented manifolds M^n and N^n are *intrinsically homologous* or *cobordant* if their difference $M^n - N^n = M^n + (-N^n)$ bounds. In particular, bounding manifolds are *intrinsically homologous to* 0. The classes of oriented closed n-manifolds intrinsically homologous to one another, form the *n-dimensional intrinsic homology group,* which we denote by $\mathfrak{O}^n$. The calculation of the group $\mathfrak{O}^n$ is a problem which includes "problem $\mathfrak{O}$". Each characteristic number determines a homorphism of the group $\mathfrak{O}^n$, $n \equiv 0 \pmod 4$, into the group of integers.

In what follows the intrinsic homology class of the oriented closed manifold M^n will be denoted by $[M^n]$, and the intrinsic homology class mod 2 of the closed manifold M^n by $[M^n]_2$.

The theorems on 1-, 2-, and 3-manifolds mentioned earlier can now be formulated as follows: the groups $\mathfrak{O}^1, \mathfrak{O}^2, \mathfrak{O}^3, \mathfrak{N}^1, \mathfrak{N}^3$ are trivial; $\mathfrak{N}^2$ is a group of order 2 with generator $[PR(2)]_2$.

What was said above about $PR(4)$ and $PC(2)$ implies that $\mathfrak{N}^4$ has at least four different elements: 0 and the classes $[PR(4)]_2$, $[PC(2)]_2$, $[PR(4)]_2 + [PC(2)]_2$. This result was obtained with the help of the residues $W_4(M^4)$ and $W_1^4(M^4)$. It turns out that none of the remaining residues can be prescribed: it follows from the Wu formula [10] that all characteristic residues of a 4-manifold are linear combinations of these two, and we have for any closed manifold M^4

$$W_2^2[M^4] = W_4[M^4] + W_1^4[M^4], \quad W_1^2 W_2[M^4] = W_1 W_3[M^4] = 0.$$

Passing to the group $\mathfrak{O}^4$ we meet for the first time with the characteristic numbers. An oriented closed 4-manifold M^4 has a characteristic number $P_4[M^4]$.

The simplest manifold for which this number is different from 0 is the complex projective plane $PC(2)$. Being a complex manifold it has a natural orientation. As Pontrjagin showed [6], $P_4[PC(2)] = 3$. Consequently, the class $[PC(2)]$ generates an infinite cyclic subgroup of $\mathfrak{D}^4$.

In 1952 I managed to show that the indicated subgroups exhaust $\mathfrak{D}^4$ and $\mathfrak{N}^4$, [16], [17]. In other words $\mathfrak{D}^4$ *is a free cyclic group with generator* $[PC(2)]$; $\mathfrak{N}^4$ *is a direct sum of groups of order* 2 *with generators* $[PC(2)]_2$ *and* $[PR(4)]_2$.

The main difficulty occurs in the proof of the first theorem (about $\mathfrak{D}^4$). The basic idea is similar to the proof of the theorem about $\mathfrak{D}^3$ but involves significantly more complicated considerations. In particular the cone which has to be corrected and turned into a manifold is 5-dimensional and lies in Euclidean 7-space. The second theorem (on $\mathfrak{N}^4$) can be deduced quite simply from the first theorem and the structure of the groups $\mathfrak{D}^3$ and $\mathfrak{N}^2$.

It follows from these theorems that *oriented closed 4-manifolds with the same characteristic number are cobordant* and that *closed 4-manifolds with the same characteristic residues are cobordant* mod 2. In particular, an oriented closed 4-manifold M^4 bounds if and only if $P_4[M^4] = 0$, and a closed manifold M^4 bounds mod 2 if and only if $W_4[M^4] = 0$, and $W_1^4[M^4] = 0$.

Of course this solution of problems $\mathfrak{D}$ and $\mathfrak{N}$ is only effective in so far as we are able to compute $P_4[M^4]$ and the residues $W_4[M^4]$, $W_1^4[M^4]$. As far as the residues are concerned, they can be considered effectively computable; moreover, by virtue of the formula of Wu [10] the characteristic residues of a manifold of any dimension are effectively computable. The problem of computing $P_4[M^4]$ was solved in my paper [16] in the following way. For any closed oriented manifold M^{4k} of dimension $4k$ we denote by $\sigma[M^{4k}]$ the signature of the quadratic form $q(U_{2k})$ defined on the $2k$-dimensional real cohomology group H^{2k} of the manifold M^{4k} by the formula $q(U_{2k}) = (U_{2k}^2, \mu^{4k})$, $U_{2k} \in H^{2k}$ (μ^{4k} is the basic homology class of the oriented $4k$-manifold M^{4k}; the signature of a real quadratic form is the number of positive squares minus the number of negative squares in the canonical form). It turns out that if M^{4k} bounds then $\sigma[M^{4k}] = 0$ and $\sigma[M^{4k} + N^{4k}] = \sigma[M^{4k}] + \sigma[N^{4k}]$. Consequently, the signature, like the characteristic number defines a homomorphism of the group $\mathfrak{D}^{4k}$ into the group of integers. In particular this is true of the cyclic group $\mathfrak{D}^4$. Since any two homomorphisms of a cyclic group into the integers differ by a constant multiple, there is a λ, such that $\sigma[M^4] = \lambda P_4[M^4]$. But $P_4[PC(2)] = 3$, and $\sigma[PC(2)] = 1$. Hence

$$P_4[M^4] = 3\sigma[M^4].$$

In particular, the characteristic number $P_4[M^4]$ is a topological (and even a homotopy type) invariant of the manifold M^4.

The note [17] also contains some theorems on the groups $\mathfrak{D}^n$ and $\mathfrak{N}^n$ for $n>4$ (cf. §8), but none of these groups were computed completely. The main difficulty lies in the fact that the method of gradually changing the cone spanned over M^n into an $n + 1$-manifold with boundary M^n, which could be carried out with difficulty for $n = 3$ and $n = 4$, does not lend itself to further generalization. In particular, it was not shown, that the groups $\mathfrak{D}^n$ and $\mathfrak{N}^n$ have a finite number of generators, although this seemed very probable.

6. **Results of Thom: the groups $\mathfrak{N}^n$.** In 1953 the brilliant paper [18] of Thom appeared, advancing the theory of intrinsic homology far ahead.

Thom established an isomorphism between the groups $\mathfrak{D}^n$ and $\mathfrak{N}^n$ and the homotopy groups of certain spaces specially constructed for this purpose and, thus, to a considerable degree, reduced the computation of the groups $\mathfrak{D}^n$ and $\mathfrak{N}^n$ to a homotopy problem. One should not think that these homotopy problems appeared easy . The success of his work is due to the fact that just at this time the French school of topologists had created (but not yet fully published) powerful new methods for the solution of homotopy problems. To this it should be added that the method with which Thom established the isomorphisms is a direct generalization of the method of ''rigged manifolds'' of Pontrjagin [19]. It is remarkable that Pontrjagin used the connection between problems in the theory of manifolds and homotopy problems to solve homotopy problems, and Thom used it to solve problems in the theory of manifolds.

Thom succeeded in completely solving the homotopy problems connected with the groups $\mathfrak{N}^n$, and thus obtained complete information on these groups. In order to formulate his results we denote by K the set of non-negative integers which are not of the form $2^m - 1$, and by $d(n)$ the number of different decompositions of n as a sum of elements of K. Thom showed that $\mathfrak{N}^n$ *is the direct sum of $d(n)$ groups of order 2* and that *two closed manifolds with the same characteristic residues are cobordant* mod 2. In particular, *a closed manifold all of whose characteristic residues are 0 bounds* mod 2.

It follows from this theorem that among the characteristic residues of n-manifolds there are precisely $d(n)$ independent ones.

Thom also investigated the connection between groups $\mathfrak{N}^n$ of different dimension arising from the fact that manifolds can be *multiplied.* Let $\mathfrak{N}$ be the direct sum of all the groups $\mathfrak{N}^n$. If we introduce in $\mathfrak{N}$ the product defined by the product of manifolds, then $\mathfrak{N}$ becomes a ring. Thom showed that *one can define closed manifolds V^n, one for each $n \in K$, such that the classes $v_n = [V^n]_2$ are a system of generators for the ring $\mathfrak{N}$, which are independent* mod 2, *i. e., $\mathfrak{N}$ becomes the ring of polynomials in the variables v_n over the prime field of characteristic* 2. In particular, the product of two manifolds which are not intrinsically homologous to

0 mod 2 is not intrinsically homologous to 0 mod 2.

Thom did not indicate a concrete system of generators for the groups $\mathfrak{N}^n$. He showed that for the manifolds V^n with even n one can take the real projective space $PR(n)$, and he indicated a concrete V^5 (cf. below) but he did not find concrete manifolds V^n with odd $n > 5$ (we note that for n odd $PR(n)$ is homologous to 0 mod 2, and if oriented remains intrinsically homologous to 0, cf. [17]). This lacuna was filled in 1956 by Dold [20]. We denote by $V(p, q)$ the manifold obtained from the product $S^p \times PC(q)$ of the sphere S^p and the complex projective space $PC(q)$ by the identification $(x, z) = (-x, \bar{z})$, where x and $-x$ are antipodal points in S^p and z and $\bar{z}$ are points in $PC(q)$ with complex conjugate projective coordinates. Dold showed that for V^n with odd n one can take the manifold $V(2^r - 1, s2^r)$, where r and s are integers defined by the relation $n + 1 = 2^r(2s + 1)$. We note that the simplest of the manifolds $V(2^r - 1, s2^r)$, namely the 5-manifold $V(1, 2)$ had already been constructed in 1950 by Wu [10] (for another purpose). It was also used by Thom for the manifold V^5 and of course served as the starting point of the more general construction of Dold.

From now on we will understand by V^n just those manifolds which were indicated by Thom and Dold. By virtue of what was said above the products

$$v_{k_1} v_{k_2} \cdots v_{k_s}, \qquad k_1 + k_2 + \ldots + k_s = n, \tag{1}$$

serve as a system of generators of the group $\mathfrak{N}^n$. The manifolds $V^n = PR(n)$ with even n are non-orientable; the manifolds $V^n = V(2^r - 1, s2^r)$, $n + 1 = 2^r(2s + 1)$, are orientable. It is not hard to verify that the symmetry of the sphere S^p, $p = 2^r - 1$, with respect to any hyperplane passing through its center, determines a smooth homeomorphism of $V(2^r - 1, s2^r)$, reversing its orientation. Consequently, the arbitrarily oriented manifold $V(2^r - 1, s2^r)$ is intrinsically homologous to $-V(2^r - 1, s2^r)$, i.e., the manifold $2V(2^r - 1, s2^r)$ is intrinsically homologous to 0. Hence for odd $n \in K$, V^n determines an element of order 2 in the group $\mathfrak{O}^n$. From now on this element will be denoted by u_n.

We denote by Z_2 the group of order 2 and on the basis of Thom's results we list the first few groups $\mathfrak{N}^n$:

	Group	Generators
	$\mathfrak{N}^1 = 0,$	
	$\mathfrak{N}^2 = Z_2,$	$[PR(2)]_2$
	$\mathfrak{N}^3 = 0,$	
	$\mathfrak{N}^4 = Z_2 + Z_2,$	$[PR(4)]_2, \quad [PR(2) \times PR(2)]_2,$
	$\mathfrak{N}^5 = Z_2,$	$[V(1, 2)]_2,$
	$\mathfrak{N}^6 = Z_2 + Z_2 + Z_2,$	$[PR(6)]_2, \quad [PR(4) \times PR(2)]_2,$
		$[PR(2) \times PR(2) \times PR(2)]_2,$
	$\mathfrak{N}^7 = Z_2,$	$[V(1, 2) \times PR(2)]_2.$

As is not hard to calculate, $PR(2) \times PR(2)$ and $PC(2)$ have the same characteristic residues. Consequently, $[PR(2) \times PR(2)]_2 = [PC(2)]_2$, and the generators of $\mathfrak{N}^4$, given here coincide with the generators given in §5.

7. **Results of Thom: the groups $\mathfrak{D}^n$.** In solving the homotopy problem connected with the computation of the groups $\mathfrak{D}^n$ Thom met with significant difficulties of an algebraic nature. Because of these difficulties he did not succeed in his study of the groups $\mathfrak{D}^n$ as fully as in the case of $\mathfrak{N}^n$. Nevertheless, here too he obtained far-reaching results.

We denote by $\mathfrak{D}^n_T$ the periodic part of $\mathfrak{D}^n$ (i.e., the subgroup of $\mathfrak{D}^n$ of elements of finite order) and by $c(k)$ the number of decompositions of the number k as the sum of non-negative integers. Thom showed that $\mathfrak{D}^n_T$ *is finite for each* n *and that* $\mathfrak{D}^n$ *coincides with* $\mathfrak{D}^n_T$ *if* $n \not\equiv 0 \pmod 4$, *and has rank* $c(k)$ *if* $n = 4k$.

Further results of Thom refer to the factor group $\mathfrak{D}^{4k}_0 = \mathfrak{D}^{4k}/\mathfrak{D}^{4k}_T$. The elements of this factor group are *weak* intrinsic homology classes defined as follows: an oriented closed manifold M^{4k} bounds *weakly* if there exists a non-negative interger m such that the manifold mM^{4k} bounds; two oriented closed manifolds *belong to the same weak intrinsic homology class* or *are weakly cobordant* if their difference bounds weakly. It follows from the theorems of Pontrjagin (§4) that *all characteristic numbers of a weakly bounding manifold are* 0 *and that weakly cobordant manifolds have the same characteristic numbers.* Thom proved the converse theorem: *if all the characteristic numbers of a closed oriented manifold are* 0 *then it bounds weakly; if two oriented closed manifolds have the same characteristic numbers then they are weakly cobordant.*

We consider for fixed k all possible manifolds of the form

$$PC(2k_1) \times PC(2k_2) \times \ldots \times PC(2k_s), \qquad k_1 + k_2 + \ldots + k_s = k. \qquad (2)$$

Thom proved that *the weak homology classes defined by them are linearly independent, i.e., generate a subgroup of* $\mathfrak{D}^{4k}_0$ *of the same rank* $c(k)$ *as* $\mathfrak{D}^{4k}_0$. *Hence, for any oriented closed manifold* M^{4k} *there is a non-negative integer* m *such that the manifold* mM^{4k} *is intrinsically homologous to an integral linear combination of products of even dimensional complex projective spaces.*

We denote be $\mathfrak{D}^{4k}_1$ the subgroup of $\mathfrak{D}^{4k}_0$ generated by the weak homology classes of the manifolds (2). We know from §5 that $\mathfrak{D}^4_1 = \mathfrak{D}^4_0 = \mathfrak{D}^4$. Thom proved that $\mathfrak{D}^8_1 = \mathfrak{D}^8_0$. Hirzebruch [21] proved that $\mathfrak{D}^{12}_1 = \mathfrak{D}^{12}_0$, but $\mathfrak{D}^{4k}_1 \neq \mathfrak{D}^{4k}_0$ for $k \geq 4$. Generators for $\mathfrak{D}^{4k}_0$ with $k \geq 4$ are not known.

Let $\mathfrak{D}$ be the direct sum of all the groups $\mathfrak{D}^n$. If we introduce a product in $\mathfrak{D}$ defined by the product of manifolds then $\mathfrak{D}$ becomes a ring. The collection $\mathfrak{D}_T$ of elements of finite order is the direct sum of the groups $\mathfrak{D}^n_T$, and in the same sense the factor ring $\mathfrak{D}_0 = \mathfrak{D}/\mathfrak{D}_T$ is the direct sum of the factor groups $\mathfrak{D}^{4k}_0 = \mathfrak{D}^{4k}/\mathfrak{D}^{4k}_T$.

Thom's results do not completely determine either the generators or the algebraic structure of the ring $\mathfrak{D}_0$. However they determine a little something in this ring. This "little something" is the tensor product

$$\mathfrak{D}_0 \otimes Q = \mathfrak{D} \otimes Q,$$

where Q is the field of rational numbers. Thom's theorems show that *this tensor product is the ring of polynomials in the variables* $u_{4k} = [PC(2k)]$ *over the field* Q.

For the groups $\mathfrak{D}^n$ with $n \leq 7$ Thom obtained complete results:

$$\mathfrak{D}^1 = \mathfrak{D}^2 = \mathfrak{D}^3 = 0, \qquad \mathfrak{D}^4 = Z,$$
$$\mathfrak{D}^5 = Z_2, \qquad \mathfrak{D}^6 = \mathfrak{D}^7 = 0.$$

Here Z is the free cyclic group. The classes u_4, u_5 serve as generators of $\mathfrak{D}^4$, $\mathfrak{D}^5$.

In summary, one can say that Thom developed powerful methods which permitted him to solve the problems of the theory of intrinsic homology to a remarkable degree. He completely determined the algebraic structure of the groups $\mathfrak{N}^n$ and the ring $\mathfrak{N}$ and indicated some of the generators (the missing generators were indicated by Dold). By determining the structure of the groups $\mathfrak{N}^n$ Thom turned out to be able to give a complete solution to problem $\mathfrak{N}$. Thom proved that the groups $\mathfrak{D}^n$ are finitely generated and completely computed the ring $\mathfrak{D} \otimes Q$. He did not find the remaining groups $\mathfrak{D}_T^n$ and ideal $\mathfrak{D}_T$ or the generators of the groups $\mathfrak{D}_0^{4k} = \mathfrak{D}^{4k}/\mathfrak{D}_T^{4k}$ or the ring $\mathfrak{D}_0 = \mathfrak{D}/\mathfrak{D}_T$. Determining the structure of the groups $\mathfrak{D}_0^{4k}$, Thom was able to solve problem $\mathfrak{D}$ partially.

8. **The homomorphism** h. The operation of depriving an oriented closed manifold of its orientation determines a natural homorphism of the group $\mathfrak{D}^n$ into the group $\mathfrak{N}^n$ and of the ring $\mathfrak{D}$ into the ring $\mathfrak{N}$. We will denote this homomorphism by h. The study of h allows one to get new information about the groups $\mathfrak{D}^n$ and the ring $\mathfrak{D}$. Moreover, it opens completely new properties of manifolds. For example, finding the image $h(\mathfrak{D}^n)$ of $\mathfrak{D}^n$ in $\mathfrak{N}^n$ means answering the question which closed manifolds M^n are cobordant mod 2 to oriented closed manifolds.

My papers [17], [22] are devoted to the study of h. In the first (1953) the kernel of h is found and the image $h(\mathfrak{D}^n)$ in $\mathfrak{N}^n$ is described; this description is applied to the study of the groups $\mathfrak{N}^n$ of lower dimension. In the second paper (1958) the image $h(\mathfrak{D}^n)$ is described in other terms and the results are applied to the study of $\mathfrak{D}^n$. The second paper is based on results of Thom expounded in §6 but apart from that is based on direct geometric methods as is the first paper.

The theorem on the kernel is very simple: *the kernel of* h *in* $\mathfrak{D}^n$ *is* $2\mathfrak{D}^n$; in other words, *an orientable closed manifold is intrinsically homologous to* 0 *mod* 2 *if and only if when it is given an orientation it is intrinsically homologous to the double of an oriented closed manifold.*

We proceed to the image $h(\mathfrak{D}^n)$. As is known, the integral $(n-1)$-dimensional homology group of a connected non-orientable closed n-manifold contains exactly one element a^{n-1} of order 2. For any manifold M^n we define a^{n-1} as the sum of the classes a^{n-1}, defined on the non-orientable components of M^n. The equation $a^{n-1} = 0$ is necessary and sufficient for M^n to be orientable. As is not hard to show a^{n-1} reduced mod 2 is dual to the Stiefel-Whitney class W_1. The $(n-2)$-dimensional homology class mod 2 dual to W_1^2 (i.e., the self-intersection in the sense of Lefschetz of a^{n-1} reduced mod 2) we denote by b^{n-2}. The theorem on $h(\mathfrak{D}^n)$ obtained in [17], asserts that: (a) *in every closed manifold M^n there is an oriented submanifold A^{n-1} which as an integral cycle belongs to a^{n-1}, and a submanifold B^{n-2}, which as a cycle* mod 2 *belongs to b^{n-2};* (b) *if M^n is intrinsically homologous* mod 2 *to an oriented closed manifold then A^{n-1}, no matter how chosen, is intrinsically homologous to 0, and B^{n-2}, no matter how chosen, is intrinsically homologous to* 0 mod 2; (c) *if for some choice of A^{n-1} and B^{n-2} the first is intrinsically homologous to* 0 *and the second is intrinsically homologous to* 0 mod 2, *then M^n is intrinsically homologous* mod 2 *to an orientable manifold.*

This theorem reduces the n-dimensional problem of interest to us to the $(n-1)$-dimensional problem $\mathfrak{D}$ and the $(n-2)$-dimensional problem $\mathfrak{N}$. For example, it follows from the relations $\mathfrak{D}^2 = 0$, $\mathfrak{N}^1 = 0$, that $h(\mathfrak{D}^3) = \mathfrak{N}^3$. This was precisely the first computation of $\mathfrak{N}^3$ and similar (though more complicated) methods gave $\mathfrak{N}^4$; cf. §5. Further, $h(\mathfrak{D}^5) = \mathfrak{N}^5$. This follows from the relations $\mathfrak{D}^4 = Z$, $\mathfrak{N}^3 = 0$ and the (almost obvious) fact that the manifold A^{n-1} always satisfies the relation $[2A^{n-1}] = 0$.

In order to formulate the theorem on the image $h(\mathfrak{D}^n)$ proved in [22], we single out among the characteristic residues $\rho[M^n]$ those such that ρ contains the factor W_1, and call them W_1*-residues.* Among the generators (1) of $\mathfrak{N}^n$ we single out those in which each even-dimensional factor v_{k_i} appears an even number of times, and call them *generators of the first kind.* The theorem we are interested in asserts that: (d) *a closed manifold M^n is intrinsically homologous* mod 2 *to an orientable manifold if and only if all its W_1-residues are* 0; (e) *the image $h(\mathfrak{D}^n)$ is the subgroup of $\mathfrak{N}^n$ generated by the generators of the first kind;* (f) *in proposition* (c) *the requirement $[A^{n-1}] = 0$ can be replaced by the weaker requirement $[A^{n-1}]_2 = 0$.* It is not hard to deduce from this theorem that any oriented manifold M^n, such that $[2M^n] = 0$, and $[M^n]_2 = 0$ is intrinsically homologous to 0. In other words, *if $u \in \mathfrak{D}^n$, $2u = 0$, and $h(u) = 0$, then $u = 0$.* Consequently, *there are no elements of order 4 in $\mathfrak{D}^n$,* and this means *there are no elements of finite order divisible by* 4.

Theorems (d) and (e) give an effective algebraic solution for our problem. Theorem (e) can clearly be formulated thus: *the image $h(\mathfrak{D})$ of the ring $\mathfrak{D}$ in the* *ring $\mathfrak{R}$ is the subring generated by the generators v_n of odd dimension and the squares of the generators v_n of even dimension.* As was shown in [22], for any n

$$[PR(n) \times PR(n)]_2 = [PC(n)]_2.$$

Hence, the generators v_{2k+1} and v_{2k}^2 of the ring $h(\mathfrak{D})$ are the images of the elements u_{2k+1} and u_{4k} of the ring $\mathfrak{D}$ defined in §§6, 7:

$$v_{2k+1} = h(u_{2k+1}) \, (2k+1 \in K), \qquad v_{2k}^2 = h(u_{4k}). \tag{3}$$

9. The groups $\mathfrak{D}_T^n$. We denote by $\mathfrak{D}_2^n$ the subgroup of $\mathfrak{D}_T^n$ consisting of those elements whose order has the form 2^m. The theorems expounded in this section completely compute the groups $\mathfrak{D}_2^n$.

Since there are no elements of finite order, a multiple of 4, in $\mathfrak{D}^n$ every element of $\mathfrak{D}_2^n$ except 0 has order 2. We have already met with some elements of this subgroup, for example $u_{2k+1} \, (2k+1 \in K)$. Clearly, all products of these classes also belong to $\mathfrak{D}_2^n$, and the same thing applies to their products with the classes u_{4k}, i.e., to products of the form

$$u_{n_1} u_{n_2} \ldots u_{n_s}, \qquad n_1 \equiv 1 \pmod 2. \tag{4}$$

It is supposed that each of the numbers $n_1, n_2, \cdots, n_s$ is either odd and in K or is divisible by 4; the classes u_n are not defined for other values of n. It turns out that *the products* (4) *with* $n_1 + n_2 + \cdots + n_s = n$ *are a system of generators for* $\mathfrak{D}_2^n$ *which are independent* mod 2.

This theorem appears here for the first time and I will give its proof here. If $u \in \mathfrak{D}_2^n$ and $h(u) = 0$, then, by virtue of §8, $u = 0$. Hence h is one-one on $\mathfrak{D}_2^n$. By (3) it carries the products (4) into generators of $\mathfrak{R}^n$ which are independent mod 2. Hence our products are also independent mod 2. The indicated generators of $\mathfrak{R}^n$ are generators of the first kind. If $n \not\equiv 0 \pmod 4$ then they are all the generators of the first kind. If $n \equiv 0 \pmod 4$ then they are only some generators of the first kind while the rest consists of those generators of the form

$$v_{2k_1}^2 v_{2k_2}^2 \ldots v_{2k_t}^2, \qquad 4k_1 + 4k_2 + \ldots + 4k_t = n. \tag{5}$$

It remains to show that in the second case the intersection of the subgroup $\mathfrak{R}_1^n$ of $\mathfrak{R}^n$ generated by the elements (5), with the subgroup $h(\mathfrak{D}_2^n)$ consists of only 0.

We consider all characteristic residues of the form

$$W_{2k_1}^2 W_{2k_2}^2 \ldots W_{2k_t}^2 [M^n], \qquad 4k_1 + 4k_2 + \ldots + 4k_t = n. \tag{6}$$

It is known [23], that W_{2k}^2 is P_{4k}, reduced mod 2. Hence, in the case of an oriented manifold M^n the residue (6) is the characteristic number

$$P_{4k_1} P_{4k_2} \ldots P_{4k_t} [M^n],$$

reduced mod 2. But this number is 0 for any weakly bounding manifold M^n. Hence, on $h(\mathfrak{D}_2^n)$ all residues (6) are identically 0. However, as is not hard to show, these residues are independent on $\mathfrak{N}_1^n$, i.e., simultaneously map only the 0 class to 0. Hence, $h(\mathfrak{D}_2^n) \cap \mathfrak{N}_1^n$ consists of only 0.

Thus the manifolds of the classes (4) and their sums exhaust all the manifolds of order 2 (i.e., all manifolds M^n for which $[2M^n] = 0$, but $[M^n] \neq 0$). We also know that no manifold can have finite order divisible by 4.

Are there manifolds of other orders, for example, of order 3, 5 or 6?

Until just recently one could only say that no such manifolds were known. Right now the situation has changed:

Averbuh [24] proved that *such manifolds do not exist*, i.e., that $\mathfrak{D}_T^n = \mathfrak{D}_2^n$. *Thus we have found all the groups* $\mathfrak{D}_T^n$, *in particular, all the groups* $\mathfrak{D}^n$ *with* $n \not\equiv 0$ (mod 4), *and the ring* $\mathfrak{D}_T$. As an ideal this ring is generated by the classes u_{2k+1}, $2k + 1 \in K$.

Averbuh's proof uses Thom's techniques and some more recent developments of algebraic topology and consists of overcoming the algebraic difficulties which were mentioned in §7.

The theorems of this section were obtained for the groups $\mathfrak{D}^8$, $\mathfrak{D}^9$, $\mathfrak{D}^{10}$, $\mathfrak{D}^{11}$, $\mathfrak{D}^{12}$ by Adachi [25] (by Thom's methods).

These results enable us to give the following remarkable solution to problem $\mathfrak{D}$: *an oriented closed manifold is intrinsically homologous to* 0 *if and only if all its characteristic numbers and residues are* 0; *two oriented closed manifolds are intrinsically homologous if and only if they have identical characteristic numbers and residues.* Proof: If all the characteristic numbers and residues of the oriented closed manifold M^n are 0, then by Thom's theorem $[M^n]_2 = 0$ and there is a non-negative m such that $[mM^n] = 0$. Since $\mathfrak{D}_T^n$ contains only elements of order 1 and 2, $[2M^n] = 0$. Since $[2M^n] = 0$ and $[M^n]_2 = 0$, then $[M^n] = 0$.

Dold's remark [20]. Every non-negative integer n different from 1, 2, 3, 6, 7 can be written as the sum of non-negative integers n_1, n_2, $\cdots$, n_s, each of which is either odd and in K or divisible by 4. The corresponding product $u_{n_1} u_{n_2} \cdots u_{n_s}$ has dimension n and is different from 0. Hence *every group* $\mathfrak{D}^n$, *except* $\mathfrak{D}^1$, $\mathfrak{D}^2$, $\mathfrak{D}^3$, $\mathfrak{D}^6$, $\mathfrak{D}^7$, *is non-trival.*

10. **Some problems.** The solutions of problems $\mathfrak{N}$ and $\mathfrak{D}$ exponded (§§6 and 9) are essentially different from one another. The solution to problem $\mathfrak{N}$ depends on the values of the characteristic residues which are topological (and even homotopy type) invariants of the manifold. In particular, the intrinsic homology class mod 2 is a topological invariant; cf. §1. The same thing applies to problem $\mathfrak{D}$ in dimensions not divisible by 4. But in dimensions divisible by 4 the solution of

problem $\mathfrak{D}$ depends on the characteristic numbers whose topological invariance
has not been proved. In this connection it is unknown whether the property of a
manifold M^{4k} bounding is a topological property. Moreover, in contrast to the
characteristic residues the characteristic numbers are not effectively computable
invariants of a manifold so that there is also a difference in the degrees of effec-
tiveness in the given solutions of problems $\mathfrak{N}$ and $\mathfrak{D}$.

Other unsolved problems in the theory of intrinsic homology are also connected
with the study of properties of characteristic numbers, the search for generators
of the factor group $\mathfrak{D}_0^{4k} = \mathfrak{D}^{4k}/\mathfrak{D}_T^{4k}$ $(k \geq 4)$ and the study of the algebraic structure
of the factor ring $\mathfrak{D}_0 = \mathfrak{D}/\mathfrak{D}_T$.

Problems $\mathfrak{N}$ and $\mathfrak{D}$ can also be set for manifolds of finite class $C^r (r = 1, 2, \cdots)$
and for combinatorial manifolds. By the general theory of Whitney [26], for any
manifold of class C^r there exists a manifold of class C^∞ which is C^r-homeomor-
phic to it and for any two C^r-homeomorphic manifolds of class C^∞ there is a C^∞-
homeomorphism. Hence the theory of intrinsic homology for manifolds of finite
class C^r is no different from the theory for manifolds of class C^∞. The theory
has not been constructed for combinatorial manifolds.

The present survey does not touch on the applications of intrinsic homology
theory. Hirzebruch's book [27] contains applications of intrinsic homology to
algebraic geometry and Milnor's work [1] is also based on intrinsic homology.

BIBLIOGRAPHY

[1] J. Milnor, *On manifolds homeomorphic to the 7-sphere*, Ann. of Math. (2) 64
(1956), 394–405.

[2] S. S. Cairns, *Triangulation of the manifold of class one*, Bull. Amer. Math.
Soc. 41 (1935), 549–552.

[3] E. Stiefel, *Richtungsfelder und Fernparallelismus in n-dimensionalen Mannig-
faltigkeiten*, Comment. Math. Helv. 8 (1936), 305–353.

[4] Hassler Whitney, *On the topology of differentiable manifolds*, Lectures in
Topology, pp. 101–141, Univ. of Michigan Press, Ann Arbor, Mich., 1941.

[5] L. S. Pontrjagin, *Characteristic cycles on differentiable manifolds*, Mat. Sb.
(N. S.) 21 (63) (1947), 233–284. (Russian)

[6] ______, *Vector fields on manifolds*, Mat. Sb. (N.S.) 24 (66) (1949), 129–162.
(Russian)

[7] Wu Wen-tsün, *Sur les classes caractéristiques des structures fibrées sphér-
iques*, Actualitiés Sci. Ind., No. 1183, Hermann, Paris, 1952.

[8] V. A. Rohlin, *Intrinsic definition of Pontrjagin's characteristic cycles*, Dokl.
Akad. Nauk SSSR 84 (1952), 449–452. (Russian)

[9] R. Thom, *Variétés plongées et i-carrés*, C. R. Acad. Sci. Paris 230 (1950)
507–508.

[10] Wu Wen-tsün, *Classes caractéristiques et i-carrés d'une variété*, C. R. Acad. Sci. Paris 230(1950), 508–511.

[11] A. Dold, *Über fasernweise Homotopieäquivalenz von Faserräumen*, Math. Z. 62(1955), 111–136.

[12] Wu Wen-tsün, *On Pontrjagin classes*. I, II, III, IV, V, Acta Math. Sinica 3 (1953), 291–315; 4(1954), 171–199; 4(1954), 323–346; 5(1955), 37–63; 5(1955), 401–410. (Chinese. English summaries)

[13] V. A. Rohlin, *On the Pontrjagin characteristic classes*, Dokl. Akad. Nauk SSSR 113(1957), 276–279. (Russian)

[14] V. A. Rohlin and A. S. Švarc, *The combinatorial invariance of Pontrjagin classes*, Dokl. Akad. Nauk SSSR 114(1957), 490–493. (Russian)

[15] V. A. Rohlin, *A three-dimensional manifold is the boundary of a four-dimensional one*, Dokl. Akad. Nauk SSSR 81(1951), 355–357. (Russian)

[16] ________, *New results in the theory of four-dimensional manifolds*, Dokl. Akad. Nauk SSSR 84(1952), 221–224. (Russian)*

[17] ________, *Intrinsic homologies*, Dokl. Akad. Nauk SSSR 89(1953), 789–792. (Russian)

[18] R. Thom, *Sous-variétés et classes d'homologie des variétés différentiables*. I, II, C. R. Acad. Sci. Paris 236(1953), 453–454, 573–575; *Variétés différentiables cobordantes*, ibid. 236(1953), 1733–1735.**

[19] L. S. Pontrjagin, *Smooth manifolds and their applications in homotopy theory*, Trudy Mat. Inst. Steklov. No. 45, Izdat. Akad. Nauk SSSR, Moscow, 1955. (Russian)

[20] A. Dold, *Erzeugende der Thomschen Algebra $\mathfrak{N}$*, Math. Z. 65(1956), 25–35.

[21] F. Hirzebruch, *Some problems on differentiable and complex manifolds*, Ann. of Math. (2) 60(1954), 213–236.

[22] V. A. Rohlin, *Internal homologies*, Dokl. Akad. Nauk SSSR 119(1958), 876–879. (Russian)

[23] Wu Wen-tsün, *Sur les puissances de Steenrod*, Colloq. de Topologie (Strasbourg, 1951), No. IX, Bibliothèque Nationale et Universitaire de Strasbourg, 1952.

[24] B. G. Averbuh, *Algebraic structure of cobordism groups*, Dokl. Akad. Nauk SSSR 125(1959), 11–14. (Russian)

[25] M. Adachi, *On the groups of cobordism Ω^k*, Nagoya Math. J. 13(1958), 135–156.

[26] H. Whitney, *Differentiable manifolds*, Ann. of Math. (2) 37(1936), 645–680.

[27] F. Hirzebruch, *Neue topologische Methoden in der algebraischen Geometrie*, Springer, Berlin, 1956.

[28] R. Thom, *Quelques propriétés globales des variétés différentiables*, Comment. Math. Helv. 28(1954), 17–86.

Translated by:
Norman Stein

*For corrections see [17].

**For a more detailed exposition see [28].

THE SEMIGROUP OF HOMEOMORPHIC MAPPINGS OF AN INTERVAL

L. M. GLUSKIN

Let S be the set of all continuous strictly monotone functions $f(\xi)$, defined on the interval $E = [0, 1]$, such that $0 \leq f(\xi) \leq 1$. It is clear that S is a semigroup with respect to superposition of functions:

$$f \cdot g = fg = f[g(\xi)].$$

The present article is devoted to the study of the semigroup S. In §2 we consider questions of divisibility in the semigroup S, and we find its one-sided and two-sided ideals. In §3 we describe two nonisomorphic subsemigroups of S that have no nontrivial homomorphisms. In §4 we find the group of automorphisms of semigroup S. §5 is devoted to problems of topology in S. In particular, we prove the existence of a finite basis in S and single out dense subsemigroups of S, which are close to S in their algebraic and topological properties. *

§1. Principal definitions and notations

1.1. A semigroup is a set with a single-valued binary operation. If the semigroup S contains an element 0 such that $x0 = 0x = 0$ for all $x \in S$, then 0 is called the zero of the semigroup S. If the semigroup S contains an element e such that $ex = xe = x$ for all $x \in S$, then e is called the unit of the semigroup S.

Let A, B be two nonvoid subsets of the semigroup S. The symbol AB denotes the set of all products ab, where $a \in A$, $b \in B$.

1.2. A nonvoid subset L of the semigroup S is called a left ideal if $SL \subseteq L$. A right ideal is defined analogously. A set which is at once a left and a right ideal is called a two-sided ideal of the semigroup S (we ordinarily call it simply an ideal of S). The semigroup S itself and the zero 0 of the semigroup S (if S contains a zero) are evidently ideals of S. Ideals of S different from 0 and S are called proper ideals of S.

Let a be an arbitrary element of the semigroup S. The set $a \cup Sa$ is the intersection of all left ideals of S that contain a, and is called the principal left ideal of S generated by the element a. Analogously, the sets $a \cup aS$ and $a \cup aS \cup Sa \cup SaS$ are called principal right and principal two-sided ideals of the semigroup

---

*A part of the results of the present article could be carried over to the semigroup of all homeomorphic mappings of a circle into itself and to the semigroup of homeomorphic mappings of an n-dimensional sphere. This, however, would require a significant increase in the size of the article.

S. If S is a semigroup with a unit, then principal left, right and two-sided ideals have, respectively, the forms Sa, aS, and SaS. a is called the generator of the ideal Sa (aS, SaS).

If a left ideal L of the semigroup S contains no left ideals of S different from 0 and L, then L is called a minimal left ideal of S.

1.3. Let a, b be arbitrary elements of the semigroup S. If the equation $xa = b$ always has a solution x in S, then S is called a semigroup with left division. In particular, for every element a of such a semigroup S, there exists at least one left unit, that is, an element e_a such that $e_a a = a$.

If for arbitrary elements a, x, $y \in S$, the equality $ax = ay$ implies $x = y$, then S is called a semigroup with left cancellation.

1.4. We will use throughout this article the following notation. If A, B are two subsets of a set S, then $A \subseteq B$ means that A is a subset of the set B; $A \subset B$ means that $A \subseteq B$ and $A \neq B$. $A \cap B$ and $A \cup B$ are respectively the intersection and the union of the sets A and B; $A \backslash B$ is the difference between the sets A and B. The void set will be denoted by $\varnothing$.

1.5. Let S be a certain semigroup of mappings of a set E into itself, $f \in S$, and A subset of E. The symbol fA will denote the set of all points $f(\xi) = f\xi$, where $\xi \in A$.

1.6. Let α, β be any two assertions. The symbol $\alpha \rightarrow \beta$ means: "if α holds, then β holds"; the symbol $\alpha \longleftrightarrow \beta$ means $\alpha \rightarrow \beta$ and $\beta \rightarrow \alpha$.

1.7. A nonvoid subset D of a semigroup S is called a subsemigroup of S, if $DD \subseteq D$. A subsemigroup D of a semigroup S is called a subgroup of S, if it is a group with respect to the operation of S.

A nonvoid subset N of elements of a semigroup S is called a normal subsemigroup [1], if for arbitrary a, $b \in S$ and $n \in N$ the following relations hold:

$$an \in N \longleftrightarrow a \in N,$$

$$nb \in N \longleftrightarrow b \in N,$$

$$anb \in N \longleftrightarrow ab \in N.$$

Normal subsemigroups of a semigroup S, and only these, are the inverse images of the identity of the semigroup ϕS under homomorphisms ϕ of the semigroup S.

1.8. Let a be an element of the semigroup S. If there exist natural numbers k, n ($k \neq n$) such that $a^k = a^n$, then a is called an element of finite order. In the opposite case a is called an element of infinite order.

Special elements of finite order of a semigroup S are its idempotents (if they exist). These are elements f such that $f^2 = f$. If $a \in S$ is an element of finite order, then there is an idempotent among the powers of the element a (see for example [2], [3]).

1.9. Throughout the remainder of this paper, S will denote the semigroup of all continuous strictly monotone functions $f(\xi) = f\xi$, defined on the closed interval $E = [0, 1]$, such that $0 \le f\xi \le 1$. We denote by e the unit of this semigroup, that is, the function $e \in S$ such that $e\xi = \xi$ for all $\xi \in E$.

Every function $f(\xi)$ obviously induces a homeomorphic mapping of the interval E into itself. Thus S can be considered as the semigroup of all homeomorphic mappings of the interval E into itself.

§ 2. Divisibility, ideals

2.1. Theorem. *Let a, b be arbitrary elements of S. In order that the equation*

$$ax = b \tag{1}$$

have a solution in S, it is necessary and sufficient that the inclusion $bE \subseteq aE$ hold. The equation (1) possesses at most one solution in S.

Proof. If there is a solution x of the equation (1) in S, then $xE \subseteq E$, $bE = (ax) E = a(xE) \subseteq aE$.

Conversely, suppose that $bE \subseteq aE$. Every point $\eta \in bE$ can be represented in the form $\eta = b\xi$, $\xi \in E$ (see 1.6). The function a^{-1} is continuous and strictly monotone, and it maps the interval aE onto E. The inclusion $bE \subseteq aE$ implies that the function $x = a^{-1}b$ is defined throughout E, and is strictly monotone and continuous. It maps E onto the interval $xE = a^{-1}(bE) \subseteq E$. Consequently $x \in S$.

Assuming the existence of two solutions x_1, x_2 of the equation (1), we obtain from $ax_1, = ax_2 = b$ the relation $x_1 = x_2 = a^{-1}b$.

2.1.1. Let G denote the set of all functions $g \in S$ such that $gE = E$. It is plain that G is a group.

Let a, b be arbitrary elements of S. In order for the solution x of the equation (1) to be contained in the group G, it is necessary and sufficient that $aE = bE$.

Proof. Suppose that $ax = b$ and that $x \in G$. Then from Theorem 2.1 we have $bE \subseteq aE$. But G is a group, and the function x^{-1} inverse to x also lies in G. From $a = bx^{-1}$ and from Theorem 2.1 we infer that $aE \subseteq bE$, so that $aE = bE$.

Conversely, suppose that $aE = bE$. From Theorem 2.1 we infer the existence in S of an element $x = a^{-1}b$ such that $ax = b$. If $xE = a^{-1}bE = E_1 \subset E$, then we would have: $bE = axE = a(a^{-1}b) E = aE_1 \subset aE$, and this contradicts the equality $bE = a(a^{-1}b) E = aE$. Hence the function $x = a^{-1}b$ maps E onto itself, that is, $x \in G$.

2.1.2. *S is a semigroup with left cancellation.*

This assertion, which incidentally is valid for an arbitrary semigroup of one-to-one mappings, follows from Theorem 2.1.

2.1.3. *Let a, b, c be arbitrary elements of S. If $caE = cbE$, then $aE = bE$.*

In fact, if $caE = cbE$, 2.1.1 implies that $ca = cbg$, where $g \in G$. Then 2.1.2 implies that $a = bg$, and again from 2.1.1 we have $aE = bE$.

2.2. Let F denote the boundary of the interval E, that is, $F = \{0, 1\}$. If a is an arbitrary element of S, then by F_a we will denote the set $a^{-1}(F \cap aE)$.

Theorem. *Let a, b be arbitrary elements of S. In order for the equation*

$$xa = b \tag{2}$$

to have a solution x, it is necessary and sufficient that $F_a \supseteq F_b$.

Proof. Suppose that the element $x \in S$ is a solution of the equation (2) for certain a, $b \in S$ and that ξ is a point in F_b. By definition, we have $\eta = b\xi \in F$. On the other hand, we also have $\eta = b\xi = xa\xi \in xE$. Since η is a point of the boundary of the interval E, it must also be a point of the boundary of the interval xE. But the function x^{-1} is monotone, and hence $x^{-1}\eta = a\xi$ is a point of the boundary of the interval E. The definition of F_a implies that $\xi \in F_a$. Thus $F_b \subseteq F_a$.

Conversely, suppose that $F_b \subseteq F_a$. The function ba^{-1} is monotone and maps the interval $aE = [\alpha, \beta]$ continuously onto bE. If $\alpha \neq 0$, then by definition, $a^{-1}\alpha \,\bar{\in}\, F_a$. It follows from the condition that $a^{-1}\alpha \,\bar{\in}\, F_b$, which means that $ba^{-1}\alpha \,\bar{\in}\, F$. Hence the mappings ba^{-1} of the interval aE onto bE can be extended in an infinite number of ways to a monotone and continuous function f that maps the interval $[0, \beta]$ onto a certain interval containing bE in such a way that $f\xi = ba^{-1}\xi$ for an arbitrary point $\xi \in aE$. If $\beta = 1$, then $f \in S$ and f is the desired solution of the equation (2). If $\beta \neq 1$, then, repeating the previous reasoning, we can extend f to a function $x \in S$ that maps the interval E continuously and strictly monotonely onto a certain interval xE ($bE \subset xE$) and that satisfies the equation (2).

2.2.1. *If $a \in G$, then for every $b \in S$, there exists one and only one solution in S of the equation (2).*

In fact, we then have $aE = E$, $a^{-1}E = E$, $ba^{-1} \in S$. Conversely, if $xa = b$, then $x = ba^{-1}$.

2.2.2. *If a, $b \in S$, $a \,\bar{\in}\, G$, and there exists at least one solution in S of the equation (2), say x_0, then there are an infinite number of solutions in S of this equation.* This assertion follows from the proof of 2.2.

2.3. *If a is an arbitrary element in S, not lying in G, then there are an infinite number of left units of the element a in S (see 1.3).*

This assertion follows from 2.2 and 2.2.2. It is also evident at once that the left units of the element a are exactly the functions $e_a \in S$ such that $e_a\xi = \xi$ for an arbitrary point $\xi \in aE$.

2.3.1. Theorem. *There are no elements of finite order in $S \backslash G$, and in particular, there are no idempotents.*

Proof. Suppose that $a \in S$, $a^2 = a$ and $aE = [\alpha, \beta]$. From 2.3 and the inclusion $\alpha, \beta \in aE$, we have $a\alpha = \alpha$, $a\beta = \beta$. Thus a is a monotone increasing function. Assuming, for example, that $\alpha \neq 0$, we would have $a\xi < a\alpha = \alpha$ for an arbitrary point $\xi (0 < \xi < \alpha)$ and $a\xi \bar{\in} aE$. Consequently $\alpha = 0$ and in the same way $\beta = 1$. Then $aE = E$, $a \in G$, and the only idempotent of the group G is its unit e.

If $f \in S$ is an element of finite order, then among the powers of the element f there is an idempotent (see 1.8). We have just shown that this idempotent is e. Suppose that $f^n = e$. It follows from Theorem 2.1 that $E = eE \subseteq fE$, that is, $f \in G$.

2.3.2. The assertion 2.3 can be strengthened in the following way:

If a, b are arbitrary elements from S and $bE \not\subseteq aE$, then among the left units of the element a, there are an infinite number that are not left units of the element b.

Proof. If $bE \not\subseteq aE$, then there is a point $\eta \in bE \backslash aE$. Among the left units e_a of the element a we can obviously choose an infinite number such that $e_a \eta \neq \eta$; then $e_a b \neq b$.

2.3.3. Assertion 2.3 implies that the right cancellation law, unlike 2.1.2, fails in S. We can, nonetheless, state the following modification of this law:

If the equality $ax = bx$ holds for certain $a, b \in S$ and every function $x \in S$ for which $F_x \neq 0$, then we also have $a = b$.

Proof. Let ξ be an arbitrary point in E, and $\{x_k\}_{k=1}^{\infty}$ a sequence of functions in S such that $F_{x_k} \neq \varnothing$ and $\bigcap_{k=1}^{\infty} x_k E$ contains only the point ξ. From the equality $ax_k = bx_k$ for all k, we infer that

$$a\xi = \bigcap_{k=1}^{\infty} a(x_k E) = \bigcap_{k=1}^{\infty} (ax_k) E = \bigcap_{k=1}^{\infty} (bx_k) E = \bigcap_{k=1}^{\infty} b(x_k E) = b\xi,$$

that is, $a = b$.

2.3.4. *If a, b are arbitrary elements of $S \backslash G$, e_a and e_b are arbitrary left units of the elements a and b, and x_0 is a solution of equation (2), then each of the elements $x_0 e_a$, $e_b x_0$ is a solution of the equation (2).*

This assertion is obvious. Using assertion 2.3.2, one can show that there are an infinite number of distinct elements among the solutions $x_0 e_a$, $e_b x_0$.

2.4. Let $\mu(a)$ denote the number of elements of the set F_a (see 2.2). By definition, $\mu(a) \leq 2$.

Theorem. *Let a, b be any elements of S. The equation $xay = b$ has a solu-*

tion $x, y \in S$ *if and only if* $\mu(a) \geq \mu(b)$.

Proof. Let $a, b, x, y \in S$ and suppose that $xay = b$. We set $xa = c$; then $cy = b$. Theorem 2.2 implies that $F_c \subseteq F_a$, and hence $\mu(c) \leq \mu(a)$. Theorem 2.1 implies that $bE \subseteq cE$, from which we infer that $\mu(b) \leq \mu(c)$. In fact, if the function b, for example, carries one of the points of F again into a point of the set F, then from $bE \subseteq cE$ we infer that the function c also carries at least one of the points of F again into a point of the boundary of the interval E.

Conversely, let $a, b \in S$ and $\mu(a) \geq \mu(b)$. Then either $F_b \subseteq F_a$ or $iF_b \subseteq F_a$, where i is the reflection of E onto itself about its center. In both cases, there exists a function $y \in S$ such that

$$yF_b \subseteq F_a. \tag{3}$$

Setting $ay = c$ and multiplying (3) on the left by c, we obtain: $aF_b \subseteq aF_a \subseteq F$, and consequently $F_b \subseteq c^{-1}(F \cap cE) = F_c$. Theorem 2.2 shows that there is a solution x in S of the equation $xc = b$. From this and the fact that $ay = c$, we have: $xay = b$.

2.5. Let $[\alpha, \beta]$ be any interval contained in $[0, 1]$. Let $R[\alpha, \beta]$ denote the set of all $x \in S$ for which $xE \subseteq [\alpha, \beta]$; let K be the set of all $x \in S$ for which $F \cap xE = \varnothing$; and let $J = S \backslash G$.

There are four different sets F_a (see 2.2): $\varnothing, \{0\}, \{1\}, \{0 \cup 1\}$. Hence there are only four principal left ideals in S, corresponding to the sets F_a of the generators a.

If $F_a = \{0 \cup 1\}$, then $a \cup Sa = S$. If $F_a = \varnothing$, then $a \cup Sa = K$. The union of the two remaining principal left ideals of the semigroup S (generated by elements $a \in S$ for which $F_a = 0$ or $F_a = 1$) forms a fifth left ideal J of the semigroup S. Since every left ideal of the semigroup S is the union of its own principal left ideals, and K is contained in all principal left ideals, there are no other left ideals in the semigroup S.

2.5.1. It follows from 2.4 that the only principal two-sided ideals of the semigroup S are S, J and K. All elements $g \in G$ are generators of the ideal S; of the ideal J, all elements $a \in S$ for which $\mu(a) = 1$; of the ideal K, all elements $a \in S$ for which $\mu(a) = 0$, that is, all elements of K.

Every two-sided ideal of the semigroup S is the union of its principal two-sided ideals. Since the set of all principal two-sided ideals of the semigroup S is linearly ordered by inclusion, all two-sided ideals of S are principal, that is, S contains no ideals different from S, J and K. K is the minimal left and minimal two-sided ideal of the semigroup S, and consequently K is a semigroup with left division (see 1.2 and 1.3).

2.6. It follows from Thoerem 2.1 that the set $R\,[\alpha,\,\beta]$ is a principal right ideal of the semigroup S, generated by an arbitrary element $a \in S$ such that $aE = [\alpha,\,\beta]$. There are no other principal right ideals in the semigroup S.

The set of all principal right ideals of the semigroup S, completed by the void set, forms a lattice under inclusion isomorphic to the lattice of all intervals contained in the interval E. In fact, if $[\alpha,\,\beta]$ and $[\gamma,\,\delta]$ are two intervals contained in E, then $R\,[\min\{\alpha,\,\gamma\},\,\max\{\beta,\,\delta\}]$ is the minimal principal right ideal of the semigroup S containing $R\,[\alpha,\,\beta]$ and $R\,[\gamma,\,\delta]$. If $[\alpha,\,\beta]\cap[\gamma,\,\delta]=\varnothing$, then $R\,[\alpha,\,\beta]\cap R\,[\gamma,\,\delta]=\varnothing$.

It is easy to verify that this lattice is complete [4], and that the set of all principal left ideals of the semigroup S also forms a lattice.

2.6.1. From 2.6 we infer:

In order that there should exist in J a common right divisor of elements of a set $B \subset S$, it is necessary and sufficient that $\bigcup_{b\in B} bE \neq E$.

§3. The semigroup D

3.1. Let D be the subset of S consisting of all monotone increasing functions $f \in S$. It is evident that D is a subsemigroup of the semigroup S. Furthermore, we have

Theorem. D *is a normal subsemigroup of the semigroup* S.

Proof. Let $n \in D$. The superposition of n and an arbitrary function $a \in S$ is a monotone increasing function if and only if $a \in D$. If $a,\,b$ are arbitrary functions in S, and $n \in D$, then the functions ab and anb are both increasing or both decreasing. By definition (1.7), D is a normal subsemigroup of the semigroup S.

3.2. Let $A = K \cap D$ (see 2.5); let B_0 be the set of all $f \in D$ such that $f(0) = 0$ and $f(1) < 1$; let B_1 be the set of all $f \in D$ such that $F(0) > 0$, $f(1) = 1$; let C_0 (and C_1, respectively) be the set of all $f \in B_0$ (or B_1) leaving invariant all points of some interval containing the point 0 (or 1). The sets $A,\,B_0,\,B_1,\,C_0,\,C_1,\,A \cup B_0,\,A \cup B_1$ are all obviously subsemigroups of the semigroup D.

The properties of the semigroup S, set forth in 2.1 – 2.3.2 and in 2.5 – 2.6.1 all hold for the semigroup D, as is easy to verify.

3.3. It follows from 2.5 that D contains four principal left ideals: $A,\,A \cup B_0,\,A \cup B_1$, and D. All elements of the ideal A are generators of A, and consequently, A is a semigroup *with left division* (see 1.3); $B_0\,(B_1)$ is the set of all generators of the ideal $A \cup B_0\,(A \cup B_1)$; $G \cap D$ is the set of generators of the ideal D.

All four of these ideals are principal two-sided ideals of the semigroup D. Furthermore, D contains also one two-sided ideal, $D \cap J = (A \cup B_0) \cup (A \cup B_1)$. A is a minimal left and minimal two-sided ideal of D. There are no other left or two-sided ideals in the semigroup D (see 2.5 and 2.5.1).

3.4. B_0 and C_0 are semigroups with left division.

Proof. Let a, b be arbitrary elements of B_0. It follows from 2.2 and 2.2.2 that there are an uncountable number of solutions of the equation (2) in S. By Theorem 3.1, all of these solutions lie in D. Since $a(0) = b(0) = 0$, we have $x(0) = 0$. We can always choose the function x in the proof of 2.2 so that $xE = [0, \gamma]$, where $\gamma < 1$, and then $x \in B_0$.

If a, $b \in C_0$ in this situation, then there is a point λ $(0 < \lambda < 1)$ such that $a\xi = b\xi = \xi$ for an arbitrary point in the interval $[0, \lambda]$. But then we also have $x\xi = \xi$ for $0 \le \xi \le \lambda$, and $x \in C_0$.

3.5. C_0 is a minimal normal subsemigroup of the semigroup B_0.

Proof. A left unit of the element $b \in B_0$ is any element $c \in D$ such that $c\xi = \xi$ for $0 \le \xi \le b$ (1). The subsemigroup $C_0 \subset B_0$ contains all left units belonging to B_0 of all elements $b \in B_0$. Every element $c \in C_0$ is a left unit of every element $b \in B_0$. Every solution lying in B_0 of the equation $xc_1 = c_2$, where c_1, $c_2 \in C_0$, is contained in C_0 (see 3.4). Since B_0 is a semigroup with left division, the main theorem of article [5] shows that C_0 is a minimal normal subsemigroup of the semigroup B_0.

3.6. A homomorphism ϕ of the semigroup S is called trivial if it is an isomorphism or a homomorphism onto the semigroup consisting of a single element. A semigroup is called simple, if it has no nontrivial homomorphisms. E. S. Ljapin [6] and the author [7] have found all commutative simple semigroups and have considered simple semigroups with zero. Up to now, no example has been given in the literature of a simple semigroup not a group containing an element of infinite order. Below (3.7 and 3.8) we shall exhibit two nonisomorphic simple subsemigroups of the semigroup S (all of their elements, in view of Theorem 2.3.1, have infinite order). *

3.7. As above (3.2), let A be the subsemigroup of D consisting of all functions $f \in D$ for which $f(0) > 0$, $f(1) < 1$.

Theorem. *The semigroup A is simple.*

Proof. Let ϕ be an arbitrary homomorphism of the semigroup A that is not an isomorphism. Then there exist elements a, $b \in A$ such that $a \ne b$ and $\phi a = \phi b$. There is at least one point $\xi \in E$ for which $a\xi \ne b\xi$.

There are disjoint intervals E_1 and E_2 such that the point $a\xi$ is in the interior of E_1 and the point $b\xi$ is in the interior of E_2. Then the point ξ is in the

---

*An example of a simple semigroup with left division, all elements of which have infinite order, has been found by Cohn (Proc. London Math. Soc. (3) 8 (1958), 466–480).

interior of both $a^{-1}E_1$ and $b^{-1}E_2$, that is, in the interior of $a^{-1}E_1 \cap b^{-1}E_2$. Consequently there is an interval $E_3 = [\alpha, \beta] \subseteq a^{-1}E_1 \cap b^{-1}E_2$ $(\alpha > 0, \beta < 1)$, which contains ξ as an interior point. Let c be an arbitrary function in A for which $cE = E_3$. Then acE and bcE have no points in common, since $acE = aE_3 \subseteq a(a^{-1}E_1) = E_1$ and similarly $bcE \subseteq E_2$. Suppose that $acE = [\alpha_1, \beta_1]$, $bcE = [\alpha_2, \beta_2]$, and for example $\alpha_2 > \beta_1$.

Now let d be any function in A such that $dE = [\alpha_3, \beta_3]$, $\alpha_3 > \beta_1$. The function $d(bc)^{-1}$ is monotone increasing and maps the interval bcE onto dE. Plainly we can construct an infinite set of functions $f \in A$ satisfying the following conditions:

a) All points of the intervals acE and dE are interior points of fE.

b) $f\xi = \xi$ for every point $\xi \in acE$.

c) $f\xi = d(bc)^{-1}\xi$ for every point $\xi \in bcE$.

From the very construction of the function f we have $fac = ac$, $fbc = d$. But then, since $\phi a = \phi b$, we have

$$\phi d = \phi(fbc) = \phi f \cdot \phi b \cdot \phi c = \phi f \cdot \phi a \cdot \phi c = \phi(fac) = \phi(ac).$$

Let t be any function in the semigroup A. One can always find a function $d \in A$ such that $dE = [\alpha_3, \beta_3]$, $\alpha_3 > \max\{\beta_1, \beta_4\}$, where $[\alpha_4, \beta_4] = tE$. From what has just been proved, we have $\phi d = \phi(ac)$. In just the same way we prove that $\phi d = \phi t$. Thus, if ϕ is not an isomorphism, it follows that ϕ maps all elements of the semigroup A into one and the same element $\phi(ac)$. The proof is completed.

3.8. Theorem. *The semigroup C_0 is simple.*

Proof. Let ϕ be an arbitrary homomorphism of the semigroup C_0 that is not an isomorphism. There are a pair of functions $a, b \in C_0$ such that $a \neq b$, $\phi a = \phi b$. Let $\xi_1 \in E$ be a point such that $a\xi_1 \neq b\xi_1$, and suppose for example that $a\xi_1 > b\xi_1$. Then, in view of continuity, there exists a neighborhood (α_1, β_1) of the points ξ_1 such that $a\xi > b\xi$ for $\alpha_1 < \xi < \beta_1$.

Let $c(\xi)$ be a function in C_0 such that $c(1) = \gamma$, where $\xi_1 \leq \gamma < \beta_1$. We introduce some notation: $a_1(\xi) = ac(\xi)$, $b_1(\xi) = bc(\xi)$, $b_1(1) = \gamma_1$. Then $a_1(1) = a[c(1)] = a(\gamma) > b(\gamma) = b[c(1)] = b_1(1) = \gamma_1$. In view of the continuity of the function a_1, there exists a point δ $(0 < \delta < 1)$ such that $a_1(\delta) = \gamma_1$. Let c_1 be a function in C_0 such that $c_1(\xi) = a_1^{-1}(\xi)$ for $0 \leq \xi \leq \gamma_1$. Suppose also that $a_2(\xi) = c_1 a_1(\xi)$, $b_2(\xi) = c_1 b_1(\xi)$. Since $b_1, c_1, a_1 \in C_0$, it follows that $a_2, b_2 \in C_0$. At the same time, for all $\xi \leq \delta$, we have $a_2(\xi) = a_1^{-1}[a_1(\xi)] = \xi$, and for all $\xi \leq 1$, in view of the monotonicity of b_1, we infer that

$$b_2(\xi) = c_1[b_1(\xi)] \leq c_1[b_1(1)] = c_1(\gamma_1) = a_1^{-1}(\gamma_1) = \delta,$$

and hence for all $\xi \in E$,

$$a_2 [b_2 (\xi)] = b_2 (\xi), \quad a_2 b_2 = b_2.$$

On the other hand, since $\phi a = \phi b$, we have

$$\phi a_2 = \phi (c_1 ac) = \phi c_1 \cdot \phi a \cdot \phi c = \phi c_1 \cdot \phi b \cdot \phi c = \phi(c_1 bc) = \phi b_2,$$

$$(\phi b_2)^2 = \phi a_2 \cdot \phi b_2 = \phi(a_2 b_2) = \phi b_2,$$

and the semigroup ϕC_0 contains the idempotent ϕb_2.

Let t be an arbitrary element of the semigroup C_0. In view of 3.4, C_0 contains elements u, v such that $u b_2 = t$, $vt = b_2$. But then:

$$\phi t \cdot \phi b_2 = (\phi u \cdot \phi b_2) \cdot \phi b_2 = \phi u \cdot \phi b_2 = \phi t, \quad \phi v \cdot \phi t = \phi b_2.$$

Thus the idempotent ϕb_2 is a right unit for all elements of the semigroup ϕC_0, and for every element of ϕC_0 there exists a left inverse element with respect to ϕb_2. As Clifford has shown [8], such a semigroup is the set-theoretic union of isomorphic disjoint groups G_i:

$$\phi C_0 = \bigcup_i G_i, \quad G_i G_j = G_i. \tag{4}$$

Assume that the semigroup (4) contains two distinct subgroups G_1 and G_2, and suppose that $\phi a_1 \in G_1$, $\phi a_2 \in G_2$. We set $\mu = \max \{a_1 (1), a_2 (1)\}$. From the fact that a_1, $a_2 \in C_0$, we infer that $\mu < 1$. There exists a function $b \in C_0$ such that $b\xi = \xi$ for $\xi \leq \mu$. Then we have

$$b a_1 = a_1, \quad b a_2 = a_2. \tag{5}$$

If $\phi b \in G_i$, then it follows from (4) and (5) that ϕa_1, $\phi a_2 \in G_i$. This contradicts the assumption that $\phi a_1, \phi a_2$ lie in distinct groups G_i. Consequently ϕC_0 contains only one subgroup G_i, that is, it is a group. The unit of this group will be the idempotent ϕb_2. But for every element $a \in C_0$, there exists an element $x \in C_0$ such that $ax = x$. In fact, there is an $\epsilon < 1$ such that $a\xi = \xi$ for $\xi \leq \epsilon$. Then an arbitrary function $x \in C_0$ for which $x(1) \leq \epsilon$ satisfies the condition pointed out above. Therefore every element ϕa of the group ϕC_0 is also a left unit for an arbitrary element $\phi x \in \phi C_0$. It is clear that such a group consists only of the unit element ϕb_2, and the homomorphism ϕ is trivial. Thus the semigroup C_0 is simple.

3.9. We introduce the following order relation in D: $a \geq b$ if $a\xi \geq b\xi$ for all $\xi \in E$. As is easy to show, D under this order relation is a lattice-ordered semigroup (l-semigroup) [4]. The group $H = G \cap D$ was studied in the works [9], [10], as well as others. This is an important example of an l-group, a lattice-ordered group (ordered by the relation $\geq$ just introduced). We note the following obvious properties of the ordering in the semigroup D:

a) the semigroups H, B_0, C_0, A, $A \cup B_0$, $H \cup B_0$ are convex subsets [4] of the semigroup D.

b) If $d \in D$, $b \in B_0$, and $d \leq b$, then $d \in B_0$.

3.10. The results of the present section give the possibility of computing explicitly all homomorphisms of the semigroups S and D. The set of these homomorphisms is finite.

§4. Automorphisms of the semigroup S

In the present section we shall consider the automorphism groups of the semigroup S and certain of its subsemigroups.

4.1. Let $f \in H$, and let x be an arbitrary element of B_0. The mapping

$$\psi x = f x f^{-1} \tag{6}$$

of each of the semigroups B_0 and C_0 onto itself is obviously an automorphism.

Theorem. *Every automorphism of the semigroups B_0 and C_0 has the form* (6).

Proof. Let ψ be any automorphism of the semigroup B_0, and let x and y be arbitrary elements of B_0 satisfying the condition $x(1) \leq y(1) = \eta$. Suppose that $\psi x(1) > \psi y(1) = \eta'$, and let a' be a function in B_0 such that $a' \xi = \xi$ for all $\xi \in [0, \eta']$ and $a' \xi \neq \xi$ for $\xi \bar{\in} [0, \eta']$. It is then obvious that $a' \cdot \psi y = \psi y$, $a' \cdot \psi x \neq \psi x$. We set $a' = \psi a$ and obtain

$$a y = y, \quad a x \neq x.$$

The first condition implies that $a \xi = \xi$ for all $\xi \in [0, \eta]$, that is, that $a x = x$, and this contradicts the second condition. Thus if $x(1) \leq y(1)$, it follows that $\psi x(1) \leq \psi y(1)$. In particular, if $x(1) = y(1)$, we have $\psi x(1) = \psi y(1)$.

If x is an arbitrary element of B_0, $x(1) = \eta$ $(0 < \eta < 1)$, then we introduce the notation

$$\psi x (1) = f (\eta).$$

It follows from our reasoning that the mapping $f(\eta)$ of the open interval $(0, 1)$ onto itself does not depend upon the choice of the element $x \in B_0$ satisfying the condition $x(1) = \eta$ for the given value of η. Since ψ is an automorphism, f is a one-to-one mapping of the open interval $(0, 1)$ onto itself. The function $f(\eta)$ is monotone increasing and has an inverse function $f^{-1}(\eta)$; consequently it is continuous.

Suppose now that x is an arbitrary element of B_0, and $\eta < 1$ an arbitrary element of E; let y be one of the elements of the semigroup B_0 for which $\eta = y(1)$. Since ψ is an automorphism of B_0, we have, upon introducing the notations $xy = z$, $\psi x = x'$, $\psi y = y'$, $\psi z = z'$, the following for all $\xi \in E$: $z'(\xi) = x'y'(\xi)$. In particular, $z'(1) = x'y'(1)$. But, by definition, $y'(1) = \psi y(1) = f(\eta)$, $z'(1) =$

$\psi z(1) = fz(1) = fxy(1) = fx(\eta)$, and for an arbitrary point $\eta \in (0,1)$, we have

$$fx(\eta) = \psi x(f\eta) \tag{7}$$

or, if we write $f\eta = \xi$,

$$\psi x(\xi) = fxf^{-1}(\xi), \tag{8}$$

where once again ξ is an arbitrary interior point of E.

The function f, defined on the open interval $(0, 1)$, admits a unique extension to a monotone increasing continuous function, defined on the closed interval $[0, 1]$: $f(0) = 0$, $f(1) = 1$. At the same time, $x(0) = \psi x(0) = 0$ for any element $x \in B_0$. Thus the relation (7) is valid for $\eta = 0$, and therefore the relation (8) is valid for $\xi = 0$. If however $\xi = 1$, then $\eta = f^{-1}(1) = 1$. Then (7), and therefore also (8), holds by the very definition of the function f: $fx(1) = \psi x(1)$. Consequently (8) holds for every point $\xi \in E$.

The assertion concerning automorphisms of the semigroup C_0 is proved similarly.

4.2. Theorem. *Every automorphism of the semigroup D has the form (6), where $f \in G$.*

Proof. Let ψ be an arbitrary automorphism of the semigroup D. As was noted in 3.3, D contains four principal left ideals with sets of generators A, B_0, B, B_1, and $H = G \cap D$. Of these only H is a group. A is the minimal ideal of the semigroup D. Hence two cases are possible:

$$\psi A = A, \quad \psi H = H, \quad \psi B_0 = B_0, \quad \psi B_1 = B_1 \tag{9}$$

or

$$\psi A = A, \quad \psi H = H, \quad \psi B_0 = B_1, \quad \psi B_1 = B_0. \tag{10}$$

Let $i(\xi) = 1 - \xi$. In case (10), we set $i^{-1} \cdot \psi x \cdot i = \psi_1 x$ for an arbitrary $x \in B_0$. In case (9), we set $\psi_1 = \psi$. In both cases ψ_1 is an automorphism of the semigroup D and $\psi_1 B_0 = B_0$.

It follows from Theorem 4.1 that the automorphism of the semigroup B_0 induced by the mapping ψ_1 has the form $\psi_1 b = fbf^{-1}$, where $b \in B_0$, $f \in H$. The mapping $\psi_2 x = f_1^{-1} \cdot \psi_1 x \cdot f_1$ $(x \in D)$ is also an automorphism of the semigroup D, and $\psi_2 b = b$ for every $b \in B_0$.

Now let a be an arbitrary element of A. We choose an element $b \in B_0$ such that $b\xi = \xi$ for all $\xi \in aE$ and $b\xi \neq \xi$ for every point $\xi \bar{\in} aE$ $(\xi \neq 0)$. Then we have: $ba = a$, $\psi_2(ba) = \psi_2(a)$, $b \cdot \psi_2 a = \psi_2 a$. It follows from the choice of the element b that $(\psi_2 a) E \subseteq aE$. Considering the automorphism ψ_2^{-1}, we obtain: $aE \subseteq (\psi_2 a) E$, and consequently $(\psi_2 a) E = aE$ for an arbitrary element $a \in A$.

Now let ξ be any interior point of E and $\{a_k\}_{k=1}^{\infty}$ a sequence of elements

$a_k \in A$ such that $\lim\limits_{k \to \infty} a_k E = \xi$. Then in view of the continuity of the mappings a

and $\psi_2 a$, $a\xi$ is the unique common point of all the intervals $aa_k E$, and $(\psi_2 a)\,\xi$ is
the unique common point of all of the intervals $(\psi_2 a)\,a_k E = \psi_2 a(\psi_2 a_k E) =$
$\psi_2\,(aa_k)\,E$. Consequently we have

$$(\psi_2 a)\;\xi = \lim_{k \to \infty}\,[\psi_2\,(aa_k)\,E] = \lim_{k \to \infty}\,[(aa_k)\,E] = a\,\xi,$$

that is, $\psi_2 a = a$ for an arbitrary element $a \in A$.

Since $ad \in A$ for arbitrary elements $a \in A$, $d \in D$, we have $ad = \psi_2(ad) =$
$a \cdot \psi_2 d$ and from 2.1.2 we have: $\psi_2 d = d$. Thus ψ_2 is the identity mapping of the
semigroup D onto itself. Consequently $\psi_1 s = fxf^{-1}$ for an arbitrary $x \in D$ ($f \in H$).
If $\psi_1 = \psi$, then this gives the relation (6). In case (10), we have $\psi_1 x = i^{-1} \cdot \psi x \cdot i =$
fxf^{-1}; from this it follows that $\psi x = (if)\,x\,(if)^{-1}$, or, if we write $if = f_1\,(f_1 \in G)$,
we obtain: $\psi x = f_1 x f_1^{-1}$, and once again we obtain the relation (6).

4.3. Theorem. *Every automorphism of the semigroup S has the form* (6),
where $f \in G$.

Proof. Let ψ be an arbitrary automorphism of the semigroup S. K is the unique
minimal ideal of S, and hence $\psi K = K$. The semigroup A contains all left units
belonging to K of an arbitrary element $k \in K$. As in 3.5, one can show that A is
the minimal normal subsemigroup of K, and hence $\psi A = A$. For elements $d \in D$
(and only for these), we have $Ad \subseteq A$. It follows from this that $\psi D = D$. ψ induces
an automorphism of the semigroup D, which in view of Theorem 4.2 has the form:
$\psi d = fdf^{-1}$ ($d \in D$, $f \in G$). Then the mapping $\psi_1 x = f^{-1} \cdot \psi x \cdot f$ is also an automorphism of the semigroup S for which $\psi_1 d = d$ for an arbitrary element $d \in D$.

G is the unique maximal subgroup of S. Hence $\psi_1 G = G$, and ψ_1 induces an
automorphism of the group G. G is a group without center, and every automorphism
of G is inner [10]. Since $\psi_1 h = h$ for an arbitrary element $h \in H$, ψ is the identity
automorphism of the group G and $\psi_1 i = i$, where $i(\xi) = 1 - \xi$ is the reflection of
the interval E about its center. Every monotone decreasing function in S can be
represented in the form $x = di$, where $d \in D$. From the relations $\psi_1 d = d$ and
$\psi_1 i = i$ it follows that $\psi_1 x = x$ for an arbitrary element $x \in S$, and this means:
$\psi x = fxf^{-1}$, which we wished to prove.

4.3.1. From the considerations of 4.1 – 4.3, it follows that:

*Every automorphism of each of the semigroups D, B_0, C_0 can be uniquely ex-
tended to an automorphism of the semigroup S.*

§5. A topology in S

5.1. We introduce in the usual way the distance between two elements of S:

$$\rho(x, y) = \max_{\xi \in E}\,|\,x(\xi) - y(\xi)\,|\quad (x, y \in S).$$

Thus S becomes a bounded metric space. The semigroup S is a topological semigroup with respect to the topology induced by the metric, or, as one says, a metric semigroup. This means that for every $\epsilon > 0$, there is a $\delta > 0$ such that if $\rho(x, x') < \delta$, $\rho(y, y') < \delta$, then $\rho(xy, x'y') < \epsilon$.

5.2. Theorem. *Every principal right ideal of the semigroup S is a closed subset of S.* [*]

Proof. Let a be an arbitrary element of S, $aE = [\alpha, \beta]$. The ideal aS is the set of all functions $x \in S$ such that $xE \subseteq [\alpha, \beta]$ (see 2.6). Let c be an arbitrary element of $\overline{aS}$ ($\overline{X}$ is the closure of the set X). This means that for every $\epsilon > 0$ there is an element $ax \in aS$ such that $\rho(c, ax) < \epsilon$. In particular, denoting by $\eta = c\xi$ an arbitrary point of cE, we have: $|c\xi - ax\xi| = |\eta - ax\xi| < \epsilon$. But $ax\xi \in aE$, and consequently $\eta \in \overline{aE} = aE$. Thus $cE \subseteq aE$, and in view of Theorem 2.1, $c \in aS$. This means that $\overline{aS} = aS$.

5.3. Theorem. *The ideal K is dense in S. Every point of K is an interior point of K.*

Proof. Let k be an arbitrary element of K (see 2.5). By definition, $F \cap kE$ is void. Every point $\xi \in kE$ is interior to E. Hence there is an $\epsilon > 0$ such that every point of the ϵ-neighborhood of the point ξ is interior to E. This means that for $\rho(k, x) < \epsilon$, the interval xE consists solely of interior points of the interval E and $x \in K$. Thus the ϵ-neighborhood of the element k is contained in K and k is an interior point of K.

Now let a be an arbitrary element of S, and $\epsilon > 0$ an arbitrarily small number, $x_\epsilon(\xi) = (1 - \epsilon)\,\xi + \epsilon/2$. Every point of $x_\epsilon E$ is interior to E, $x_\epsilon \in K$; then we also have $x_\epsilon a \in K$. Furthermore,

$$\rho(a,\ x_\epsilon a) = \max_{\xi \in E} |a\xi - x_\epsilon a\xi| \leqslant \max_{\eta \in E} |\eta - x_\epsilon \eta| < \epsilon.$$

Thus every ϵ-neighborhood of an arbitrary point $a \in S$ has nonvoid intersection with K, $\overline{K} = S$. The theorem is proved.

5.3.1. *Let aS be an arbitrary principal ideal of the semigroup S. The semigroup aK is dense in aS. All points of the semigroup aK are interior points of aK.*

Proof. It follows from 2.1.2 and 5.1 that the mapping $x \to ax$ is a homeomorphism of S onto aS. Under a homeomorphism an interior point goes into an interior point and a dense subset into a dense subset.

5.4. Let T be a topological semigroup and B a certain subset of element of T. If the subsemigroup $\{B\} \subseteq T$, generated by all of the elements $b_i \in B$ is dense in T, then B is said to be a basis of the semigroup S. If the set B is finite, then

---

[*]It is assumed that the reader is familiar with the basic concepts of topology.

S is called a semigroup with a finite basis [11].

It follows from Theorem 5.3 that every basis of the semigroup K (and in particular K itself) is also a basis of the semigroup S. It is nonetheless interesting to look for less trivial bases of the semigroup S, and in particular to look for finite bases in S.

5.4.1. *The semigroup J admits no finite basis.*

Proof. Let $B = \{b_1, b_2, \cdots, b_k\}$ be an arbitrary finite set of elements in J. Then $\{B\} \subset B' = \bigcup_{i=1}^{k} b_i S$. In view of Theorem 5.2, each of the principal right ideals $b_i S$ is closed: hence we have $\overline{B'} = B'$ and $\{\overline{B}\} \subseteq B'$. By hypothesis all $b_i \in J$ and hence $b_i S \subseteq J$, and $B' \subseteq J$. On the other hand, it is easy to construct an interval xE ($xE \neq E$) contained in none of the sets $b_i E$. Then the function $x \in J$ is contained in none of the ideals $b_i S$ and consequently $x \,\overline{\in}\, B'$. Hence $J \not\subseteq B'$, and also $J \not\subseteq \{\overline{B}\}$. B is not a basis of the semigroup J.

5.4.2. From 5.4.1, we infer:

J contains no finite basis of the semigroup S.

On the other hand it is evident that G contains no basis whatsoever of the semigroup S (finite or not).

5.5 Theorem. *Let B be an arbitrary basis of the group G, and let k be any element of K. Then the set $B \cup k$ is a basis of the semigroup S.*

Proof. Let k be an element of K, ϵ an arbitrary positive number, and s an arbitrary element of S. Theorem 5.3 implies that there is an element $a \in K$ such that $\rho(s, a) < \frac{\epsilon}{2}$. In view of Theorem 2.2, the equation $xk = a$ is solvable in S, while from the construction of this solution it is evident that as x one can take an element $g \in G$.

By hypothesis B is a basis of the group G. Hence there is an element b in the semigroup $\{B\}$ such that $\rho(b, g) < \frac{\epsilon}{2}$. Then we have

$$\rho(bk, \ gk) = \max_{\xi \in E} |bk\xi - gk\xi| \leqslant \max_{\eta \in E} |b\eta - g\eta| = \rho(b, \ g) < \frac{\varepsilon}{2} \, ;$$

recalling that $gk = a$, we obtain:

$$\rho(s, \ bk) \leqslant \rho(s, \ a) + \rho(gk, \ bk) < \frac{\varepsilon}{2} + \frac{\varepsilon}{2} = \varepsilon.$$

5.5.1. In the article [11], it was proved that the group G admits a finite basis. From this and from 5.5 we obtain

Theorem. *The semigroup S admits a finite basis. One of the elements of this basis is an arbitrary element of K, and the remaining elements form a basis of the*

group G.

In the articles [11], [12], it is proved that there is a finite basis in the semi-group of all continuous functions defined on the interval $[0, 1]$ and such that $0 \leq f(\xi) \leq 1$. However, the bases found in these papers contain no basis of the semigroup S.

5.6. In the following two subheads we will construct certain classes of sub-semigroups of the semigroup S, dense in S, for which many of the properties established above for the semigroup S remain valid.

Let $a(\xi)$ be a monotone increasing continuous function defined on the interval $\Gamma_1(a) = [\alpha_a, \beta_a]$, with range of values $\Gamma_2(a) = [\gamma_a, \delta_a]$. $a(\xi)$ is a homeomorphic partial transformation of the interval E [13]. Let o be the partial transformation for which $\Gamma_1(o) = \Gamma_2(o) = \varnothing$. The operation of multiplication is defined for partial mappings in the usual way, as follows: if $\xi \in \Gamma_1(b)$ and $b\xi \in \Gamma_1(a)$, then $\xi \in \Gamma_1(ab)$ and $ab\xi = a(b\xi)$. Under this operation the set $W^*(E)$ of all homeomorphic partial transformations described above forms a semigroup (if we adjoin the element o), and the subset $W_1(E) \subset W^*(E)$ consisting of o and all $a \in W^*(E)$ for which $\alpha_a = \beta_a$ is an ideal of the semigroup $W^*(E)$.

However, in the set $Y(E) = [W^*(E) \backslash W_1(E)] \cup o$ one can define multiplication as follows: $a * b = o$ if $\Gamma_2(b) \cap \Gamma_1(a)$ contains no more than one point; and $a * b = ab$ otherwise. It is easy to see that $Y(E)$ is a semigroup isomorphic to the "difference" factor-semigroup $W^*(E) - W_1(E)$ (see [3]).

In the set $Y(E)$, we define the following partial operation of "addition": suppose that $a, b \in Y(E)$ and $\beta_a = \alpha_b$, $\delta_a = \gamma_b$; then $c = a + b$ if $c(\xi) = a(\xi)$ for $\xi \in \Gamma_1(a)$ and $c(\xi) = b(\xi)$ if $\xi \in \Gamma_1(b)$.

Now let T_Λ be a dense subset of the interval E that contains 0 and 1; suppose further that Λ is a subsemigroup of $Y(E)$ closed under the operation $+$ and satisfying the following conditions:

a) $\alpha_a, \beta_a \in T_\Lambda$ for every function $a \in \Lambda$.

b) If $a \in \Lambda$ and $\xi \in \Gamma_1(a) \cap T_\Lambda$, then $a\xi \in T$.

c) If $\gamma_a = 0$, then $\alpha_a = 0$; if $\delta_a = 1$, then $\beta_a = 1$.

d) For arbitrary $\alpha, \beta, \gamma, \delta \in T_\Lambda$ ($\alpha < \beta$, $\gamma < \delta$), satisfying condition c), there is a function $a \in \Lambda$ such that $\Gamma_1(a) = [\alpha, \beta]$, $\Gamma_2(a) = [\gamma, \delta]$.

e) If $c \in \Lambda$ and $c = a + b$, where $\beta_a \in T_\Lambda$, then $a, b \in \Lambda$.

f) If the function $a^{-1}(\xi)$ inverse to $a \in \Lambda$ satisfies condition c), then $a^{-1} \in \Lambda$.

Let $T = \{t_0, t_1, \cdots, t_n\}$ be an arbitrary finite subset of T_Λ; $t_0 = 0$, $t_n = 1$, $t_{k-1} < t_k$; $d(\xi)$ the sum as defined above of the functions $a_k(\xi)$, where $\alpha_k = t_{k-1}$,

$\beta_k = t_k$. By construction, $d(\xi)$ is continuous and monotone on the same interval. Consequently $d(\xi) \in D$. Let $D(\Lambda)$ denote the set of all functions $d(\xi)$ constructed in this way, and by $S(\Lambda)$ the subsemigroup of S, generated by all elements of $D(\Lambda)$ and the function $i(\xi) = 1 - \xi$.

$D(\Lambda)$ *is a dense subsemigroup of* D.

We will not give the proof of this assertion: it is easy to establish , as is also the validity for $D(\Lambda)$ of almost all of the properties of the semigroup D established above. The law of left cancellation in $D(\Lambda)$ follows from the inclusion $D(\Lambda) \subseteq D$. Problems of divisibility, which lie at the bottom of all of our considerations, are solved in $D(\Lambda)$ just as they are in D. For this it suffices to note that for arbitrary a, $b \in D(\Lambda)$, the functions ba^{-1} and $s^{-1}b$ (where they are defined) are sums of functions in Λ, and ba^{-1} can be extended by adding summands from Λ.

If we also require that $1 - \xi$ lie in T_Λ whenever $\xi \in T_\Lambda$, we find that if $d(\xi) \in D(\Lambda)$, then all functions $i^{-1} di$ are also in $D(\Lambda)$. Then all properties of the semigroup S considered in the present work remain valid for the dense subsemigroup $S(\Lambda)$ of S.

5.6.1. For T_Λ one can take all of the interval E, and for Λ arbitrary functions from $Y(E)$. Then $S(\Lambda) = S$. However, one can give less trivial examples (in these examples, we imply where not stated explicitly the validity of the conditions of 5.6):

1) P is a real field, $T_\Lambda = P \cap E$, Λ consists of functions that are sums in the sense of 5.6 of a finite number of linear functions $a(\xi) = a\xi + b$, where a, $b \in P$, $a > 0$. All of the functions in $D(\Lambda)$ are piecewise linear.

2) P is a real field, $T_\Lambda = P \cap E$; the generators of Λ (as in example 1)) are functions $a(\xi)$ that are fractional-linear:

$$a(\xi) = \frac{a\xi + b}{c\xi + d}, \text{where } a,\ b,\ c,\ d \in P, \ \begin{vmatrix} a & b \\ c & d \end{vmatrix} > 0.$$

3) P is an algebraically closed field, $T_\Lambda = P \cap E$, and $a(\xi)$ is a branch of an algebraic function.

4) T_Λ is an arbitrary dense subset of E symmetric about its center; $a(\xi)$ has a continuous nonvanishing derivative in $\Gamma_1(a)$.

In the case where the field P is countable, the semigroup $S(\Lambda)$ in examples 1) – 3) consists of a countable set of elements. In particular, the semigroups $K \cap S(\Lambda)$, $A \cap S(\Lambda)$ and all of their right ideals are countable semigroups with left division and left cancellation. *

____

* An example is given in the article [14] of a countable semigroup with left division and left cancellation. This semigroup coincides with the semigroup $A \cap D(\Lambda)$ of example 1), if P is taken as the field of rational numbers.

In conclusion we note that one can construct semigroups like the semigroups $S(\Lambda)$ that are dense in S, the elements of which are all possible "sums" of a countable set of functions $a(\xi) \in \Lambda$ under various hypotheses concerning the limit points of the sets T.

BIBLIOGRAPHY

[1] E. S. Ljapin, *Normal complexes of associative systems*, Izv. Akad. Nauk SSSR Ser. Mat. 14 (1950), 179–192. (Russian)

[2] A. K. Suškevič, *Theory of generalized groups*, Kharkov, 1937.

[3] D. Rees, *On semi-groups*, Proc. Cambridge Philos. Soc. 36 (1940), 387–400.

[4] G. Birkhoff, *Lattice theory*, Amer. Math. Soc., New York, 1948.

[5] L. M. Gluskin, *Homomorphisms of unilaterally simple semigroups on groups*, Dokl. Akad. Nauk SSSR 102 (1955), 673–676. (Russian)

[6] E. S. Ljapin, *Simple commutative systems*, Izv. Akad. Nauk SSSR Ser. Mat. 14 (1950), 275–282. (Russian)

[7] L. M. Gluskin, *Simple semigroups with zero*, Dokl. Akad. Nauk SSSR 103 (1955), 5–8. (Russian)

[8] A. H. Clifford, *A system arising from a weakened set of group postulates*, Ann. of Math. (2) 34 (1933), 865–871.

[9] C. J. Everett and S. Ulam, *On ordered groups*, Trans. Amer. Math. Soc. 57 (1945), 208–216.

[10] N. J. Fine and G. E. Schweigert, *On the group of homeomorphisms of an arc*, Ann. of Math. (2) 62 (1955), 237–253.

[11] J. Schreier and S. Ulam, *Über topologische Abbildungen der euklidischen Sphären*, Fund. Math. 23 (1934), 102–118.

[12] W. Sierpiński, *Sur l'approximation des fonctions continues par les superpositions de quatre fonctions*, Fund. Math. 23 (1934), 119–120.

[13] V. V. Vagner, *The theory of generalized heaps and generalized groups*, Mat. Sb. (N.S.) 32 (74) (1953), 545–632. (Russian)

[14] N. N. Vorob'ev, *On the theory of ideals of associative systems*, Leningrad. Gos. Ped. Inst. Uč. Zap. 103 (1955), 31–73. (Russian)

Translated by:
Edwin Hewitt

ON CLOSED MAPPINGS

A. D. TAĬMANOV

1. The following theorem of Katetov is well known: *For each n-dimensional metric space R, the set of all bounded, uniformly zero-dimensional* mappings of the space R into E^n is everywhere dense in the space $C(R, E^n)$ of all bounded mappings of the space R into E^n.*

From this the theorem of Šersnev easily follows: *For each n-dimensional metric space R the set of all $(n-k)$-dimensional mappings of the space R into E^k, where $0 \leq k \leq n$, is everywhere dense in the space $C(R, E^k)$.*

In connection with this theorem P. S. Aleksandrov posed the following problem: *On each such n-dimensional space is it possible to find an $(n-k)$-dimensional, closed mapping into E^n where $k \leq n$?*

In the present note we answer this question negatively and give the following new characterization of closed mappings.

2. **Theorem 1.** *Suppose f is a continuous mapping from the metric space X onto the metric space Y, which can be extended to a continuous mapping $\tilde{f}$ from a compact extension $\tilde{X}$ of X onto a compact extension $\tilde{Y}$ of Y; the mapping f is closed if and only if for each $y \in Y$ the set $\tilde{f}^{-1}(y) \backslash X$ does not contain limit points of the set $X \backslash \tilde{f}^{-1}(y)$.***

Necessity: Suppose f is closed and that there exists a point $y_0 \in Y$ such that the set $\tilde{f}^{-1}(y_0) \backslash X$ contains a point x_0 which is a limit point for the set $X \backslash \tilde{f}^{-1}(y_0)$. We will show that in the set $X \backslash \tilde{f}^{-1}(y_0)$ there exists a closed set A, the closure (in $\tilde{X}$) of which contains a point x_0 of the inverse image $\tilde{f}^{-1}(y_0)$.

According to a lemma of I. A. Vaĭnšteĭn [4] the complete inverse image $f^{-1}(y_0)$ of the point y_0 has a compact boundary $B = \operatorname{Br} f^{-1}(y_0)$. The point

---

* A mapping of a metric space X into a metric space Y is called uniformly *zero-dimensional* if for each $\epsilon > 0$, there exists a $\delta > 0$ such that each set M of the space X, having diameter less than δ, can be decomposed into the union of disjoint, open sets in M, each of which has diameter less than ϵ.

** If we consider completely regular spaces X and Y, then from the conditions of the theorem the closedness of the mapping follows. From the proof of the theorem it is clear that the conditions of the theorem are also necessary, if we consider a closed mapping in which the complete inverse image $f^{-1}(y)$ of each point has a compact boundary.

However, the following example, due to Ju. M. Smirnov, shows that the conditions of the theorem are not necessary for closed mappings of arbitrary completely regular spaces.

Suppose X is $(-\infty, +\infty)$, and the space Y is obtained from the space X by identifying all integers with one point y_0. The natural mapping f from X into Y is closed, continuous and if it is extended to a mapping $\tilde{f}$ on the Stone-Čech compactification βX of the space X, then there are limit points of the nonintegral reals in the inverse image $\tilde{f}^{-1}(y_0)$ of the point y_0.

$x_0 \in \tilde{f}^{-1}(y_0)\backslash X$ does not belong to the compact set B and hence there is a neighborhood U of the point x_0 which does not intersect B. Suppose V is a neighborhood of x_0 such that $\overline{V} \subset U$. Then the point x_0 is a limit point for the set $A = \overline{V} \cap (X\backslash \tilde{f}^{-1}(y_0))$. Now it suffices to show that the set A is closed in $X\backslash \tilde{f}^{-1}(y_0)$. For if A is not closed in $X\backslash \tilde{f}^{-1}(y_0)$, then there exists a point x_1 in $\tilde{f}^{-1}(y_0)$, which is a limit point for A and consequently belongs to B, which contradicts the choice of U. The set A does not intersect $\tilde{f}^{-1}(y_0)$ and $f(A)$ does not contain the point $y = \tilde{f}(x_0)$. In other words, the point y_0 must be a limit point of $f(A)$. Thus, $f(A)$ is not closed in $f(X) = Y$, which contradicts our assumption.

Sufficiency: Suppose f is a mapping such that for each $y \in Y$ the set $\tilde{f}^{-1}(y)\backslash X$ does not contain limit points of the set $X\backslash \tilde{f}^{-1}(y)$. Now let A be a closed subset of X. If A is closed in $\tilde{X}$ then $f(A)$ is closed in $\tilde{Y}$ and, consequently, in Y.

Suppose now that A is closed in X, but not closed in $\tilde{X}$. It is necessary to show that in this case $f(A)$ is closed in Y. Obviously, $\tilde{f}(\overline{A}^*) = \tilde{f}(A)$. Assume that $f(A)$ is not closed in Y, i.e., there exists a point, belonging to the closure of $f(A)$ in Y and not belonging to $f(A)$. Now consider the set $\alpha = \tilde{f}^{-1}(y_0)$.

From $y_0 \in Y$ it follows that $\alpha \cap X \neq \Lambda$.

From $y_0 \in f(A)$ it follows that $\alpha \cap X = \Lambda$.

From $y_0 \in \tilde{f}(A)$ it follows that $\alpha \cap \overline{A}^* \neq \Lambda$.

This means that $\tilde{f}^{-1}(y_0)\backslash X$ contains limit points of A. Then the set $\tilde{f}^{-1}(y_0)\backslash X$ contains limit points of $X\backslash \tilde{f}^{-1}(y_0)$.

Thus the theorem is proved.

Lemma 1. *Suppose f is a continuous, closed mapping from the metric space X onto the metric space Y and F is some compact subset of Y such that the inverse image $f^{-1}(y)$ for $y \in F$ is compact; then the set $f^{-1}(F)$ is compact.*

This lemma follows from Theorem 2.2 of [3].

Lemma 2. *Let f be a continuous, closed k–dimensional mapping from the metric space X of dimension n onto the set Y; then X is the union of two sets X_1, X_2, so that $X_1 \cap X_2 = \Lambda$ and the mapping on the set X_1 is compact (i.e., $f^{-1}(y) \cap X$ is compact for all $y \in Y$) and each point $x \in X_2$ has a neighborhood of dimension k.*

Proof. If all sets $f^{-1}(y)$ are compact, then we can put $X = X_1$, and $X_2 = \Lambda$. If there are points $y \in Y$, whose inverse images are noncompact, then denote the set of such points by N. If $y \in N$, then $\alpha = f^{-1}(y)$ is noncompact and according to Theorem 1 the set $\alpha_0 = \alpha \cap \overline{(X\backslash \alpha)}^*$ is compact and $\alpha_0 \subset \alpha$ and $\alpha\backslash \alpha_0$ are open in X. The set $X_2 = \bigcup_{y \in N} (\alpha \backslash \alpha_0)$ is open in X, and for each point $x \in \alpha \backslash \alpha_0$ there is a neighborhood U of dimension k such that $U \cap \overline{(X \backslash \alpha)}^* = \Lambda$. By putting $X_1 = X \backslash X_2$, we complete the proof.

3. **Construction of an example.** From the closed square $Q = \{0 \leq x \leq 1;$ $0 \leq y \leq 1\}$ we delete all interior points with an irrational abscissa. The remaining set R is a connected one-dimensional set of type F_σ.

We will show that there does not exist a zero-dimensional, continuous, closed mapping of R into the interval $[0, 1]$. Let us suppose there is such a mapping f. Then $f(R)$ will be a connected set and will contain an interval $[\alpha, \beta] \subset f(R)$. If all $f^{-1}(z)$ where $z \in [\alpha, \beta]$ were compact, then according to Lemma 1 the set $f^{-1}([\alpha, \beta])$ would be a compact set, containing an interior point of R, which is impossible. Therefore, the mapping f is not compact. Then according to Lemma 2 the set R is the sum of two sets R_1 and $R_2 \neq \Lambda$ such that each point x of R has a zero-dimensional neighborhood U in the set R. But by the construction of the set R this is clearly impossible. Therefore the mapping f is not closed, which completes the proof.

4. **Note:** If we delete from Q all interior points with a rational abscissa, then we obtain a one-dimensional set S of type G_δ, on which there does not exist a closed, continuous and zero-dimensional mapping into $[0, 1]$.

From Lemma 2 it immediately follows that there is no k-dimensional, closed mapping from a universal n-dimensional set into E^{n-k} where $0 \leq k < n$. A set X of dimension n is called uniformly n-dimensional if there do not exist points in X having neighborhoods of dimension $\leq n$.

From Lemma 2 it easily follows that each k-dimensional, closed mapping of a uniformly n-dimensional set into E^{n-k} is a compact mapping provided $0 \leq k < n$.

The following question, posed by L. V. Keldyš is open: Do there exist n-dimensional sets of type F_σ or G_δ on which there are zero-dimensional, closed mappings into E^n?

The author wishes to express his gratitude to Ju. M. Smirnov for advice concerning this note.

BIBLIOGRAPHY

[1] V. Katetov, *On the dimension of non-separable spaces.* I, Czechoslovak. Math. J. 2 (77) (1952), 333-368.

[2] M. Šersnev, *The characterization of the dimension of metric spaces in terms of continuous mappings into Euclidean spaces*, Uspehi Mat. Nauk 12 (1957), no.5 (77), 245-247. (Russian)

[3] M. Henriksen and J. Isbell, *Some properties of compactifications*, Duke Math. J. 25(1957), 83-105.

[4] I. A. Vaĭnšteĭn, *On closed mappings*, Moskov. Gos. Univ. Uč. Zap. 155 Mat. 5 (1952), 3-53. (Russian)

Translated by:

Jack Ceder

SOME GENERALIZATIONS OF IMBEDDING THEOREMS

S. L. SOBOLEV

Introduction

Functions of several variables encountered in analysis are usually treated as functions of an n-dimensional vector in Euclidean space. However, this point of view often does not correspond to the essence of the matter. There are cases when the roles of different independent variables are entirely different and consideration of all of them assimilated in one space is not suitable. A simple example of exactly this kind is, for instance, the coefficients of a quasi-linear partial differential equation of 2nd order

$$\Sigma \, A_{ij} \left[u, \frac{\partial u}{\partial x_1}, \frac{\partial u}{\partial x_2}, \cdots, \frac{\partial u}{\partial x_n}, x_1, x_2, \cdots, x_n \right] \frac{\partial^2 u}{\partial x_i \partial x_j} = F.$$

It is clear that the spatial variables x_k play one role here and the variables $u, \frac{\partial u}{\partial x_1}, \cdots, \frac{\partial u}{\partial x_m}$ an entirely different one. Correspondingly, the character of the dependence of A_{ij} on these variables is supposed ordinarily to be different.

Let the function

$$\Phi(u_1, u_2, \cdots, u_m \mid x_1, x_2, \cdots, x_n)$$

be a function of two such groups of variables.

For fixed $x_1, x_2, \cdots, x_n$ it becomes a function of m variables and consequently to each vector x of the Euclidean space R_n corresponds some element of the corresponding functional space X of functions of the m variables $u_1, u_2, \cdots, u_m$. Thus it is entirely natural to consider this function of $m + n$ variables as a function of m variables with values in the functional space X.

The fundamental operations of analysis, differentiation, integration, etc., for abstract functions are natural generalizations of the ordinary operations for functions of numerical variables. However, as explained in the investigations of various authors [4, 5], such a generalization is often accomplished in a non-trivial manner. In the present article we shall develop successively concepts of the integral of abstract functions in a somewhat different manner than the authors cited.

The fundamental goal of our investigation is the translation to abstract functions of certain questions of functional analysis, of the type of so-called *imbedding theorems*, establishing the behavior of functions whose derivatives are integrable to some power $p \geq 1$, i.e., belonging to the space $W_p^{(l)}$ (see [1, 2]).

As we have already remarked, constructing the operation of differentiation for abstract functions is not difficult. However, the construction of normed Banach spaces of the type L_p and $W_p^{(l)}$, as we shall see, is less trivial and may be accomplished in various ways.

In order to clarify the essence of the matter here, we shall make a few remarks. To each summable numerical point function $\phi(x)$ in the n-dimensional domain Ω may be made to correspond a certain set function, its indefinite integral

$$\phi(E) = \int_E \phi(x)\, dx. \tag{0.1}$$

The function $\phi(x)$ thus defines a set function, i.e., a certain mapping of the set of all L-measurable sets into the Euclidean line.

Conversely, having an additive, absolutely continuous numerical set function $\phi(E)$, we can always construct its derivative $\phi(x)$, related to $\phi(E)$ by formula (0.1).

Therefore, considerations of $\phi(E)$ and of $\phi(x)$ are equivalent.

Additive, absolutely continuous numerical set functions and the corresponding summable point functions form two isomorphic Banach spaces L_1. For the construction of these spaces it is sufficient to define a norm in them: for point functions this will be

$$\| \phi(x) \|_{L_1} = \int_\Omega |\phi|\, dx, \tag{0.2}$$

and correspondingly for set functions

$$\| \phi(E) \|_{L_1} = \sup_{E_1, E_2} [\phi(E_1) - \phi(E_2)], \tag{0.3}$$

where E_1 and E_2 run through arbitrary measurable sets.

Analogously, we may norm the subspaces $L_p (p \geq 1)$, important in applications, of the space L_1 by means of the formulas

$$\| \phi(x) \|_{L_p} = \left[\int_\Omega |\phi|^p\, dx \right]^{1/p} \tag{0.4}$$

or

$$\| \phi(E) \|_{L_p'} = \sup \frac{|\int \psi(x)\, d\phi(E)|}{\| \psi \|_{L_{p'}}}, \quad \frac{1}{p} + \frac{1}{p'} = 1, \tag{0.5}$$

where $\psi(x)$ runs through all possible step functions assuming only a finite number of values (for such functions the integral in the numerator has an obvious meaning) and $\| \psi \|_{L_{p'}}$ is defined by formula (0.4).

Isomorphism of the two spaces L_p and L_p' under the law of correspondence expressed by formula (0.1) is first established directly not for all functions, but only for $\phi(x)$ belonging to a certain everywhere dense set, namely, for step functions assuming only a finite number of values.

The right-hand side of formula (0.1) thus represents some operator acting in the space of numerical, summable point functions and having values in a space of set functions. This operator, defined on some set of numerical functions, is extended to arbitrary summable functions. This incidentally enables us to approach from a new point of view the concept of integral, coinciding now with $\phi(E)$.

For abstract functions formulas (0.2) and (0.3) as well as (0.4) and (0.5) generalize in a completely natural way but, as we shall see below, they will not be equivalent, since the norms defined by them will not coincide and will not even be equivalent. These norms are related to two aspects of absolute convergence. An important property of absolutely convergent integrals of numerical functions, i.e., those for which their absolute values are summable, as well as a property of absolutely convergent numerical series, is the fact that their values (or sums of the series) are independent of the order of integration or summation. As soon as the set E_n is sufficiently close to E (in measure, for instance, in the Lebesgue case) or as soon as sufficiently many terms of the series are added, the integral or sum will differ as little as desired from its limiting value. This independence of order for numerical functions or series is equivalent to convergence of the sums of the absolute values. However, even series whose terms are elements of the Banach space X may have this property without the series of their norms converging. Let us indicate a very simple example:

Let the space X be l_2, i.e., the space of sequences with convergent sums of squares. Let us denote by $i_1, i_2, \cdots, i_k, \cdots$ vectors of the form $(0, 0, 0, \cdots \overset{k}{\cdots}, 0, 1, 0, \cdots)$.

The series

$$\Sigma \frac{i_k}{k} \tag{0.6}$$

will not converge absolutely in the ordinary sense since the sum of the norms of its terms increases unboundedly. However, this sum does not depend on the order of the terms since the norm of the sum of any number of its members situated sufficiently far out will be as small as desired:

$$\left\| \sum_{k_s > N} \frac{i_{k_s}}{k_s} \right\| \leq \left[\sum_{t=N}^{\infty} \frac{1}{t^2} \right] < \epsilon.$$

In the following sections we shall study integrals and set functions with the norms (0.3) and (0.5), which are generalizations of such "absolutely convergent" integrals.

Our investigation into parts of the theory of integration is related to certain earlier papers. In his article devoted to integrals of abstract functions S. Bochner, using the norm

$$\| \phi \|_{B_1} = \int \| \phi(x) \|_X \, dx,$$

which generalizes (0.2), constructs the integral of abstract point functions $\phi(x)$, thus extending the integration operator (0.1).

Another method of constructing the integral, which is more general and makes use of another idea, was proposed by I. M. Gel'fand [5].

By means of the formula

$$\phi(E) = \int_E \phi(x) \, dx \tag{0.8}$$

one may construct a copy of the function space B_1 obtained by Bochner in the collection of abstract set functions $\phi(E)$. Now let us consider this question differently. Let us introduce into the set of functions $\phi(E)$ the norm $\| \; \|_{\Phi_1}$ by means of the formula

$$\| \phi \|_{\Phi_1} = \sup_{E_1, E_2} \| \phi(E_1) - \phi(E_2) \|_X, \tag{0.9}$$

which is a natural generalization of (0.3).

Thus we separate out the Banach space Φ_1 from the collection of additive abstract set functions. It turns out that this space does not coincide with the copy of B_1 given by formula (0.8). For functions $\phi(E)$ and $\phi(x)$ related by (0.8) we shall have the inequality

$$\| \phi(E) \|_{\Phi_1} \leq \| \phi(x) \|_{B_1}. \tag{0.10}$$

With corresponding definitions of the norms B_p and Φ_p by the formulas

$$\| \phi(x) \|_{B_p} = \left[\int \| \phi(x) \|^p \, dx \right]^{1/p}, \tag{0.11}$$

$$\| \phi(E) \|_{\Phi_p} = \sup \frac{\| \int \psi(x) \, d\phi(E) \|_X}{\| \psi(x) \|_{L_{p'}}}, \quad \frac{1}{p} + \frac{1}{p'} = 1, \tag{0.12}$$

the analogous inequality

$$\| \phi(E) \|_{\Phi_p} \leq \| \phi(x) \|_{B_p} \tag{0.13}$$

is established for these norms also.

In the present article it will be shown that the image of the function space B_p ($p \geq 1$) defined by (0.8) is essentially a space of set functions from Φ_p, i.e., that Φ_p contains elements that are not the image of any elements from B_p.

Moreover, if we introduce a norm for point functions $\phi(x)$ by means of the mapping (0.8), carrying over the norm of Φ_p to them (as before, this may always be done for step functions), then we can not obtain the entire space from these functions $\phi(x)$ without introducing ideal elements. Hence it is immediately apparent that the natural class of objects for which we shall construct our theorem is precisely the class of abstract set functions.

The task of the present article is the construction of the functional spaces Φ_p and $\Psi_p^{(l)}$ of abstract set functions, analogous to the spaces L_p and $W_p^{(l)}$, whose norm is constructed according to formulas (0.9) and (0.12), and the establishment for them of imbedding theorems analogous to theorems for numerical functions.

First we shall attend to the analysis of some important properties of abstract set functions.

§1. The fundamental functional space

1. **Some theorems on continuous abstract functions.** Suppose given a metric space P in which is defined a distance between elements

$$\rho(\alpha, \beta) = \rho(\beta, \alpha). \tag{1.1}$$

The function $\rho(\alpha, \beta)$ is assumed to be positive and to satisfy the triangle condition:

$$\rho(\alpha, \gamma) \leq \rho(\alpha, \beta) + \rho(\beta, \gamma). \tag{1.2}$$

Also given is an abstract function $F(\alpha)$ defined on a set Ω of elements of P with values from the Banach space X.

The function $F(\alpha)$ is called *uniformly continuous* on Ω if for any $\epsilon > 0$ one may determine $\delta(\epsilon)$ such that for

$$\rho(\alpha, \beta) < \delta(\epsilon)$$

we have the inequality

$$\| F(\alpha) - F(\beta) \|_X < \epsilon. \tag{1.3}$$

Any such function $\delta(\epsilon)$ is called a *modulus of continuity* for $F(\alpha)$. Several theorems ordinarily proved for continuous numerical functions are valid for abstract, uniformly continuous functions. We shall indicate these theorems here without carrying out their proofs, which are completely repetitious of well-known ones.

We shall call a set of functions $\{F(\alpha)\}$ *equicontinuous* if they have a common modulus of continuity, by which we understand an arbitrary function $\delta(\epsilon)$ satisfying condition (1.3) for all $F(\alpha) \in \{F(\alpha)\}$.

For a uniformly convergent sequence uniform continuity of the limit function and equicontinuity of the sequence are equivalent.

Theorem 1. *The limit function $F_0(\alpha)$ of a uniformly convergent sequence $F_k(\alpha)$ of equicontinuous functions is uniformly continuous.*

Theorem 2. *A uniformly convergent sequence $F_k(\alpha)$ of uniformly continuous functions is equicontinuous. From convergence everywhere on a compact set and equicontinuity follows uniform convergence.*

Theorem 3. *An equicontinuous sequence $F_k(\alpha)$ converging everywhere on a set E compact in itself, i.e., closed and containing a finite ϵ-net for any ϵ, converges uniformly.*

Theorem 4. *A continuous function $F(\alpha)$ on a closed compact set E is bounded on it and attains the maximum of its norm there.*

Theorem 5. *A function $F(\alpha)$ continuous on a closed compact set E is uniformly continuous there.*

Theorem 6. *A function $F(\alpha)$ defined on an everywhere dense set E in a metric space P and uniformly continuous on it may be continuously extended to the whole space and, moreover, in a unique manner.*

The proof of this theorem presents no difficulty. It is sufficient to establish that for convergence of the sequence of points α_k to α_0 the set of values $F(\alpha_k)$ converges and defines $F(\alpha_0)$. Continuity of the function constructed is established in an elementary manner.

2. Let us consider some measurable set Ω of n-dimensional Euclidean space and let $\phi(E)$ be some abstract set function with values from a vector Banach space X, defined on all sets E which are Lebesgue measurable and which are subsets of Ω. The function $\phi(E)$ will be considered to be *finitely additive*. This means that for two arbitrary sets E_1 and E_2 without common points we have the equality

$$\phi(E_1) + \phi(E_2) = \phi(E_1 \cup E_2). \tag{1.4}$$

From (1.4) it follows that for the empty set O

$$\phi(O) = \theta$$

where θ denotes the zero element in the space X.

Below we shall consider not one individual set function but a set of them. Moreover, in order that misunderstandings will not arise, we shall denote by $\phi(E)$ the element of the space X which is the value of ϕ for a definite or a variable E.

By the symbol ϕ we shall denote the given set function as a whole, i.e., a given mapping of all Lebesgue measurable sets into the space X.

For the set of functions ϕ which we are considering, addition and multiplication by a constant are defined in the obvious manner.

The manifold Φ of all additive abstract set functions is thus a vector manifold. As the zero in this manifold will serve the function identically equal to the zero element in X.

From the manifold Φ is determined a normed manifold Φ^* if we introduce into it some norm $\|\phi\|_{\Phi^*}$ obeying the usual axioms of a norm:

1. $\|a\phi\|_{\Phi^*} = |a| \, \|\phi\|_{\Phi^*}.$

2. $\|\phi_1 + \phi_2\|_{\Phi^*} \leq \|\phi_1\|_{\Phi^*} + \|\phi_2\|_{\Phi^*}.$

3. $\|\phi\|_{\Phi^*} > 0$ if ϕ is not trivially a constant equal to zero.

4. From $\|\phi\|_{\Phi^*} = 0$ follows $\phi \equiv \theta.$ $\qquad\qquad\qquad$ (1.5)

Let us suppose that for the norm chosen in this manner the *completeness condition* is fulfilled, i.e., every sequence ϕ_k which converges in norm $\|\phi\|_{\Phi^*}$ has a limit: in order words, from

$$\|\phi_k - \phi_l\|_{\Phi^*} < \epsilon \text{ for } k, l > N_1(\epsilon)$$

follows the existence of a ϕ_0 such that

$$\|\phi_k - \phi_0\|_{\Phi^*} < \epsilon \text{ for } k > N_1(\epsilon). \qquad\qquad (1.6)$$

Then Φ^* will in turn be a Banach space. As one of the simplest norms Φ^* may serve the uniform norm

$$\|\phi\|_M = \sup_E \|\phi(E)\|_X. \qquad\qquad (1.7)$$

The manifold of all additive abstract set functions with a finite norm $\|\phi\|_M$ will be a linear Banach space.

Indeed, fulfillment of the four axioms (1.5) is obvious. Let us demonstrate the completeness of the space M with this norm.

For any set E_1 from the convergence of ϕ_k in the norm $\|\phi\|_M$ and completeness of the Banach space X follows the existence of the limit

$$\phi_0(E_1) = \lim_{k \to \infty} \phi_k(E_1)$$

as well as its boundedness in norm.

The limit $\phi_0(E_1)$, as is easily shown, will be an additive function of E_1. On some linear manifolds Φ^* from Φ may be defined some other norm $\|\ \|_{\Phi^*}$.

Let us agree to call the norm $\|\phi\|_{\Phi^*}$ *subordinate to the norm* $\|\phi\|_M$ if there exists a constant A such that

$$\|\phi\|_M \leq A\|\phi\|_{\Phi^*}. \tag{1.8}$$

If $\|\phi\|_\Phi$ is subordinate to $\|\phi\|_M$, then from the convergence of a sequence in the norm Φ follows its convergence in the norm M. The subordinate norm is defined, generally speaking, on a more restricted manifold.

Remark. Any manifold Ψ^* from Φ on which is everywhere defined a norm $\|\ \|_\Phi$ subordinate to $\|\ \|_M$ may be closed with respect to the norm $\|\ \|_\Phi$ and extended to a linear space $\Psi \subset \Phi$. As is well known, the manifold Ψ thus obtained may not be closed with respect to the norm $\|\ \|_M$.

Let us prove these assertions. First of all, let us show that for any sequence ϕ_k converging in the sense of the norm of Φ there exists a limit element in the sense of the norm M. This follows from the estimate

$$\|\phi_k - \phi_l\|_M \leq A\|\phi_k - \phi_l\|_\Phi$$

and from the completeness of the space Φ with respect to the norm M.

For such a limit element ϕ_0 we may in turn define the norm $\|\phi_0\|_\Phi$. Indeed we have from the triangle axiom

$$\|\phi_k\| \leq \|\phi_k - \phi_l\|_\Phi + \|\phi_l\|_\Phi,$$

from which, along with the symmetric inequality, it follows that

$$|\ \|\phi_k\| - \|\phi_l\|\ | < \epsilon.$$

Thus $\|\phi_k\|_\Phi$ must converge to a finite limit which must be the same for all convergent ϕ_k which may be assimilated into one sequence. Let us write for ϕ_0 the norm

$$\|\phi_0\|_\Phi = \lim \|\phi_k\|_\Phi. \tag{1.9}$$

Elements of the form $a_1 \phi_0^{(1)} + a_2 \phi_0^{(2)}$ where $\phi_0^{(1)}$ and $\phi_0^{(2)}$ are limits of two sequences convergent in $\|\ \|_\Phi$ are always limits of convergent sequences themselves. For them is defined the norm

$$\| a_1 \phi_0^{(1)} + a_2 \phi_0^{(2)} \| = \lim \| a_1 \phi_k^{(1)} + a_2 \phi_k^{(2)} \|. \qquad (1.10)$$

Consequently, the extension of Ψ^* which has been constructed will be linear itself.

A passage to the limit gives us the triangle axiom and the other properties of a norm for Ψ.

Finally, some sequences which converge in the sense of $\| \ \|_M$ may not be convergent in the sense of $\| \ \|_\Phi$.

While completing Ψ we will not thereupon introduce any limit elements for them and so their limits, existing in the sense of $\| \ \|_M$, may not belong to Ψ. From this follows the last of our assertions, according to which the set with norm $\| \ \|_\Phi$ may not be closed in $\| \ \|_M$. Later we shall introduce several such norms $\| \ \|_\Phi$ subordinate to the norm $\| \ \|_M$ and study some of their properties.

The set of all sets E which are Lebesgue measurable and lie in Ω may be considered as a metric space by introducing into it an appropriate metric.

Set $\rho(E_1, E_2) = m(E_1 \leftrightarrow E_2) \equiv m(E_1 \cup E_2 \setminus E_1 \cap E_2)$.

We shall consider $\rho(E_1, E_2)$, namely, the measure of the so-called symmetric difference of the two sets, as the distance between these sets. It is easily verified that all of the axioms of a distance will be fulfilled for it. We may thus determine among the functions $\phi(E)$ from Φ the set of continuous functions.

Closure of the linear manifold of continuous functions with respect to the norm $\| \ \|_M$ clearly leads again to continuous functions. This follows from Theorem 1 since convergence with respect to the norm $\| \ \|_M$ is uniform convergence. It is obvious that closure with respect to any norm subordinate to M also leads to continuous functions.

3. Let us consider some normed space of set functions $\phi(E)$. Let us agree again to denote every such function, considered as an element of Φ^*, by the symbol ϕ, reserving the notation $\phi(E)$ for denoting that element of X which corresponds to a given concrete or variable E_0.

Corresponding to each function ϕ we shall define a new abstract function $\phi(E, E_1)$ of two variable sets by the formula

$$\phi(E, E_1) = \phi(E \cap E_1). \qquad (1.11)$$

The values of this function for fixed E_1 will be functions of E, i.e., elements of Φ^*. We shall denote this new function of the set E_1 with values in Φ^* by $\psi_\phi(E_1)$ and we shall say that $\phi(E)$ *generates* $\psi_\phi(E_1)$ *by intersection.*

As we have agreed earlier, $\psi_\phi(E_1)$ will denote the element of the space Φ^* obtained for a definite or variable E_1 by means of the function ϕ from formula (1.11). In case we are interested in not the individual dependence on E_1 of some element of Φ^* but the whole set of such dependences or mappings of the set of measurable sets into Φ^*, we shall use simply the symbol ψ_ϕ.

Theorem 7. *The set Φ^* generates by intersection a set of abstract functions $\psi_\phi(E_1)$ defined for any measurable E_1 from Ω and satisfying the two conditions:*

(a) $\psi_\phi(E_1)$ *are finitely additive for variable E_1;*

(b) $\psi_\phi(E, E_1) = \theta$, *where θ is the zero element of X, on all E for which $E \cap E_1$ is empty.*

Conversely, any function $\psi(E_1)$ with values in Φ^* which satisfies (a) *and* (b) *is itself generated by intersection by the set function $\phi(E)$ from Φ^* which is equal to the value of $\psi_\phi(E_1)$ for $E_1 = \Omega$*

$$\phi = \psi_\phi(\Omega). \tag{1.12}$$

Let us prove this theorem. The proof of the direct assertion is obvious. Additivity in E_1 follows from the relation

$$E \cap (E_2 \cup E_3) = (E \cap E_2) \cup (E \cap E_3)$$

and the additivity of the function $\phi(E)$.

From the remark that $\phi(E) = \theta$ for empty $E \cap E_1$ follows the assertion (b).

Let us prove the converse assertion of the theorem. Let $\psi(E_1)$ be given. Consider the function $\phi(E)$ defined by (1.12) and let us construct by formula (1.11) the function $\psi_\phi(E_1)$ corresponding to it.

For convenience we shall denote as earlier the value of the function $\psi_\phi(E_1)$ for a given E by $\phi(E, E_1)$. This will be an element of Φ^* if E is considered as a variable.

Let us show that $\psi_\phi(E_1)$ coincides with $\psi(E_1)$.

Indeed, for any E

$$\psi(E, E_1) = \psi(E, E_1 \cap E) + \psi(E, E_1 \setminus E_1 \cap E) =$$

$$= \psi(E_1 \cap E, E_1 \cap E) + \psi(E \setminus E_1 \cap E, E_1 \cap E) + \psi(E, E_1 \setminus E_1 \cap E).$$

However, $\psi(E, E_1 \setminus E_1 \cap E) = \theta$ and $\psi(E_1 \cap E, E \setminus E_1 \cap E) = \theta$, for $E \cap (E_1 \setminus E_1 \cap E)$ and $(E \setminus E_1 \cap E) \cap (E_1 \cap E)$ are empty sets.

Consequently,

$$\psi(E, E_1) = \psi(E_1 \cap E, E_1 \cap E). \tag{1.13}$$

Moreover,

$$\psi(E_1 \cap E, \Omega) = \psi(E_1 \cap E, \Omega \setminus E_1 \cap E) + \psi(E_1 \cap E, E_1 \cap E) =$$

$$= \psi(E_1 \cap E, E_1 \cap E)$$

and so

$$\psi(E_1, E) = \psi(\Omega, E_1 \cap E). \tag{1.14}$$

Together with (1.13) this proves the theorem.

The function $\psi_\phi(E_1)$ is an abstract function with values in some normed linear manifold Φ^* defined in the metric space Y.

We shall say that $\phi(E)$ is *absolutely continuous in the norm* $\|\ \|_\Phi$ if $\|\psi_\phi(E_1)\|_\Phi$ may be made as small as desired for sufficiently small $m(E_1)$. By the *modulus* of absolute continuity of $\phi(E_1)$ we shall mean a function $\delta(\epsilon)$ satisfying the condition:

from $m(E_1) < \delta(\epsilon)$ follows

$$\|\psi_\phi(E_1)\|_\Phi < \epsilon. \tag{1.15}$$

Theorem 8. *In order that the function* $\psi_\phi(E_1)$ *be continuous on* Y *it is necessary and sufficient that the function* $\phi(E)$ *generating it by intersection be absolutely continuous.*

Let us first prove sufficiency of the condition of the theorem.

Assume that the condition is satisfied. For any two sets E_1 and E_2 such that $\rho(E_1, E_2) < \delta(\frac{1}{2}\epsilon)$ we will obtain by (1.15)

$$\|\psi_\phi(E_1) - \psi_\phi(E_2)\|_\Phi = \|\psi_\phi(E_1 \setminus E_1 \cap E_2) - \psi_\phi(E_2 \setminus E_1 \cap E_2)\|_\Phi \le$$

$$\le \|\psi_\phi(E_1 \setminus E_1 \cap E_2)\|_\Phi + \|\psi_\phi(E_2 \setminus E_1 \cap E_2)\|_\Phi \le$$

$$\le \frac{1}{2}\epsilon + \frac{1}{2}\epsilon = \epsilon.$$

The continuity of $\psi_\phi(E_1)$ is proved.

Conversely, if $\psi_\phi(E_1)$ is continuous, then $\|\psi_\phi(E_1) - \psi_\phi(0)\| \le \epsilon$ for any E_1 such that $m(E_1) < \delta(\epsilon)$ and any ϵ. As was proved in the previous theorem, $\psi_\phi(0) = 0$; this means that $\|\psi_\phi(E_1)\|_\Phi < \epsilon$ whenever $m(E_1) < \delta(\epsilon)$. This establishes the absolute continuity of ϕ. The theorem is proved.

Suppose further that the norm $\|\ \|_\Phi$ of functions $\phi(E)$ that we are considering possesses the following property. For any measurable sets E_1 and E_2 without common points we have the equality

$$\|\psi_\phi(E_1) + \psi_\phi(E_2)\|_\Phi = \|\psi_\phi(E_1) - \psi_\phi(E_2)\|_\Phi. \tag{1.16}$$

We shall call such a norm *symmetric*.

It is obvious, for instance, that for indefinite integrals of numerical functions integrable in the sense of Lebesgue, $\phi(E) = \int_E \phi(x)\,dx$, the norm

$$\|\phi\|_\Phi = \int_\Omega |\phi(x)|\,dx$$

will be symmetric, since for such functions

$$\| \psi_\phi(E_1) \pm \psi_\phi(E_2) \|_\Phi = \int_{E_1} |\phi(x)|\, dx + \int_{E_2} |\phi(x)|\, dx.$$

Let us furthermore call a norm in Φ *monotone* if it follows from $E_2 \supset E_1$ that

$$\| \psi_\phi(E_2) \|_\Phi \geq \| \psi_\phi(E_1) \|_\Phi. \tag{1.17}$$

A symmetric norm is monotone.

Indeed, corresponding to E_2 and E_1, where $E_2 \supset E_1$, consider the functions ϕ_1 and ϕ_2 defined by the formulas

$$\phi_1 = \psi_\phi(E_1) - \psi_\phi(E_2 \setminus E_2 \cap E_1),$$

$$\phi_2 = \psi_\phi(E_2) = \psi_\phi(E_1) + \psi_\phi(E_2 \setminus E_2 \cap E_1).$$

By the property of symmetry $\| \phi_1 \|_\Phi = \| \phi_2 \|_\Phi = \| \psi_\phi(E_2) \|_\Phi$.
On the other hand, $\psi_\phi(E_1) = \frac{1}{2} \phi_1 + \frac{1}{2} \phi_2$ and consequently

$$\| \psi_\phi(E_1) \|_\Phi \leq \frac{1}{2} \| \phi_1 \|_\Phi + \frac{1}{2} \| \phi_2 \|_\Phi = \| \psi_\phi(E_2) \|_\Phi,$$

as was to be proved.

It is useful to note that in the norm $\| \ \|_M$ absolute continuity coincides with continuity of $\phi(E)$ as an abstract function on the metric space Y.

Indeed, the condition that from $m(E) < \delta(\epsilon)$ follows $\| \phi(E) \|_M < \epsilon$, as is easily seen, is equivalent to continuity on Y of the function $\phi(E)$. This in turn, by monotonicity of the norm $\| \ \|_M$, may be replaced by the condition $\| \phi(E \cap E_1) \|_M < \epsilon$ for $m(E_1) < \delta(\epsilon)$, and this last condition expresses absolute continuity in the metric of M.

For monotone norms $\| \ \|_\Phi$ we have the theorem:

Theorem 9. *Any sequence of functions $\psi_{\phi_k}(E_1)$ which converges everywhere on Y converges uniformly.*

The proof of this theorem follows from the fact that

$$\| \psi_{\phi_k}(E_1) - \psi_{\phi_l}(E_1) \|_\Phi \leq \| \psi_{\phi_k}(\Omega) - \psi_{\phi_l}(\Omega) \|_\Phi$$

so that for sufficiently large k and l the uniform smallness of the left-hand side for any E_1 follows from the convergence of the sequence $\psi_{\phi_k}(\Omega)$.

Now we can prove several theorems applying the theory of abstract functions on metric spaces to functions $\psi_\phi(E_1)$ which are generated by intersection by functions $\phi(E)$. Any manifold Φ^* of functions ϕ may be closed by adjoining to it the limits of any convergent sequences. A passage to the limit in each such sequence gives us again some set function, since from the convergence of ϕ_k in norm follows convergence in the norm of X of the sequence $\phi_k(E)$ for every E.

Theorem 10. *The closure in norm of a linear manifold of absolutely continuous (in the norm of Φ) set functions also consists of absolutely continuous functions.*

In fact a convergent sequence ϕ_k generates by intersection a sequence $\psi_{\phi_k}(E_1)$ which converges uniformly.

By Theorem 1 the limit function $\psi_0(E_1)$, which obviously exists, will be uniformly continuous. Moreover, conditions (a) and (b) of Theorem 8 are obviously fulfilled for it. Consequently, it is generated by some absolutely continuous function $\phi_0(E)$, which is obviously the limit of the convergent sequence.

The theorem is proved.

We shall say that the sequence $\phi_k(E)$ is *absolutely equicontinuous* if the functions $\phi_k(E)$ have a common modulus of continuity.

Theorem 11. *A sequence of elements ϕ_k from Φ^* converging with respect to the norm of Φ to a function ϕ_0 is absolutely equicontinuous.*

In fact, convergence of ϕ_k to ϕ_0 in norm implies uniform convergence of $\psi_{\phi_k}(E_1)$ to $\psi_{\phi_0}(E_1)$ and by Theorem 2 equicontinuity of ϕ_k. This indicates absolute equicontinuity.

The theorem is proved.

4. Let us extend the domain Ω to the entire space R_n and extend the definition of the function $\phi(E)$ to any measurable sets E in R_n by means of the formula

$$\phi(E) = \phi(E \cap \Omega). \tag{1.18}$$

The right-hand side of (1.18) is defined for any E and serves as the definition of the left-hand side. Instead of the manifold Y of sets in Ω we shall define ϕ on the wider manifold of all measurable sets $\tilde{Y}$. To the set E, considered as a set of vectors, let us add the constant vector y. We shall obtain a new set $E + y$. We shall say that the function $\phi(E)$ is *continuous with respect to translation* in the norm Φ if the following is true: there exists a function $\delta(\epsilon)$, positive for all positive ϵ and called the *modulus of continuity*, such that from $|y| < \delta(\epsilon)$ follows

$$\|\phi(E + y) - \phi(E)\|_\Phi < \epsilon. \tag{1.19}$$

Corresponding to a given set function $\phi(E) \in \Phi$ we shall construct an abstract point function $Z_\phi(y)$ with values in Φ in the following way.

Let

$$Z_\phi(E \mid y) = \phi(E + y) \tag{1.20}$$

be an abstract function of the measurable set E and the variable point y. For fixed y it represents an element of Φ. By varying y we shall obtain different elements of Φ. This function of the point y with values in Φ will be called $Z_\phi(y)$. We shall say that the function $\phi(E)$ *generates the function* $Z_\phi(y)$ *by translation.*

Having an abstract function $Z_\phi(y)$ we can always recover a function of two variables $Z_\phi(E \mid y)$ as the values assumed by $Z_\phi(y)$ for a given E. The following theorem holds.

Theorem 12. *For the abstract function* $Z_\phi(y)$ *generated by the function* $\phi(E)$ *by translation we have the equalities*

$$Z_\phi(E - y \mid y) = Z_\phi(E \mid 0), \quad Z_\phi(E \mid 0) = Z_\phi(\Omega \cap E \mid 0). \tag{1.21}$$

Conversely, any function $Z(y)$ *satisfying conditions* (1.21) *is generated by translation by the function*

$$\phi(E) = Z(E \mid 0). \tag{1.22}$$

The proof of this theorem is obvious. Indeed, calculating $Z_\phi(E - y \mid y)$ we obtain by definition from (1.20)

$$Z_\phi(E - y \mid y) = Z_\phi(E \mid 0).$$

The first formula of (1.21) is proved. Furthermore, the second formula of (1.21) is obvious. Conversely, suppose $Z(E \mid y)$ satisfies (1.21). We define $\phi(E)$ by formula (1.22) and construct $Z_\phi(y)$.

By definition, using (1.22), we obtain

$$Z_\phi(E \mid y) = Z(E + y \mid 0). \tag{1.23}$$

However, by (1.21)

$$Z(E + y \mid 0) = Z(E \mid y). \tag{1.24}$$

From (1.23) and (1.24) follows

$$Z_\phi(E \mid y) = Z(E \mid y). \tag{1.25}$$

Property (1.21) says that $\phi(E)$ satisfies (1.18). Formulas (1.22) and (1.25) prove our theorem.

It may happen again that some norm $\| \phi \|_\Phi$ is defined in the space of $\phi(E)$.

We shall call this norm *invariant* if it can be extended to set functions defined on $\widetilde{Y}$ in such a way that two conditions are satisfied:

Condition 1.

$$\| \phi(E) \|_{\widetilde{\Phi}} = \| \phi(E) \|_\Phi \tag{1.26}$$

for all E such that $\phi(E) = \phi(\Omega \cap E)$, where $\widetilde{\Phi}$ is the new extended norm.

Condition 2.

$$\| \phi(E + y) \|_{\widetilde{\Phi}} = \| \phi(E) \|_{\Phi}. \qquad (1.27)$$

In the following we shall consider expressly an invariant norm.

Let $\| \phi(E) \|_{\Phi}$ denote precisely $\| \phi(E) \|_{\widetilde{\Phi}}$.

The norm introduced by us determines directly the set of functions $Z_\phi(y)$ continuous in the sense of this norm. As always, the function $Z_\phi(y)$ is continuous at the point y_0 if there exists a function $\delta(\epsilon)$ called the *modulus of continuity* such that

$$\| Z_\phi(y_0 + \Delta y_0) - Z_\phi(y_0) \|_\Phi < \epsilon \ \text{ for } \ |\Delta y_0| < \delta(\epsilon). \qquad (1.28)$$

Theorem 13. *A function $Z_\phi(y)$ generated by translation by the function $\phi(E)$ which is continuous at one point is uniformly continuous.*

In fact we have by the property of invariance of the norm

$$\| Z_\phi(y + \Delta y) - Z_\phi(y) \| = \| Z_\phi(\Delta y) - Z_\phi(0) \|$$

and so for $|\Delta y| < \delta(\epsilon)$ the norm of the increase of the function will be the same for all y. The theorem is proved.

We shall call a function which generates by translation a continuous function $Z_\phi(y)$ *continuous with respect to translation.* From Theorem 13 follows the corollary.

Corollary. *For the continuity of the function $Z_\phi(y)$ in its entirety it is necessary and sufficient that the function $\phi(E)$ generating it be continuous with respect to translation.*

We shall omit the proof, which is obvious.

The theorems about continuous abstract functions make it possible to obtain some results about the theory of set functions continuous with respect to translation.

Theorem 14. *If the sequence $\phi_k(E)$ converges to $\phi_0(E)$ in norm, then the sequence of abstract functions $Z_{\phi_k}(y)$ generated by translation by $\phi_k(E)$ converges uniformly to $Z_{\phi_0}(y)$.*

The proof of this theorem follows from the fact that

$$\| Z_{\phi_k}(y) - Z_{\phi_0}(y) \| = \| \phi_k(E) - \phi_0(E) \| \ \text{ for any } \ y.$$

Theorem 15. *Convergence of a sequence $Z_k(y)$ of functions satisfying (1.21) at least at one point implies uniform convergence for all y.*

The proof follows from (1.21) and invariance of the norm.

Theorem 16. *The closure in norm Φ of a manifold Φ^* of functions $\phi(E)$ continuous with respect to translation consists of functions continuous with* *respect to translation.*

This theorem is proved by a simple application of Theorem 15 and Theorem 2.

5. Of course, if is possible to introduce a norm into the space of functions $\phi(E)$ in a different way. We have already mentioned that it is possible, for instance, following Bochner, first to define the manifold Φ^* as the manifold of integrals $\int_E \phi(x)\,dx$ of step functions of a point $\phi(x)$ with summable norm. Then we may take

$$\| \phi(E) \|_B = \int \| \phi(x) \|_X \, dx \tag{1.29}$$

and close this manifold.

However, we shall use other methods of norming.

By the norm $\| \phi(E) \|_{\Phi_1}$ we shall understand

$$\| \phi(E) \|_{\Phi_1} = \sup \{ \| \phi(E_1) - \phi(E_2) \| \} \tag{1.30}$$

such that $E_1 \cup E_2 = E$, $E_1 \cap E_2$ is empty.

It is obvious that the norm $\| \; \|_{\Phi_1}$ is subordinate to the uniform norm $\| \; \|_M$ and even is equivalent to it, for

$$\| \phi \|_M \leq \| \phi \|_{\Phi_1} \leq 2 \| \phi \|_M. \tag{1.31}$$

Let us prove the theorem:

Theorem 17. *All of the axioms of a norm are fulfilled for the norm* $\| \phi(E) \|_{\Phi_1}$.

The equality

$$\| a\phi \| = | a | \cdot \| \phi \|$$

is obvious. Furthermore,

$$\| \phi_1 + \phi_2 \| = \sup \| \phi_1(E_1) + \phi_2(E_1) - \phi_1(E_2) - \phi_2(E_2) \| \leq$$

$$\leq \sup [\| \phi_1(E_1) - \phi_1(E_2) \| + \| \phi_2(E_1) - \phi_2(E_2) \|] \leq$$

$$\leq \sup \| \phi_1(E_1) - \phi_1(E_2) \| + \sup \| \phi_2(E_1) - \phi_2(E_2) \| =$$

$$= \| \phi_1 \| + \| \phi_2 \|.$$

No less evident is the third property.

From $\| \phi \|_{\Phi_1} = 0$ obviously follows $\phi = 0$. The theorem is proved.

It is obvious that such a norm will be homogeneous in the sense defined above. Let us indicate one more method of introducing a norm for set functions.

Again let $\phi(E)$ be an element and let ω be a numerical step function, $\omega(x) = \alpha_i$ for $x \in E_i$. We introduce the function

$$\omega\phi(E) = \Sigma a_i \phi(E \cap E_i). \tag{1.32}$$

Consider $\|\omega(x)\|_{L_{p''}}$, where $p' = p/(p-1)$, $p > 1$ and hence $p' > 1$. It may happen that

$$\frac{\|\omega\phi(\Omega)\|_X}{\|\omega\|_{L_{p'}}} < A$$

for any step functions ω, where A is some positive number not depending on ω.

In this case we shall say that the function $\phi(E)$ *belongs to the abstract space* Φ_p. Consider

$$\sup \frac{\|\omega\phi(\Omega)\|_X}{\|\omega\|_{L_{p'}}} = \sigma(\phi). \tag{1.33}$$

The number σ is a positive functional of ϕ. Let us verify for this functional the fulfillment of the basic axioms of a norm.

Theorem 18. *The properties*

$$\sigma(a\phi) = |a|\,\sigma(\phi); \tag{1.34}$$

$$\sigma(\phi_1 + \phi_2) \le \sigma(\phi_1) + \sigma(\phi_2) \tag{1.35}$$

hold.

Property (1.34) is obvious. Let us prove property (1.35).

Let

$$\sigma(\phi_1 + \phi_2) = \frac{\|\omega_3[\phi_1(\Omega) + \phi_2(\Omega)]\|_X}{\|\omega_3\|_{L_{p'}}} + \epsilon.$$

Then

$$\sigma(\phi_1 + \phi_2) - \epsilon = \frac{\|\omega_3[\phi_1(\Omega) + \phi_2(\Omega)]\|_X}{\|\omega_3\|_{L_{p'}}} \le \frac{\|\omega_3\phi_1(\Omega)\|_X}{\|\omega_3\|_{L_{p'}}} + \frac{\|\omega_3\phi_2(\Omega)\|_X}{\|\omega_3\|_{L_{p'}}} \le$$

$$\le \sup \frac{\|\omega\phi_1(\Omega)\|_X}{\|\omega\|_{L_{p'}}} + \sup \frac{\|\omega\phi_2(\Omega)\|_X}{\|\omega\|_{L_{p'}}} = \sigma(\phi_1) + \sigma(\phi_2).$$

From the validity of this relation for arbitrary ϵ we deduce our inequality. The theorem is proved.

Using $\sigma(\phi)$ let us introduce a new space with the norm

$$\|\phi(E)\|_{\Phi_p} = \sigma(\phi). \tag{1.36}$$

Remark. It is useful to note that formula (1.33) may also be used for the definition of $\|\ \|_{\Phi_1}$. For this it is necessary to set

$$\| \omega \|_{L_\infty} = \sup | \omega(x) |, \tag{1.37}$$

where only values assumed by the step function $\omega(x)$ on sets of positive measure enter into the calculation of sup. The proof is based on the following lemma.

Lemma 1. *Let* $\omega(x) = \alpha_i$ *for* $x \in E_i$ $(i = 1, 2, \cdots, N)$ *or*

$$\omega(x) = \sum_{i=1}^{N} \alpha_i \xi_i(x),$$

where $\xi_i(x)$ *is the characteristic function of the set* E_i.

Let $| \alpha_i | \leq 1$. *In the positive function*

$$\lambda(\alpha_1, \alpha_2, \cdots, \alpha_N) = \| \sum \alpha_i \phi(E_i) \|_X$$

of the variables $\alpha_1, \alpha_2, \cdots, \alpha_N$ *one may replace each of the* α_i *by* ± 1 *in such a way that it does not decrease.*

For the proof of the lemma it is sufficient to note that the function $\| \sum \alpha_i \phi(E_i) \|_X$ is convex in each α_i and therefore assumes its largest value at an end of the interval of variation of this variable.

Using Lemma 1 we see that the expression $\| \sum \alpha_i \phi(E_i) \|_X$ in the numerator of formula (1.33) for $| \alpha_i | \leq 1$, and hence for all step functions such that $\| \omega(x) \|_{L_\infty} = 1$ can only increase if we replace the function by some other one which assumes only the two values ± 1. Hence it is clear that it is sufficient to calculate sup only for such $\omega(x)$ and this proves that formula (1.33) is equivalent to (1.30).

Let us establish the relation between $\| \phi \|_{\Phi_1}$ and $\| \phi \|_{\Phi_p}$.

We have the inequality

$$\| \phi \|_{\Phi_p} \geq \| \phi \|_{\Phi_1} | \Omega |^{-1/p'}, \tag{1.38}$$

where $| \Omega |$ is the volume of the domain Ω. Indeed, let

$$\| \phi \|_{\Phi_1} = \| \omega \phi(\Omega) \|_X + \epsilon,$$

where ω is some step function equal to $+ 1$ or $- 1$ and ϵ is an arbitrarily small positive constant.

But for such an ω we have $\| \omega \|_{L_{p'}} = | \Omega |^{1/p'}$, where $| \Omega |$ is the volume of the domain Ω, and so

$$\| \omega \phi \|_X = \frac{\| \omega \phi \|_X}{\| \omega \|_{L_{p'}}} | \Omega |^{1/p'},$$

i.e.,

$$\| \omega \phi \|_X \leq | \Omega |^{1/p'} \sup \frac{\| \omega \phi \|_X}{\| \omega \|_{L_{p'}}}.$$

From this and from (1.36) follows our theorem.

From inequality (1.38), in particular, it follows that for $\sigma(\phi) = 0$ the function ϕ is equal to zero. By the same inequality $\|\phi\|_{\Phi_p}$ is subordinate to $\|\phi\|_{\Phi_1}$.

§2. Integrals of abstract functions

1. Among the set functions $\phi(E)$ a special role in our investigation will be played by those that are indefinite integrals of abstract functions. Let us dwell somewhat more in detail on this question.

Consider the set $\mathfrak{M}$ of abstract step functions, i.e., those functions $\phi(x)$ which assume only a finite number of values:

$$\phi(x) = \alpha_k, \quad x \in E_k, \quad \text{where} \quad \alpha_k \in X \ (k = 1, 2, \cdots, N).$$

A mapping of $\mathfrak{M}$ into additive set functions called the *operator of indefinite integration* is represented by the formula

$$\phi(E) = \int_E \phi(x)\, dx = \Sigma \ \alpha_k\, m(E \cap E_k). \tag{2.1}$$

The set $\mathfrak{M}^*$ of functions $\phi(E)$ obtained in this way will, of course, not be closed in the given norm B_p, $p \geq 1$, or Φ_p, $p \geq 1$. The closure of $\mathfrak{M}^*$ contains many elements not in $\mathfrak{M}^*$.

Let us construct the closure of the operator of indefinite integration. For this it is first necessary to make some choice of convergence in the vector space of abstract point functions. Let such a convergence be given. Call it R-*convergence*. The integration operator can be extended to those elements $\overline{\mathfrak{M}}$ of the closure of $\mathfrak{M}$ for which we have R-convergence of a sequence $\phi_k(x)$ of elements of $\mathfrak{M}$ and at the same time convergence of the corresponding set functions $\phi_k(E)$ in the norm Φ in the Banach space of abstract set functions.

Closure of such an operator is possible only in case, under such a definition, the unique limit for the image $\phi_k(E)$ of a sequence $\phi_k(x)$ converging to zero must be zero. It is necessary to verify this property for various R-convergences and various norms Φ.

The simplest R-convergence is uniform convergence. It is obvious that if the sequence $\phi_k(x)$ converges uniformly to zero, then the sequence $\phi_k(E)$ also converges to zero in any of the norms Φ_p or B_p.

Therefore, for bounded measurable functions the operator of integration as closure of the operator (1.36) always has meaning. Now let us consider convergence almost everywhere as the property of R-convergence. Functions obtained by closing $\mathfrak{M}$ in this convergence are called measurable. Let us verify that it follows from convergence almost everywhere to zero of the sequence $\phi_k(x)$ from $\mathfrak{M}$ and convergence of $\phi_k(E)$ in the norm Φ_p or B_p that the limit of $\phi_k(E)$

can only be zero.

Lemma. *An almost everywhere convergent sequence of measurable point functions* $\phi_k(x)$ *will converge uniformly on a system of closed sets with measure as near as desired to the measure of* Ω.

We shall carry over the proof repeating word for word the well-known proof for numerical functions (see [3]). Let the limit function for $\phi_k(x)$ be $\phi_0(x)$. Consider the system of sets $E_\nu(\epsilon)$ such that for $x \in E_\nu(\epsilon)$

$$\| \phi_\nu(x) - \phi_0(x) \| < \epsilon \quad \text{for all} \quad \nu > N.$$

The system $E_\nu(\epsilon)$ obviously forms an expanding system for given ϵ:

$$E_{\nu+1}(\epsilon) \supset E_\nu(\epsilon). \tag{2.2}$$

It is further obvious that

$$\lim_{\nu \to \infty} mE_\nu(\epsilon) = m\Omega.$$

If this were not so for some ϵ, then there would be a set $\Omega \setminus E_\nu$ with measure greater than zero on which convergence of $\phi_k(x)$ to $\phi_0(x)$ would not hold.

Each of the sequences $E_\nu(1/2^k)$ $(\nu = 1, 2, \cdots)$ is covered by the preceding ones, i.e., for all $E_\nu(1/2^k)$ and any $s > 0$ there is a set $E_m(1/2^{k-s})$ such that $E_\nu(1/2^k) \subset E_m(1/2^{k-s})$.

In addition, there exists a sequence covered by all of the $E_m(1/2^s)$ with measure converging to the measure of Ω (see [3], Lemma 7, lecture VI). This means that by choosing a sufficiently large member E_0 of this sequence, with measure larger than $m\Omega - \delta$, we will find for it a sufficiently large N such that on E_0: $\| \phi_k - \phi_0 \| < 1/2^s$, for $k > N(s)$, as was to be proved.

Now we note that, if $\phi_k(E)$ converges to $\phi(E)$ in the norm Φ_1 (and Φ_1 is the broadest of the convergences in norm Φ_p or B_p), then by Theorems 10 and 11 equicontinuity holds for $\phi_k(E)$ as well as the limit function. Having chosen δ so as to have $\| \phi(E) \|_X < \epsilon$ for $mE < \delta$, we see that

$$\| \phi_k(E) - \phi(E) \|_X \leq \| \phi_k(E_0) - \phi(E_0) \|_X +$$

$$+ \| \phi_k(\Omega - E_0) - \phi(\Omega - E_0) \|_X \leq 2\epsilon,$$

which proves our theorem.

The class of functions $\phi(x)$ on which is defined the integration operator obtained by taking the closure of the operator (2.1) will be the class of respectively Φ_1, Φ_p, B_1 or B_p summable abstract functions.

It is useful to prove that to those summable in any sense belong the functions continuous on a closed and bounded domain. Such functions form a Banach space K with norm $\|\phi(x)\|_K = \max \|\phi(x)\|_X$.

For the proof it is sufficient to construct partitions of the domain into sub-domains Δ_i on which the oscillation of the function $\phi(x)$ converges to zero and take as approximating functions ones that have constant values on each cell of a certain partition. Such functions converge to $\phi(x)$ uniformly. By what has been said above the closure of the integration operator is defined on them. Note that the definition of Bochner in our interpretation is the closure of the integration operator in the norm B_1 for almost everywhere converging functions.

We have used step functions as the point of departure. A second way of constructing indefinite integrals starting from abstract, continuous point functions is possible. It is useful to state its fundamental features.

Let x be a point of the Euclidean space R_n. We shall consider abstract, continuous functions defined in a domain Ω and assuming values in a Banach space X.

The property of continuity means that

$$\|\phi(x + \Delta x) - \phi(x)\|_X < \epsilon \quad \text{for} \quad |\Delta x| < \delta(\epsilon). \tag{2.3}$$

We shall define the concept of integral of a continuous function.

Consider first a set E_1 consisting of a finite number of cells of some cubical net. We shall call such sets *grid sets*.

For each cube lying in Ω we shall consider some division Ξ of it into smaller cubes by dividing each side into 2^s parts.

Let Q_i^{Ξ} be the cells of such a subdivision ($i = 1, 2, \cdots$).

Consider all values of $\phi(x)$ in the cube Q_i^{Ξ}. These values form in the space X some set $M_{i\Xi}$ with minimal convex hull N_i^{Ξ}. Let us multiply according to Minkowski each such convex set by the volume of the cube Q_i^{Ξ}, i.e., multiply each vector by this volume, and add all of the convex sets, corresponding to every Q_i^{Ξ} (form the set consisting of all sums of one vector from each). The sum of these sets forms a new convex set Λ_{Ξ} in the space X.

Now suppose that some new subdivision H is a daughter subdivision of Ξ, i.e., consists of cells each of which lies only in one of the cells of Ξ.

We shall prove that the convex set Λ_H is then a subset of the set Λ_{Ξ}.

The set Λ_{Ξ} is the closure of a set of elements of the form

$$\xi^{\Xi} = \sum_{i=1}^{N} Q_i^{\Xi}(a_1^{(i)}\xi_1^{(i)} + a_2^{(i)}\xi_2^{(i)} + \cdots + a_N^{(i)}\xi_N^{(i)}), \tag{2.4}$$

obtained for all possible positive $a_s^{(i)}$ such that

$$\sum_{i=1}^{N} a_s^{(i)} = 1.$$

In the parentheses in (2.4) let us group those terms that correspond to different cells Q_j^Ξ into which Q_i^Ξ divides. We will obtain

$$\xi = \sum Q_i^\Xi \left[(a_1^{(i)} \xi_1^{(i)} + a_2^{(i)} \xi_2^{(i)} + \cdots + a_{k_1}^{(i)} \xi_{k_1}^{(i)}) + \right.$$

$$+ (a_{k_1+1}^{(i)} \xi_{k_1+1}^{(i)} + \cdots + a_{k_2}^{(i)} \xi_{k_2}^{(i)}) + (a_{k_2+1}^{(i)} \xi_{k_2+1}^{(i)} + \cdots + a_{k_3}^{(i)} \xi_{k_3}^{(i)}) + \cdots$$

$$\left. \cdots + (a_{k_2+1}^{(i)} \xi_{k_2+1}^{(i)} + \cdots + a_N^{(i)} \xi_N^{(i)}) \right].$$

Introducing $a_{k_j+1}^{(i)} + \cdots + a_{k_j+1}^{(i)} = \beta_j$ we can take out of each parentheses the multiplier β_s, after which the sum takes the form

$$\xi^\Xi = \sum_i [\beta_1 Q_i^\Xi (\beta_1^{(i)} \xi_1^{(i)} + \cdots + \beta_{k_1}^{(i)} \xi_{k_1}^{(i)}) +$$

$$+ \beta_2 Q_i^\Xi (\beta_{k_1+1}^{(i)} \xi_{k_1+1}^{(i)} + \cdots + \beta_{k_2}^{(i)} \xi_{k_2}^{(i)}) +$$

$$+ \beta_3 Q_i^\Xi (\beta_{k_2+1}^{(i)} \xi_{k_2+1}^{(i)} + \cdots + \beta_{k_3}^{(i)} \xi_{k_3}^{(i)}) + \cdots$$

$$\cdots + \beta_r Q_i^\Xi (\beta_{k_r+1}^{(i)} \xi_{k_r+1}^{(i)} + \cdots + \beta_{k_r+1}^{(i)} \xi_{k_r+1}^{(i)})].$$

Equating this vector with the vector

$$\xi^H = \sum_j Q_j^H (\beta_1^{(j)} \xi_1^{(j)} + \beta_2^{(j)} \xi_2^{(j)} + \cdots + \beta_r^{(j)} \xi_r^{(j)}),$$

we see that any vector ξ^H is found among the vectors ξ^Ξ and is obtained from this formula for $\beta_i Q_i^\Xi = Q_j^H$.

The converse is, generally speaking, not true, since not every vector ξ^Ξ is contained among the ξ^H.

Consequently, the set ξ^H is a subset of ξ^Ξ and so its closure Λ_H is also a subset of Λ_Ξ the closure of ξ^Ξ, as was to be proved.

For any continuous function $\phi(x)$, as the diameter of the largest cell Q_i of the subdivision Ξ decreases indefinitely, the diameter of the body Λ_Ξ decreases indefinitely. In fact, this diameter does not exceed

$$\sum m Q_i \max_{x_1, x_2 \in Q_i} \| \phi(x_1) - \phi(x_2) \| < \epsilon.$$

Consequently, as the diameter converges indefinitely to zero, we will obtain a system of nested, closed, convex sets Λ_Ξ. Obviously, there is one and only one point common to all Λ_{Ξ}. We shall call this point the *integral* of $\phi(x)$ over Ω.

Thus we can define the integral over any grid set, which gives us the function

$$\phi(E) = \int\limits_E \phi(x)\,dx, \tag{2.5}$$

where E is any set consisting of a sum of cubes.

Let us prove an important property of the function $\phi(E)$ defined in this manner:

The function $\phi(E)$, defined on grid sets, is absolutely continuous with respect to the norm Φ_p $(p \geq 1)$.

The proof is elementary since the norm $\|\phi\|_{B_p}$ is subordinate to the norm $\|\phi\|_{\Phi_p}$ and the norm $\|\phi\|_{\Phi_p}$ is monotone. Thus

$$\|\Psi_\phi(E_1)\|_{\Psi_p}^p \leq \int\limits_{E_1} \|\phi(x)\|_X^p\,dx \leq m\|\phi(x)\|_X^p\,mE_1.$$

For any Lebesgue set E_0 we can determine a grid set E_1 such that

$$m(E_0 \setminus E_0 \cap E_1) < \epsilon, \quad m(E_1 \setminus E_0 \cap E_1) < \epsilon;$$

from which it follows that the system of grid sets is dense in the metric space Y of Lebesgue sets in Ω, which we introduced above.

The function $\phi(E)$, uniformly continuous on an everywhere dense set in Y, can be uniquely extended to a continuous function on all of Y by Theorem 6.

The $\phi(E)$ so obtained is the indefinite integral of ϕ over any measurable set.

All of the rest of the construction is carried out in exactly the same way as for step functions. We will again obtain the same class of measurable functions as the closure of the continuous functions under almost everywhere convergence.

The Bochner integral is the closure of integrals of step functions in the sense of the metric $\|\phi\| = \int \|\phi\|_X\,dx$. The metric of Bochner is subordinate to the one considered by us, i.e., convergence in our sense follows from convergence in the sense of Bochner.

Thus it follows that the domain of definition of the integration operator according to Bochner is a subset of the domain of definition of integration in the sense Φ_1. We shall show that it is a proper subset, i.e., that an abstract function may fail to have an integral in the sense of Bochner and nevertheless be integrable in our sense. With the aim of proving this let us consider a small example.

Example 1. As the domain Ω we take the interval $0 \leq x \leq 1$ and let the space X represent the l_2-space of numerical sequences with convergent sum of squares.

The elements of l_2 are $a(a_1, a_2, \cdots)$ with

$$\| a \| = \sqrt{\sum_{k=1}^{\infty} | a_k |^2}. \tag{2.6}$$

Let i_k be the element of l_2 in which the kth component equals 1 and all remaining components are zero. Then

$$a = \sum_{k=1}^{\infty} a_k i_k. \tag{2.7}$$

We define a function $\phi(x)$ with values in l_2 by the formulas

$$\phi(x) = c_k i_k \text{ for } 2^{-k} < x \le 2^{-k+1},$$

$$\phi(0) = 0, \text{ where } c_k > 0. \tag{2.8}$$

Let us find out what c_k must be in order that the function $\phi(x)$ should be an element of various spaces.

a. In order to have $\phi(x) \in B_1$ it is necessary and sufficient that

$$\int \| \phi(x) \| \, dx < \infty;$$

this gives

$$\text{for } \Sigma \frac{c_k}{2^k} < + \infty \quad \phi(x) \in B_1, \tag{2.9}$$

$$\text{for } \Sigma \frac{c_k}{2^k} \longrightarrow + \infty \quad \phi(x) \notin B_1. \tag{2.10}$$

b. Analogously, we obtain for belonging to B_2:

$$\text{for } \Sigma \frac{c_k^2}{2^k} < + \infty \quad \phi(x) \in B_2, \tag{2.11}$$

$$\text{for } \Sigma \frac{c_k^2}{2^k} \longrightarrow + \infty \quad \phi(x) \notin B_2. \tag{2.12}$$

c. For the function $\phi(x)$ to belong to Φ_1 it is obviously necessary that the norms

$$\left\| \int_{2^{-s}}^{1} \phi(x) \, dx \right\| = \sqrt{\sum_{k=1}^{s} \frac{c_k^2}{2^{2k}}} \tag{2.13}$$

be bounded, and hence

$$\phi(x) \notin \Phi_1 \quad \text{for} \quad \Sigma \, \frac{c_k^2}{2^{2k}} \longrightarrow +\infty. \tag{2.14}$$

We can easily see that boundedness of the indicated norms is also a sufficient condition for $\phi(x)$ to belong to Φ_1;

$$\text{for} \quad \Sigma \, \frac{c_k^2}{2^{2k}} < +\infty \quad \phi(x) \in \Phi_1. \tag{2.15}$$

We need to prove that, when (2.15) is satisfied, the operator of indefinite integration can be extended to $\phi(x)$ by continuity from the set of step functions. If we replace $\phi(x)$ by the function $\phi^{(s)}(x)$, namely, the projection of $\phi(x)$ onto the s-dimensional space with basis vectors $(i_1, i_2, \cdots, i_s)$, we see that

$$\| \phi^{(m+p)}(x) - \phi^{(m)}(x) \|_{\Phi_1} \le \sqrt{\sum_{k=m}^{m+p} (c_k^2/2^{2k})}, \tag{2.16}$$

from which convergence of $\phi^{(m)}(x)$ to $\phi(x)$ is clear, Q.E.D.

d. Now let us consider the conditions for $\phi(x)$ to belong to Φ_2. Let $\omega(x)$ be the numerical step function defined by the equations

$$\omega(x) = g_k, \quad 2^{-k} < x \le 2^{-k+1},$$

$$\omega(0) = 0, \quad \text{where} \quad g_k > 0,$$

and let

$$\| \omega(x) \|_{L_{p'}} = \| \omega(x) \|_{L_2} = 1,$$

i.e.,

$$\Sigma \, \frac{g_k^2}{2^k} = 1.$$

Let us introduce the set function

$$\phi(E) = \sum_{k=1}^{\infty} m(E \cap [2^{-k} < x \le 2^{-k+1}]) \, c_k i_k. \tag{2.17}$$

This set function will serve as the indefinite integral of $\phi(x)$ in the sense B_1 in the case $\Sigma \, (c_k/2^k) < +\infty$.

We shall look at some sort of condition of boundedness of the norm of this function in Φ_2. Using the fact that

$$\| \int \omega(x) \phi^{(m)}(x) \, dx \|^2 = \left[\sum_{k=1}^{\infty} \frac{g_k^2 \, c_k^2}{2^{2k}} \right], \tag{2.19}$$

we will give to this condition the form

$$\sum_{k=1}^{\infty} \frac{g_k^2 c_k^2}{2^{2k}} < + \infty.$$
(2.20)

It is easy to see that condition (2.20) is equivalent to boundedness of the relation

$$c_k^2 / 2^k < M.$$
(2.21)

Indeed, for any sequence

$$c_{k_j}^2 / 2^{k_j}$$
(2.22)

converging to ∞, we can always select a convergent series

$$\sum \frac{g_{k_j}^2}{2^{k_j}} = 1,$$

such that termwise multiplication of its terms by the sequence (2.22) makes it divergent. Consequently, (2.21) is necessary for the validity of (2.20). Its sufficiency is obvious.

Nevertheless, condition (2.21) in the given case is not sufficient in order to continuously extend the indefinite integration operator to $\phi(x)$ in the metric Φ_2. In other words, $\phi(x)$ will not, generally speaking, be the limit of step functions in this metric. Let us call Ψ_2 the domain of definition of the operator of indefinite integration in the norm Φ_2. By what was proved above, every function

$$\phi(E) = \int_E \phi(x)\, dx, \quad \phi \in \Psi_2,$$

must be absolutely continuous in the metric of Φ_2.

We shall prove that such absolute continuity will hold if and only if

$$\lim_{k \to \infty} \frac{c_k^2}{2^k} = 0.$$
(2.23)

In fact, if $c_{k_j}^2 / 2^{k_j} > \epsilon_0$, then for an infinite set of values of k_j there are intervals $\Delta_k (2^{-k_j} < x \le 2^{-k_j+1})$ such that $m\Delta_k = 2^{-k_j}$ and

$$\| \psi_\phi(\Delta_{k_j}) \|_\Phi > \epsilon_0.$$
(2.24)

Inequality (2.24) can be obtained from (2.19) by setting

$$g_k = \begin{cases} 0 & \text{for } k \ne k_j, \\ 2^{k_j} & \text{for } k = k_j. \end{cases}$$

Consequently, (2.23) is necessary.

Now if we estimate the norm of $\phi^{(m+p)}(E) - \phi^{(m)}(E)$ by means of formula (2.19), we will have after easy calculations

$$\| \phi^{(m+p)}(E) - \phi^{(m)}(E) \| \leq \sup \frac{c_k^2}{2^k}.$$

Hence follows the sufficiency of condition (2.23) for $\phi(x)$ to belong to Ψ_2.

The conditions that we have derived prove that:

a. *The space B_1 is a proper subspace of Φ_1.*

It is sufficient to set $c_k = 2^k/k$. Then we will have

$$\phi(x) \in \Phi_1, \quad \phi(x) \notin B_1.$$

b. *The space B_2 is a proper subspace of Φ_2.*

It is sufficient to set $c_k = 2^{k/2}/\sqrt{k}$. Then we obviously will have

$$\phi(x) \in \Phi_2, \quad \phi(x) \notin B_2.$$

c. *The space Ψ_2 is a proper subspace of Φ_2. There are elements in Φ_2 that are not absolutely continuous in the sense of Φ_2.*

It is sufficient to set $c_k = 2^{k/2}$. We will obtain

$$\phi(E) \in \Phi_2, \quad \phi(E) \notin \Psi_2.$$

d. *There exist functions of B_1 that do not belong to Φ_2.*

Let $c_k = 2^k/k^2$. Then

$$\phi(x) \in B_1, \quad \phi(x) \notin \Phi_2.$$

The following example shows that among the elements of Φ_2 are contained ones not belonging to B_1.

Example 2. Let Ω again be $[0, 1]$ and X be l_2. We construct the function $\phi(x)$ by the formulas

$$\phi(x) = c_k i_k, \quad \frac{1}{\sqrt{k+1}} < x \leq \frac{1}{\sqrt{k}},$$

$$\phi(0) = 0 \quad \text{where} \quad c_k > 0.$$

Similarly to Example 1 we can obtain the conditions:

$$\text{for } \Sigma \frac{c_k}{k^{3/2}} < \infty \quad \phi(x) \in B_1, \tag{2.25}$$

$$\text{for } \Sigma \frac{c_k}{k^{3/2}} \to +\infty \quad \phi(x) \notin B_1, \tag{2.26}$$

$$\text{for } \Sigma \frac{c_k^2}{k^{3/2}} < +\infty \quad \phi(x) \in B_2, \tag{2.27}$$

$$\text{for } \Sigma \frac{c_k^2}{k^{3/2}} \to +\infty \quad \phi(x) \notin B_2, \tag{2.28}$$

$$\text{for } \Sigma \frac{c_k^2}{k^3} < \infty \qquad \phi(x) \in \Phi_1, \; \phi(x) \in \Psi_1, \tag{2.29}$$

$$\text{for } \Sigma \frac{c_k^2}{k^3} \to +\infty \qquad \phi(x) \notin \Phi_1, \; \phi(x) \notin \Psi_1, \tag{2.30}$$

and furthermore

$$\text{for } \frac{c_k^2}{k^{3/2}} < M < \infty \quad \phi(E) \in \Phi_2, \tag{2.31}$$

$$\text{for } \frac{c_{k_j}^2}{k_j^{3/2}} \to +\infty \qquad \phi(E) \notin \Phi_2, \tag{2.32}$$

$$\text{for } \frac{c_k^2}{k^{3/2}} \to 0 \qquad \phi(E) \in \Psi_2, \tag{2.33}$$

$$\text{for } \frac{c_{k_j}^2}{k_j^{3/2}} > \epsilon_0 \qquad \phi(E) \notin \Psi_2. \tag{2.34}$$

Set

$$c_k = k^{5/8}. \tag{2.35}$$

Then

$$\Sigma \frac{c_k}{k^{3/2}} = \Sigma \, k^{-1/6} \to +\infty$$

and so

$$\phi(x) \notin B_1. \tag{2.36}$$

On the other hand, $c_k^2/k^{3/2} = k^{-1/4} \to 0$ and so

$$\phi(x) \in \Phi_2. \tag{2.37}$$

The closure in the metric of Φ_p of the manifold of continuous functions represents some subspace Ψ_p, which may be proper or improper (i.e., coincide with Φ_p).

We shall now show that this manifold Ψ_p can be broader than the set of integrals of all summable functions. This is shown by the following examples.

Example 3. Again let Ω be $[0, 1]$ and let X coincide with l_2. The functions $\chi_s(x)$ are defined by the formulas

$$\chi_0(x) = i_1, \quad \chi_1(x) = \begin{cases} i_2 & \text{for } 0 \le x \le \dfrac{1}{2}, \\[2mm] i_3 & \text{for } \dfrac{1}{2} < x \le 1, \end{cases}$$

$$\chi_2(x) = \begin{cases} i_4 & \text{for } 0 \le x \le \dfrac{1}{4}, \\[2mm] i_5 & \text{for } \dfrac{1}{4} < x \le \dfrac{1}{2}, \\[2mm] i_6 & \text{for } \dfrac{1}{2} < x \le \dfrac{3}{4}, \\[2mm] i_7 & \text{for } \dfrac{3}{4} < x \le 1. \end{cases} \tag{2.38}$$

Let

$$\chi_s(E) = \int_E \chi_s(x)\, dx.$$

We construct the series

$$\phi(E) = \sum_{s=1}^{\infty} \frac{2^{s/2}}{s} \chi_s(E). \tag{2.39}$$

We shall show that this series converges in the sense of the norm $\|\phi\|_{\Phi_2}$. Indeed,

$$\chi_s(\Omega) = \int_\Omega \chi_s(x)\, dx = \sum_{k=1}^{2^s} \frac{i_{jk}}{2^s}. \tag{2.40}$$

Calculating the norm of $\chi_s(\Omega)$, we will have

$$\|\chi_s(\Omega)\|_{l_2} = \sqrt{\sum_{k=1}^{2^s} (1/2^{2s})} = 2^{-s/2}.$$

It is obvious that all of the $\chi_s(\Omega)$ are orthogonal. Therefore, convergence of the series $\|\chi_s\|^2\, 2^s/s^2$ is sufficient for convergence of the series for ϕ. It is easily seen that this series converges. Its limit is $\|\phi(E)\|^2$. On the other hand, it is obvious that the series of point functions $\sum \chi_s(x)$ does not converge.

We shall show that there exists no function $\phi(x)$ for which $\phi(E)$ is the indefinite integral. But first it is useful to make a more general remark.

Let the abstract function $\phi(x)$ with values in l_2, defined on the interval

[0, 1], be integrable, i.e., let the integral

$$\phi(E) = \int_E \phi(x)\, dx$$

exist over any measurable set. This means, by our definition, that there exists a sequence of step functions $\phi_k(x)$ converging to $\phi(x)$ almost everywhere and such that the sequence $\phi_k(E)$ converges.

In other words, for any $\epsilon > 0$ we can determine $N(\epsilon)$ such that

$$\|\phi_m - \phi_p\|_{\Phi_1} < \epsilon \quad \text{for} \quad m,\, p > N(\epsilon).$$

We shall show that it is possible in this case, instead of the indicated sequence $\phi_k(x)$ in the definition of the integral, to consider the sequence

$$\phi^{(r)}(x) = \sum_{s=1}^{r} i_s \widetilde{\phi}^{(s)}(x)$$

of partial sums of the orthogonal expansion of the function $\phi(x)$. By $\widetilde{\phi}^{(s)}(x)$ we shall denote the components of this expansion of the function $\phi(x)$. This sequence $\phi^{(r)}(x)$ will obviously converge to $\phi(x)$ almost everywhere (everywhere that $\phi(x)$ has meaning).

Note that each function $\phi^{(r)}(x)$ is integrable, for its values are finite-dimensional and each component is a measurable, summable function.

It remains to show that in the integral

$$\int_E \phi^{(r)}(x)\, dx$$

we can pass to the limit as $r \to \infty$, i.e.,

$$\int_E \phi(x)\, dx = \lim \int_E \phi^{(r)}(x)\, dx = \sum_{s=1}^{\infty} i_s \int_E \widetilde{\phi}^{(s)}(x)\, dx. \tag{2.41}$$

We may establish by classical considerations that the sequence $\phi^{(r)}(x)$ converges to $\phi(x)$ uniformly on sets F_h with measure as near to unity as desired.

From the absolute continuity of $\phi(x)$ (see Theorem 9) it follows that for any ϵ there exists a $\delta(\epsilon)$ such that

$$\left\| \int_E \phi(x)\, dx \right\|_X < \epsilon \quad \text{if} \quad mE < \delta(\epsilon).$$

However,

$$\left\| \int_E \phi^{(r)}(x)\, dx \right\|_X \le \left\| \int_E \phi(x)\, dx \right\|_X$$

and so for sufficiently small δ the integrals of $\phi^{(r)}(x)$ taken over the set

$[0, 1] \setminus F_h$ will be as small as desired.

But on F_h all of the $\phi^{(r)}(x)$ converge to $\phi(x)$ uniformly. From this follows formula (2.41).

But this means that for our function $\phi_0(x)$, the integral of which would coincide with $\phi(E)$, all components in the space l_2 would have to coincide with $\chi_s(E) 2^{s/2}/s$ and the series (2.39) would have to converge to $\phi(E)$ almost everywhere, which is impossible.

One more example of the same kind as just considered has an important meaning for the following.

Example 4. Let x and y be two vectors in n-dimensional Euclidean space R_n and let r be the modulus of their difference, $r = |x - y|$. We shall assume that they belong to some bounded domain Ω. For fixed x the function $1/r^\lambda$ as a function of y represents an element of the space $L_{p'} \, (p' > 1)$ for any λ satisfying the inequality

$$\lambda < n/p'. \tag{2.42}$$

Consequently, $1/r^\lambda$ is then a point function taking values in $L_{p'}$. This function is continuous, for

$$\left| \frac{1}{r^\lambda(x_1, y)} - \frac{1}{r^\lambda(x, y)} \right| < |x_1 - x| \, (r + r_1)^{\lambda - 1} \tag{2.43}$$

and the norm of this last fraction in $L_{p'}$ with respect to the variable y is as small as desired. The modulus of continuity of $1/r^\lambda$ is equal to ϵ^β, where $\beta > 0$.

Now consider this same function $1/r^\lambda$ for $n/p' < \lambda < n$.

This function will not be for any $x \in \Omega$ an element of $L_{p'}$ in the domain Ω of variation of the point y, since $1/r^{\lambda p'}$ will not be summable. However, if we stop considering $1/r^\lambda$ as a function of the point x and go over to set functions, the situation changes. Let S_s be some linear manifold in Ω whose variable vector x' is defined by the equations $x'_{s+1} = \mathrm{const}, \ x'_{s+2} = \mathrm{const}, \cdots, x'_n = \mathrm{const}$.

In this manifold let us consider the sets J'_s that are Lebesgue measurable (in the variables $x'_1, x'_2, \cdots, x'_s$). Instead of a function of the point x^* we shall consider $1/r^\lambda(x^*, y)$ as a function of the set J'_s and of the vector $x_2(x_{s+1}, \cdots, x_n)$ according to the formula

$$\frac{1}{r^\lambda}(J'_s, x_2 \mid y) = \int_{J'_s} \frac{1}{r^\lambda(x^*, y)} \, dx'_1 \cdots dx'_n,$$

where the components of x^* are $(x'_1, x'_2, \cdots, x'_s, x_{s+1}, x_{s+2}, \cdots, x_n)$.

With this understanding, as we shall prove later, the function $1/r^\lambda$ will be an abstract function of J_s' and x_2 with values in $L_{p'}$, which is continuous in the metric Φ_{q^*} on S_s, where $q^* \le q$; $s/q = n/p - (n - \lambda)$.

Let us introduce another vector x_1, lying in the subspace of $x_1, x_2, \cdots, x_s$. Then $x = x_1 + x_2$ represents an arbitrary n-dimensional vector. The function

$$\frac{1}{r^\lambda} (J_s' + x_1, x_2 \mid y)$$

will be an abstract function of the point x and the set J_s' with values in $L_{p'}$ with respect to the point y, which is continuous in x for $q^* < q$ and absolutely continuous in J_s' in the metric of Φ_{q^*} on S_s.

The proof of this assertion follows from the imbedding theorem given, for instance, in the author's book [1] (see [2, 6]).

Indeed, in the book cited it is established that

$$\left[\int_{S_s} \left| \int \frac{\phi(y)}{r^\lambda} \, dy \right|^{q^*} dS_s \right]^{1/q^*} \le A \| \phi(y) \|_{L_p} \tag{2.44}$$

for any $\phi(y) \in L_p$. Consequently, the form

$$A_\lambda(\omega, \phi) = \int_{S_s} \omega(x') \left(\int_\Omega \frac{\phi(y)}{r^\lambda} \, dy \right) dx' \tag{2.45}$$

satisfies the inequality $| A_\lambda(\omega, \phi)| \le A \| \phi(y) \|_{L_p} \| \omega(x) \|_{L_{q^{*'}}}$.

Thus it follows that this form is a linear functional of $\phi(y)$ in L_p and the integral

$$\int \frac{\omega(x_1')}{r^\lambda} \, dx_1'$$

is an element of $L_{p'}$ for any $\omega(x')$ satisfying the inequality

$$\left\| \int \frac{\omega(x_1')}{r^\lambda} \, dx_1' \right\| \le A \| \omega(x_1') \|_{L_{q^{*'}}}.$$

Our assertion follows from (57) if we recall the definition of the norm Φ_{q^*} on S_s. In exactly the same way the continuity in x of the function $1/r^\lambda$ is proved using a theorem of Kondrašov (see [1]).

According to the theory developed in §1, the closure of the set of integrals of continuous functions is the set functions $\phi(E)$ that are absolutely continuous and continuous with respect to translation. The question arises whether these two properties are independent of each other. For numerical functions both of these properties are consequences of the fact of existence of the integrals, but for abstract set functions the two continuities prove to be not coincident and not consequences of the existence of $\phi(E)$. We shall now cite an example of a set function which is absolutely continuous with respect to the metric Φ_1 but is not continuous with respect to translation in the same metric.

Example 5. Let X be the space of bounded sequences $(a_1, a_2, \cdots)$.

We shall denote as before by i_k the element of this space having all coordinates zero except the kth, which equals one. Any element ϕ from m can be written conditionally in the form of the series

$$\phi = \sum_{k=1}^{\infty} a_k i_k \tag{2.46}$$

(this series will not converge in the metric of m), where all of the a_k are bounded. As the norm in m we take as usual $\max |a_k|$. Let us divide the segment $0 \le x \le 1$ into 2^s parts of equal length and consider on this segment the abstract function $\chi_s(x)$ which assumes the value i_s on all even-numbered segments and $-i_s$ on all odd-numbered segments:

$$\chi_1(x) = \begin{cases} i_1 & \text{for } 0 \le x \le \frac{1}{2}, \\ -i_1 & \text{for } \frac{1}{2} < x \le 1, \end{cases}$$

$$\chi_2(x) = \begin{cases} i_2 & \text{for } 0 \le x \le \frac{1}{4}, \\ -i_2 & \text{for } \frac{1}{4} < x \le \frac{1}{2}, \\ i_2 & \text{for } \frac{1}{2} < x \le \frac{3}{4}, \\ -i_2 & \text{for } \frac{3}{4} < x \le 1; \end{cases} \tag{2.47}$$

further, let us form

$$\psi_N(x) = \sum_{s=1}^{N} \chi_s(x). \tag{2.48}$$

Corresponding to each $\psi_N(x)$ we construct the integral

$$\psi_N(E) = \int_E \psi_N(x)\, dx.$$

We shall prove that the functions $\psi_N(E)$ converge for each E, in the norm of X, i.e., m (not with respect to $\|\ \|_{\Phi_p}$), to some limit function $\phi_0(E)$ which is absolutely continuous.

First note that the series representing these functions on all cell sets with the end points of the cells expressed by finite binary fractions terminate in a finite number of terms, since integrals of functions defined on the cells of a finer subdivision will be zero over the cells of a coarser subdivision. Thus $\phi_0(E)$ is defined on all grid sets. The functions $\phi_N(E)$ are absolutely equicontinuous, being integrals of functions bounded in norm.

Grid sets of this type are everywhere dense in the space Y. Therefore, corresponding to any measurable set we can produce a grid set E_δ such that

$$m(E \leftrightarrow E_\delta) < \delta.$$

By the absolute equicontinuity of the ϕ_N we have

$$\| \phi_m(E) - \phi_p(E) \| \leq$$

$$\leq \| \phi_m(E) - \phi_m(E_\delta) \| + \| \phi_m(E_\delta) - \phi_p(E_\delta) \| + \| \phi_p(E_\delta) - \phi_p(E) \| <$$

$$< \frac{1}{3}\epsilon + \frac{1}{3}\epsilon + \frac{1}{3}\epsilon = \epsilon \tag{2.49}$$

for sufficiently small δ and sufficiently large m and p chosen corresponding to E_δ. From this follows the convergence of $\phi_N(E)$ for all sets E and, by Theorem 2, the continuity of $\phi_0(E)$, i.e., absolute continuity in the norm of Φ_1.

On the other hand the function $Z_\phi(y)$ generated by translation from $\phi_0(E)$ is not continuous, since upon translation by 2^{-k} along the x-axis

$$\phi_0(E + 2^{-k}) - \phi_0(E) = 2 \int_E (\pm i_k) \, dx, \tag{2.50}$$

$$\| \phi_0(E + 2^{-k}) - \phi_0(E) \| = 2. \tag{2.51}$$

This proves our assertion.

§3. The fundamental theorem about Ψ_1 and Ψ_p

1. In §2 we considered integration of abstract functions with the aid of the closure of the operator of indefinite integration. Below we shall study another operator acting in the space of abstract set functions, the operator of multiplication by a numerical function.

For numerical step functions $\omega(x) = \alpha_i$, $x \in E_i$, and an arbitrary abstract, additive set function we define the integral

$$\psi(E) = \int_E \omega(x) \, d\phi(E) \tag{3.1}$$

by the formula

$$\psi(E) = \Sigma \, \alpha_i \phi(E \cap E_i). \tag{3.2}$$

If $\phi(E)$ is in turn the integral of a step function,

$$\phi(E) = \int_E \phi(x)\, dx, \quad \text{where} \quad \phi(x) = \beta_j, \quad x \in \hat{E}_j, \tag{3.3}$$

then, as is easily seen, the integral (3.1) can be expressed by the formula

$$\psi(E) = \int_E \omega(x)\,\phi(x)\, dx. \tag{3.4}$$

Thus (3.1) is the natural generalization of multiplication of an abstract function by a numerical function. The function $\psi(E)$ represents a bilinear operator over the numerical function ω and the abstract set function $\phi(E)$ with values in the space of abstract set functions.

Lemma I. *Let*

$$\phi(E) \in \Phi_p, \quad \omega(x) \in L_q, \tag{3.5}$$

where

$$p > 1, \quad q \geq 1, \quad \frac{1}{p} + \frac{1}{q} = \frac{1}{s} \leq 1. \tag{3.6}$$

(The values $p = \infty$ *and* $q = \infty$, *and also* $s = \infty$, *are not excluded.) The operator (3.1) is a bounded and consequently continuous operator in the indicated spaces (3.6) with values in* Φ_s.

The inequality

$$\|\psi\|_{\Phi_s} \leq \|\omega\|_{L_q} \|\phi\|_{\Phi_p} \tag{3.7}$$

is valid.

By proving inequality (3.7) we shall prove all of the lemma.

By definition

$$\|\psi\|_{\Phi_s} = \sup \frac{\|\int \omega'\, d\psi(E)\|_X}{\|\omega'\|_{L_{s'}}}, \quad \text{where} \quad \omega'(x) = \beta_j, \quad x \in \hat{E}_j.$$

Transforming this expression, we give it the form

$$\|\psi\|_{\Phi_s} = \sup \frac{\|\Sigma\, \alpha_i \beta_j \phi(E_i \cap \hat{E}_j)\|_X}{\|\omega\omega'\|_{L_{p'}}} \cdot \frac{\|\omega\omega'\|_{L_{p'}}}{\|\omega'\|_{L_{s'}}}. \tag{3.8}$$

But $\omega\omega'$ is again a step function, defined by the formula

$$\omega\omega' = \alpha_i \beta_j, \quad x \in E_i \cap \hat{E}_j.$$

Hence

$$\frac{\|\Sigma\, \alpha_i \beta_j \phi(E_i \cap \hat{E}_j)\|_X}{\|\omega\omega'\|_{L_{p'}}} \leq \|\phi(E)\|_{\Phi_p}. \tag{3.9}$$

Furthermore,

$$\frac{1}{s'} = 1 - \frac{1}{s} = 1 - \frac{1}{p} - \frac{1}{q} = \frac{1}{p'} - \frac{1}{q}$$

and so

$$\frac{1}{p'} = \frac{1}{s'} + \frac{1}{q}.$$

Note further that from the last inequality follows

$$\| \omega \omega' \|_{L_{p'}} \leq \| \omega \|_{L_q} \| \omega' \|_{L_{s'}}, \tag{3.10}$$

from which with (3.8) our lemma follows. With $q = p'$, i.e., for $s = 1$ inequality (3.7) gives

$$\frac{\| \omega \phi \|_{\Phi_1}}{\| \omega \|_{L_{p'}}} \leq \| \phi \|_{\Phi_p},$$

from which it follows that for the definition of $\| \ \|_{\Phi_1}$ by formula (1.36) we can consider the function ω in that formula running through not only the step functions, but also any bounded measurable functions.

From the lemma that has been proved and the results of the previous section it follows that, if $\phi(E)$ belongs to Ψ_p, i.e., is the limit of a sequence of step functions, then the function $\psi(E)$ will belong to Ψ_s. This follows from a passage to the limit in formula (3.1) and from the fact that the product of two step functions is in turn a step function.

2. Now we need to study in more detail the concept of convolution of functions as applied to abstract set functions.

Let $\omega(x)$ be some finite, numerical, measurable function, i.e., a function equal to zero outside of some finite domain. For simplicity we shall restrict ourselves here to the case where $\omega(x)$ is bounded, because for our purposes this will be sufficient.

Moreover, let $\phi(E)$ be an abstract additive set function, which for simplicity is also finite, i.e., equal to zero for all E not containing points of some finite set. Consider the integrals

$$\psi(y) = \int \omega(y - x) \, d_x \phi(E) \tag{3.11}$$

over the entire space, where under the integral y is considered to be fixed and the function $\omega(y - x)$ is considered as a function of the point x.

We shall consider this integral (3.11) as a new abstract function of the point y with values in X and we shall call $\psi(y)$ the *convolution* of the set function $\phi(E)$ with the numerical point function ω. Let us indicate still another formula for the convolution, now not for set functions, but for abstract functions. With $\phi(E) = \int_E \phi(x) \, dx$ formula (3.11) can be rewritten

$$\psi(y) = \int \omega(y - x) \phi(x) \, dx. \tag{3.12}$$

We can carry out a change of variables here by introducing $y - x = z$. Then $x = y - z$ and (3.12) can be rewritten as

$$\psi(y) = \int \omega(z)\,\phi(y - z)\,dz. \tag{3.13}$$

Let us prove another lemma:

Lemma II. *In the triple convolution*

$$\psi(y) = \int \omega_1(z)\,[\int \omega_2(y - z - x)\,d_x\phi(E)]\,dz, \tag{3.14}$$

where ω_2 and ω_1 are bounded measurable functions and $\phi(E) \in \Phi_p$, $p > 1$, is continuous with respect to translation, the associative law holds, i.e., we can change the order of carrying out the operations in it.

We have the formula

$$\psi(y) = \int [\int \omega_1(z)\,\omega_2(y - x - z)\,dz]\,d_x\phi(E). \tag{3.15}$$

Let us prove this lemma. The function $\chi(y - z) = \int \omega_2(y - x - z)\,d_x\phi(E)$ for any fixed y is a continuous function of the variable z.

Indeed, let us estimate the norm of the difference

$$\int [\omega_2(y - x - z - \Delta z) - \omega_2(y - x - z)]\,d_x\phi(E) =$$
$$= \int \omega_2(y - x - z)\,\{d_x[\phi(E + \Delta z) - \phi(E)]\}. \tag{3.16}$$

But $\phi(E)$ is continuous with respect to translation by assumption and consequently the norm of the function $\phi(E + \Delta z) - \phi(E)$ is as small as desired. By inequality (3.16) we obtain $\|\Delta \chi\| < \epsilon$. From this follows the continuity of $\chi(y - z)$.

Now let us define a new step function $\chi^{(N)}(y, z)$ in the following way:

Partition the domain of variation of z into subdomains Δ_i $(i = 1, 2, \cdots, N)$ such that the diameter of each such subdomain converges to zero as $N \to \infty$.

Set

$$\chi^{(N)}(y, z) = \chi(y - z_i), \quad z \in \Delta_i,$$

where z_i is some value of z from the domain Δ_i.

By the continuity of $\chi(y - z)$, the function $\chi^{(N)}(y, z)$ will converge uniformly to $\chi(y - z)$. Therefore, by Lemma I we can pass to the limit in the integral

$$\int \omega_1(z)\,\chi^{(N)}(y, z)\,dz$$

and write instead of (3.14) the formula

$$\psi(y) = \lim_{\substack{N \to \infty \\ d\Delta_i \to 0}} \int \omega_1(z)\,\chi^{(N)}(y, z)\,dz. \tag{3.17}$$

By expanding the expression on the right-hand side and using the representation of $\chi^{(N)}$ in each Δ_i, we transform it into the form

$$\psi(y) = \lim_{\substack{N \to \infty \\ d\Delta_i \to 0}} \sum_{i=1}^{N} \int_{\Delta_i} \omega_1(z)\, dz\, \chi(y - z_i) =$$

$$= \lim_{\substack{N \to \infty \\ d\Delta_i \to 0}} \sum_{i=1}^{N} \int_{\Delta_i} \omega_1(z)\, dz [\int \omega_2(y - x - z_i)\, d_x \phi(E)] =$$

$$= \lim_{\substack{N \to \infty \\ d\Delta_i \to 0}} \int [\sum_{i=1}^{N} \int_{\Delta_i} \omega_1(z)\, dz\, \omega_2(y - x - z_i)]\, d_x \phi(E). \qquad (3.18)$$

Now let us consider the inner sum under the last integral sign.

The function $\omega_2(y - x - z)$ is an abstract function of the point z with values in the space of functions f of the difference $y - x$ which are integrable to the power $p' < \infty$.

This function is continuous with respect to z since the norm of the difference

$$\| \omega_2(y - x - z - \Delta z) - \omega_2(y - x - z) \|_{L_{p'}},$$

will be as small as desired for sufficiently small Δz.

Now let us construct the step function $\omega_2^{(N)}(y - x, z) = \omega_2(y - x - z)$, $z \in \Delta_i$. This function will converge to $\omega_2(y - x - z)$ uniformly. The sum in the right-hand member of (3.18) may be rewritten with the aid of $\omega_2^{(N)}$ in the form of an integral and we will obtain

$$\psi(y) = \int [\int \omega_1(z)\, \omega_2^{(N)}(y - x, z)\, dz]\, d_x \phi(E). \qquad (3.19)$$

By the uniformity of convergence of $\omega_2^{(N)}(y - x, z)$ to the limit we see that the function

$$\zeta^{(N)} = \int \omega_1(z)\, \omega_2^{(N)}(y - x, z)\, dz$$

with values in $L_{p'}$ for the argument $y - x$ also converges to a limit in the norm $\| \ \|_{L_{p'}}$ and

$$\lim_{L_\infty} \zeta^{(N)} = \zeta = \int \omega_1(z)\, \omega_2(y - x - z)\, dz.$$

Therefore, from (3.19) follows

$$\psi(y) = \lim \int \zeta^{(N)}(y - x)\, d_x \phi(E) = \int \zeta(x - y)\, d_x \phi(E),$$

as was to be proved.

3. The main goal of the present section is the following theorem:

Theorem 20. *Any function $\phi(E)$ which generates by translation a continuous function $Z_\phi(y)$ is the limit of a sequence $\phi_m(E)$ of integrals of continuous functions.*

By the corollary to Theorem 13 we may assume in the formulation of this theorem that the function ϕ is continuous with respect to translation.

From this theorem and from Theorem 16 it follows that Ψ_p not only consists of functions that generate continuous $Z_\phi(y)$ by translation, but contains all such functions. Before proving this theorem, we shall have to give some preliminary definitions.

Let $\phi(E)$ be an element of Φ_p. Consider the convolution of $\phi(E)$ with the kernel

$$\omega_h(\xi) = \frac{1}{h^n}\,\omega\left(\frac{\xi}{h}\right),$$

where $\omega(\xi)$ differs from zero only in the sphere $(\xi) \leq 1$ and is such that

$$\int \omega(\xi)\,d\xi = 1. \tag{3.20}$$

We shall call such a convolution $\phi_h(y) = \int \omega_h(y - x)\,d_x \phi(E)$ a *mean function* for $\phi(E)$. In case the kernel $\omega(\xi)$ is infinitely differentiable, the mean function will also be infinitely differentiable.

Indeed, the function

$$\Delta\omega_h = \frac{\omega_h(y + \Delta y - x) - \omega_h(y - x)}{\Delta y}.$$

converges uniformly to $\partial\omega_h/\partial y$ as $\Delta y \to 0$ maintains a constant direction.

Hence it follows that

$$\frac{\partial\phi_h}{\partial y} = \lim \int \Delta\omega_h(y - x)\,d_x\phi(E) = \int \frac{\partial\omega_h}{\partial y}\,d_x\phi(E), \tag{3.21}$$

as was to be proved.

Lemma III. *If $\phi(E)$ is the integral of a continuous abstract function $\phi(x)$, then the mean function $\phi_h(y)$ will be as close as desired to $\phi(x)$ with respect to the norm $\|\phi\| = \sup \|\phi(x)\|$.*

Proof. Note that in this case $\phi_h(y)$ may be written in the form

$$\phi_h(y) = \int \omega_h(y - x)\,\phi(x)\,dx \tag{3.22}$$

and furthermore

$$\phi_h(y) - \phi(y) = \int \omega_h(y - x)\,[\phi(x) - \phi(y)]\,dx. \tag{3.23}$$

But the kernel $\omega_h(y - x)$ differs from zero only in the sphere $|y - x| \leq h$.

By choosing h so small as to have $\| \phi(x) - \phi(y) \| < \epsilon$ in this sphere, we will obtain

$$\| \phi_h(y) - \phi(y) \| \le \epsilon \int \omega_h(y - x)\, dx < \epsilon,$$

as was to be proved.

Corresponding to the mean function which we have defined as an abstract point function $\phi_h(y)$, we shall construct the set function which is its indefinite integral

$$\phi_h(E) = \int_E \phi_h(y)\, dy. \tag{3.24}$$

Theorem 20 will be proved if we establish that, in the case when $\phi(E)$ is continuous with respect to translation, the function $\phi_h(E)$ converges in the norm of Φ_p to $\phi(E)$.

First of all, let us prove another lemma.

Lemma IV. *The function $Z_{\phi_h}(y)$ generated by translation by the mean function $\phi_h(E)$ is equal to the mean function of the function $Z_\phi(y)$ generated by translation by the function $\phi(E)$ itself. In other words, the averaging operator*

$$\int_{E_0} [\int \omega_h(y - x)\, d_x \phi(E)]\, dy$$

may be transposed with the operator of generation by translation.

Let us prove this lemma. Consider the value of the function $Z_{\phi_h}(z)$ on some set E_0. This value will be

$$Z_{\phi_h}(E_0 \mid z) = \int_{E_0 + z} [\int \omega_h(y - x)\, d_x \phi(E)]\, dy.$$

By introducing $\chi_{E_0}(z)$, namely, the characteristic function of the set E_0, this value may be written

$$Z_{\phi_h}(E_0 \mid z) = \int \chi_{E_0 + z}(y) [\int \omega_h(y - x)\, d_x \phi(E)]\, dy.$$

But it is well known that

$$\chi_{E_0 + z}(y) = \chi_{E_0}(y - z)$$

and so

$$Z_{\phi_h}(E_0 \mid z) = \int \chi_{E_0}(y - z) [\int \omega_h(y - x)\, d_x \phi(E)]\, dy.$$

In the last integral it is convenient to introduce a new variable of integration $y - z = u$, after which it can be rewritten in the form

$$Z_{\phi_h}(E_0 \mid z) = \int \chi_{E_0}(u) [\int \omega_h(u + z - x)\, d_x \phi(E)]\, du.$$

χ_{E_0} and ω_h are bounded measurable functions. Consequently, we can apply Lemma II on the associativity of convolution.

This lemma gives

$$Z_{\phi_h}(E_0 \mid z) = \int [\int \chi_{E_0}(u) \, \omega_h(u + z - x) \, du] \, d_x \, \phi(E).$$

In the inner integral we can again change the variables of integration, setting $u = v + x - z$. We get

$$Z_{\phi_h}(E_0 \mid z) = \int [\int \chi_{E_0}(v + x - z) \, \omega_h(v) \, dv] \, d_x \, \phi(E).$$

But to the last integral we can apply Lemma II for a second time, after which we will have

$$Z_{\phi_h}(E_0 \mid z) = \int \omega_h(v) \, [\int \chi_{E_0}(v + x - z) \, d_x(E)] \, dv.$$

Let us again change the variable v to $y = z - v$ in the inside integral:

$$Z_{\phi_h}(E_0 \mid z) = \int \omega_h(z - y) \, [\int \chi_{E_0}(x - y) \, d_x \phi(E)] \, dy =$$
$$= \int \omega_h(z - y) \, \phi(E_0 + y) \, dy. \tag{3.25}$$

This last formula is nothing else than the value of the mean function of $Z_\phi(z)$ on the set E_0.

Formula (3.25) proves the assertion of the lemma.

Our theorem follows without difficulty from the lemmas that have been established. In fact, by Lemma III the mean function $Z_{\phi \mid h}(y)$ of the function $Z_\phi(y)$ converges to $Z_\phi(y)$ uniformly in the norm of Φ_p. But this means that for each point y this function, being at the same time, by Lemma IV, the function generated by translation by $\phi_h(y)$, converges to $Z_\phi(y)$.

This means that for all y, and among them for $y = 0$, $\phi_h(E + y)$ converges to $\phi(E + y)$, i.e., $\phi_h(E) \to \phi(E)$.

The theorem is proved.

Theorems 16 and 20 thus give us a characteristic property in the space Φ_p of set functions of the subspace Ψ_p obtained by closure of the manifold of indefinite integrals of continuous abstract functions with respect to the norm of Φ_p.

It will be the absolutely continuous abstract set functions which generate continuous abstract functions by translation. Along the way we have obtained another property of abstract set functions.

Theorem 21. *Every function that is continuous with respect to translation in the norm of* Φ_p *is absolutely continuous in that same metric.*

Indeed, we established that a function continuous with respect to translation is the limit of a sequence of indefinite integrals of continuous functions.

But by Theorem 11 this limit is absolutely continuous, Q. E. D.

§4. Integrals of potential type and imbedding theorems

1. For abstract set functions, just as for numerical point functions, it is possible to prove a series of theorems about integrals of potential type.

Theorem 22. *The integral*

$$U(x) = \int_{\Omega} \omega(x, y)\, d_y\, \phi(E), \tag{4.1}$$

where $\phi(E)$ is an absolutely continuous function from Φ_p and $\omega(x, y)$ is a function of the point x with values in $L_{p'}$ as a function of y, continuous in the metric of $L_{p'}$, is itself a continuous function of the point x with values in X.

If $\omega(x, y)$ is uniformly continuous, then the function $U(x)$ is also uniformly continuous.

In fact by Lemma I (the case $s = 1$)

$$\left\| \int_{\Omega} \omega(x + \Delta x, y)\, d_y\, \phi(E) - \int_{\Omega} \omega(x_0, y)\, d_y\, \phi(E) \right\|_X =$$

$$= \left\| \int_{\Omega} [\omega(x + \Delta x, y) - \omega(x, y)]\, d_y\, \phi(E) \right\|_X \le$$

$$\le \| \omega(x + \Delta x, y) - \omega(x, y) \|_{L_{p'}(y)} \cdot \| \phi(E) \|_{\Phi_p} \le \epsilon \| \phi(E) \|_{\Phi_p}. \tag{4.2}$$

Inequality (4.2) proves our theorem.

It is useful to note that for $\omega(x, y)$ we can take, for instance, the function $\omega(x, y) = r^{-\lambda} K(x, y)$, where $K(x, y)$ is an everywhere bounded, measurable function of both variables and $\lambda < n/p'$. Continuity everywhere of such a function follows from Example 4, §2 (p. 325).

Theorem 23. *Let $\omega(x, y)$ satisfy the condition*

$$| \omega(x + \Delta x, y) - \omega(x, y) | \le \frac{| \Delta x | (r + r_1)^{\lambda - 1}}{r^{\lambda} r_1^{\lambda}}. \tag{4.3}$$

Then for $q \le p' < n/\lambda$

$$\| \omega(x + \Delta x, y) - \omega(x, y) \|_{L_q(y)} \le | \Delta x |^{\beta}, \quad \| \omega(x, y) \|_{L_q(y)} < A,$$

$$where\ \beta > 0. \tag{4.4}$$

The derivation of formula (4.4) is essentially contained in the author's book [1].

Applying (4.4) to (4.1) we obtain an obvious expression for the modulus of continuity of the function $U(x)$.

2. Now let us consider some s-dimensional Euclidean manifold S_s and suppose given on it a numerical, additive function of sets J, also depending on the point y in n-dimensional space: $\omega(J, y)$, $J \subset S_s$. We shall denote points of the set J by x.

Suppose that for any y this function belongs to Φ_q: $\omega(J, y) \in \Phi_q$.

Let $\tau(x) \in L_{q'}$ $(x \in S)$. Denote

$$\sigma(y) = \int \tau(x)\, d_x\, \omega(J, y)$$

and for some p' let

$$\|\sigma(y)\|_{L_{p'}} \le A \|\tau(x)\|_{L_{q'}}$$

where A does not depend on the choice of τ.

To each additive set function $\phi(E)$ of measurable sets E in n-dimensional space we can make correspond a function $\chi(J)$ of measurable sets J in the s-dimensional manifold according to the formula

$$\chi(J) = \int \omega(J, y)\, d_y\, \phi(E), \tag{4.5}$$

where y is a vector in the space in which the sets E vary.

Theorem 24. $\chi(J)$ *is an additive abstract function with values in* X.

Let us estimate the norm of this function in Φ_q on S_1.

By definition of the norm we have

$$\|\chi(J)\|_{\Phi_q} = \sup \frac{\left\| \int \tau(x)\, d_x\left[\int \omega(J, y)\, d_y\, \phi(E) \right] \right\|_X}{\|\tau(x)\|_{L_{q'}}}, \tag{4.6}$$

where $\tau(x)$ is a step function.

Let us estimate the numerator of formula (4.6). Using the additivity of the integral and the fact that $\tau(x)$ is a step function: $\tau(x) = \beta_j$, $x \in J_j$, we will get

$$\left\| \tau(x)\, d_x\left[\int \omega(J, y)\, d_y\, \phi(E) \right] \right\|_X = \left\| \Sigma\, \beta_j \left[\int \omega(J_j, y)\, d_y\, \phi(E) \right] \right\|_X,$$

$$\left\| \int \Sigma\, \beta_j\, \omega(J_j, y)\, d_y\, \phi(E) \right\|_X = \left\| \int \left[\int \tau(x)\, d_x\, \omega(J, y) \right] d_y\, \phi(E) \right\|_X$$

and consequently

$$\left\| \int \tau(x)\, d_x\left[\int \omega(J, y)\, d_y\, \phi(E) \right] \right\|_X \le \left\| \int \tau(x)\, d_x\, \omega(J, y) \right\|_{L_{p'}} \cdot \|\phi(E)\|_{\Phi_p}. \tag{4.7}$$

However, the first factor of the right-hand side of (4.7) does not exceed $A\|\tau(x)\|_{L_{q'}}$ and consequently

$$\|\chi(J)\|_{\Phi_q} \le A \|\phi(E)\|_{\Phi_p}. \tag{4.8}$$

If we take $\dfrac{1}{r^\lambda}(J + x_1, x_2, y)$ for the function $\omega(J, y)$, as shown in

Example 4 ($\S$2), we will obtain a theorem which it is natural to call a theorem about integrals of potential type.

In some cases, as for instance in the example being considered, the function $\omega(J, y)$ also depends on an n-dimensional vector x by which we can displace the set J together with a surface on which it lies.

We have the theorem:

Theorem 25. *If*

$$\left\| \int \tau(x)\, d_x [\omega(J + \Delta x, y) - \omega(J, y)] \right\|_{L_{p'}} < A\eta(|\Delta x|) \cdot \|\tau(x)\|_{L_{q'}}, \qquad (4.9)$$

for any $\tau(x)$, then $\chi(J + x)$ is continuous in the metric of Φ_q with respect to translation. The modulus of continuity of $\chi(J + x)$ is equal to $B\eta(|\Delta x|)$.

The proof of this assertion is obvious. For this it is sufficient to repeat the reasoning which led us to formula (4.8), replacing $\omega(J, y)$ everywhere by the difference $\omega(J + x, y) - \omega(J, y)$.

We will obtain

$$\|\chi(J + x) - \chi(J)\|_{\Phi_{q'}} \le A\|\phi(E)\|_{\Phi_p} \eta(|\Delta x|),$$

which proves the theorem.

It is easily proved (see [1]) that condition (4.9) will be satisfied by any function $\omega(J, y)$ that besides (4.7) also satisfies the condition

$$|\omega(J + x, y) - \omega(J, y)| \le \int_J K(x, y)\, \frac{(r + r_1)^{\lambda - 1}|x|}{r^\lambda r_1^\lambda}\, dx.$$

The conditions of Theorems 23 and 24 essentially give us a generalization of theorems about the complete continuity of the integral operator by establishing the modulus of continuity of the functions $U(x)$ and $\chi(J)$.

3. Let us consider some set function $\phi(E) \in \Psi_p$.

By what has been stated above, we can for this function form the integral

$$\int \omega(x)\, d\phi(E),$$

where $\omega(x)$ is an arbitrary smooth, finite function. Let L be some differential operator with sufficiently smooth, for instance constant, coefficients. Form the expression $M\omega(x)$ where M is the operator adjoint to L in the sense of ordinary adjointness of differential operators.

Suppose that there exists a set function $\psi(E) \in \Psi_p$ satisfying the identity

$$\int \omega(x)\, d\psi(E) = \int M\omega\, d\phi(E)$$

for all sufficiently smooth finite ω.

Then we shall say that the function $\psi(E)$ is the value of the operator L

on $\phi(E)$:

$$\psi(E) = L\,\phi(E).$$

The operator $L\phi$ is uniquely determined since, if for some ϕ there existed two functions ψ_1, ψ_2, their difference ψ would satisfy the condition

$$\int \omega(x)\,d\psi(E) = 0$$

for all sufficiently smooth finite ω. From this it would follow that $\psi_h = 0$ and since, by what has been proved, $\|\psi_h - \psi\|_{\Phi_p} \to 0$, we would have to have $\psi = 0$, i.e., $\psi_1 = \psi_2$.

The simplest differential operators are operators of differentiation.

Let us consider the collection of set functions $\phi(E)$ from Φ_1 admitting all derivatives of order

$$\frac{\partial^l \phi(E)}{\partial x_1^{a_1} \cdots \partial x_n^{a_n}} = \psi_{a_1, \cdots, a_n}(E),$$

where all $\psi_a(E)$ belong to some Ψ_p.

By the introduction of a corresponding norm this collection can be turned into a Banach space $\Psi_p^{(l)}$ analogous to the space $W_p^{(l)}$ for numerical functions considered in [1]. It is convenient first to introduce the space $\widetilde{\Psi}_p^{(l)}$ consisting of classes of functions $\{\phi(E)\}$ such that all derivatives of order l are the same for any two functions from the same class.

We may define a norm in this space, for instance, by the formula

$$\| \phi(E) \|_{\widetilde{\Psi}_p^l}^p = \Sigma \left\| \frac{\partial^l \phi(E)}{\partial x_1^{a_1} \cdots \partial x_n^{a_n}} \right\|_{\Phi_p}, \tag{4.10}$$

where Σ runs over all derivatives of order l.

We have the theorem:

Theorem 26. *The set of functions from $\Psi_p^{(l)}$ equivalent to zero in the metric of $\widetilde{\Psi}_p^{(l)}$, i.e., those functions for which all derivatives of order l are zero, is the set of indefinite integrals of polynomials in $x_1, x_2, \cdots, x_n$ of degree less than l with coefficients in X.*

We shall carry out the proof of this theorem later along with several others.

As for numerical functions we shall introduce the space Σ_l of all such polynomials with norm, for instance, in the form of the sum of the squares of the norms of all of the coefficients. To each projection Π_l of $\Psi_p^{(l)}$ into Σ_l corresponds the complementary projection $E - \Pi_l$ which determines in $\Psi_p^{(l)}$ a complementary space isomorphic to the space of classes $\widetilde{\Psi}_p^{(l)}$.

Using the two projections Π_1 and $\Pi_1^* = I - \Pi_1$, we can introduce a norm into $\Psi_p^{(l)}$ as any convex function of $\| \Pi_1 \phi(E) \|_{\Sigma_l}$ and $\| \phi(E) \|_{\widetilde{\Psi}_p^{(l)}}$.

The set of projections Π_1 decomposes into classes, leading to equivalent norms, among which it is easy to determine one regular class (see [1]). With the norms introduced we have the theorems:

4. **Theorem 27.** *For $lp > n$ the function $\phi(E)$ is the integral of some continuous point function $\phi(x)$. The modulus of continuity of $\psi(x)$ equals $A|\Delta x|^{\beta}$, where $\beta > 1$,*

$$\| \phi(x) \|_X \leq A \| \phi(E) \|_{\Psi_p^{(l)}},$$

A being some constant which does not depend on ϕ.

This theorem is a generalization to abstract functions of the imbedding theorems for smooth functions.

Theorem 28. *For $lp < n$ the function $\phi(E)$ is defined on all smooth sets of dimension s, where $s > n - lp$, and represents a set function $\phi(J)$ belonging to L_q where*

$$\frac{s}{q} = \frac{n}{p} - l.$$

Moreover,

$$\| \phi(J) \|_{\Psi_q} \leq A \| \phi(E) \|_{\Psi_p^{(l)}}.$$

For Ψ_{q*} with $q < q$ we have continuity in the sense of Example 4 (§2).*

The proof of all of these theorems will be based on some simple lemmas which we shall present here:

Lemma I. *The differentiation operator $\partial^k / \partial x_1^{a1} \cdots \partial x_n^{an}$ commutes with the averaging operator. If*

$$\phi_h(x) = \frac{1}{h^n} \int \omega \left[\frac{r}{h} \right] d_y \phi(E) \tag{4.11}$$

and the generalized derivative

$$\frac{\partial^k \phi(E)}{\partial x_1^{a1} \cdots \partial x_n^{an}} = \psi(E)$$

exists, then

$$\frac{1}{h^n} \int \omega \left[\frac{r}{h} \right] d\psi(E) = \frac{\partial^k \phi_h}{\partial x_1^{a1} \cdots \partial x_n^{an}}, \tag{4.12}$$

where the derivative is calculated in the usual sense.

The proof of Lemma I is exactly the same as the proof which is ordinarily used for numerical functions. Differentiating (4.11) we have

$$\frac{\partial^k \phi_h(x)}{\partial x_1^{a_1} \cdots \partial x_n^{a_n}} = \int \frac{\partial^k \omega(r/h)}{\partial x_1^{a_1} \cdots \partial x_n^{a_n}} \, d_y \phi(E),$$

where $x_1, \cdots, x_n$ are the coordinates of the point x.

Using the definition of the generalized derivative and noting that the derivatives with respect to $x_1, x_2, \cdots, x_n$ of $\omega(r/h)$ differ by a multiple of $(-1)^k$ from the derivatives with respect to $y_1, \cdots, y_n$, we obtain the proof of our lemma.

Lemma II. *For continuously differentiable abstract point functions $\phi_h(x)$ we have the identity*

$$\phi_h(x) = \int_\Omega K(x, y) \phi_h(y) \, dy + \int_\Omega \Sigma \frac{K_{a_1 \cdots a_n}}{r^{n-l-1}} \frac{\partial^l \phi_h}{\partial y_1^{a_1} \cdots \partial y_n^{a_n}} \, dy, \qquad (4.13)$$

where K and $K_{a_1 \cdots a_n}$ are the same kernels through which is obtained the analogous identity with numerical functions.

The proof of Lemma II coincides word for word with the proof of the same identity for numerical functions.

Recall that the kernel $K(x, y)$ is a polynomial in the variables $x_1, \cdots, x_n$ (namely, the coordinates of x) with coefficients infinitely differentiable with respect to $y_1, \cdots, y_n$ (the coordinates of y).

The kernels $K_{a_1 \cdots a_n}(x, y)$ are bounded and

$$\left| \frac{K_{a_1 \cdots a_n}(x + \Delta x, y)}{r_1^{n-l-1}} - \frac{K_{a_1 \cdots a_n}(x, y)}{r^{n-l-1}} \right| \le \frac{A|\Delta x| (r + r_1)^{n-l-2}}{r_1^{n-l-1} r^{n-l-1}}. \qquad (4.14)$$

We rewrite formula (4.13) in the form

$$\phi_h(x) = \Pi_1 \phi_h(x) + \Pi_1^* \phi_h(x). \qquad (4.15)$$

The operator Π_1 is obviously the projection of $\phi_h(x)$ into Σ_l since upon substituting an element ϕ_0 from Σ_l into (4.15) we obtain

$$\phi_0(x) = \Pi_1 \phi_0(x), \qquad (4.16)$$

from which it follows that $\Pi_1^* = I - \Pi_1$ is also a projection.

From formulas (4.13) and (4.15) we can obtain the proof of our theorems. Let us investigate Theorem 27 first.

The operator Π_1^* is defined on the set of smooth l-times continuously differentiable functions while, by Theorem 22, its values will be continuous abstract point functions.

Let us again consider the space K of continuous abstract point functions with the metric defined by formula (2.2).

The domain of values of the operator Π_1^* will lie in this space and by the same theorem it will be bounded as an operator acting from $\widetilde{\Psi}_p^{(l)}$ into K.

But then it can be continuously extended from the everywhere dense set of differentiable functions to the entire space $\widetilde{\Psi}_p^{(l)}$.

Therefore, we shall agree to consider it as a bounded operator defined directly on the whole space $\widetilde{\Psi}_p^{(l)}$ with values in K.

In the same way the operator Π_1 may be considered as a bounded operator defined on the entire space Φ_1 with values in K.

Now we can define a metric in the space $\Psi_p^{(l)}$ by means of our projection Π_1 as we indicated in part 3 of this section.

With such a definition both operators Π_1 and Π_1^* will be continuous on $\Psi_p^{(l)}$.

Now consider the operators

$$I_1 = \int_E \Pi_1 \phi(x)\, dx, \quad I_2 = \int_E \Pi_1^* \phi(x)\, dx \tag{4.17}$$

with values in $\Psi_p^{(l)}$. The sum

$$I = I_1 + I_2 \tag{4.18}$$

is the identity operator on an everywhere dense set in $\Psi_p^{(l)}$. Its continuous extension is simply the identity operator on $\Psi_p^{(l)}$.

As we see, it pairs with any function $\phi(E)$ from $\Psi_p^{(l)}$ the indefinite integral of a continuous point function.

Theorem 27 is proved.

Theorem 28 is proved analogously.

Consider the s-dimensional manifold S_s and let J be a set in S_s which is Lebesgue measurable.

Integrate both sides of (4.13) over the set J. We will have

$$\phi_h(J) = \int_\Omega \widetilde{K}(J, y)\,\phi_h(y)\, dy + \int_\Omega \Sigma\, \widetilde{K}_{a_1 \cdots a_n}(J, y)\, \frac{\partial^l \phi_h}{\partial x_1^{a_1} \cdots \partial x_n^{a_n}}\, dy, \tag{4.19}$$

where

$$\widetilde{K}(J, y) = \int_J K(x, y)\, dx$$

$$\widetilde{K}_{a_1 \cdots a_n}(J, y) = \int_J \frac{K_{a_1 \cdots a_n}(x, y)}{r^{n-\lambda-1}}\, dx. \tag{4.20}$$

Applying the same argument as for the proof of Theorem 27 and using the fact that the kernels $\widetilde{K}_{a_1 \cdots a_n}(J, y)$ also behave like the functions $\dfrac{1}{r^\lambda}(J, y)$ in

Example 5 ($\S 2$), we prove that the identity operator in $\Psi_p^{(l)}$ transforms any element from $\Psi_p^{(l)}$ into that set function $\phi(E)$ which is represented in the form

$$\phi(E) = \int\limits_{x_2} \phi(J_{x_2}, x_2)\, dx_2, \qquad (4.21)$$

where the J_{x_2} are sections of the set by the s-dimensional surfaces $x_2 = \text{const.}$

Moreover, the norm of the set function $\phi(J_{x_2}, x_2)$ in Φ_q is estimated through the norm of $\phi(E)$ in $\Psi_p^{(l)}$ by the formula

$$\| \phi(J_s) \|_{\Phi_q} \le A \| \phi \|_{\Psi_p^{(l)}}.$$

Theorem 28 is proved.

5. Still other theorems, analogous to those that have been proved, appear to me to be of substantial interest. This remark applies, for instance, to the imbedding theorems of Gagliardo-Nirenberg reported at the International Mathematical Congress at Edinburgh in August 1958. These theorems make precise the properties of the derivatives of ϕ of order $1, 2, \cdots, l-1$ for functions from $W_p^{(l)}$ if it is known that ϕ itself is summable to some power greater than that coming from the imbedding theorems.

We have not attempted here to carry out a complete proof for exponent q bounded by an arbitrary $q^* < q$. It is thought that the theorem is true in the broadest formulation.

Finally, it would be very interesting in the applications to establish a theorem analogous to the theorem about composite functions used for the construction of the theory of quasi-linear hyperbolic equations. This theorem is easily established for the metric of B_p, but whether it can be carried over to the norm of Φ_p and under what conditions remains unclear.

The author wishes to express his gratitude to Professor Vitol′d Bortaševič, who has made a number of very useful remarks regarding the present article.

BIBLIOGRAPHY

[1] S. L. Sobolev, *Some applications of functional analysis to mathematical physics*, Izdat. Leningrad. Gos. Univ., Leningrad, 1950. (Russian)

[2] ――――, *On a theorem of functional analysis*, Mat. Sb. (N.S.) 4 (46) (1938), 471–498. (Russian)

[3] ――――, *The equations of mathematical physics*, 3rd ed., GITTL, Moscow, 1954. (Russian)

[4] S. Bochner, *Integration von Funktionen deren Werte die Elemente eines Vektorraumes sind*, Fund. Math. 20 (1933), 262–276.

[5] I. M. Gel'fand, *Abstrakte Funktionen und lineare Operatoren*, Mat. Sb. (N.S.) 4 (46) (1938), 235–286.

[6] V. P. Il'in, *On a theorem of inclusion for a limiting exponent*, Dokl. Akad. Nauk SSSR 96 (1954), 905–908. (Russian)

[7] S. L. Sobolev, *Sur la théorie des équations hyperboliques aux dérivées partielles*, Mat. Sb. (N.S.) 5 (47) (1939), 71–99. (Russian. French summary)

[8] Ja. S. Bugrov, *On imbedding theorems*, Dokl. Akad. Nauk SSSR 116 (1957), 531–534. (Russian)

[9] A. A. Vašarin, *The boundary properties of functions having a finite Dirichlet integral with a weight*, Dokl. Akad. Nauk SSSR 117 (1957), 742–744. (Russian)

[10] L. D. Kudrjavcev, *On extension of functions and imbedding of classes of functions*, Dokl. Akad. Nauk SSSR 107 (1956), 501–504. (Russian)

[11] S. M. Nikol'skiĭ, *Inequalities for entire functions of finite degree and their application in the theory of differentiable functions of several variables*, Trudy Mat. Inst. Steklov. 38 (1951), 244–278. (Russian)

[12] ———, *Properties of certain classes of functions of several variables on differentiable manifolds*, Mat. Sb. (N.S.) 33 (75) (1953), 261–326. (Russian)

[13] ———, *An imbedding theorem for functions with partial derivatives considered in different metrics*, Izv. Akad. Nauk SSSR Ser. Mat. 22 (1958), 321–336. (Russian)

Translated by:
J. R. Brown

EMBEDDING OF LOCALLY COMPACT ABELIAN TOPOLOGICAL GROUPS IN EUCLIDEAN SPACES

M. BOGNAR

The structure of locally compact, abelian topological groups was worked out by L. S. Pontrjagin in 1934 [3], and is well known. Regarding the question of embedding the space of this group in an Euclidean space, there exists up to now only one result, that of the Japanese mathematicians K. Kodaira and M. Abe. They proved in 1940 that an n-dimensional, compact, connected, abelian group with a countable base can be embedded in $(n+1)$-dimensional Euclidean space if and only if it is an n-dimensional toroidal group [2].

We will prove the following theorem.

Each n-dimensional, locally compact, abelian topological group with a countable base can be embedded in $(n-2)$-dimensional Euclidean space.

We will prove this theorem only for compact groups of positive dimension.

The definition of the group of characters of a locally compact, abelian topological group G, with a countable base, is the following: we denote by Ξ the factor group of the additive group of the real numbers by the subgroup of integers. Elements of the group Ξ will be denoted by ξ. If x is a real number then the coset of the group Ξ which contains x, will be denoted by $\bar{x}$. A neighborhood of zero in the space Ξ will be denoted by U. Any homomorphism γ of the topological group G into the topological group Ξ is called a character of the topological group. We define the set Γ of characters as a topological group in the following manner: the sum of 2 characters γ and γ^1 is that character $\gamma + \gamma^1$ such that for each element $g \in G$:

$$(\gamma + \gamma^1) g = \gamma (g) + \gamma^1 (g).$$

Neighborhoods V of the zero element of Γ are defined in terms of neighborhoods U of zero in Ξ and compact subsets Φ of the space G by putting $\gamma \in V (\Phi, U)$ if $\gamma (\Phi) \subset U$.

In such a way we obtain a topological group Γ which is abelian, locally compact and has a countable base. The basic theorem of Pontrjagin is that, if Γ is the group of characters of the group G and G^* is the group of characters of the group Γ, then G and G^* are isomorphic.

We note that the group of characters of a compact group is discrete, and that the group of characters of a discrete group is compact. Moreover, the dimension of a compact group coincides with the rank of its discrete group of characters and conversely.

In view of this, we can reformulate our theorem as follows.

The space of the group of characters of each discrete, countable, abelian group G of rank n $(n \geq 1)$ may be embedded in $(n + 2)$-dimensional Euclidean space.

We will prove the theorem in this form.

First of all we give a construction, similar to the construction of van Dantzig [1] for the realization of the solenoid of Vietoris [4] in 3-dimensional Euclidean space.

Let K_n be a full circle in the plane R^2 where n is the rank of the group G. We will further assume that each K_l $(1 \leq l \leq n)$ is compact and lies in $(2+n-l)$-dimensional R^{2+n-l}. We take a $(1 + n - l)$-dimensional plane S^{1+n-l} in R^{2+n-l} such that K_l lies inside one side of the plane S^{1+n-l}. Next we embed the space R^{2+n-l} isometrically in $3 + n - l = 2 + n - (l - 1)$-dimensional Euclidean space and orientate the rotation of R^{3+n-l} around S^{1+n-l}. Define the rotation $T_{l,\xi}$ $(\xi \in \Xi)$ as the rotation of angle $2\pi x$, where $\bar{x} = \xi$, around S^{1+n-l} in R^{3+n-l}. Observe that if $\xi \neq \xi^1$, then $T_{l,\xi}(K_l)$ does not intersect $T_{l,\xi^1}(K_l)$. Let

$$K_{l-1} = \bigcup_{\xi \in \Xi} T_{l,\xi}(K_l).$$

It is not difficult to show that K_{l-1} is compact in $R^{2+n-(l-1)}$.

Thus we define the compact sets $K_n, K_{n-1}, \cdots, K_0$, where $K_0 \subset R^{2+n}$, and the rotations $T_{n,\xi}, T_{n-1,\xi}, \cdots, T_{1,\xi}$.

Now we give an orientation to the plane R^2. Let C be a point in R^2 and $\xi \in \Xi$. We define the rotation $T_{C,\xi}$ of the plane R^2 as the rotation of angle $2\pi x$ where $\bar{x} = \xi$, around the center C.

We define the full circles $K_n, K_{n+1}, \cdots, K_l, \cdots$ with centers $C_n, C_{n+1}, \cdots, C_l, \cdots$ as follows: K_n has already been defined. Suppose that the circle K_{l-1} $(l > n)$ with center C_{l-1} has been defined. Now take the circle K_l inside K_{l-1} such that for $\xi \neq \xi^1$ and $l(\xi - \xi^1) = \bar{0}$ the circles $T_{C_{l-1},\xi}(K_l)$ and $T_{C_{l-1},\xi^1}(K_l)$ do not intersect.

We denote the rotation $T_{C_{l-1},\xi}$ by $T_{l,\xi}$ $(l > n)$. Then

a) $T_{l,\xi}(K_l) \subset K_{l-1}$ for all l and ξ $(l \geq 1, \xi \in \Xi)$.

b) If $\xi \neq \xi^1$ and $l \leq n$ or $l(\xi - \xi^1) = \bar{0}$, then $T_{l,\xi}(K_l)$ and $T_{l,\xi^1}(K_l)$ do not intersect.

c) $\delta(K_l) \to 0$, if $l \to \infty$ where $\delta(K_l)$ is the diameter of the compact set K_l.

Now we define the compact set $M_{\xi_1, \xi_2, \cdots, \xi_l}$. Let $\xi_1, \xi_2, \cdots, \xi_l$ be any finite sequence in Ξ. Let

$$M_{\xi_1, \xi_2, \cdots, \xi_l} = T_{1, \xi_1}(T_{2, \xi_2}(\cdots(T_{l, \xi_l}(K_l))\cdots)).$$

We note that

$$M_{\xi_1, \xi_2, \cdots, \xi_l, \xi_{l+1}} \subset M_{\xi_1, \xi_2, \cdots, \xi_l}.$$

Now we take the infinite sequence $\xi_1, \xi_2, \cdots, \xi_l, \cdots$ and define a point $s(\xi_1, \xi_2, \cdots, \xi_l, \cdots)$ so that:

$$s(\xi_1, \xi_2, \cdots, \xi_l, \cdots) = M_{\xi_1} \cap M_{\xi_1, \xi_2} \cap \cdots \cap M_{\xi_1, \xi_2, \cdots, \xi_l} \cap \cdots .$$

Then it is obvious that if $s(\xi_1, \xi_2, \cdots, \xi_l)$ and $s(\xi_1^1, \xi_2^1, \cdots, \xi_l^1, \cdots)$ are

two points for which $\xi_i = \xi_i^1$ where $i < l$ and $\xi_l \neq \xi_l^1$ where $l \leq n$ or $l(\xi_l - \xi_l^1) = \bar{0}$, then

$$s(\xi_1, \xi_2, \cdots, \xi_l, \cdots) \neq s(\xi_1^1, \xi_2^1, \cdots, \xi_{ll}^1, \cdots).$$

Now we prove our theorem.

First of all we note that if G is an abstract countable, abelian group of rank n $(n \geq 1)$ and $g_1, \cdots, g_n$ are n linearly independent elements from the group G, then we can enumerate the remaining elements of the group G as $g_{n+1}, \cdots, g_l, \cdots$ such that for each l

$$l_{g_l} \in G_n$$

where G_n is the subgroup of the group G generated by the elements $g_1, \cdots, g_n$.

Suppose now that γ is any character of the group G. Suppose

$$\phi(\gamma) = s(\gamma(g_1), \gamma(g_2), \cdots, \gamma(g_l), \cdots).$$

We obtain a mapping ϕ of the space of the group of characters Γ of the group G into R^{n+2}.

First we prove that the mapping ϕ is one to one.

Suppose $\gamma \neq \gamma'$. Then there exists some g_k such that $\gamma(g_k) \neq \gamma'(g_k)$. Assume g_l is the first such g_k in the sequence $g_1, \cdots, g_k, \cdots$. If $l \leq n$ then obviously

$$\phi(\gamma) \neq \phi(\gamma').$$

If $l > n$, then

$$\gamma(g_1) = \gamma'(g_1), \cdots, \gamma(g_n) = \gamma'(g_n),$$

and consequently

$$(\gamma - \gamma')(g_1) = \cdots = (\gamma - \gamma')(g_n) = \bar{0} .$$

Therefore

$$(\gamma - \gamma')(G_n) = \overline{0}$$
$$(\gamma - \gamma')(lg_l) = \overline{0}$$
$$l(\gamma(g_l) - \gamma'(g_l)) = \overline{0}$$

that is,

$$\phi(\gamma) \neq \phi(\gamma').$$

Consequently the mapping ϕ is one to one.

Now we show that the mapping ϕ is continuous. Let $\gamma \in \Gamma$ and W be any neighborhood of the point $\phi(\gamma) = s(\gamma(g_1), \gamma(g_2), \cdots, \gamma(g_l), \cdots)$ in R^{n+2}. Then there obviously exists some l so that

$$T_{1,\,\gamma(g_1)}(T_{2,\,\gamma(g_2)}(\ldots(T_{l,\,\gamma(g_l)}(K_l))\ldots)) = M_{\gamma(g_1),\,\gamma(g_2),\,\ldots,\,\gamma(g_l)} \subset W.$$

It is not difficult to show that for some l we can find a neighborhood U of zero in Ξ so that if $\xi_1 \in \gamma(g_i) + U$ $(i = 1, 2, \cdots, l)$, then

$$M_{\xi_1,\,\xi,\,\ldots,\,\xi_l} = T_{1,\,\xi_1}(T_{2,\,\xi_2}(\ldots(T_{l,\,\xi_l}(K_l))\ldots)) \subset W.$$

$\Phi = \{g_1, g_2, \cdots, g_l\}$ is a compact subset of the discrete space G. Let $V(\Phi, U)$ be the neighborhood of zero in the group Γ. Let $\gamma' \in \gamma + V$. Then $\gamma'(g_i) - \gamma(g_i) \in U$ $(i = 1, 2, \cdots, l)$ i.e., $\gamma'(g_i) \in \gamma(g_i) + U$ $(i = 1, 2, \cdots, l)$. Consequently

$$\phi(\gamma') = s(\gamma'(g_1), \gamma'(g_2), \ldots, \gamma'(g_l), \ldots) \in M_{\gamma'(g_1),\,\gamma'(g_2),\,\ldots,\,\gamma'(g_l)} \subset W,$$

i.e., $\phi(\gamma') \in W$. Therefore $\phi(\gamma + V) \subset W$. ϕ is thus a one to one continuous mapping of the compact set Γ into the Hausdorff space R^{n+2}, and therefore the mapping ϕ is also a homeomorphism of the space Γ onto the subspace $\phi(\Gamma)$ of the space R^{n+2}. The proof is now finished.

We have proved that each n-dimensional $(n \geq 1)$ compact, abelian topological group with a countable base can be embedded into $(n + 2)$-dimensional Euclidean space. We note that for $n = 0$ our theorem is also valid.

Now we prove our initial theorem, about embedding a locally compact, abelian group with a countable base.

According to Pontrjagin's theorem each locally compact, abelian group Γ with a countable base is the direct sum of two groups V and Γ_1, where V is a factor-group and Γ_1 has some compact subgroup Γ_2, such that the factor-group $\Gamma_1 - \Gamma_2$ is discrete and countable. It is not difficult to show that if Γ is n-dimensional and V is a k-dimensional group, then Γ_1 and also Γ_2 are $(n-k)$-dimensional groups. Thus, both Γ_1 and Γ_2 can be embedded in R^{n-k+2}. The group V can be embedded in R^k and therefore can be embedded in $n - k + 2 + k = n + 2$-dimensional Euclidean space. Thus the theorem is proved.

BIBLIOGRAPHY

[1] D. van Dantzig, *Über topologisch homogene Kontinua*, Fund. Math. **15**
(1930), 102-125.

[2] K. Kodaira and M. Abe, *Über zusammenhängende kompakte abelsche Gruppen*,
Proc. Imp. Acad. Tokyo **16** (1940), 167-172.

[3] L. Pontrjagin, *The theory of topological commutative groups*, Ann. of Math.
(2) **35** (1934), 361-388.

[4] L. Vietoris, *Über den höheren Zusammenhang kompakter Räume und eine
Klasse von zusammenhangstreuen Abbildungen*, Math. Ann. **97** (1927),
454-472.

Translated by:
Jack Ceder

ON THE COMPLETENESS OF A PARTIALLY ORDERED SPACE

D. A. VLADIMIROV

In the present article three problems are considered which were formulated in the work of L. V. Kantorovič, B. Z. Vulih and A. G. Pinsker [1]. In what follows we will make use without reservation of the terminology and notations in [1] and [2].

Let X be a K-space, $\{x_n\}$ $(n = 1, 2, \cdots)$ a sequence of elements of X. We will say that the sequence x_n is (o)-*fundamental* if

$$n_i, \ m_i \longrightarrow \infty \tag{1}$$

implies that

$$x_{n_i} - x_{m_i} \xrightarrow{(o)} 0. \tag{2}$$

If from (1) it always follows that

$$x_{n_i} - x_{m_i} \xrightarrow{(t)}$$

then the sequence is called (t)-fundamental. It is easy to see that a sequence (o)-convergent $((t)$-convergent) to a finite limit is always (o)-fundamental $((t)$-fundamental, respectively). Whether the converse is true, i.e., whether an (o)- (or (t)-) fundamental sequence of elements of a K-space always has a finite (o)- (or, respectively (t)-) limit, constitutes problems 1 and 2[1] considered. Below a negative answer is given to these questions. Indeed, an example of a K-space will be given which is not (t)-complete. Also, it will be shown that the hypothesis of (o)-completeness of any K-space contradicts the continuum hypothesis (even weaker hypothesis).

In what follows by T we will denote the set of all sequences $\tau_\xi = (n_1^{(\xi)}, n_2^{(\xi)}, \cdots, n_k^{(\xi)}, \cdots)$ of natural numbers that increase without bounds. T is partially ordered by the following: $\tau_\alpha \succ \tau_\beta$ if for $k \geq k_0 (\alpha, \beta)$ we have $n_k^{(\alpha)} \geq n_k^{(\beta)}$.

It is not difficult to convince oneself that any countable subset of T is both bounded above and below in the sense of our ordering (the proof can be done by the usual diagonal method). We will denote by a the smallest cardinal of sets not bounded above in the sense of our order[2]. Clearly

---

1)See the list of problems at the end of [1].

2)Here, of course, we make use of the axiom of choice.

$$\aleph_0 < a \leqslant 2^{\aleph_0} \ .$$

Lemma 1. *Let* $T' \subset T$, $\overline{\overline{T}}' < a$[1]. *Then there exists a sequence* $\{c_m\}$ $(m = 1, 2, \cdots)$ *of non-negative real numbers for which*

a) $\sum\limits_{m=1}^{\infty} c_m = + \infty$;

b) *for any* $\tau_\beta = (n_1^{(\beta)}, n_2^{(\beta)}, \cdots, n_k^{(\beta)}, \cdots) \in T'$

$$\sigma_k(\beta) = \sum_{m=k+1}^{k+n_k^{(\beta)}} c_m \to 0 \ \ for \ \ k \to \infty.$$

Proof. Because the cardinality of T' is less than a then this set is bounded above in the sense of our ordering. This means that there exists a sequence $\tau = (n_1, n_2, \cdots, n_k, \cdots)$ such that $\tau \succ \tau_\beta$ for all β. Now let us suppose: $m^{(1)} = n_1 + 1$; if $m^{(1)}$, $m^{(2)}$, $\cdots$, $m^{(s)}$ are already constructed, let $m^{(s+1)} = m^{(s)} + n_m(s)$.

Define the sequence $\{c_m\}$ by the following conditions:

$$c_m = \begin{cases} \dfrac{1}{s}, & \text{if } m = m^{(s)} \\ 0 & \text{in other cases.} \end{cases}$$

This is the required sequence.

Lemma 2. *Let a be a regular cardinal number. This means that the following equality is impossible:*

$$a = \sum_{\gamma \in \Gamma} b_\gamma,$$

where $a > \overline{\overline{\Gamma}}$ *and* $a > b_\gamma$ *for all* $\gamma \in \Gamma$.

This lemma is an evident consequence of the transitivity of $\succ$.

We now define a partially ordered linear set X as follows: Let $A = \{\alpha\}$ be the set of all ordinal numbers of cardinality less than a: $\alpha < \omega_a$ where ω_a is the first ordinal number of cardinality a.

We consider the K-space F of all real functions $f(\alpha)$ defined on A. We denote by X its factor-space, obtained by identifying those functions which are different on sets of cardinality less than a; such sets will be called *null-sets*.

---

[1] $\overline{\overline{T}}'$ denotes the cardinality of T'.

X is partially ordered in the natural way: $f \geq g$ if $f(\alpha) \geq g(\alpha)$ everywhere except on some null-set [1]. X is a K^--space, i.e., in X there exists a supremum of any countable set bounded above. To prove this, observe that by virtue of Lemma 2 the union of a countable set of null-sets is again a null-set. Therefore the bounds of countable sets can be computed "point-wise". Precisely: let be given a countable set $G = \{f_n\}$ ($n = 1, 2, \cdots$) and let $f_n \leq f$ for all n. We select in an arbitrary way "representatives" of the classes f_n and f. This gives us functions $f_n(\alpha)$, $f(\alpha)$ where $f_n(\alpha) \leq f(\alpha)$ except for a certain null-set. Let $\bar{f}(\alpha) = \sup_n f_n(\alpha)$ [2]. Then the class $\bar{f}$, consisting of all functions equivalent to $\bar{f}(\alpha)$[3], will be the least upper bound of the set. The verification of this fact does not present any difficulty. For lower bounds everything is analogous.

In a K^--space the concept of (o)- convergence has meaning. In our case this will be convergence "almost everywhere". More precisely: $f_n \xrightarrow{(o)} 0$ means that for arbitrary choice of "representatives" $f_n(\alpha)$ we have $f_n(\alpha) \longrightarrow 0$ for all except perhaps of a null-set. For the verification of this we make use of the possibility to compute the exact bound "point-wise" and the fact that the class of null-sets is closed with respect to countable unions. All these questions concerning the convergence of certain sequences of elements of X, can be answered by considering sequences of arbitrarily chosen "representatives".

A fundamental theorem was given by A. I. Judin, according to which every Archimedian K-linear (and consequently all K^--spaces) can be imbedded with the preservation of order in some K-space ($[3, 2, 4]$). The imbedding is realized by the method of sections. Here, if some set had an exact bound before imbedding then it will not be changed after imbedding. Therefore, if this process is applied to a K^--space X, then the meaning of (o)- and (t)-convergence is preserved.

We denote by $\bar{X}$ the completion of X. The remark made above signifies that if $x, x_n \in X$ then the statements a) $x_n \xrightarrow{(o)} x$ in X and b) $x_n \xrightarrow{(o)} x$ in $\bar{X}$ are equivalent (not excluding $x = \infty$).

Now, we can, finally, realize the indicated construction. We bring an example of a sequence in the space $\bar{X}$ which will be (t)-fundamental but which will have limit $+ \infty$.

Let $T_\alpha = \{\tau_\alpha\}$, $\alpha \in A$, be some set of sequences $\tau_\alpha = (n_1^{(\alpha)}, n_2^{(\alpha)}, \cdots, n_k^{(\alpha)}, \cdots)$ that is not bounded above. For each $\alpha_0 \in A$ we construct with the

---

1) The functions $f(\alpha)$, $g(\alpha)$ are arbitrary "representatives" of the classes f, g which are the true elements of X.

2) Where this supremum is infinite we can consider $\bar{f}(\alpha) = 0$. This may happen only on null-sets.

3) As usual, functions are called *equivalent* if they are different on null-sets.

help of Lemma 1 a series $\sum\limits_{m=1}^{\infty} c_m(\alpha_0)$ with non-negative terms, such that

a) $\sum\limits_{m=1}^{\infty} c_m(\alpha_0) = +\infty$,

b) if $\alpha \leqslant \alpha_0$, then $\sum\limits_{m=k+1}^{k+n_k^{(\alpha)}} c_m(\alpha_0) \to 0$ for $k \to \infty$.

Consider a series in the space $\bar{X}$: $\sum\limits_{m=1}^{\infty} c_m$ where c_m is a class containing the function $c_m(\alpha_0)$. Because for each α we have $\sum\limits_{m=1}^{\infty} c_m(\alpha) = +\infty$ then also $\sum\limits_{m=1}^{\infty} c_m = +\infty$ in our space. We convince ourselves that the sequence of partial sums of this series is (t)-fundamental, i.e., any arbitrary sequence $\{\sigma_k = \sum\limits_{n=m_k+1}^{m_k+p_k} c_n\}$ contains a subsequence $\{\sigma_{k_s}\}$, (o)-convergent to zero.

For this it is sufficient to single out from the sequence

$$\{\sigma_k(\alpha) = \sum_{n=m_k+1}^{m_k+p_k} c_n(\alpha)\}$$

the subsequence $\{\sigma_{k_s}(\alpha)\}$ which would converge to zero everywhere except on some null-set.

And so, let be given a sequence of natural numbers

$$\{m_k\} \qquad (k = 1, 2, \ldots), \qquad m_k \to \infty.$$

Single it out to obtain $m_{k_s} + 1 > s$. Define $\bar{p}_s = \max\limits_{i \leq s} \{m_{k_i} + p_{k_i}\}$. We have the evident inequality

$$0 \leqslant \sigma_{k_s}(\alpha) = \sum_{n=m_{k_s}+1}^{m_{k_s}+p_{k_s}} c_n(\alpha) \leqslant \sum_{n=s+1}^{n=s+\bar{p}_s} c_n(\alpha). \tag{3}$$

Because the set T_α is not bounded, the sequence $\tau = (\bar{p}_1, \bar{p}_2, \cdots, \bar{p}_s, \cdots)$ is not its upper bound and for some α_0 for an infinite sequence of indices $\{s_j\}$ $(j = 1, 2, \cdots)$ the following inequalities will be true

$$\bar{p}_{s_j} < n_{s_j}^{(\alpha_0)}, \tag{4}$$

$$\sum_{n=s_j+1}^{s_j+p_{s_j}} c_n(\alpha) \leqslant \sum_{n=s_j+1}^{s_j+n_{s_j}^{(\alpha_0)}} c_n(\alpha) = \bar{\sigma}_{s_j}(\alpha). \tag{5}$$

The sequence $\{\bar{\sigma}_s(\alpha)\}$ converges to zero for all $\alpha \geq \alpha_0$. This follows from its very way of construction. In other cases $\bar{\sigma}_s(\alpha) \longrightarrow 0$ everywhere except on a null-set, because the set of numbers smaller than α_0 has cardinality smaller than a and is consequently a null-set. Hence $\{\bar{\sigma}_s\}$ (o)-converges to zero in X and consequently in $\bar{X}$ too. The same is true for the subsequence $\{\bar{\sigma}_{s_j}\}$. By virtue of the inequalities (3), (5)

$$0 \leqslant \sigma_{k_{s_j}} \leqslant \bar{\sigma}_{s_j}.$$

Thus, $\sigma_{k_{s_j}} \xrightarrow{(o)} 0$ in X and therefore $\sigma_k \xrightarrow{(t)} 0\,^{1)}$, which implies the following theorem.

Theorem 1. *The space $\bar{X}$ contains a (t)-fundamental sequence converging to $+\infty$ and, consequently, it is not (t)-complete.*

If we accept the continuum hypothesis then $a = 2^{\aleph_0}$. Then we can assume that T_a contains, in general, all monotone increasing sequences of natural numbers approaching infinity. Hence it follows easily that the sequence constructed by us is (o)-fundamental and the space $\bar{X}$ is not (o)-complete. Therefore we have

Theorem 2. *If $\aleph_1 = 2^{\aleph_0}$ then the space $\bar{X}$ is not (o)-complete.*

We note that the construction of the same space X does not require any kind of hypothesis.

Theorem 3. *If $2^{\aleph_0} < \aleph_\omega$, then there exists a K-space which is not (o)-complete.*

We will say that $T_1 = \{\tau_\xi\}$ is a majorizing set if for arbitrary $\tau \in T$ we can find $\tau_{\xi_0} \in T_1$ such that $\tau_{\xi_0} > \tau$. The least of the cardinalities of majorizing sets will be denoted by a_1. There exists a majorizing set T^* of cardinality a_1. Let $a_1 = \aleph_k$. We assume that $2^{\aleph_0} < \aleph_\omega$; consequently $k < \omega$, and therefore all cardinalities not exceeding a_1 are regular because $\aleph_\omega$ is the first non-regular cardinal $^{2)}$.

The main points of the proof are given as a lemma:

Lemma 3. *Let for every set $T' \subset T$ and for every $\bar{\bar{T}}' < b < \aleph_\omega$ there be a K-space X' containing a non-negative sequence $\{x'_s\}$ such that*

$^{1)}$We showed that an arbitrary sequence $\{\sigma_k\}$ contains an (o)-convergent subsequence.

$^{2)}$In the proof we make use only of the inequality $a_1 < \aleph_\omega$.

a) $\quad \sum_{s=1}^{\infty} x_s' = +\infty,$

b) $\quad$ *if* $\quad \tau = (n_1', n_2', \ldots, n_k', \ldots) \in T'$, *then* $\quad \sum_{s=m}^{m+n_m'} x_s' \xrightarrow[m\to\infty]{(o)} 0.$

Let further, $b \geq \aleph_1$. *Then for an arbitrary set* T'' *of cardinality* b *we can find a K-space* X'' *and a non-negative sequence* $\{x_s''\} \subset X''$ *for which conditions* a) *and* b) *are satisfied (of course, replacing* x_s' *by* x_s'', *etc.).*

We index the elements of T'' by transfinite numbers, less than ω_b (first number of cardinality b). Let us suppose that

$$T_\gamma = \{\tau_\alpha\}, \qquad \alpha \leqslant \gamma, \qquad \text{for all} \quad \gamma < \omega_b.$$

According to the hypothesis, there exist K-spaces X_γ and, in these, non-negative sequences $\{x_s^{(\gamma)}\}$ for which we have

a) $\quad \sum_{s=1}^{\infty} x_s^{(\gamma)} = +\infty,$

b) $\quad$ if $\quad \tau = (n_1, n_2, \ldots, n_k, \ldots) \in T_\gamma$, then $(o) \lim_m \sum_{s=m}^{m+n_m} x_s^{(\gamma)} = 0.$

Now we define a K-space X'' similar to the one constructed in the proof of Theorem 1. Namely, in the space $F = SX_\gamma$ [1] consisting of functions $y(\gamma)$ with values in corresponding spaces X_γ, identifying functions that are different only on sets with cardinality less than b. The factor space obtained this way is a K-space. The proof of this differs little from the corresponding reasoning in the proof of Theorem 1. This time it is essential that b is non-denumerable regular cardinal and consequently the set of cardinality b is not a countable union of sets of smaller cardinalities. Therefore the class of "null-sets" will be closed with respect to countable unions, and the exact bounds of countable sets may be computed "point-wise". Complete the constructed factor-space by the method of sections; thus we obtain the required K-space X''. The required sequence $\{x_s''\}$ is generated by the functional sequence $\{x_s(\gamma) = x_s^{(\gamma)}\}$. The proof that this sequence satisfies conditions a) and b) is analogous to the proof of Theorem 1.

Because for $b \leq \aleph_1$ the existence of the corresponding space and sequence was established in the proof of Theorem 1, then Lemma 3 gives the possibility to establish by induction, the existence of similar spaces and sequences for all $b \leq a_1$.

---

[1] SX_γ is the complete union of the spaces X_γ (see [2], p. 69).

In particular, we can find for the set T^* a K-space X^* and a sequence $\{x_s\} \subset X^*$ such that

a) $\displaystyle\sum_{s=1}^{\infty} x_s = +\infty,$

b) $(o)\lim \displaystyle\sum_{s=m}^{m+n_m} x_s = 0,$ if $\tau = (n_1, n_2, \ldots, n_k, \ldots) \in T^*.$

We show that the sequence of partial sums of the divergent series $\displaystyle\sum_{s=1}^{\infty} x_s$ is (o)-fundamental.

Let $\{k_i\}$, $\{m_i\}$ be two arbitrary sequences of natural numbers, tending to infinity.

We can assume that $k_i \leq m_i$. Let

$$\bar{k}_i = \min_{s \geq i} \{k_s\}, \quad \bar{m}_i = \max_{s \leq i} \{m_s\}, \quad \bar{\bar{m}}_i = \max_{\bar{k}_j = i} \{\bar{m}_j\}.$$

Because the set T^* is majorizing, then there exists

$$\tau_1 = (m_1', m_2', \ldots, m_k', \ldots) \in T^* \text{ and for } i \geq i_0, \quad m_i' > \bar{\bar{m}}_i - i,$$

$$\sum_{s=i}^{\bar{\bar{m}}_i} x_s \leq \sum_{s=i}^{i+m_i'} x_s = \sigma_i.$$

Moreover, $\sigma_i \xrightarrow{(o)} 0$ and then as can easily be seen $\displaystyle\sum_{s=k_i}^{m_i} x_s \xrightarrow{(o)} 0$ also.

Theorem 3 is proved.

The given cardinal number a turns out to be useful for considering still another problem in the theory of K-spaces — the seventh problem mentioned in paper [1]. We may ask whether we could give a proof — without the use of the continuum hypothesis — of the regularity of a K-space in which the theorem about the diagonal sequence[1] holds. The answer to this question is as follows: the continuum hypothesis could be replaced by the weaker assumption that $a = \aleph_1$; but this assumption cannot be dropped.

Let us consider two assertions:

(A) A K-space in which the theorem about the diagonal sequence holds is regular.

(B) $a = \aleph_1.$

---

1) See [2], p. 176.

Theorem 4. *Assertions* (A) *and* (B) *are equivalent.*

It is possible to prove that (B) implies (A) with an insignificant change in the argument of A. G. Pinsker's proof of Theorem 1.48 from Vol. V of his monograph [2] (making use of the existence of a subset T' of T having cardinality a, that is, unbounded from below).

We show that (A) implies (B). We presuppose that $a > \aleph_1$. Let F be the K-space of all real functions on the set of transfinite numbers of the first of two classes. Consider the set $\{f_n^{(m)}\} \subset F$ such that $f_n^{(m)} \searrow 0$ as $n \to \infty$, $m = 1, 2, \cdots$. For every α there exists a diagonal sequence

$$f_{\substack{n(\alpha) \\ m}}^{(m)}(\alpha) \xrightarrow[m \to \infty]{} 0.$$

Because $\alpha < \Omega$ then there is an increasing sequence of natural numbers $(n_1, n_2, \cdots, n_k, \cdots)$, which "exceeds" every one of the sequences $(n_1^{(\alpha)}, n_2^{(\alpha)}, \cdots, n_k^{(\alpha)}, \cdots)$. Clearly, $f_{\substack{n \\ m}}^{(m)} \xrightarrow{(o)} 0$ for $m \to \infty$. But then[1] in F the theorem about the diagonal sequence holds, although F is a non-regular space.

Theorem 4 shows that the use of set-theoretical hypotheses in the theory of partially ordered spaces is not only convenient in the proofs but they are deeply connected with the essence of this theory.

The author thanks B. Z. Vulih for his valuable suggestions.

BIBLIOGRAPHY

[1] L. V. Kantorovič, B. Z. Vulih and A. G. Pinsker, *Partially ordered groups and linear partially ordered spaces*, Uspehi Mat. Nauk 6 (1951), no. 3 (43), 31-98. (Russian)

[2] ———, *Functional analysis in partially ordered spaces*, GITTL, Moscow, 1950. (Russian)

[3] Garrett Birkhoff, *Lattice theory*, Amer. Math. Soc., New York, 1940.

[4] A. I. Judin, *Sur le complètement des ensembles semiordonnés*, Leningrad State Univ. Annals Math. Ser. 12 83 (1941), 57-61. (Russian. French summary)

Translated by:
Tamas Bartha

---

1) See [2], p. 171.